BIOLOGY OF *PEROMYSCUS*
(RODENTIA)

SPECIAL PUBLICATIONS

This series, published by the American Society of Mammalogists, has been established for papers of monographic scope concerned with some aspect of the biology of mammals.

Correspondence concerning manuscripts to be submitted for publication in the series should be addressed to the chairman of the Committee on Special Publications (now J. Knox Jones, Jr., address below).

Copies of special publications may be ordered from the Secretary-treasurer of the Society, Dr. Bryan P. Glass, Department of Zoology, Oklahoma State University, Stillwater, Oklahoma 74074.

Price of this issue $15.00

COMMITTEE ON SPECIAL PUBLICATIONS

WILLIAM H. BURT, Chairman and Editor
Museum of Zoology,
University of Michigan,
Ann Arbor, Michigan 48104.

J. KNOX JONES, JR.
Museum of Natural History,
University of Kansas,
Lawrence, Kansas 66044.

JAMES N. LAYNE
Archbold Biological Station,
Route 2, Box 380,
Lake Placid, Florida 33852.

M. RAYMOND LEE
Museum of Natural History,
University of Illinois,
Urbana, Illinois 61803.

Photograph by Lee R. Dice

BIOLOGY OF *PEROMYSCUS* (RODENTIA)

EDITED BY

JOHN A. KING

SPECIAL PUBLICATION NO. 2

THE AMERICAN SOCIETY OF MAMMALOGISTS

PUBLISHED DECEMBER 20, 1968

Contributors

ROLLIN H. BAKER, Professor of Zoology and Director
  The Museum
  Michigan State University, East Lansing

W. FRANK BLAIR, Professor of Zoology
  Department of Zoology
  University of Texas, Austin

LEE R. DICE, Professor Emeritus of Zoology
  University of Michigan, Ann Arbor

JOHN F. EISENBERG, Resident Scientist
  National Zoological Park
  Smithsonian Institution, Washington, D. C.

BASIL E. ELEFTHERIOU, Assistant Professor of Biology
  Division of Biology
  Kansas State University, Manhattan

J. BRUCE FALLS, Professor of Zoology
  Department of Zoology
  University of Toronto

CLAUDE W. HIBBARD, Professor of Geology
  Museum of Palaeontology
  University of Michigan, Ann Arbor

EMMET T. HOOPER, Professor of Zoology
  Museum of Zoology
  University of Michigan, Ann Arbor

JOHN A. KING, Professor of Zoology
  Department of Zoology
  Michigan State University, East Lansing

DAVID KLINGENER, Assistant Professor of Zoology
  Department of Zoology
  University of Massachusetts, Amherst

JAMES N. LAYNE, Director
  Archbold Biological Station
  American Museum of Natural History
  Lake Placid, Florida

DAVID I. RASMUSSEN, Associate Professor of Zoology
Department of Zoology
Arizona State University, Tempe

LUCILLE F. STICKEL, Research Biologist
U. S. Fish and Wildlife Service
Patuxent Wildlife Research Center
Laurel, Maryland

C. RICHARD TERMAN, Associate Professor of Biology
Department of Biology
College of William and Mary
Williamsburg, Virginia

JOHN O. WHITAKER, JR., Associate Professor of Life Sciences
Department of Life Sciences
Indiana State University, Terre Haute

PREFACE

THE PURPOSES of this book are to gather the previous research on *Peromyscus* into one volume, to organize and evaluate this research in order to direct future investigations, and to illustrate the applicability of species-level comparisons to many scientific disciplines. These purposes could not readily be achieved by a single investigator for a large and complex genus like *Peromyscus*. The integration of material and the unity of style possible by one author has been sacrificed for the knowledge of experts in each discipline. The book is, therefore, a compilation of separate treatises with the genus *Peromyscus* as the common subject. In the future, some single student of the genus may find the material contained in this volume an incentive for preparing a definitive and interpretative analysis of *Peromyscus* biology.

I have limited my editorial responsibility to bringing the contributions together and attempting some consistency in form and terminology. While these goals are not the most laudable for a unified and integrated volume, they do leave the responsibility for the content where it belongs—with the expert.

A certain amount of repetition is unavoidable because many of the studies cited include several disciplines. Whereas this repetition may at first suggest impoverishment in the area of *Peromyscus* biology, careful reading will reveal that each author has used often-cited studies to illustrate different points or he has selected different material from the study. For example, the chapters on behavior and psychology draw upon many of the same studies, but present different information from them. Perhaps one function of this book will be to encourage specific and definitive studies of *Peromyscus*, which will be less amorphous, diffuse, and prone to repetitious citations than the currently reviewed material.

Omission of material from the book is a more serious shortcoming than repetition. Each chapter is of necessity a summary of the material reviewed and only a fraction of the available material has been presented. Furthermore, certain topics, like pathology and physiology, have been almost completely excluded because contributors could not be located. The organization of the book also provided room for some omissions, both of material and competent contributors. The most serious omission, however, results from the absence of sound research, which this book may stimulate.

Proper nomenclature is essential to a book using the comparative technique. However, consistency and agreement in nomenclature is difficult to achieve in an area as dynamic as the systematics of *Peromyscus*. Therefore, the nomenclature of the cited study has been retained even where synonyms are applicable or the systematics has been revised. Although some authors have indicated synonyms in their chapters, the species index can be consulted for synonymy and systematic discrepancies.

The initiation of this book was urged upon me several years ago by C. R. Terman and B. E. Eleftheriou while we were together at The Jackson Laboratory in Bar Harbor. We later met again at Purdue University to prepare an outline and list of contributors. Most of the contributors we invited accepted the responsibility of preparing a chapter and most stayed throughout the preparation of the book. At various stages during its preparation, W. H. Burt, E. T. Hooper, and L. R. Dice provided incentives and advice. Sketches throughout were made by David I. Rasmussen (DIR). Many of the contributions to this book have been supported by federal grants, primarily from the National Science Foundation and the National Institutes of Health. The recently initiated special publication series of the American Society of Mammalogists has provided an appropriate vehicle for the publication of this scholarly work.—J. A. K.

CONTENTS

INTRODUCTION

W. Frank Blair

THE DIVERSITY of life, of which his population is a part, has intrigued man as far back as his written, or even crudely drawn, records reveal. Pictographs of ancient man attest to this interest. There is no reason to doubt that this interest will continue or that challenging questions about this diversity will remain as long as civilized man exists on the earth. Regardless of how sophisticated our knowledge becomes concerning the chemical basis of life or of molecular and submolecular organization of living materials, the organization of life into populations of individuals, each adapted to reproduce itself, to carry out the physiological and mechanical functions necessary to its continued existence, and necessary to its coexistence with other populations, will continue to be a superbly challenging biological phenomenon. How is this diversity achieved? How is it maintained? What are the virtually endless ways in which populations adapt to their physical and biotic environments? Why do populations become extinct?

The diversity of life is so great that we can never hope to answer all of the questions we might pose for all of the kinds of organisms, so we look to detailed studies of representative taxa for elucidation of the principles governing diversification and maintenance of diversity. Many taxa are relatively unsuitable for this purpose, but in most of the major branches that organic evolution has produced there are outstandingly useful taxa for such investigations.

Ideally, the taxon should have a good fossil record that we might know something of its evolutionary history. Ideally, it should have a considerable amount, but not an impossibly great amount, of diversification in living forms so that various stages of the evolutionary process are adequately represented, e.g., from the level of genus through subgenus, species group, and species to populations representing incipient species. Ideally, the taxon should not be one

1

that has run its evolutionary course, but one that is actively evolving so that the mechanisms of evolution may be investigated as dynamic processes. Ideally, the organisms should be obtainable with relative ease and in adequate numbers, and they should be amenable to laboratory culture.

The North American genus of cricetine rodents, the white-footed or deer mice of the genus *Peromyscus*, about which this book is written, come as near to fulfilling these requirements as one might hope for in a taxon of small mammals. The fossil record leaves much to be desired (Chap. 1), but the included species indicate a range from Lower Pliocene to Recent, with some species groups appearing to have radiated as recently as the late Pleistocene. Diversity of living types within the genus is indicated by the 40 to 50 biological species of some 15 to 20 cenospecies classified in 7 subgenera (Chap. 2). These 40 to 50 species show a wide range of adaptation to their environments, from desert to taiga, from deciduous forest to prairie grassland and tropical scrub, from terrestrial-fossorial to saxicoline and arboreal (Chap. 4). Their geographic ranges vary from the restricted, as exemplified by *P. floridanus,* to the enormously broad as exemplified by *P. maniculatus.* Furthermore, the latter species, and to a lesser extent such species as *P. leucopus,* ranges through a great diversity of biomes and of local environmental variations so that its subpopulations reflect the effects of regional and local selection on the genotype. Such a species as *P. maniculatus,* ranging through most North American biomes, provides a living system for investigation of the counter-effects of local selection and gene dispersal in a huge biological species. As might be deduced from the short geologic history and from the diversity of types, the taxon is an evolutionarily active one. Present distributions reflect the effects of Wisconsin glacial-stage climates and shifts of biota. Within some species groups, e.g., the *maniculatus, leucopus, boylei,* and *truei* groups, populations occur in stages of geographic speciation ranging from disjunct, allopatric isolates with minimal morphological differentiation to populations at or near the level of distinct species that are interacting with one another in zones of secondary contact.

Various attributes of *Peromyscus* contribute to their usefulness for investigating the dynamics and behavior of natural populations. They are small, but not too small. Most are readily trapped in live traps and hence provide material for investigation of a number of populational parameters. Some live readily in nest boxes, thus making it possible to record such reproductive data as litter size and frequency, and to construct geneologies for natural populations. Several of the species are very abundant in nature, and one or more species are available for study in most parts of North America below the Arctic Circle.

Lastly, and very significantly, several of the species adapt readily to laboratory life and thrive and reproduce in captivity. Some species have failed to breed in the laboratory, but fortunately the members of the widely distributed and variable *maniculatus* and *leucopus* groups have proved to be splendid laboratory animals. This adaptability to laboratory life has made it possible to investigate modes of inheritance of genetic traits, physiological attributes, and behavioral characteristics of various species. Among all of the species, however, the huge, widely distributed and highly variable *P. maniculatus* has proved the most useful and consequently has been the object of probably more research than all other species of *Peromyscus* combined.

The history of any research trend is the history of the men whose vision and research initiated and influenced its development. Three names stand out in the history of the development of research with *Peromyscus* as it exists today and as it is reviewed in the following pages of this book.

The stage was set by Wilfred H. Osgood, who published in 1909 a monographic revision of the genus *Peromyscus*. Osgood did a monumental task of sorting out a confusing mass of names and specimens and succeeded in erecting a classification, mostly from straight morphological evidence, that has stood the test of time remarkably well. A major factor in Osgood's success was probably his ability to see beyond the typological concept of species that dominated the thinking of most taxonomists of his day (see Chap. 2). Emergence of the biological species concept and the utilization of many kinds of information other than that derived from straight morphology have resulted in changes in Osgood's classification, but the relatively small magnitude of these changes is a tribute to Osgood as a classifier of animal diversity.

Credit for demonstrating the feasibility of using *Peromyscus* as a laboratory animal for genetic and evolutionary studies belongs to

Francis B. Sumner. Sumner established laboratory colonies of *Peromyscus* and pioneered the work of studying geographic variation in a quantitative way. For measurement of pelage color, he devised a crude but effective colorimeter that has since been replaced for such work by commercially available electronic instruments. Sumner devoted most of his energies through the 1920's to geographic variations in pelage color and particularly to the striking parallels between coat color and such backgrounds as basaltic lava flows and white sands. As an interesting footnote, Sumner entered on this work with rather strong leanings toward a neo-Lamarckian explanation for the resemblance between coat color and background. Before his researches on *Peromyscus* were completed, Sumner came around to realization of the genetic basis and Darwinian explanation for coat resemblance to background. Sumner ultimately found his research on *Peromyscus* incompatible with his employment in an institution devoted to marine science. He summarized his *Peromyscus* work in an article in Bibliographia Genetica in 1932 and turned his energies to experimental studies of protective coloration in fishes.

When Sumner discontinued his *Peromyscus* work he gave his stocks, his colorimeter, and the benefits of his experience to Lee R. Dice, one of the contributors to this book. At the University of Michigan in the 1930's and 1940's, Dice transformed what was formerly a *Mus* genetics laboratory into a major center for the investigation of many aspects of *Peromyscus* biology and evolution. He perfected methods of raising and maintaining *Peromyscus* in the laboratory and the stocks he developed from his own field collecting and that of his students soon engulfed those he had inherited from Sumner. Dice, himself, worked on many aspects of *Peromyscus* biology, but among his many research publications those reporting interspecific hybridization experiments and those reporting his extensive studies of the genetic basis for geographic differentiation seem the most important. More importantly, during this period Dice trained a large number of students, and there are few facets of *Peromyscus* biology which have not been affected by this man either directly or through his students. Populational studies of *Peromyscus* through mark and recapture methods have their beginning in his group. The first nest-box studies of field populations of *Peromyscus* were done there. The first experimental studies of behavioral isolating mechanisms in *Peromyscus* were done there. Large breeding stocks of *Peromyscus* provided material for conventional genetic analysis of various traits of *Peromyscus,* for the studies of geographic variation, and for physiological, behavioral, and other studies. With Dice's retirement the work on *Peromyscus* became dispersed to many institutions and workers around the country. The following pages of this book will reveal the stage to which the biology of *Peromyscus* has progressed.

1
PALAEONTOLOGY

Claude W. Hibbard

Introduction

MANY OF THE first species of the New World living cricetines to be named were assigned to the genus *Mus* Linnaeus. Waterhouse (1839) erected the genus *Hesperomys* "to separate the New World *Murinae,* collectively, from typical *Mures* of the Old World, upon the broad characters of the tubercula-tion of the molars" (Coues and Allen, 1877:43).

E. D. Cope (1874) described a fossil cricetine rodent in North America as *Hesperomys loxodon* (= *Eumys loxodon,* J. A. Allen, 1877) from the Santa Fe Marls of New Mexico and another in 1879, as *Hesperomys nematodon,* from the John Day beds of Oregon. These were placed in the genus *Peromyscus* by O. P. Hay (1902). Later, A. E. Wood (1936) assigned each to a new genus.

Wheatley (1871) was the first to questionably assign the remains of a fossil cricetine to that of a Recent species. This was in regard to a lower jaw with M_1 and M_2 recovered from the Port Kennedy bone deposit. Cope (1899) made the following comment concern-ing the above specimen, "agreeing in detail of structure with the group with which our recent *H. leucopus* is type, and of the size of that species, not certainly referable to the latter without further comparison." O. P. Hay (1902) assigned it to *Peromyscus leucopus* (Rafinesque). If the specimen examined by Cope was a fossil associated with the fauna, it is most certain that it does not belong to a Recent species of *Peromyscus.*

Louise Kellogg (1910) described the first extinct species of a cricetine rodent, *Peromyscus antiquus,* that is presently assigned to the genus *Peromyscus.*

Many fossil cricetines have been incorrectly assigned to the genus *Peromyscus.* Fossil species considered by some workers as not belonging to the genus *Peromyscus, s. s.,* are:

6

Peromyscus loxodon (Cope, 1874) ; Hay, 1902; late Miocene or lowermost Pliocene of New Mexico = *Copemys loxodon* (Cope) , A. E. Wood, 1936.

Peromyscus nematodon (Cope, 1879) ; Hay, 1902; uppermost Oligocene or lowermost Miocene, John Day of Oregon = *Leidymys nematodon* (Cope) , A. E. Wood, 1936.

Peromyscus parvus Sinclair (1905) ; uppermost Oligocene or lowermost Miocene, John Day of Oregon = cf. *?Paciculus parvus* (Sinclair) , Wilson, 1949: 56 = *Leidymys parvus* (Sinclair) , Clark, Dawson, and Wood, 1964: 63.

Peromyscus brachygnathus Gidley (1922) , middle Pleistocene, Curtis Ranch local fauna, Arizona = *Baiomys brachygnathus* (Gidley) ; Hibbard, 1941: 352; Packard, 1960.

Peromyscus minimus Gidley (1922) , late Blancan, Benson local fauna, Arizona = *Baiomys minimus* (Gidley) ; Hibbard, 1941: 352; Packard, 1960.

Peromyscus dentalis Hall (1930a) , lower Pliocene (early to middle Clarendonian) , Fish Lake Valley fauna, Nevada = *Copemys dentalis* (Hall) ; Clark, Dawson, and Wood, 1964.

Peromyscus longidens Hall (1930b) , late Miocene, Barstow beds, California = *Copemys longidens* (Hall) ; Clark, Dawson, and Wood, 1964.

Peromyscus eliasi Hibbard (1937b) , upper Pliocene (early Blancan) Kansas = *Bensonomys eliasi* (Hibbard) ; Hibbard, 1956.

Peromyscus martinii Hibbard (1937a) , middle Pliocene (Hemphillian) , Edson beds, Kansas = *Onychomys martini* (Hibbard) ; Hoffmeister, 1945.

Peromyscus cf. *nuttalli* (Harlan) , Olson, 1940; late Pleistocene of Missouri = *Ochrotomys* cf. *nuttalli* (Harlan) , Hooper, 1958; Hooper and Musser, 1964b.

Peromyscus kelloggae Hoffmeister (1959) , early Pliocene (Clarendonian) Niobrara River fauna Nebraska = *Copemys kelloggae* (Hoffmeister) , Clark, Dawson, and Wood, 1964.

Peromyscus russelli James (1963) , late Barstovian to late Clarendonian, California = *Copemys russelli* (James) , Clark, Dawson, and Wood, 1964.

Peromyscus sawrockensis Hibbard (1964) , late Hemphillian, Saw Rock Canyon fauna, Kansas. Not *Peromyscus s. s.*; more closely related to the *Bensonomys–Symmetrodontomys* group.

Pliocene Records

The following fossil records are considered in this paper as belonging to the genus *Peromyscus*. They are arranged from the oldest occurrence to the youngest.

Peromyscus dentalis Hall, 1930
(Fig. 1, No. 1)

Peromyscus dentalis Hall, 1930a, Univ. Calif. Publ., Bull. Dept. Geol. Sci., 19 (12) :306, Figs. 15–16. Fish Lake Valley fauna, Nevada.

Peromyscus dentalis Hall. Wilson, 1939, Carnegie Inst. Wash. Publ. 514:38. Avawatz fauna, late Clarendonian, California.

Peromyscus cf. *P. dentalis* Hall. Shotwell and Russell, 1963, Trans. Amer. Phil. Soc., n.s., 53 (1) :43.
Copemys dentalis (Hall). Clark, Dawson, and Wood, 1964. Bull. Mus. Comp. Zool., 131 (2) :47–49.

HOLOTYPE.—No. 29635, Univ. Calif. Mus. Paleo.; fragment of right lower jaw with incisor, M_1–M_2.

TYPE LOCALITY AND AGE.—Seven miles north of Chiatovich Ranch, Esmeralda County, Nevada, Esmeralda Formation, Fish Lake Valley local fauna, early Pliocene (Clarendonian).

DIAGNOSIS.—Hall, 1930*a*: "Size about as in *Peromyscus maniculatus gambeli*; masseteric fossa well defined; space between coronoid process and M_3 wide and including deep fossa; horizontal ramus thick, especially at diastema; lower teeth wide, without accessory tubercles, and principal cusps low, M_3 large; M_2 large relative to M_1."

DISCUSSION.—I have not examined the holotype of *Peromyscus dentalis* Hall, nor the specimens referred to this species. Hall's description restricts it to the genus *Peromyscus*. To my knowledge it is the only fossil cricetine in which the size relationship of M_2 to M_1 can be considered as ancestral to the subgenus *Megadontomys* (ex. *Peromyscus thomasi*).

This species is known from parts of five lower jaws. The upper molars and M_3 are unknown.

Peromyscus pliocenicus Wilson, 1937
(Fig. 1, No. 2)

Peromyscus pliocenicus Wilson, 1937, Carnegie Inst. Wash., Publ. 487:15, Pl. 3, Figs. 2–4.

HOLOTYPE.—No. 1966 C. I. T. Coll. Vert. Paleo. (now part of the Los Angeles County Museum Collection), fragmentary right lower jaw with M_1 and M_2. This species is known from the holotype and a fragmentary right maxillary with M^1 and incomplete M^2.

TYPE LOCALITY AND AGE.—C. I. T. loc. 49, Kern County, California; Kern River beds, Kern River fauna, middle Pliocene (Hemphillian).

DIAGNOSIS.—Wilson, 1937: "Cheek-teeth hypsodont, but crowns show tendency to wear to flat surfaces; without accessory folds although intermediate tubercles may be present or absent. M_1 with divided antero-median cusp; tip of anterointernal re-entrant angle becomes isolated with wear leaving a broad, shallow re-entrant. M_3 relatively unreduced. Pit or foramen in maxillary bone lateral to anterior root of M^1. Size large, slightly exceeding any known fossil species, but approximately that of *Peromyscus nesodytes*." Occlusal length of M_1–M_2 is 4.4 mm. This specimen has not been studied.

Peromyscus antiquus Kellogg, 1910
(Fig. 1, No. 6)

Peromyscus antiquus Kellogg, 1910, Univ. Calif. Publ., Bull. Dept. Geol., 5 (29) : 432, Fig. 16.
Peromyscus near *P. antiquus* Kellogg. Wilson, 1936, Smith Valley local fauna, Nevada. Carnegie Inst., Wash., Publ. 473:33.
Peromyscus antiquus Kellogg. Hoffmeister, 1949. Chalk Springs local fauna, Nevada, late Clarendonian. Univ. Calif. Publ., Bull. Dept. Geol. Sci., 28 (7) : 178, Fig. 4.

HOLOTYPE.—No. 12571, Univ. Calif. Mus. Paleo.; fragmentary left lower jaw with M_1–M_3.

TYPE LOCALITY AND AGE.—UCMP loc. 1103, Humboldt County, Nevada, Alturus Formation, Thousand Creek local fauna, late Hemphillian.

GEOLOGIC RANGE.—Late Clarendonian to late Hemphillian.

DIAGNOSIS.—Kellogg, 1910: "Its large size distinguishes it from *Peromyscus californicus* but the tooth pattern is practically identical, except that M_2 and M_3 have a ridge on the anterior face which is present in M_2 of *P. californicus* but not in M_3."

DISCUSSION.—A lower jaw with M_1–M_3 and another with M_1 and M_2 have been assigned to this species. Specimens have not been seen.

Peromyscus kansasensis Hibbard, 1941
(Fig. 1, No. 8)

Peromyscus kansasensis Hibbard, 1941, Amer. Midland Nat., 26 (2) :352, Fig. 11.
Peromyscus kansensis Hibbard, 1962, Jour. Mamm., 43 (4) :484 = *P. kansasensis* Hibbard, 1941.

HOLOTYPE.—No. 4597, Univ. Kans. Mus. Coll., part of right lower jaw with incisor, M_1–M_3. This species is known only from the holotype.

TYPE LOCALITY AND AGE.—Univ. Kans. Meade County, Kansas loc. 3, W ½ SW ¼ Sec 22, T. 33 S, R. 29 W, Rexroad Ranch; Rexroad Formation, upper Pliocene (early Blancan).

DIAGNOSIS.—A form the size of *Peromyscus c. californicus* (Gambel) with low open cusps and wide open valleys, between cusps; no lophids or stylids. M_2 not compressed by crowding, with rather narrow transverse width in proportion to anteroposterior length. Length of M_1–M_3 is 4.4 mm.

Peromyscus baumgartneri Hibbard, 1954
(Fig. 1, No. 9)

Peromyscus baumgartneri Hibbard, 1954, Trans. Kans. Acad. Sci., 57 (2) :232, Figs. 3F–H.

HOLOTYPE.—No. 29677, Univ. Mich. Mus. Paleo., fragmentary right lower jaw with incisor and M_1–M_3.

TYPE LOCALITY AND AGE.—Univ. Kansas loc. 3, W ½ SW ¼ Sec 22, T. 33 S, R. 29 W, Rexroad Ranch; Meade County, Kansas; Rexroad Formation, upper Pliocene (early Blancan).

GEOLOGIC RANGE.—Rexroad and Fox Canyon local faunas, also known from Univ. Mich. loc. UM-Kl-47 (Fox Canyon), Meade County, Kansas.

DIAGNOSIS.—Smallest of the known Pliocene species of *Peromyscus*. No anteromedian groove on face of anteroconid of M_1; no lophids or stylids. Anteroposterior length of M_1–M_3 is 3.4 mm.

DISCUSSION.—*Peromyscus baumgartneri* was recovered from the same deposits as *P. kansasensis*. Besides the holotype there are four paratypes all of which are fragmentary jaws.

Peromyscus hagermanensis Hibbard, 1962
(Fig. 1, No. 7)

Peromyscus hagermanensis Hibbard, 1962. Jour. Mamm., 43 (4) :484, Fig. 2.

HOLOTYPE.—No. 34441, Univ. Mich. Mus. Paleo., fragmentary right lower jaw with M_2. The holotype is the only known specimen of this species.

TYPE LOCALITY AND AGE.—U. S. Geol. Survey Cenozoic loc. 20765 (D. W. T. loc. 540) SW ¼ Sec 28, T. 7 S, R. 13 E, Twin Falls County, Idaho; Glenns Ferry Formation, Upper Pliocene (early Blancan) K/A date 3,500,000 years (Evernden, Curtis, Savage, and James, 1964: 164).

DIAGNOSIS.—Size larger than *Peromyscus maniculatus* (Wagner) and *P. crinitus* (Merriam) now living in Idaho. *P. hagermanensis* is smaller than *P. antiquus* Kellogg, *P. irvingtonensis* Savage, *P. nesodytes* Wilson, *P. pliocenicus* Wilson, or *P. kansasensis* Hibbard. It is near the size of *P. dentalis* Hall from which it is distinguished by the absence of the deep fossa between the coronoid process and M_3. It is larger than *P. baumgartneri* Hibbard and *P. cragini* Hibbard. Furthermore, thin stylids close the base of the re-entrant valleys between the metaconid and entoconid and between the entoconid and posterior cingulum. Anteroposterior length of M_2 is 1.35 mm.

Pleistocene Records

Peromyscus irvingtonensis Savage, 1951
(Fig. 1, No. 3)

Peromyscus (*Haplomylomys*) *irvingtonensis* Savage, 1951, Univ. Calif. Publ., Bull. Dept. Geol. Sci., 28 (10) :228, Fig. 5.

HOLOTYPE.—No. 38781, Univ. Calif. Mus. Paleo., fragmentary right lower jaw with M_1 and M_3. Known only from the holotype.

TYPE LOCALITY AND AGE.—UCMP loc. V-3604, Irvington Site 2, 0.8 mile SE from the center of Irvington, Alameda County, California; Pleistocene, Irvingtonian age.

DIAGNOSIS.—Savage, 1951: "Length of cheek-tooth row (5.2 mm) intermediate between *P. californicus californicus* (Gambel) and *P. nesodytes* Wilson; teeth relatively massive when compared with all species of the subgenus *Haplomylomys* except *P. antiquus* Kellogg, *P. pliocenicus* Wilson, and *P. nesodytes*; M_3 relatively large (anteroposterior diameter 1.4 mm) as compared to other species except *P. antiquus* and possibly *P. pliocenicus*; M_3 slightly more reduced relative to other cheek teeth than in type of *P. antiquus*."

Peromyscus cragini Hibbard, 1944
(Fig. 1, No. 10)

Peromyscus cragini Hibbard, 1944, Geol. Soc. Amer., Bull. 55 (6) :724, Fig. 8.
Peromyscus cragini Hibbard. Paulson, 1961, Papers Mich. Acad. Sci., Arts, and Letters, 46:140.

HOLOTYPE.—No. 6618, Univ. Kans. Mus. Coll., fragmentary left lower jaw with incisor and M_1–M_3.

TYPE LOCALITY AND AGE.—Univ. Kans. Meade County, Kansas loc. 17, Pearlette ash mine in NW ¼ SE ¼ Sec 26, T. 32 S, R. 28 W,

John J. Isaacs Ranch; Crooked Creek Formation, Pleistocene, late Kansan, early Irvingtonian.

GEOLOGIC RANGE.—Known only from the Cudahy fauna, KU localities 10 and 17, Meade County, Kansas.

DIAGNOSIS.—A form slightly smaller than *Peromyscus maniculatus nebrascensis* (Coues) and with the valleys between the cusps broader and more open, not so deep. No occurrence of lophids, a small mesostylid rarely present. Anteroposterior length of M_1–M_3 is 3.5 mm.

DISCUSSION.—The specimens of this species were taken from below the Pearlette volcanic ash. Besides the holotype there are parts of nine other specimens consisting of lower and upper dentitions.

Peromyscus berendsensis Starrett, 1956
(Fig. 1, No. 13)

Peromyscus berendsensis Starrett, 1956, Jour. Paleo., 30 (5) :1188, Fig. 1a.
Peromyscus berendsensis Starrett. Hibbard, 1963. Mt. Scott local fauna, late Illinoian, Meade County, Kansas. Papers Mich. Acad. Sci., Arts, and Letters, 48:207, Fig. 4c.

HOLOTYPE.—No. 31789, Univ. Mich. Mus. Paleo., fragmentary right lower jaw with M_1. Parts of two other jaws were recovered in Meade County, Kansas, in association with the Mt. Scott local fauna.

TYPE LOCALITY AND AGE.—Berends ranch, SE corner of Sec 6, T. 5 N, R. 28 E. C. M., Beaver County, Oklahoma; Pleistocene, Illinoian, early Rancholabrean.

DIAGNOSIS.—M_1 (greatest length, 1.4 mm; greatest width, 0.9 mm) shorter and narrower than those of *Peromyscus baumgartneri* Hibbard or *P. cochrani* Hibbard. A small mesolophid is present. M_1 closely resembles that of *P. maniculatus,* which generally has a more complicated cusp pattern. The M_1 of *P. maniculatus bairdi* (Hoy and Kennicott) differs from that of *P. berendsensis* by being longer and wider. Also the posterior slopes of the cusps are steeper than those of *P. berendsensis.*

Peromyscus oklahomensis Stephens, 1960
(Fig. 1, No. 14)

Peromyscus oklahomensis Stephens, 1960, Bull. Geol. Soc. Amer., 71 (11) :1692, Fig. 7B.

Peromyscus oklahomensis Stephens. Dalquest, 1962. Easley Ranch local fauna, Foard County, Texas. Jour. Paleo., 36 (3) :576.

HOLOTYPE.—No. 38672, Univ. Mich. Mus. Paleo., a right M_2.

TYPE LOCALITY AND AGE.—UMMP loc. 2, NE ¼, SW ¼ Sec 10, T. 27 N, R. 24 W, Dees Ranch, Harper County, Oklahoma; Doby Springs lake beds, Pleistocene, Illinoian, early Rancholabrean.

GEOLOGIC RANGE.—Early to mid-Rancholabrean.

DIAGNOSIS.—Stephens, 1960: "A mouse tooth the size of M_2 (length, 1.68 mm) of *Peromyscus floridanus* Bangs, but with extremely broad re-entrant valleys between the cusps. The mesostylid is well developed, but there is no evidence of a mesolophid. The ectostylid is very rudimentary."

DISCUSSION.—Other specimens have been recovered by Dalquest in Foard County, Texas.

Peromyscus progressus Hibbard, 1960
(Fig. 1, No. 11)

Peromyscus progressus Hibbard, 1960, Contrib. Mus. Paleo. Univ. Mich., 16 (1) : 171, Figs. 13A and 13H.

Peromyscus progressus Hibbard, 1963. Mt. Scott local fauna, late Illinoian, Meade County, Kansas. Papers Mich. Acad. Sci., Arts, and Letters, 48:207.

HOLOTYPE.—No. 35637, Univ. Mich. Mus. Paleo., fragmentary right lower jaw with M_1–M_3.

TYPE LOCALITY AND AGE.—Univ. Kans. Meade County, Kansas, loc. 6, Cragin Quarry, SW ¼ Sec 17, T. 32 S, R. 28 W, Big Springs Ranch; Kingsdown Formation, Pleistocene, Sangamon, Rancholabrean.

GEOLOGIC RANGE.—Known from the Mt. Scott local fauna (late Illinoian) and the Cragin Quarry local fauna (Sangamon), Meade County, Kansas.

DIAGNOSIS.—A *Peromyscus* the size of *P. cochrani* Hibbard; larger than *P. cragini* Hibbard or *P. berendsensis* Starrett. Styles, stylids, lophs, and lophids present or absent. Internal and external re-entrant valleys between the cusps broader than in *P. maniculatus* and most specimens of *P. leucopus*. The anteroposterior length of M_1–M_3 is 3.9 mm.

DISCUSSION.—Parts of ten lower jaws and eight maxillaries with teeth were taken in association with the holotype. These have been considered as members of a single species. The characters of the teeth would allow the placement of some of these specimens in the subgenus, *Haplomylomys* and others in the subgenus *Peromyscus*.

Peromyscus cochrani Hibbard, 1955
(Fig. 1, No. 12)

Peromyscus cochrani Hibbard, 1955, Contrib. Mus. Paleo. Univ. Mich., 12 (10) : 210, Fig. 5A.
Peromyscus cochrani Hibbard. Slaughter and Ritchie, 1963. Clear Creek local fauna, Denton County, Texas. So. Methodist Univ. Grad. Research Center Jour., 31 (3) :123.

HOLOTYPE.—No. 27542, Univ. Mich. Mus. Paleo., right lower jaw with incisor and M_1–M_3.

TYPE LOCALITY AND AGE.—Univ. Mich. loc. UM-K2-47, Jinglebob pasture, SW ¼ Sec 32, T. 33 S, R. 29 W, XI Ranch, Meade County, Kansas. Kingsdown Formation, Pleistocene, late Sangamon, Rancholabrean.

GEOLOGIC RANGE.—Known from type locality, Jinglebob local fauna, late Sangamon and the Clear Creek local fauna, Wisconsin, Denton County, Texas.

DIAGNOSIS.—A mouse the size of *Peromyscus leucopus novebora-censis* (Fischer). Dental characters intermediate between *Peromyscus progressus* Hibbard and the Recent species, *P. leucopus* and *P. maniculatus*. Internal and external re-entrant valleys between the cusps broader in *P. cochrani* than in *P. leucopus* and *P. maniculatus*. Mesostylid, ectostylid, mesostyle, and enterostyle not as well developed in most specimens as in Recent species of the subgenus *Peromyscus*. Lophids rarely present on the lower teeth; mesolophs on upper molars either slightly developed or absent. Anteroposterior length of M_1–M_3 is 3.9 mm when measured with the same ocular micrometer as was the holotype of *P. progressus*. Previously published (Hibbard, 1955) as 3.76 mm.

DISCUSSION.—Five other lower jaws were recovered with the holotype, a lower jaw from a Wisconsin deposit in Denton County, Texas, is assigned to this species.

Peromyscus imperfectus Dice, 1925
(Fig. 1, No. 4)

Peromyscus imperfectus Dice, 1925, Carnegie Inst. Wash., Publ. 349:123, Figs. 1–2.
Peromyscus gambeli (?) Kellogg, 1912, Univ. Calif. Publ. Bull. Dept. Geol., 7:166.

HOLOTYPE.—No. 21879, Univ. Calif. Mus. Paleo., a skull with right M^1 and M^2, and left M^2 and M^3; nasals, parietals, occipitals, bullae, and mastoids missing. This species is known only from the holotype.

TYPE LOCALITY AND AGE.—UCMP loc. 20-N-1, 2051, Rancho La Brea deposits, Los Angeles County, California; Wisconsin, late Rancholabrean.

DIAGNOSIS.—Dice, 1925: "Size about that of *Peromyscus maniculatus gambelii*. Small accessory tubercules present on M^1 and M^2. Palatine slits proportionally longer and shelf of bony palate proportionally shorter than in the skulls of other species of the genus. The palatine slits are wide in front as in *P. m. gambelii* and not narrowed anteriorly as in *P. eremicus fraterculus* and in *P. e.* [*sic*] *crinitus*."

Peromyscus nesodytes Wilson, 1936
(Fig. 1, No. 5)

Peromyscus nesodytes Wilson, 1936, Jour. Mamm., 17 (4) :408, Fig. 1.

HOLOTYPE.—No. 1780, C. I. T. Coll. Vert. Paleo. (now part of the Los Angeles County Museum Collection) ; part of right lower jaw with M_1–M_3. This species is known only from the holotype.

TYPE LOCALITY AND AGE.—C. I. T. loc. 106, Santa Rosa Island, Santa Barbara County, California; Pleistocene, late Rancholabrean.

DIAGNOSIS.—Wilson, 1936: "Partially bipartite antero-median cusp in M_1. Antero-internal valley in M_1, V-shaped. No accessory tubercles on cheek-teeth. Size very large, slightly larger than *Peromyscus antiquus*. Length of tooth row M_1–M_3, 5.9 mm."

The distribution of the following late Pleistocene fossils that have been assigned to living species are shown in Figure 1, A–H.

Peromyscus boylei Baird
(Fig. 1, A)

Stock, C. C., 1918, Univ. Calif. Publ. Bull. Dept. Geol., 10 (24) :468. Hawver Cave local fauna, Eldorado County, California. Wisconsin. Number of specimens recovered not reported.

Jakway, G. E., 1958, Trans. Kans. Acad. Sci., 61 (3) :320. San Josecito Cave, Nuevo Leon, Mexico. Wisconsin. Jakway reports two fragmentary skulls and twelve dentaries with complete dentition.

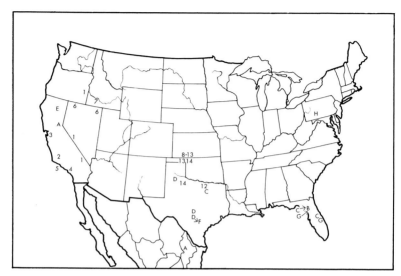

FIG. 1.   Geographic location of remains of fossil *Peromyscus*. 1, *P. dentalis;*
2, *P. pliocenicus;* 3, *P. irvingtonensis;* 4, *P. imperfectus;* 5, *P. nesodytus;* 6, *P.
antiquus;* 7, *P. hagermanensis;* 8, *P. kansasensis;* 9, *P. baumgartneri;* 10, *P.
cragini;* 11, *P. progressus;* 12, *P. cochrani;* 13, *P. berendsensis;* 14, *P. oklahomensis.* A, *P. boylei;* B, *P. floridanus;* C, *P. gossypinus;* D, *P. leucopus;* E, *P.
maniculatus gambeli;* F, *P.* cf. *nasutus;* G, *P. polionotus;* H, *P. maniculatus
gracilis.*

Peromyscus cf. *californicus* (Gambel)

Schultz, J. R., 1938, Carnegie Inst. Wash., Publ. 487:208. McKittrick local fauna,
Kern County, California. Wisconsin. Approximately 100 left lower jaws and
nearly 90 right lower jaws are questionably referred to *Peromyscus* cf. *californicus.*

Peromyscus floridanus (Chapman)
(Fig. 1, B)

Sherman, H. B., 1952, Quart. Jour. Florida Acad. Sci., 15 (2) :93. Number of
specimens not reported.
Ray, C. E., 1958, Bull. Mus. Comp. Zool., 119 (7) :432, footnote. Reddick local
fauna, Marion County, Florida. Number of specimens not reported.

Peromyscus gossypinus (Le Conte)
(Fig. 1, C)

Ray, C. E., 1958, Bull. Mus. Comp. Zool., 119 (7) :432. Reddick local fauna,
Marion County, Florida, and Melbourne local fauna, Brevard County, Florida.
The latter record is based on two lower jaws with teeth.

Crook, W. W., Jr., and R. K. Harris, 1958, Amer. Antiquity, 23 (3) :241. Lewisville site local fauna, Denton County, Texas. Wisconsin, greater than 37,000 B. P. years. Number of specimens not given.

Gut, J. H., and C. E. Ray, 1963, Quart. Jour. Fla. Acad. Sci., 26 (4) :325. Reddick local fauna, Marion County, Florida. Late Pleistocene. Number of specimens not reported.

Peromyscus leucopus (Rafinesque)
(Fig. 1, D)

Dalquest, W. W., 1962, Jour. Paleo., 36 (3) :576. Easley Ranch local fauna, Foard County, Texas. Early Wisconsin. Number of specimens not given.

Patton, T. H., 1963, Bull. Texas Memorial Museum, No. 7:27. Miller's Cave local fauna, Llano County, Texas. Late Wisconsin. Seven lower jaws recovered.

Peromyscus cf. *P. leucopus* (Rafinesque)

Gidley, J. W., and C. L. Gazin, 1938, Bull. U. S. Natl. Mus., 171:59. Cumberland Cave local fauna, Allegany County, Maryland. Late Pleistocene; known from eight fragmentary lower jaws with incisors. Seven jaws have one or more cheek teeth each.

Pruitt, W. O., Jr., 1954, Papers Mich. Acad. Sci., Arts, and Letters, 39:255. Wisconsin. Part of three lower jaws lacking molar teeth.

Tamsitt, J. R., 1957, Texas Jour. Sci., 9 (3) :362. Friesenhahn Cave local fauna, Bexar County, Texas. Wisconsin. Based on a study of 86 lower jaws.

Guilday, J. E., 1962, Ann. Carnegie Mus., 36 (9) :93. Natural Chimneys local fauna, Augusta County, Virginia. Wisconsin. Parts of 12 individuals.

Guilday, J. E., P. S. Martin, and A. D. McCrady, 1964, Bull. Natl. Speleological Soc., 26 (4) :158. New Paris No. 4 local fauna, Bedford County, Pennsylvania. Late Wisconsin. C^{14} date, 11,300 B. P. years. Parts of 2 left, and 1 right maxilla; 7 left, and 11 right lower jaws.

Dalquest, W. W., 1965, Jour. Paleo., 39 (1) :70. Howard ranch local fauna, Grosebeck Formation, Hardeman County, Texas. Late Wisconsin, C^{14} dates are $16,775 \pm 565$, and $19,098 \pm 1,074$ B. P. years. Based on two jaws with one tooth each; reported as *Peromyscus* cf. *maniculatus*.

Peromyscus maniculatus gambeli (Baird)
(Fig. 1, E)

Kellogg, Louise, 1912, Univ. Calif. Publ. Bull. Dept. Geol., 7 (8) :158. Samwel Cave, Eldorado County, California. Wisconsin. Known from part of a skull with left upper tooth row.

Peromyscus maniculatus gracilis (Le Conte)
(Fig. 1, H)

Peterson, O. A., 1926, Ann. Carnegie Mus., 16:276. Reported as *Peromyscus canadensis* Miller. Specimens consist of three lower jaws and some fragmentary upper jaws and isolated teeth from the Frankstown Cave local fauna, Blair County, Pennsylvania.

Peromyscus cf. *P. maniculatus* (Wagner)

Tamsitt, J. R., 1957, Texas Jour. Sci., 9 (3) :362. Friesenhahn Cave local fauna, Bexar County, Texas. Wisconsin. Number of specimens not given.

Guilday, J. E., 1962, Ann. Carnegie Mus., 36 (9) :93. Natural Chimneys local fauna, Augusta County, Virginia. Wisconsin. Parts of 35 individuals.

Slaughter, B. H., and R. Ritchie, 1963, So. Methodist Univ. Grad. Research Center Jour., 31 (3) :124. Clear Creek local fauna, Denton County, Texas. Wisconsin, C^{14} age 28,840 ± 4,740 B. P. years. Known from a left lower jaw and maxillary.

Guilday, J. E., P. S. Martin, and A. D. McCrady, 1964, Bull. Natl. Speleological Soc., 26 (4) :158. New Paris No. 4 local fauna, Bedford County, Pennsylvania. Late Wisconsin, C^{14} date 11,300 B. P. years. Parts of four skulls; 72 left, and 93 right maxillae; 128 left, and 117 right lower jaws.

Dalquest, W. W., 1965, Jour. Paleo., 39 (1) :70. Howard ranch local fauna, Grosebeck Formation, Hardeman County, Texas. Late Wisconsin, C^{14} dates are 16,775 ± 565, and 19,098 ± 1,074 B. P. years. A lower jaw with teeth and two isolated teeth reported as *Peromyscus* cf. *leucopus*.

Peromyscus cf. *P. nasutus* (J. A. Allen)
(Fig. 1, F)

Tamsitt, J. R., 1957, Texas Jour. Sci., 9 (3) :362. Friesenhahn Cave local fauna. Bexar County, Texas. Wisconsin. Number of specimens not given. Hoffmeister and de la Torre (1961) consider *P. nasutus* as a synonym of *P. difficilis*.

Peromyscus polionotus (Wagner)
(Fig. 1, G)

Bader, R. S., 1959, Jour. Paleo., 33 (5) :968. Stratum No. 2, Vero Beach local fauna, Indiana River County, Florida. Wisconsin. One lower jaw with M_1–M_3.

Gut, H. J., and C. E. Ray, 1963, Quart. Jour. Fla. Acad. Sci., 26 (4) :325. Reddick local fauna, Marion County, Florida (tentative Illinoian age). Number of specimens not given.

Geographic and Geologic Distributions

These distributions are shown in Figures 1 and 2. All fossil specimens assigned to the genus *Peromyscus,* and of Lower (Clarendonian) and Middle (Hemphillian) Pliocene age, have been taken from deposits in western North America.

The Late Pliocene *Peromyscus hagermanensis* from Idaho appears to be more advanced in tooth structure than *P. kansasensis* or *P. baumgartneri* in the development of stylids which have not been observed in the teeth of *Peromyscus* from the Late Pliocene of the Plains region. *Peromyscus hagermanensis* was recovered well below the basalt that yielded a K/A date of 3.48 ± 0.27 × 10^6 yrs. (Evernden *et al.,* 1964:191; K/A sample 1173).

The earliest records of *Peromyscus, s. s.,* east of the Rocky Mountains are known from Upper Pliocene deposits of southwestern

Kansas. It is interesting to note that *Peromyscus* is by far the rarest of all cricetines taken from these deposits in association with *Neotoma, Onychomys, Symmetrodontomys, Bensonomys, Reithrodontomys, Baiomys,* and *Sigmodon.*

In the past, the *Peromyscus* from the Upper Pliocene of southwestern Kansas were assigned to the subgenus *Haplomylomys* because of the absence of stylids and lophids. From the study of later Pleistocene populations it appears that the absence of stylids, lophids, styles, and lophs represents a morphological grade in some populations. These populations do not belong to the subgenus *Haplomylomys* as based upon Recent species (see Hibbard and Taylor, 1960:175) of this subgenus.

A single tooth of *Peromyscus* was recovered from deposits considered to be Nebraskan in age (Dixon local fauna, Hibbard, 1956: 164). No remains of *Peromyscus* are known from deposits of Aftonian (first interglacial) age, though remains of *Bensonomys* and *Sigmodon* have been recovered with the Sanders local fauna (Hibbard, 1956:179).

Peromyscus remains are unknown from Yarmouth (second interglacial) deposits. These deposits at the site of the Borchers local fauna have yielded remains of *Reithrodontomys,* abundant jaws of *Onychomys* and *Sigmodon,* as well as numerous teeth of *Neotoma.*

Peromyscus from the Plains region during Illinoian (third glacial) and Sangamon (third interglacial) show an increase in the development of styles, stylids, lophs, and lophids from the earlier to the later deposits.

There is one isolated tooth known from the Mt. Scott local fauna (late Illinoian, Hibbard, 1963:209) that is as well developed as any tooth in the *truei* or the *boylei* group. If one considered only the present range of the two species it would be assigned to *Peromyscus boylei,* but pollen studies show that juniper was common in this region during part of Illinoian time (Kapp, 1965). Hoffmeister (1951:vii) states that "Throughout much of this range, the mice [*Peromyscus truei*] occur only in association with piñons, or junipers where these trees replace the piñon, and where the piñon-juniper grows among rocks." Rocks are not abundant in the Kansas area though a number of trees are known to have extended their ranges eastward from the Rocky Mountain region into southwestern Kansas during Illinoian time.

Another such specimen, a right lower jaw with M_1–M_3, was recovered with the Jinglebob local fauna of late Sangamon age (Hibbard, 1955:213).

Most of the specimens of *Peromyscus* recovered from Wisconsin deposits are referable to living species. *Peromyscus* does not appear as a common member of a fauna in the United States until the late Pleistocene.

The fossil record indicates that the species *maniculatus, polionotus, leucopus,* and *gossypinus* reached their present dental grade in the Wisconsin.

It appears that the genus *Peromyscus* was able to take advantage of the diversified climate of the Pleistocene to reinhabit much of southwestern United States that was occupied during the late Pliocene by a number of genera of cricetines.

Ancestral Stock

It has been shown by Rinker (1954), Hooper (1958 and 1960), Burt (1960), Hershkovitz (1962), and Hooper and Musser (1964a) that there are two distinct groups of Recent genera of Cricetinae in North America and South America. One consists of *Oryzomys, Sigmodon,* and related genera, and the second of *Neotoma, Onychomys, Reithrodontomys, Baiomys, Peromyscus,* and related genera.

If these two groups of genera were derived from a common ancestral stock, they must have separated by Oligocene or earliest Miocene. I consider *Onychomys, Neotoma, Baiomys, Reithrodontomys,* and *Peromyscus* to have been derived from a common ancestor.

The earliest record of *Onychomys* is from the Edson Quarry fauna of Middle (Hemphillian) Pliocene age.

The earliest *Baiomys* are known from the Saw Rock Canyon fauna of late Hemphillian age from Kansas. Specimens of *Baiomys* were recovered with those of *Onychomys* and two teeth of *Neotoma* (Hibbard, 1967).

The earliest *Reithrodontomys* are known from the Rexroad fauna, Upper Pliocene (early Blancan) of Kansas. Here they are found with remains of *Baiomys, Peromyscus, Neotoma, Onychomys,* and other extinct cricetines.

It appears more logical to derive the above five genera from a common cricetine with a simple tooth pattern than from *Leidymys* with a complicated tooth pattern which in turn is derived from *Eumys* with a more complicated pattern, as suggested by Clark, Dawson, and Wood (1964:43). If *Peromyscus* is derived as suggested by Clark, Dawson, and Wood, some of the Pleistocene species regained lophs, styles, lophids, and stylids that their ancestors once possessed but lost.

Areas for Future Study

Pliocene and early Pleistocene deposits of Mexico should be carefully searched for remains of *Peromyscus* and associated vertebrates.

The specimen from Port Kennedy, Pennsylvania, and those from Cumberland Cave, Maryland, should be carefully re-studied and compared with large series of Recent specimens.

Specimens from the Reddick local fauna of Florida that are identified as Recent species of the *maniculatus* and *leucopus* groups of *Peromyscus,* and assigned an Illinoian age, should be compared with a large series of Recent specimens taken both north and west of Florida. Such a comparison would show where the Floridian specimens fit in regard to the present cline for the development of lophs, styles, lophids, and stylids as shown by Hooper (1957). If the age and specific assignment of these specimens are correct, it appears that these species groups may have developed in Florida and spread westward and northward.

I found that there is greater variation in Sangamon specimens from southwestern Kansas, in the development of lophs, styles, lophids, and stylids, than there is in Recent series of *maniculatus* and *leucopus* from Kansas and Michigan. Is it possible that these species groups possessed more complicated teeth during Illinoian time in Florida than in Kansas, and that the fossil specimens from Kansas that I considered to be related to the *maniculatus* and *leucopus* groups are not closely related, but belong to an extinct line of *Peromyscus?*

Brown (1908) reported that the remains of *Peromyscus* from the Conard Fissure in Arkansas were the most abundant of all the animals found. He reported 700 lower jaws as well as other elements. There is always the problem regarding the age of all of the specimens, but it is the only collection reported that would provide material for a careful detailed study of the variation of

EPOCHS	MAMMALIAN AGES	GLACIAL AGES	EXTINCT SPECIES OF PEROMYSCUS
PLEISTOCENE	RANCHOLABREAN	WISCONSIN	P. nesodytes P. imperfectus
		SANGAMON	P. cochrani P. progressus
		ILLINOIAN	P. oklahomensis P. berendsensis
	IRVINGTONIAN	YARMOUTH	P. irvingtonensis P. cragini
		KANSAN	
ca. 2 x 10^6 yrs.	BLANCAN	AFTONIAN NEBRASKAN	
			P. hagermanensis P. baumgartneri P. kansasensis
PLIOCENE	HEMPHILLIAN		P. antiquus P. pliocenicus
ca. 12 x 10^6 yrs.	CLARENDONIAN		P. dentalis

Fig. 2. Geological occurrence of the extinct species of *Peromyscus*.

dental characters in a large fossil sample. This material has never been studied. It probably includes a number of species.

In the study of specimens from late Pleistocene deposits it is most important to compare them with a large series of Recent specimens taken throughout the present range of the species because of the possible shift of ranges in the past. In such studies I have disregarded subspecies and looked for clinal characters through the entire range of the species as in the study of *Synaptomys cooperi* (Hibbard, 1963). This allows for a better interpretation of the fossil specimens.

Summary

The genus *Peromyscus* has been used as a "catch-all" name to which 25 extinct forms of fossil cricetids have been assigned. Of these 25 species, eleven seem to belong to other genera within the subfamily Cricetinae. These are: *Copemys loxodon* (Cope), *Leidymys nematodon* (Cope), *Leidymys parvus* (Sinclair), *Baiomys brachygnathus* (Gidley), *Baiomys minimus* (Gidley), *Copemys longidens* (Hall), *Bensonomys eliasi* (Hibbard), *Onychomys martini* (Hibbard), *Copemys kelloggae* (Hoffmeister), *Copemys russelli* (James), and *Peromyscus sawrockensis* Hibbard, which is more closely related to the *Bensonomys–Symmetrodontomys* group.

The earliest fossil records now included within the concept of the genus were taken from the Early Pliocene deposits of western United States. The fossil record is poorly known, in part because of the small size of *Peromyscus,* and in part because this mouse was not a common member of the Pliocene and early Pleistocene faunas. Few specimens have been recovered in association with other cricetids. Unfortunately, fossils of this genus are unknown from deposits of these ages in Canada, Mexico, and Middle America.

Six Recent species of *Peromyscus* are known from Wisconsin deposits. Evidence from the fossil record indicates that the *maniculatus* and *leucopus* groups are products of late Pleistocene adaptive radiation.

A diagnosis is given for each of the extinct species and a map is given to show their geographic occurrence. Their range through geologic time is shown on a chart (Fig. 2). Some of the problems for future study are summarized.

Literature Cited

ALLEN, J. A. 1877. Synoptical list of the fossil rodentia of North America. Monographs of North American Rodentia. Rept. U. S. Geol. and Geog. Surv. Terr., F. V. Hayden, U. S. Geologist in charge, Vol. XI: Appendix A: 934–950.

BROWN, BARNUM. 1908. The Conard fissure, a Pleistocene bone deposit in northern Arkansas: with descriptions of two new genera and twenty new species of mammals. Mem. Amer. Mus. Nat. Hist., 9:157–208.

BURT, WILLIAM H. 1960. Bacula of North American mammals. Misc. Publ. Mus. Zool. Univ. Mich., 113:1–76, 25 pls.

CLARK, J. B., MARY R. DAWSON, AND A. E. WOOD. 1964. Fossil mammals from the Lower Pliocene of Fish Lake Valley, Nevada. Bull. Mus. Comp. Zool., 131:29–63.

COPE, E. D. 1874. Notes on the Santa Fe marls and some of the contained vertebrate fossils. Proc. Acad. Nat. Sci. Phila., 26:147–152.

———— 1899. Vertebrate remains from Port Kennedy bone deposit. Jour. Acad. Nat. Sci. Phil., 11:193–267.

COUES, ELLIOTT, AND J. A. ALLEN. 1877. Monographs of North American Rodentia. Rept. U. S. Geol. and Geog. Surv. Terr., F. V. Hayden, U. S Geologist, in charge, Vol. XI:1091 pp.

EVERNDEN, J. F., et al. 1964. Potassium-argon dates and the Cenozoic mammalian chronology of North America. Amer. Jour. Sci., 262:145–198.

GIDLEY, JAMES W. 1922. Preliminary report on fossil vertebrates of the San Pedro Valley, Arizona. U. S. Geol. Survey, Prof. Pap., 131-E:119–130.

HALL, E. R. 1930a. Rodents and lagomorphs from the later Tertiary of Fish Lake Valley, Nevada. Univ. Calif. Publ., Bull. Dept. Geol. Sci., 19: 295–312.

———— 1930b. Rodents and lagomorphs from the Barstow Beds of southern California. Ibid., 19:313–318.

HAY, O. P. 1902. Bibliography and catalogue of the fossil vertebrata of North America. Bull. U. S. Geol. Survey, 179:9–868.

HERSHKOVITZ, PHILIP. 1962. Evolution of Neotropical cricetine rodents (Muridae) with special reference to the Phyllotine group. Fieldiana Zool., 46:1–524.

HIBBARD, CLAUDE W. 1937a. Additional fauna of Edson Quarry of the Middle Pliocene of Kansas. Amer. Midland Nat., 18:460–464.

———— 1937b. An Upper Pliocene fauna from Meade County, Kansas. Trans. Kans. Acad. Sci., 40:239–265.

———— 1941. New mammals from the Rexroad fauna, Upper Pliocene of Kansas. Amer. Midland Nat., 26:337–368.

———— 1955. The Jinglebob interglacial (Sangamon?) fauna from Kansas and its climatic significance. Contrib. Mus. Paleo. Univ. Mich., 12: 179–228.

———— 1956. Vertebrate fossils from the Meade Formation of southwestern Kansas. Papers Mich. Acad. Sci., Arts, and Letters, 41 (1955):145–203.

———— 1963. A late Illinoian fauna from Kansas and its climatic significance. *Ibid.*, 48 (1962) :187–221.

———— 1964. A contribution to the Saw Rock Canyon local fauna of Kansas. *Ibid.*, 49 (1963) :115–127.

HIBBARD, CLAUDE W., AND DWIGHT W. TAYLOR. 1960. Two late Pleistocene faunas from southwestern Kansas. Contrib. Mus. Paleo. Univ. Mich., 16 (1) :1–223.

HOFFMEISTER, DONALD F. 1945. Cricetine rodents of the Middle Pliocene of the Mulholland Fauna, California. Jour. Mamm., 26 (2) :186–191.

———— 1951. A taxonomic and evolutionary study of the piñon mouse, *Peromyscus truei.* Ill. Biol. Monog., 21:1–104.

———— 1959. New cricetid rodents from the Niobrara River fauna, Nebraska. Jour. Paleo., 33:696–699.

HOFFMEISTER, DONALD F., AND L. DE LA TORRE. 1961. Geographic variation in the mouse *Peromyscus difficilis.* Jour. Mamm., 42 (1) :1–13.

HOOPER, EMMET T. 1957. Dental patterns in mice of the genus *Peromyscus.* Misc. Publ. Mus. Zool. Univ. Mich., 99:1–59, 24 figs.

———— 1958. The male phallus in mice of the genus *Peromyscus. Ibid.,* 105:1–24, 14 pls., 1 fig.

———— 1960. The glans penis in *Neotoma* (Rodentia) and allied genera. Occ. Pap. Mus. Zool. Univ. Mich., 618:1–21.

HOOPER, EMMET T., AND GUY G. MUSSER. 1964*a.* The glans penis in Neotropical cricetines (Family Muridae) with comments on classification of muroid rodents. Misc. Publ. Mus. Zool. Univ. Mich., 123:1–57, 9 figs.

———— 1964*b.* Notes on classification of the rodent genus *Peromyscus.* Occ. Pap. Mus. Zool. Univ. Mich., 635:1–13, 2 figs.

JAMES, GIDEON T. 1963. Paleontology and nonmarine stratigraphy of the Cuyama Valley badlands California. Part I. Geology, faunal interpretations, and systematic descriptions of Chiroptera, Insectivora, and Rodentia. Univ. Calif. Publ. Geol. Sci., 45:1–154.

KAPP, RONALD O. 1965. Illinoian and Sangamon vegetation in southwestern Kansas and adjacent Oklahoma. Contrib. Mus. Paleo. Univ. Mich., 19:167–255.

KELLOGG, LOUISE. 1910. Rodent fauna of the late Tertiary beds at Virgin Valley and Thousand Creek, Nevada. Univ. Calif. Publ., Bull. Dept. Geol., 5:421–437.

OLSON, E. C. 1940. A late Pleistocene fauna from Herculaneum, Missouri. Jour. Geol., 48:32–56.

PACKARD, ROBERT L. 1960. Speciation and evolution of the pygmy mice, genus Baiomys. Univ. Kans. Publ. Mus. Nat. Hist., 9:579–670, 4 pls., 12 figs.

RINKER, GEORGE C. 1954. The comparative myology of the mammalian genera *Sigmodon, Oryzomys, Neotoma,* and *Peromyscus* (Cricetinae), with remarks on their intergeneric relationship. Misc. Publ. Mus. Zool. Univ. Mich., 83:1–124.

SINCLAIR, W. J. 1905. New or imperfectly known rodents and ungulates from the John Day series. Univ. Calif. Publ., Bull. Dept. Geol., 4:125–143.

WATERHOUSE, G. R. 1839. The zoology of the voyage of H. M. S. Beagle, Pt. II. Mammalia:1–97.

WHEATLEY, C. M. 1871. The bone cave of eastern Pennsylvania. Amer. Jour. Sci. and Arts, 3rd Ser., 1 (5) :384.

WILSON, ROBERT W. 1949. Rodents and lagomorphs of the upper Sespe. Publ. Carnegie Inst. Wash., 584:51–65.

WOOD, ALBERT E. 1936. The cricetid rodents described by Leidy and Cope from the Tertiary of North America. Amer. Mus. Novitates, No. 822:1–8.

ADDENDUM

ALVAREZ, TICUL. 1966. Roedores fosiles del Pleistoceno de Tequesquinahua, Estado de Mexico, Mexico. Acta Zool. Mexicana, 8 (3) :1–16. (New *Peromyscus maldonadoi*)

GUILDAY, JOHN E., H. W. HAMILTON, AND A. D. McCRADY. 1966. The bone breccia of Bootlegger Sink. Ann. Carnegie Mus., Art. 8, 38:145–163.

GUILDAY, JOHN E., AND C. O. HANDLEY, JR. 1967. A new *Peromyscus* (Rodentia: Cricetidae) from the Pleistocene of Maryland. Ann. Carnegie Mus., Art. 6, 39:91–103.

HIBBARD, CLAUDE W. 1967. New rodents from the late Cenozoic of Kansas. Papers Mich. Acad. Sci., Arts and Letters, 52 (1966) :115–131.

MARTIN, ROBERT A. 1967. A comparison of two mandibular dimensions in *Peromyscus*, with regard to identification of Pleistocene *Peromyscus* from Florida. Tulane Studies in Zool., 14:75–79.

SHOTWELL, J. ARNOLD. 1967. *Peromyscus* of late Tertiary in Oregon. Bull. Mus. Nat. Hist., 5:1–35.

WHITE, JOHN A. 1966. A new *Peromyscus* from the late Pleistocene of Anacapa Island, California, with notes on variation in *Peromyscus nesodytes* Wilson. Los Angeles Co., Mus., Contrib. in Sci., No. 96:2–8.

2

CLASSIFICATION

Emmet T. Hooper

Introduction

THE PRINCIPAL objectives here are to indicate highlights in the taxonomic history of the genus, to discuss current classifications of its Recent species, and to point out systematic problems fruitful for further investigation. I have intentionally avoided ample discussion of concepts and methodologies of systematics, of which classification is  an ordered critically summated part, because these are fully and skillfully covered in many publications (e.g., Simpson, 1961; Mayr, 1963; Hull, 1964; Sattler, 1964; and other reports cited therein). Furthermore, I have made no attempt to present a classification which takes into account any and all relationships among the species of the genus. Because this book is rich in that kind of information, it should set the stage for a comprehensive critical summary based on these and all other systematic data.

The classifications discussed herein derive from systematic neomammalogy. The species discussed, all of Recent geologic time, are defined principally on anatomical and ecological grounds, but other kinds of data are also employed in their recognition. It is inferred that these taxonomic species correspond to similarly circumscribed populations in nature that exhibit the traits seen in the samples. It is further assumed that there is actual or potential genetic continuity within a species, and that between species gene flow is reduced or absent. For a few species there are genetic data to bolster this assumption. Intergradation of characters as between two samples or among series of samples in general is taken to mean that the sampled populations interbreed, and conversely, the absence of intergradation may signify reproductive isolation. Intraspecific variants recognized by formal name are termed subspecies. They are also referred to as geographic or ecologic races.

There is no basic difference between subspecies, geographic races, and ecologic races.

The unit group of animals in any category of the systematic heirarchy is termed a taxon; a given taxonomic category, thus, consists of all the taxa placed at that level in the hierarchy. Superspecific taxa are delimited on the principal that all members of each taxon have a single phylogenetic origin. The superspecific taxa treated in this chapter are subgenera and genera. In addition, an informal, loosely defined category termed species group is employed. A species group is an assemblage of similar species within a subgenus. Some of these groups conform in definition to cenospecies and some to superspecies (see Blair, 1943*a* and 1950, and Mayr, 1963, for discussions of these concepts and terms). Eventually the recognition of cenospecies or superspecies may supplant the use of the species group category in classification, but there are not yet enough facts about all kinds in the genus to warrant this action now.

History of the Genus

GENERIC NAMES APPLIED.—While undoubtedly mice of this genus were known to New World natives and to early white settlers, references to them do not appear in the scientific literature until the latter part of the eighteenth century. At that time the state of knowledge of small mammals of the world was meager. Most of them were referred to under the generic term *Mus,* a name which Linnaeus had applied in 1758 to various kinds of mice. Moreover, in comparing different forms, attention was given mostly to external appearances, and the mice of Europe tended to be used as standards for comparisons. Thus, when the first specimens received from North America (probably examples of *P. leucopus noveboracensis*) were observed to resemble the European woods mouse and field mouse, as might be expected they were designated by the names of those European forms, *Mus agrarius* and *Mus sylvaticus.* For a number of years nearly all American murines were included in the genus *Mus.* In 1832 the name *Arvicola* was employed in the description of a new form now known by the name *Ochrotomys nuttalli.* In 1839 the name *Hesperomys* replaced the name *Mus* for many American rodents. This was the result of work by Waterhouse who drew attention to the dental characters which distinguished the American species from those in the genus *Mus.* Then for various biological or procedural reasons there followed a series of names which are applied to mice now known by the name *Peromyscus.* These names are *Vesperimus,* suggested in 1874; *Calomys,* suggested in 1888; *Sitomys,* in 1892; and finally *Peromyscus* in 1894. *Peromyscus* was first used in 1841 but it was subsequently overlooked until shown by Thomas in 1894 to be applicable to the species. Thenceforth its validity and application have not been questioned. Its scope—the species comprising it—however, has been modified.

SCOPE OF THE GENUS.—As knowledge of mammalian faunas has increased, the scope of the generic category typically has been reduced. Forms that superficially resemble each other were later found to be dissimilar in many features which were considered taxonomically more revealing of evolutionary relationships and for that reason they were removed to other usually newly erected genera, thus reducing the morphological limits of the earlier named genera. This has been the trend in the history of the genus *Peromyscus.* Murine mice were known by the generic name *Mus* for many years until Waterhouse in 1839 proposed the name *Hesperomys* to include all New World rodents having two rows of cusps in the upper cheek teeth. The genus *Hesperomys* then comprised several subgenera, most of which are now recognized as genera, for example *Oryzomys, Onychomys,* and *Phyllotis.* There were few forms in any of these genera or subgenera throughout much of the 19th century. For example, in Baird's Mammals of North America (1859), the first comprehensive treatise on North American mammals, only 12 forms of *Peromyscus* (as now understood) were listed. These were contained within the subgenus *Hesperomys.* Similarly, Coues' Monograph of North American Rodentia (1877) includes only nine forms of *Hesperomys.* Like Baird, he also included *Onychomys* and *Oryzomys* as subgenera of *Hesperomys.* After 1885 collections of small mammals began to increase rapidly, many new forms were described, and increased attention was given to phyletic relationships of the species. As a result, various groups which long had been included as subgenera (for example *Onychomys, Oryzomys, Tylomys,* and various other South American groups) were eliminated and given generic rank.

By 1898 the genus *Peromyscus* had been restricted to forms confined to North America, and soon after the turn of the century when Osgood began his studies of this group of rodents, the genus was rid of many species which were dissimilar to the species now comprising the genus. The small-mammal fauna of North America also was becoming much better known, a host of new forms had been described, and collections of specimens in museums had swelled tremendously. The task of analyzing all of the material which had been accumulated and of arriving at a reasonable, data-based estimate of characteristics and interrelationships of the populations they represented was difficult. W. H. Osgood, who undertook the task, did an excellent job in coping with the mass of specimens, scramble of scientific names, and encrusted viewpoints of some taxonomists of his time. His monograph of 1909, the latest thorough revision of the genus, must be considered high-grade, certainly one of the best systematic reviews to that date. It is a quality item not only because it evidences careful scholarly work, but also because it reflects modern viewpoints. To Osgood, unlike many taxonomists both before and after him, the specimens before him were really not what he was attempting to classify. The living populations they represented were his prime target, and it is this viewpoint coupled with accuracy and fair documentation of his analyses which has allowed his classification to stand the test of time remarkably well.

The following figures provide a brief sketch of Osgood's view of *Peromyscus*. Out of 181 supposedly recognizable forms which had been described, he considered that 143 represented distinguishable populations which warranted formal recognition by specific or subspecific name. Of this 143, forty-two were listed as monotypic; they were considered true biologic species or at least "tentative" species pending accumulation of additional information about them. These 143 forms were distributed in six subgenera as follows: *Haplomylomys* (16 forms contained in three species); *Peromyscus* (114 forms in 32 species); *Megadontomys* (three nominal species); *Podomys* (1 species); *Ochrotomys* (two forms of one species); and *Baiomys* (seven forms in two species).

The next restriction in scope of the genus involved the deletion of the species of *Baiomys*. *Baiomys* was proposed by True (1894) as a subgenus which with *Vesperimus* and *Onychomys* formed three sections of the genus *Sitomys*. The subgenus *Baiomys* served to recognize the peculiar characters of the species *taylori*. In discussing *Sitomys musculus*, as well as *S. taylori*, Mearns (1907) indicated that the two species are not closely related to any of the other described species from the United States; he raised *Baiomys* to generic rank. Osgood (1909) considered that the characters of *Baiomys* were comparable to those of other subgenera of *Peromyscus*, so he reduced *Baiomys* to a subgenus. Miller (1912), however, again elevated it to generic rank where it has remained. The more complete information which has accumulated since 1909 has served to corroborate Mearns' view that the species of *Baiomys* in many features of their lives are indeed set well apart from the other species of *Peromyscus* and deserve generic recognition. Information accruing to 1960 is reviewed by Packard (1960) in his systematic analysis of the species of *Baiomys*. Additional materials pertaining to relationships of the genus are presented by Hooper and Musser (1964*a*) and Arata (1964).

The final step to date in the restriction of *Peromyscus* involved the deletion of the species *nuttalli*. The subgenus *Ochrotomys* was created by Osgood (1909) in recognition of characters of *nuttalli* which indicated a more distant kinship than that between most other species of the genus. In discussing affinities within *Peromyscus*, Blair (1942) pointed out that the baculum of *O. nuttalli* is strikingly different from that of species of the subgenera *Peromyscus, Haplomylomys,* and *Podomys*. He stated that "*Ochrotomys* is so different from the other *Peromyscus* as to raise the question as to whether its relationships might be better shown by elevating *Ochrotomys* to generic rank." Hooper (1958) indicated that on the basis of the peculiar characters of skin, skull, and male phallus of *nuttalli, Ochrotomys* should be raised to generic rank. Supportive evidence for this view is provided by a number of authors, including McCarley (1959), Rinker (1960), Manville (1961), and Arata (1964).

The last major modifications suggested for the genus took place in 1964 when Hooper and Musser (1964*b*) proposed three new subgenera, as follows: *Habromys,* for the species *lepturus, lophurus,* and *simulatus; Osgoodomys,* for *banderanus;* and *Isthmomys,* for *flavidus* and *pirrensis*. These suggested taxonomic changes pertain to relationships of species; they do not alter the scope of the genus.

Relatives of Peromyscus

The closest relatives of *Peromyscus* probably are other genera of the subfamily Cricetinae. In summarizing and paraphrasing opinions of the period 1900–1950 regarding affinities of *Peromyscus*, one might write as follows: *Peromyscus* is a genus of cricetine rodent apparently closely related to *Baiomys, Reithrodontomys, Onychomys, Nyctomys, Oryzomys, Neotoma,* and other New World cricetines. It is somewhat farther removed phyletically from *Microtus* and other microtines, and is more distantly related to murine rodents of the Old World. What these statements say is that anatomically *Peromyscus* resembles *Baiomys* and the other listed genera, and is less similar to microtines and murines. The statements imply that (1) when properly appraised, anatomical resemblances indicate kinship and (2) these familial and sub-familial categories indicate levels in kinship, i.e., that all cricetines are anatomically more similar and therefore, more closely related to each other than any one of them is to any microtine (Subfamily Microtinae, Family Cricetidae) or murine (Subfamily Murinae, Family Muridae).

The first implication listed above makes good sense; it accords with biological facts. There are no potent reasons to doubt that resemblance in form, function, or other dynamic aspects indicates kinship, provided that due attention is given to parallelism and other homoplastic phenomena (resemblances not due to inheritance from a common ancestry).

The second inference is less secure. It is by no means certain that the families Cricetidae and Muridae and the subfamilies Cricetinae, Microtinae, and Murinae, as now understood, all correctly express propinquity of the species contained within them. That the cricetids warrant familial separation from the murids has been debated for a number of years (see review by Simpson, 1945: 206). In addition, there is evidence now which strongly suggests that the subfamilial lines have been erroneously drawn; they may not fit the facts for Recent species at least. The evidence suggests that the subfamily Cricetinae as heretofore understood (e.g., Simpson, 1945) is polyphyletic, some of its species more closely related to microtines and murines than to other cricetines. Further discussion of classification of relationships of Old World rodents would be out of place here. Suffice it to say that several kinds of evidence (some of it mentioned below) indicate that in the New World there may be three major phyletic groups of murid rodents. These are the microtines, South American cricetines, and neotomines-peromyscines. These three are of about equal taxonomic rank, and may be approximately equidistantly related to the Old World cricetines and groups of murines. The neotomines consist of the genera *Neotoma, Xenomys, Tylomys,* and *Nelsonia.* These woodrat-like forms probably represent a minor phyletic line which evolved from that which gave rise to the peromyscines. This latter group includes the following genera: *Peromyscus, Reithrodontomys, Neotomodon, Ochrotomys, Onychomys, Baiomys,* and *Scotinomys.* Authors agree that all of these genera are probably closely related, but some (e.g., Hershkovitz, 1944, 1960, 1962 and Vorontsov, 1959) might add other genera to the group. That *Peromyscus* likely is more closely related to these genera than to woodrats and other North American groups is supported by evidence from the external parts, skeleton, musculature (Rinker, 1963), and reproductive structures (Hooper and Musser, 1964a; Arata, 1964), to mention a few sources.

There is not much point in further discussing the interrelationships of these genera, for to do so would require delving into the mass of information about both Recent and fossil forms. *Peromyscus* appears to have affinities with all genera of the peromyscine group, and various times in the past the species of three of these genera, namely, *Baiomys, Ochrotomys,* and *Onychomys,* have been included in *Peromyscus.* As indicated above, however, there are now good reasons for excluding them because they appear no more closely related to *Peromyscus* than are *Reithrodontomys, Neotomodon,* and *Scotinomys. Baiomys,* together with *Scotinomys,* appears to be set apart from *Ochrotomys, Neotomodon,* and *Reithrodontomys,* the three genera which may be closest to *Peromyscus.*

Classifications of the Species

A taxonomic classification is a tentative thing; it is not sacred. It is a systematist's best summary estimate of relationships among the organisms with which he is concerned, the organisms grouped and provided with names in accordance with taxonomic principles and rules. His conclusions are expressed as clearly as is permitted by the inadequate, cumbersome, hierarchal system with which he has to work. But since the classification is an estimate, it obviously will be modified from time to time to accord with new data. It is thus basically a test system. It is a reasonable yet tentative framework against which new systematic data may be tried for fit. So long as old and new information about the species suits the framework, the classification holds up and is retained as the best estimate of phyletic relationships of the species. As new information accumulates, however, various parts of the framework may appear to be contorted or misplaced, and eventually so many inconsistencies of data and framework may have accrued to make patching of the old scheme undesirable. Scrapping the old and starting afresh is called for.

Classifications also vary with viewpoints of the classifier. One author may attach particular importance and give special weight to a particular set of characters, and in addition may overestimate the distinctions of groups of species. As a result, he may employ numerous genera and subfamilies in his classification. Another student working with the same data may be more conservative in viewpoint and require fewer subfamilies, genera, and subgenera to present his views of the interrelationships of the species. Thus, a given classification is to be taken as a temporary model representing a classifier's view at one time of some relationships of the organisms. The classifications which follow are examples in point.

Three taxonomic arrangements of the species of *Peromyscus* are presented here. The one by Osgood (1909) is the product of the most recent, detailed, systematic analysis of all species of the genus known at that time. The other two are based on Osgood's work and in general serve to bring his classification up to date by incorporating new forms which have been described. But each also introduces new viewpoints. For example, in the arrangement by Hall and Kelson (1959) the placement of *Haplomylomys* at the head of their list of forms incorporates the idea that the "simple-tooth" condition may be primitive in the genus and the evolutionary trend has been toward increased complexity of the molar teeth. And again, the arrangement by Hooper and Musser (1964*b*) attempts to emphasize the observed reticular pattern in differentiation in the genus; few species have diverged from the main cluster.

The list by Hall and Kelson is essentially a projection of Osgood's classification to accommodate additional forms described since 1909

TABLE 1

COMPARISON OF ARRANGEMENT BY HALL AND KELSON (1959)
WITH OSGOOD'S CLASSIFICATION (1909)

Osgood's *Haplomylomys* has been shifted to the top of his list to facilitate
comparison of species; he placed it between *Podomys* and *Baiomys*

Osgood (1909)	Hall and Kelson (1959)
Genus *Peromyscus* Gloger	Genus *Peromyscus* Gloger
Subgenus *Haplomylomys* Osgood	Subgenus *Haplomylomys* Osgood
crinitus (Merriam)	*crinitus* (Merriam)
	collatus Burt
	pseudocrinitus Burt
californicus (Gambel)	*californicus* (Gambel)
eremicus (Baird)	*eremicus* (Baird)
[incl. *merriami* Mearns]	*merriami* Mearns
goldmani Osgood	[incl. *goldmani*]
	caniceps Burt
	guardia Townsend
	dickeyi Burt
	pembertoni Burt
	stephani Townsend
Subgenus *Peromyscus* Gloger	Subgenus *Peromyscus* Gloger
maniculatus group	*maniculatus* group
maniculatus (Wagner)	*maniculatus* (Wagner)
	sejugis Burt
	slevini Mailliard
sitkensis Merriam	*sitkensis* Merriam
polionotus (Wagner)	*polionotus* (Wagner)
melanotis Allen & Chapman	*melanotis* Allen & Chapman
leucopus group	*leucopus* group
leucopus (Rafinesque)	*leucopus* (Rafinesque)
gossypinus (Le Conte)	*gossypinus* (Le Conte)
boylei group	*boylei* group
boylei (Baird)	*boylei* (Baird)
[incl. *aztecus* (Saussure)	[incl. *aztecus* (Saussure)
and *evides* Osgood]	and *evides* Osgood]
	perfulvus Osgood
oaxacensis Merriam	*oaxacensis* Merriam
hylocetes Merriam	*hylocetes* Merriam
pectoralis Osgood	*pectoralis* Osgood
	polius Osgood
truei group	*truei* group
truei (Shufeldt)	*truei* (Shufeldt)
nasutus (Allen)	*nasutus* (Allen)
polius Osgood	
difficilis (Allen)	*difficilis* (Allen)
bullatus Osgood	*bullatus* Osgood

TABLE 1 (Continued)

Osgood (1909)	Hall and Kelson (1959)
melanophrys group	*melanophrys* group
melanophrys (Coues)	*melanophrys* (Coues)
xenurus Osgood	[incl. *xenurus* Osgood]
mekisturus Osgood	*mekisturus* Merriam
lepturus group	*lepturus* group
lepturus Merriam	*lepturus* Merriam
lophurus Osgood	*lophurus* Osgood
	hondurensis Goodwin
simulatus Osgood	*simulatus* Osgood
nudipes (Allen)	*guatemalensis* Merriam
furvus Allen & Chapman	*nudipes* (Allen)
guatemalensis Merriam	*altilaneus* Osgood
altilaneus Osgood	*furvus* Allen & Chapman
	latirostris Dalquest
	ochraventer Baker
	mexicanus group
mexicanus (Saussure)	*mexicanus* (Saussure)
allophylus (Osgood)	*allophylus* Osgood
banderanus Allen	*banderanus* Allen
yucatanicus Allen & Chapman	*yucatanicus* Allen & Chapman
	stirtoni Dickey
megalops group	*megalops* group
megalops Merriam	*megalops* Merriam
	sloeops Goodwin
melanocarpus Osgood	*melanocarpus* Osgood
zarhynchus Merriam	*zarhynchus* Merriam
	grandis Goodwin
Subgenus *Megadontomys* Merriam	Subgenus *Megadontomys* Merriam
thomasi Merriam	*thomasi* Merriam
nelsoni Merriam	*nelsoni* Merriam
flavidus (Bangs)	*flavidus* (Bangs)
	pirrensis Goldman
Subgenus *Ochrotomys* Osgood	Subgenus *Ochrotomys* Osgood
nuttalli (Harlan)	*nuttalli* (Harlan)
Subgenus *Podomys* Osgood	Subgenus *Podomys* Osgood
floridanus (Chapman)	*floridanus* (Chapman)
Subgenus *Baiomys* True	Genus *Baiomys* True
taylori (Thomas)	*taylori* (Thomas)
musculus (Merriam)	*musculus* (Merriam)

and to incorporate other pertinent information published to the year 1959. The two lists are compared in Table 1. Excepting the exlusion of *Baiomys* from *Peromyscus* and a few changes in arrangement of the species, the principal difference between the two lists is in quantity of forms. The number of taxa has increased considerably. Osgood recognized 143 forms (species and subspecies), seven of which are *Baiomys*. Hall and Kelson list 235 forms which in the opinion of various authors are formally recognizable. The increase is approximately 75 per cent, and at least another half dozen forms have been described subsequent to 1959. The largest increment is in the *maniculatus* group in which the count to date jumps from 43 to 83. Many of the post-1909 taxa, particularly of *P. maniculatus,* are insular populations of the Pacific and Atlantic coasts.

Table 2 presents a third arrangement of the species. This classification is essentially that suggested by Hooper and Musser (1964*b*), modified slightly to accord with more recent information. It is in harmony with most of Osgood's data together with later information on morphology, anatomy, ecology, and other aspects of the biology of the animals. It is suggested as a tentative framework against which any and all data on relationships—whether anatomical, behavioral, biochemical, or other—may be tried for fit.

The following discussion of this classification indicates the state of knowledge of the various species and suggests some of the many systematic problems in the genus yet to be solved. The distribution maps are adapted from Hall and Kelson (1959). For listings of synonymies, subspecies, and details of geographic ranges, see Hall and Kelson (1959), Miller and Kellogg (1955), and subsequent reports mentioned on following pages.

Before proceeding to the classification, however, I wish to comment on the number of species of *Peromyscus*; 57 are listed. Certainly not all of these are species in the sense of the term intended in the classification. That is, not all of them are biological species, defined as "groups of actually or potentially interbreeding natural populations which are reproductively isolated from other such groups" (Mayr, 1963). Some, when better known, probably will prove to be either geographic races or synonyms of earlier named taxa. Taking into account all known forms and any which possibly have not yet been discovered, how many biological species are there

TABLE 2

A Classification of *Peromyscus* Modified from that
Suggested by Hooper and Musser (1964*b*)

Allocation of some forms* to subgenera or species group is tentative

Subgenus *Haplomylomys* Osgood
  eremicus (Baird)
  merriami Mearns
  caniceps Burt*
  guardia Townsend
  interparietalis Burt*
  collatus Burt*
  dickeyi Burt
  pembertoni Burt*
  stephani Townsend*
  californicus (Gambel)
Subgenus *Peromyscus* Gloger
  maniculatus group
    polionotus (Wagner)
    maniculatus (Wagner)
    sejugis Burt
    slevini Mailliard*
    sitkensis Merriam
    melanotis Allen & Chapman
  leucopus group
    leucopus (Rafinesque)
    gossypinus (Le Conte)
  crinitus group
    crinitus Merriam
    pseudocrinitus Burt*
  boylei group
    pectoralis Osgood
    boylei (Baird)
    polius Osgood
    evides Osgood
    aztecus (Saussure)
    hondurensis Goodwin
    oaxacensis Merriam
    hylocetes Merriam
  truei group
    truei (Shufeldt)

  bullatus (Osgood)
  difficilis (Allen)
melanophrys group
  melanophrys (Coues)
  mekisturus Merriam
  perfulvus Osgood*
mexicanus group
  ochraventer Baker
  stirtoni Dickey
  yucatanicus Allen & Chapman
  allophylus Osgood*
  mexicanus (Saussure)
  furvus Allen & Chapman
  latirostris Dalquest
  melanocarpus Osgood
  zarhynchus Merriam
  grandis Goodwin
  altilaneus Osgood
  guatemalensis Merriam
  nudipes (Allen)
  megalops Merriam
Subgenus *Osgoodomys* Hooper & Musser
  banderanus Allen
Subgenus *Habromys* Hooper & Musser
  simulatus Osgood
  lophurus Osgood
  lepturus Merriam
  ixtlani Goodwin
Subgenus *Podomys* Osgood
  floridanus (Chapman)
Subgenus *Megadontomys* Merriam
  thomasi Merriam
Subgenus *Isthmomys* Hooper & Musser
  flavidus (Bangs)
  pirrensis Goldman

in the genus? My estimate is 40, the number of superspecies or cenospecies in the order of 15. The count of biological species certainly will not exceed 50 unless the scope of *Peromyscus* is enlarged or another concept of the species is employed.

Discussion of the Classification

SUBGENUS *Haplomylomys*

This subgenus was first employed by Osgood in 1904 to set apart the simple-toothed *P. eremicus* and *P. californicus* from most of the other forms of the genus. Later (1909) he gave a more nearly complete diagnosis of the subgenus and added *P. goldmani* and *P. crinitus* to it, remarking however that *crinitus* approaches the subgenus *Peromyscus* in regard to presence of "accessory cusps" (styles and stylids) in the molar teeth. Subsequently a number of newly named forms have been included. In the present classification, *P. crinitus* has been excluded. Characters in which *crinitus* approaches or resembles species of the subgenus *Peromyscus* are pointed out by Osgood (1909), Hall and Hoffmeister (1942), Hooper (1957*b*, 1958), and Hooper and Musser (1964*b*). Probably *P. crinitus* is more closely related to members of the *boylei* group of the subgenus *Peromyscus*. It is placed near this group in the list (Table 2).

Affinities of some forms of the subgenus are not clear. *P. merriami* and *P. goldmani* are cases in point. Recent authors (e.g., Hall and Kelson, 1952; Commissaris, 1960; Hoffmeister and Lee, 1963) have associated these with *P. eremicus,* as Osgood did, but concluded on morphological and ecological grounds that one or both of those forms is, or are, specifically distinct from *eremicus.* Hall and Kelson and Hoffmeister and Lee treat them as races of the species *P. merriami,* and I have followed their recommendation in the current classification. The procedure of considering *merriami* and *goldmani* conspecific (Fig. 4) and of listing them near *P. eremicus,* however, should be considered tentative, for these forms require additional study. The possibility that *goldmani* and *merriami,* one or both, are related to *P. pectoralis* or other Mexican forms should be explored with the fact in mind that the molar teeth are more variable than formerly was supposed (Hooper, 1957*b*) and that presence or absence of "accessory" styles and lophs

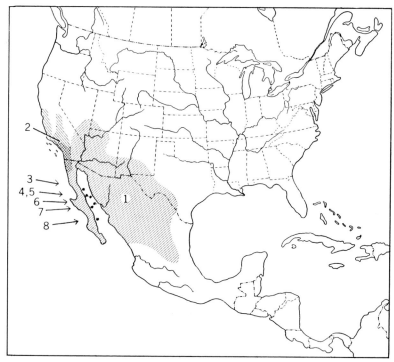

Fig. 1. 1, *P. eremicus;* 2, *P. californicus;* 3, *P. guardia;* 4, *P. interparietalis* and *P. stephani;* 5, *P. collatus;* 6, *P. pembertoni;* 7, *P. dickeyi;* 8, *P. caniceps.*

in those teeth are not totally reliable key characters for distinguishing and ordering these species.

There is little information on the forms *caniceps, guardia, interparietalis, collatus, dickeyi, pembertoni,* and *stephani,* all of which are restricted in distribution to islands in the Gulf of California (Fig. 1). Both *P. guardia* and *P. interparietalis* are represented by three subspecies (Banks, 1967); the other forms are monotypic. *P. dickeyi* and *P. guardia* probably belong in *Haplomylomys* (Burt, 1932; Hooper and Musser, 1964b). *P. stephani* is reported to fit near *guardia,* and *pembertoni* near *dickeyi* (Burt, 1932). *P. collatus* was considered to be close to *P. crinitus* (Burt, 1932), but the glandes of *collatus* examined by Hooper and Musser (1964b) clearly were like those of *eremicus* or *californicus,* not of *crinitus.* In short, questions concerning kinships, immigration routes, stages in speciation, and other systematic aspects of these forms remain to be answered. Any of eight mainland species—*eremicus, merriami, californicus, crinitus, maniculatus, pectoralis, boylei,* and *truei*—may have been intimately involved in the history of these insular populations.

The polytypic species *P. californicus* poses no major taxonomic problems. Its characters suggest kinship with *P. eremicus,* but the two species are strongly differentiated and there is no question of their distinctness in anatomical and ecological characters, and in behavior (Eisenberg, 1962, 1963). Attempts to hybridize individuals of the two species were unsuccessful (Dice, 1933).

SUBGENUS *Peromyscus*

Though most of the species are assignable to the subgenus *Peromyscus,* in effect they constitute the residue remaining after a few specialized forms have been siphoned off from the main mass and segregated in other subgenera. Another way to describe the situation is as follows: most of the species are morphologically similar to one another; they cluster. Only a few of them are differentiated in a special way from the main cluster and thus are set apart by means of subgeneric categories. This is not to say, however, that each of the 38 species listed in the subgenus *Peromyscus* is like each other species of the subgenus, for that array exhibits diversity in form, function, and other aspects. Rather it is to say that though there is diversity among the species of the subgenus, the entire series constitutes essentially a continuum of all known facets of the species' biologies, and the few kinds listed in other subgenera do not fit within that continuum. They appear to be peripheral to the mosaic core of differentiation represented by the subgenus *Peromyscus.* Thus it is that there are 38 species listed in the subgenus *Peromyscus* and only 19 in the other six subgenera, the maximum number in any one of those being 10 (*Haplomylomys*).

MANICULATUS GROUP.—There is no sound basis for disagreement with Osgood's view (1909) that *polionotus, maniculatus, sitkensis,* and *melanotis* fit together morphologically; these constitute his *maniculatus* group. The abundant information now available on the biology of these species has served to bolster, not contradict, the view that these are closely related, derived from the same parental stock. The data are entirely in harmony with the view that the geographically peripheral species, *polionotus, sitkensis,* and *melanotis,* and probably also *sejugis* and *slevini* which have been named subsequent to Osgood's revision, evolved as a result of geographic isolation of peripheral stocks (Dice, 1940*a*, 1940*b*; Blair, 1943*a*, 1950, 1953, 1958). Ecologic changes in the past are probably responsible for the disjunct distribution—insularity—of these geographically peripheral forms, separating them from the widespread *maniculatus* which is distributed over most of North America (Blair, 1950, 1958).

While there seems to be little doubt that *polionotus, maniculatus, sitkensis,* and *melanotis* are closely related and logically constitute the *maniculatus* group, there are questions as to the status of some taxa which are here included in that group. Some of these problems are pointed up in the following synopsis.

P. polionotus: That this species is more closely related to *P. maniculatus* than to any other species of the genus seems clear. *P. polionotus* (Fig. 2) consists of several geographic races (11 listed by Hall and Kelson, 1959). According to Schwartz (1954) these fall in two groups, one restricted to beach habitats along the ocean and the other occurring principally in situations back of the beach at higher elevations. Some of these races are rather highly differentiated in anatomy and habits (Blair and Howard, 1944; Blair, 1954) and there is some evidence that there are varying degrees of fertility among them, also between them and the races of *P. maniculatus* (Blair, 1943*a*). In regard to affinities with *P. maniculatus,* morphologically *P. polionotus* is more similar to the short-tailed prairie forms, e.g., *pallescens* and *bairdi,* than to the long-tailed forest forms of the species. Moreover, the morphological resemblance to *maniculatus* is closer by way of the mainland, upland forms of *polionotus* rather than through the more highly differentiated beach and island forms of that species.

P. maniculatus: This is perhaps the best known species of the genus. Published accounts of various aspects of its biology would fill several volumes. There is information on morphology, life history, geographic and ecologic variation, population dynamics, behavior, genetics, cytology, and other facets of its biology. It is a favorite species for study in its native environments, and it rivals the house mouse and the Norway rat in usefulness as an experi-

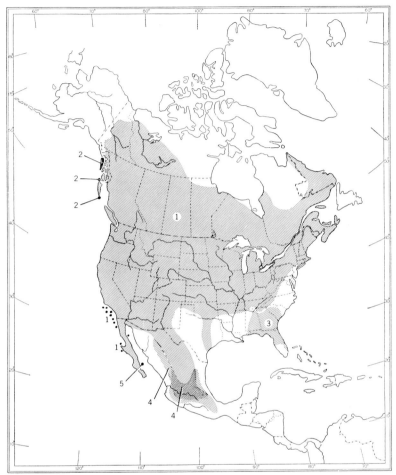

FIG. 2. 1, *P. maniculatus*; 2, *P. sitkensis*; 3, *P. polionotus*; 4, *P. melanotis*; 5, *P. sejugis* and *P. slevini*.

mental animal. References to various facets of its biology will be found in other chapters of this book.

It is a ubiquitous animal in North America, probably the most widely distributed native small mammal on the continent (Fig. 2). It occurs in a wide variety of environments and has undergone much geographic differentiation. Currently some 27 or 28 continental geographic races and some 40 insular subspecies are recognized. Some of these indicate the presence of long series of intergrading and interbreeding populations between which gene flow may vary in rate but is not completely discontinuous. In addition there are instances wherein the populations of two morphologic forms, different subspecies, occur in juxtaposition or sympatrically and apparently do not interbreed. It is said that in the latter instances though interbreeding does not occur directly between contiguous or overlapping populations, it does occur by way of a circuitous route through other populations. The inference is that the highly complex distributional pattern of *P. maniculatus* consists of at least two chains of intergrading forms: a large-body, long-tailed group of forest inhabitants and a smaller short-tailed group that lives predominantly on prairies and in other open situations.

The matter of conspecificity of all races of *P. maniculatus* needs to be further explored by means of detailed studies of all aspects of the biology of the animals. The most critical areas for such studies are those in which long-tailed forest populations closely approach or come into direct contact with short-tailed prairie populations. Few detailed studies of the sort needed have been conducted in these critical areas. Examples of these areas reported in the literature are as follows: the coastal region of Washington, Glacier Park of Montana, various parts of the Upper and Lower peninsulas of Michigan, and parts of the state of New York. These are only a few of the many areas in which two morphologically different populations of *P. maniculatus* come together and apparently do not intergrade and, inferentially, interbreed. The number and location of them probably has changed considerably in this century as a result of man's drastic altering of ranges of forests and grasslands. Range limits of many of the taxa certainly have been modified since 1909 when Osgood revised the genus. Nonetheless the long-tailed, forest-inhabiting *maniculatus* today still occupy the same region which is roughly the shape of an inverted U, extending from the Rocky Mountains and coastal forests of western states eastward across Canada and southward in the Appalachian chain. It is possible that the short-tailed *maniculatus* which now occupy the central grasslands and deserts of the west, evolved principally in southern regions, and following Pleistocene glaciation moved northward to make secondary contact with the forest forms in various parts of northern United States and southern Canada.

Some of the interdigitating populations that have been studied are in the coastal region of Washington. They are treated under the subspecific names *oreas, austerus,* and *gambeli.* The populations of these three forms are distributed as follows: *austerus,* in a narrow belt along the coast from Puget Sound, Washington, north to southern British Columbia; *oreas,* mountains and coastal areas of western Washington and southern British Columbia south to the Columbia River; *gambeli,* more arid areas to the southwest and south of the range of the other two forms. Osgood (1909) pointed out that *oreas* and *austerus* occur together at a number of localities and apparently maintain their identities. He suggested that further collecting may show that they are sympatric over a wide area. Though some specimens seemed to him to be intermediate between the two forms he suggested that in reality perhaps this apparent anatomical intermediacy represents special differentiations of one or the other of the forms; the inference was that it might not result from interbreeding of the two taxa. Dice (1949) also pointed out that *oreas* and *austerus* occur together without intergradation and that there is no apparent ecologic barrier between them. He indicated that interbreeding of the two forms had not yet been demonstrated. Fox (1948) found that *oreas* was sympatric with *austerus* and *gambeli* in certain localities. He suggested on the basis of several specimens that *oreas* behaves as a distinct species toward *austerus* and *gambeli,* with intermittent hybridization taking place. Liu (1954) accumulated evidence in the laboratory of highly incomplete interfertility of the two forms. Johnson and Ostenson (1959) and Sheppe (1961) indicated that *oreas* is ecologically and morphologically distinct from *austerus* and *gambeli,* and that its range widely overlaps that of the other two forms. All of this seems to point to the fact that knowledge of the relationships of the long-tailed forest-dwelling and short-tailed prairie chains of populations is not as complete and clear-cut as the classifications of the species would seem to indicate. To treat *oreas* as a separate species, as has been suggested, would solve no biological problem. Rather it would tend to obscure any possible relationships which that form might have with other long-tailed forest populations to the north and east. What is required is additional work with *P. maniculatus* to throw more light on the subject.

P. sejugis, P. slevini, and *P. sitkensis*: These three are insular forms, the first two occurring on islands in the Gulf of California, and *sitkensis* on islands in southern Alaska and northern British Columbia (Fig. 2). They are all large bodied, a fact which may have physiological significance and should be taken into account in the detailed studies which are needed. In external and cranial characters, *sejugis* is clearly a member of the *maniculatus* group (Burt, 1932) ; data on the male phallus support this viewpoint (Hooper and Musser, 1964*b*). Affinities of *slevini* are less certain. In morphology of the skin, skull, and baculum *slevini* belongs in the subgenus *Peromyscus* (Burt, 1934), not in *Haplomylomys* as indicated by Maillaird (1924), but where it fits within the subgenus *Peromyscus* yet remains to be precisely determined.

The scanty information available on *sitkensis* indicates that it is similar to *maniculatus* and probably stems from the same stock that gave rise to the mainland populations of that species. Anatomically it is close enough to *maniculatus* that no violence would be done if it were included within that species. The level in speciation that it has reached, however, still has to be determined. It has a peculiarly disrupted distribution on the islands. It apparently occurs on some of the islands but not on others or on the mainland, while populations of *P. maniculatus* occur on intervening islands. Populations of *maniculatus* and *sitkensis* are not known to occupy the same island.

P. melanotis: Available information indicates that *melanotis* and *maniculatus* are closely related but are separate gene pools. In the highlands of Mexico, the home of *melanotis* (Fig. 2), both species are inhabitants of open situations such as are provided in grasslands, meadows, and second growth forests; each thus utilizes much the same type of plant cover. The two species are mostly allopatric. *P. melanotis* lives predominantly in the fir and upper pine belts and on grass-studded slopes and plains above tree line, while *P. maniculatus* usually occurs in lower vegetational belts. In many areas of the central highlands of Mexico the ranges overlap or interdigitate. In these areas the two species live sympatrically and apparently do not interbreed. One such area is near the town of Perote, Veracruz (Hooper, 1957*a*). There, where pine forests give way to open grassy plains, the two species occur in the same ravines and same clumps of grass and, indeed, the ravines likely are the

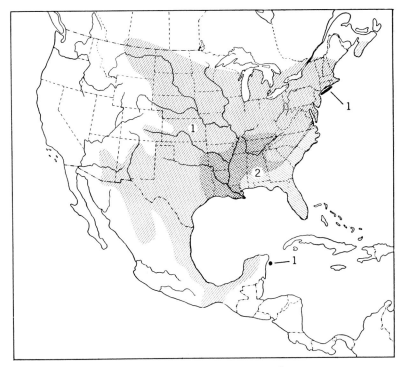

Fig. 3.   1, *P. leucopus*; 2, *P. gossypinus*

principal contact areas for the two species. They constitute avenues of *melanotis*-suited habitat that extend from the fir belt down through the inhospitable open pine forests to the oak belt and open plains where *maniculatus* lives. Insofar as now known the two species do not interbreed anywhere in natural environments; however, at least one litter of fertile young from a *maniculatus* × *melanotis* mating has been produced in the laboratory (Clark, 1966).

Leucopus Group.—*P. leucopus* and *P. gossypinus,* the two species of this group, rival *P. maniculatus* and *P. polionotus* in regard to amount of available information about them. They have been the subjects of many studies in both field and laboratory, and as a result there is now a fair picture of the biology of each of these, particularly of *leucopus.* They are similar in form and habits. Both are inhabitants of deciduous forests and are semi-arboreal. *P. leucopus* has an extensive range in Mexico and eastern and southern United States, whereas *gossypinus* is restricted to the southeastern tier of states (Fig. 3). Both have differentiated geographically; some fifteen geographic races of *leucopus* and some seven races of *gossypinus* are currently recognized.

These two forms recall *P. maniculatus* and *P. polionotus* not only because both pairs of these species are fairly well known but also because the pairs are somewhat alike in geographic relationships and in level attained in speciation. *P. polionotus* and *P. gossypinus* have small geographic ranges, both situated in approximately the same part of southeastern United States. Just as *maniculatus* and *polionotus* probably were derived from the same parental stock and have barely reached the specific level in speciation, so also have *leucopus* and *gossypinus* apparently diverged recently and scarcely are completely separate gene pools. Environmental factors which have led to the separation and differentiation of *leucopus* and *gossypinus* probably are the same ones, at least in part, which have been involved in the evolution of *polionotus* (Blair, 1943*a*, 1950, 1958).

Though the two species are for the most part allopatric, their ranges overlap in the northern part of the range of *gossypinus*. Interspecific competition is evident (Bradshaw, 1965), and in most localities where both species have been observed or sampled each appears to maintain its identity (Osgood, 1909). In the Dismal Swamp, *gossypinus* and *leucopus* frequently occur together in the same habitat, and there is no evidence of hybridization in nature (Dice, 1940*c*). That the two species may be separated ecologically, however, is suggested by McCarley (1954*a*, 1963) who indicated that in eastern Texas *gossypinus* occurs predominantly in lowland forests while *leucopus* is principally in the uplands. That the two species sometimes may hybridize or interbreed in nature is suggested by McCarley (1954*b*) who presents some evidence of hybridization between the two forms at several localities in Texas and Louisiana. The two are interfertile in the laboratory (Dice, 1937, 1940*b*). They may also hybridize in nature, but since the principal morphological difference between the two is size, it is difficult to recognize hybrids—to be sure that the observed intermediacy results from interbreeding of the two forms. Detailed

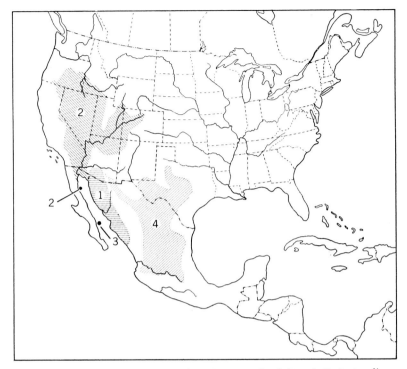

Fig. 4. 1, *P. merriami;* 2, *P. crinitus;* 3, *P. pseudocrinitus;* 4, *P. pectoralis.*

studies of several aspects of the biology of each of these species are needed to throw further light on the level of speciation attained by these two groups of populations.

CRINITUS GROUP.—*P. crinitus* is an inhabitant of rocky situations in arid environments of the Great Basin intermountain region of the western United States and northwestern Mexico (Fig. 4). Eight geographic races are currently recognized.

As indicated in the discussion of the *maniculatus* group, the affinities of *P. crinitus* are not entirely clear. In some morphological characters (e.g., of molar teeth) *crinitus* seems to be between the subgenera *Peromyscus* and *Haplomylomys,* and considering the complex pattern of differentiation observed in the genus, *P. crinitus* perhaps does fit in the gray area between the two subgenera. In architecture of skull and of the male phallus and in some other features, however, *crinitus* is at least as close to *pectoralis* of the *boylei* group of subgenus *Peromyscus* as to *eremicus* in *Haplomylomys*. The possibility that in its relationships it belongs with species of the subgenus *Peromyscus* should be further explored from the point of view of all aspects of its biology. It seems clear that in estimates to date of its affinities, attention has been focused on the principal diagnostic characters of the two subgenera and too little consideration has been given to other aspects of anatomy and biology of the species. A fresh approach is called for in which there is no a priori assumption that the species belongs in one or the other of the subgenera. Osgood (1909) and Hall and Hoffmeister (1942) treat *crinitus* as an "aberrant member" of the subgenus *Haplomylomys*. Burt (1960) states that the baculum of *crinitus* is similar to that of *eremicus* not to that of species of the subgenus *Peromyscus*. Further detailed studies are clearly needed.

P. pseudocrinitus: Relationships of this form are obscure. In the present classification, *pseudocrinitus* is listed with *crinitus* because the characters ascribed to it (Burt, 1932) fit better with *crinitus* in the subgenus *Peromyscus* than in *Haplomylomys* when features of the molar teeth are appropriately discounted as diagnostic characters (Hooper, 1957*b*). *P. pseudocrinitus* is apparently restricted to Coronados Island in the Gulf of California, Mexico (Fig. 4). It is one of a series of at least nine distinct forms restricted to islands in that gulf. The status of each of those taxa is unclear (see discussion of the subgenus *Haplomylomys* and of the *maniculatus* group of the subgenus *Peromyscus*).

BOYLEI GROUP.—Though the list of species in the current *boylei* group is different from that given by Osgood (1909), the changes that have been made in the list were anticipated by that author. The definition and limits of the group thus remain essentially unchanged. The species are inhabitants predominantly of temperate climates and rocky habitats, though they also occur in other situations. The range of *P. boylei* extends from Honduras northward to Oregon and Utah and eastward to the Ozark Mountains and overlaps that of each other species of the group. Though the species differ from one another anatomically and otherwise, they constitute a reasonably compact group which likely stemmed from a single ancestral stock.

The level in speciation attained by the various forms listed here varies. Some pairs of forms clearly do not intergrade and probably do not interbreed with each other under natural conditions. Other forms may prove to be geographic races of one or another of the species in the group.

P. pectoralis: This species inhabits the tableland of Mexico and adjoining parts of the southwestern United States (Fig. 4). Its range complements that of *P. crinitus,* a fact which should be kept in mind in subsequent studies of these species. Four geographic races are currently recognized, two of which (*eremicoides* and *pectoralis*) resemble *P. eremicus* in external appearance and in habits. The other two (*collinus* and *laceianus*) show striking resemblances to populations of *P. boylei* which occur sympatrically with them. The resemblance to *eremicus* may be owing to similar evolutionary responses to similar environments. The resemblances to *boylei* probably reflect close kinship, for the two species are quite similar in many respects, and in some localities individuals are identified as to species with difficulty. Examples from southern Texas (Hooper, 1952), San Luis Potosí (Dalquest, 1953), and southern Arizona (Hoffmeister and Goodpastor, 1954) are cases in point. The two taxa seem to be mostly separated ecologically, for example in western Texas (Blair, 1940) and in northern Tamaulipas (Hooper, 1952). That their ecological preferences are different at least where they occur sympatrically should be further explored. The baculum provides a character which may be used for distinguishing males of the two species; that of *pectoralis* bears a long, attenuate, cartilaginous tip (Clark, 1953; Hooper, 1958). The similarity of, and the difficulty of distinguishing, the two sets of populations in some areas suggests that the two species may sometimes interbreed in nature. Evidence of such, however, has not yet been presented.

P. boylei: This species has the largest geographic range (Fig. 5) of any in the group, and considering its characteristics it may well be most like the parental form which gave rise to the group. Thirteen geographic races are listed by Hall and Kelson (1959), and since the appearance of that publication, two additional races have been described and one form has been raised from synonymy. Thus, some fifteen or sixteen subspecies are recognized. Greatest differentiation has occurred in the highlands of Mexico, a region

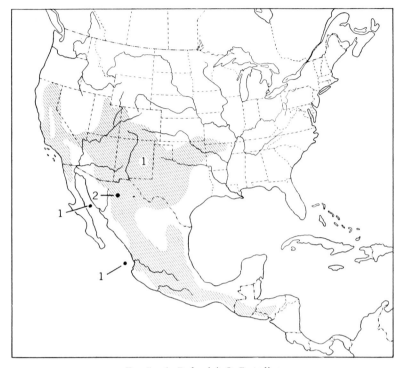

Fig. 5.   1, *P. boylei*; 2, *P. polius*

of high density of mammalian species (Simpson, 1964), and it is in that region that some of the major problems with respect to relationships of various forms of the species are to be found. In fact, it is extremely doubtful that all of the populations known by the name *P. boylei* are conspecific. Evidence that has accumulated to date indicates that some of the forms occur sympatrically without losing their respective identities. This suggests that either more than one species is represented in *P. boylei,* or that there are complex interrelationships of an intergrading chain of forms in which interbreeding between morphologically diverse populations occurs in some places but not in others. Populations known by the names *levipes, aztecus, evides, spicilegus,* and *simulus* are cases in point.

The status of the populations known as *P. b. levipes* is at the core of the problem of the interrelationships of the *boylei*-like populations in the highlands of Mexico. The form *levipes* occurs in the mountains of central Mexico southward to Honduras. In the central highlands its range is bordered on the west by *evides* (Fig. 6) and *spicilegus,* on the northwest by *rowleyi* and on the east by *aztecus* (Fig. 6). The form occurs in a variety of ecologic situations and it is morphologically quite variable over much of its range (Osgood, 1909). It occurs sympatrically with *aztecus* (Alvarez, 1961; Musser, 1964), *evides* (Hooper, 1961; Musser, 1964), *spicilegus* or *evides* (Hooper, 1955), and with the larger and otherwise different forms, *oaxacensis* and *hylocetes.* These questions immediately arise: How does *levipes* tie in with *rowleyi,* and through *rowleyi* with the other populations of *P. boylei* to the north? Secondly, what are the relationships among the other *"boylei*-like" forms which are sympatric with *levipes*? Those forms are *aztecus, evides,* and *spicilegus.*

Some *levipes* populations probably interbreed with those of *rowleyi,* but this is not yet certain. If the two forms do interbreed the interbreeding likely takes place in the southern part of the Mesa Central rather than in the mountains of northeastern Mexico, because in the northeastern segment there appears to be a large hiatus between the range of *levipes* and that of any other populations of *P. boylei* to the north or northwest (Baker, 1956). Few specimens are known from near the southern limits of the range of *rowleyi.* Those from the State of Aguascalientes, near the southern limits, retain most or all of the characters seen in *rowleyi* from the southern United States (Hooper, 1955). Detailed studies are needed of the populations in the southern part of the range of *rowleyi* to determine whether they do in fact interbreed with those of *levipes.*

Populations of *levipes* may merge with those of *spicilegus* to the northwest, but the scanty evidence available to date on this point is unconvincing. Its populations clearly are morphologically and ecologically separate from those of *aztecus* and *evides* in some areas (see above). Whether in other areas there is reproductive continuity between *aztecus* and *levipes* or *evides* and *levipes* is unknown, and it will remain unknown until the extent of the ranges of the *aztecus* and *evides* populations is determined. In short, *levipes* lives sympatrically without losing its identity with *aztecus* in parts of eastern Mexico and with *evides* and *spicilegus* in parts of western Mexico. It possibly merges with *rowleyi,* but this is not yet certain. The entire complex of characters gives one the impression that (a) the isolation of eastern and western arms of the range of *boylei*-like forms in Mexico has been removed and (b) the large gap between them has been filled by morphologically different populations which moved in from the south and now are incompletely inter-fertile with those western and eastern series of populations.

There are problems, too, with respect to *simulus* and *spicilegus*. There is some evidence (Hooper, 1955) that in the lowlands of western Mexico there are two morphological types of *P. boylei* masquerading under the name *simulus*. It is possible that these types represent reproductively isolated populations. One of these may not be reproductively connected to *spicilegus* which occurs in the mountains to the east (Baker and Greer, 1962). In the mountains of Durango the characters of *rowleyi* apparently grade into those of *spicilegus* but probably not into those of *simulus* (Baker and Greer, 1962).

What about the relationships of *evides* and *aztecus* which are separated from each other by the range of *levipes*? Although Osgood (1909) included *aztecus* as a subspecies of *P. boylei* he could find no evidence of intergradation of *aztecus* and *levipes* and stated it would not be surprising to find both of these forms at the same locality. Furthermore, he indicated that his samples of *aztecus*, *evides*, and *spicilegus* constituted a morphological group separable from both *levipes* and the other races of *P. boylei*. His prediction has been borne out, for now there are clear instances of sympatry of *aztecus*, *evides*, and *spicilegus* on the one hand with *levipes* on the other. In recognition of this it has been suggested that *aztecus* be treated as a separate species (Alvarez, 1961). If *aztecus* is to be regarded as a species, so should *evides*; but it should be remembered that the two are similar to each other and to *spicilegus*. Further studies are clearly needed to work out the relationships of this complex mosaic of morphological types, to assess the level of specia-tion of each, and to determine whether and where reproductive barriers between them have been let down or are being erected.

P. polius: This form is known only from its type locality in the state of Chihuahua, Mexico (Fig. 5). Little is known about it. Osgood (1909) included it in his *truei* group. Hoffmeister referred it tentatively to the *boylei* group, at the same time commenting that it might fit in the *melanophrys* group.

P. oaxacensis and *P. hylocetes*: Recognizing that *oaxacensis* bears

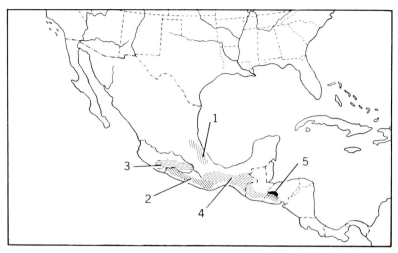

Fig. 6. 1, *P. aztecus*; 2, *P. evides*; 3, *P. hylocetes*; 4, *P. oaxacensis*; 5, *P. hondurensis*.

somewhat the same relationships ecologically and anatomically to *levipes* in southern Mexico as *aztecus* and *evides* does to *levipes* farther north, Osgood intimated that *oaxacensis* might be conspecific with *aztecus, evides,* and possibly *spicilegus*. *P. hylocetes* was also considered as a likely possibility for inclusion in that series. Whether *oaxacensis, hylocetes, aztecus, evides,* and perhaps *spicilegus* do prove to be disjunctly separated populations of one species has yet to be demonstrated. At least two of the forms, namely *oaxacensis* and *hylocetes,* likely will be found to be conspecific. These two have complementary ranges (Fig. 6), they are similar cranially and externally and they apparently are ecologic counterparts.

P. hondurensis: This taxon is known from a few localities in Honduras (Fig. 6). Closely similar to *P. evides* and *P. oaxacensis,* it likely will prove to be no more than a geographic variant of *oaxacensis*. It clearly belongs in the *boylei* group.

TRUEI GROUP.—This well-defined morphological group consists of *P. truei, P. difficilis,* and a third form (*P. bullatus*) which likely will prove to be a geographic race of *P. truei*. Over much of their ranges *truei* and *difficilis* occur sympatrically in the same habitats —predominantly rocky, arid situations—and in many respects *truei* is a smaller counterpart of *difficilis*. *P. truei* ranges from the Pacific

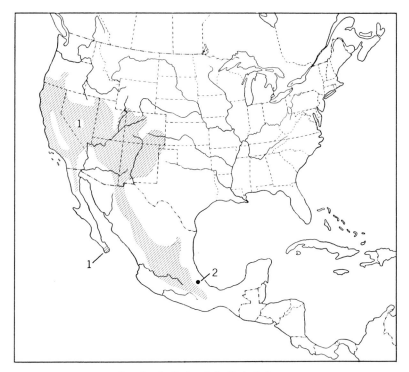

Fig. 7. 1, *P. truei*; 2, *P. bullatus*

Coast region of the United States (southern Oregon into Baja California) eastward to extreme western Oklahoma and southward into the State of Oaxaca, Mexico (Fig. 7). The range of *difficilis* extends from northern Colorado southward into Oaxaca (Fig. 8). The southern limits of the two species are essentially identical. Both species have differentiated geographically; fourteen races of *truei* and eight of *difficilis* are currently recognized. The available evidence indicates that the two species do not interbreed under natural conditions. Cross-breeding of examples of their northern populations has been obtained in the laboratory, however. The male hybrids are sterile while females are not (Dice and Liebe, 1937).

While the species *truei* seems to be a clear-cut case of a series of geographically intergrading populations, the situation with respect to *difficilis* is not so clear. Osgood (1909) suggested that

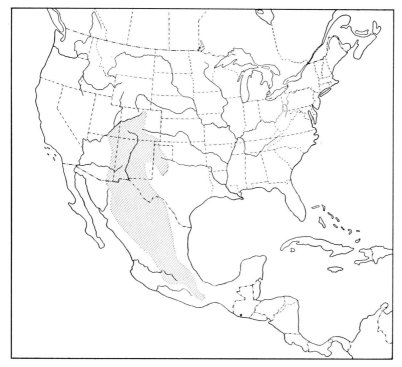

Fig. 8.   *P. difficilis*

the northern populations (formerly known as *P. nasutus*) likely are conspecific with the southern populations, but the available evidence to support this thesis is scanty. There remains a large hiatus in the known ranges of the southern and northern forms in which no specimens have yet been obtained to indicate more clearly whether the southern and northern populations could or do interbreed.

Biologically more important, perhaps, is the question of the status of *P. comanche*. This form is closely similar to *P. nasutus* and it has been treated as conspecific with *nasutus* by Hoffmeister (1951) and Hoffmeister and de la Torre (1961). Though it is but slightly differentiated externally and cranially from *nasutus* there is evidence that its populations are only partially interfertile with those of that species. Laboratory testing of stocks of the two forms indicates incomplete fertility between them (Blair, 1943*b*; Dice,

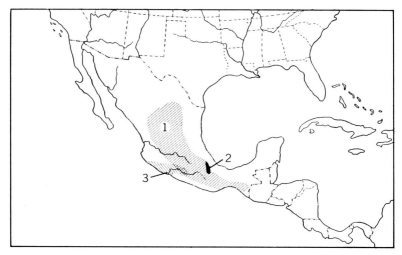

Fig. 9.   1, *P. melanophrys*; 2, *P. mekisturus*; 3, *P. perfulvus*

1952). Though the populations of *nasutus* and *comanche* now appear to be geographically isolated from each other, special attempts should be made to find and study geographically intermediate populations to see what happens between the two under natural conditions.

Melanophrys Group.—As defined by Osgood (1909), the *melanophrys* group includes *P. melanophrys* together with the forms *xenurus* and *mekisturus*. The form *xenurus* is viewed by Baker (1952) as a subspecies of *P. melanophrys,* making a total of six subspecies of that species. *P. mekisturus* is known from few specimens. It occurs in rocky situations and arid environments as does *melanophrys* (Fig. 9), and notwithstanding anatomical differences *melanophrys* may be its closest relative. Both species appear to have affinities with temperate forms, particularly with some of the *boylei* group, and a case could be made for assigning *melanophrys* and *mekisturus* to that group. Since the two species are rather strongly differentiated from those taxa, it seems appropriate to retain them as a separate group pending the accumulation of additional information.

Little is known of the biology of *P. perfulvus,* and just where it fits in a classificatory scheme is uncertain. It is recorded from a small segment of west-central Mexico where it occurs in arid tropical situations in foothills and lowlands near the Pacific Ocean (Fig. 9). Though, as Osgood stated (1945), it is unlikely that this species is wholly distinct from any previously described, nonetheless its affinities continue to remain obscure. Osgood indicated that it contrasted rather strongly with *P. melanophrys* and that many of its features resembled those of species in the *boylei* group, particularly *evides, aztecus, spicilegus,* and *simulus.* He suggested that closest relationship may be with *simulus.* A later study (Hooper, 1955) of samples from various parts of the range of *perfulvus* points out trenchant similarities of *perfulvus* and *melanophrys* and suggests that those two species may not be as distantly related as Osgood intimated. In the present classification it is tentatively included in the *melanophrys* group because that group in a sense is transitional between temperate *boylei*-like forms and tropical *mexicanus*-like species. *P. perfulvus* is unlike *P. banderanus,* with which it is ecologically associated.

MEXICANUS GROUP.—That fourteen nominal species are included in this group acknowledges the facts that many of these forms are poorly known and their affinities are unclear. With further study some of them will prove to be geographic races or to rank in synonymy of other species. All are inhabitants of tropical or subtropical environments in Middle America and almost all of them exhibit external and cranial features which seem to be characteristic of tropical species of the genus (Hooper and Musser, 1964b). As here defined, the group differs in several respects from the *mexicanus* group of Osgood (1909). It brings together some, not all, species of his *lepturus, megalops,* and *mexicanus* groups and includes other taxa which have been described subsequently.

P. ochraventer: This form is known from a few localities in arid tropical or subtropical foothills of northeastern Mexico (Fig. 11). Although it has been suggested that *ochraventer* is related to *furvus* and *latirostris* (Baker, 1951), it resembles *P. mexicanus* in several respects. It belongs in the *mexicanus* group and it may fit with the *P. mexicanus* complex of forms (see below).

P. stirtoni: This little known form is recorded from a few localities in El Salvador and Honduras (Fig. 14) (Goodwin, 1942; Burt, 1961). Though it is reported "to be an isolated species with

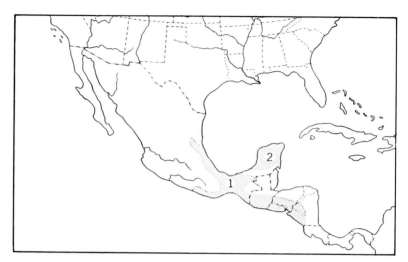

Fig. 10. 1, *P. mexicanus*; 2, *P. yucatanicus*

no close geographically related forms" (Goodwin, 1942), when better known it likely will be found to tie in with another previously described species.

P. yucatanicus: As now known, *P. yucatanicus* is restricted to the Yucatan Peninsula of Mexico where it occurs principally in forests and thickets of brush. Although it has differentiated slightly geographically, it is monotypic; none of the geographic variants appear worthy of formal recognition (Lawlor, 1965). The level in speciation attained by this form is unknown. It resembles *P. mexicanus* in several respects, and that species may be its nearest relative, but there is no known intergradation or interbreeding between those two sets of populations. A gap exists between their known geographic ranges. The possibility of character displacement in the area where the ranges of the two might meet has been suggested (Lawlor, 1965).

P. allophylus: Knowledge of *allophylus* remains as incomplete as when the form was described by Osgood in 1904; no new information indicating any relationship of it has been published, to my knowledge. Osgood's estimate (1904) of possible affinities of the form follows: "but for the size of its ears and shortness of its tail, it might well pass for an *Oryzomys* of the *O. chapmani* group. Its dark, scaly tail immediately suggests *Oryzomys*, and the character and color of its pelage bear out the resemblance. Its skull, however, is that of an ordinary type of *Peromyscus* without any striking characters. It seems probable that its closest relationship is with the *mexicanus* group, though it might easily be a northern member of some Central American group not yet known." It is known only from Chiapas, Mexico (Fig. 14).

P. mexicanus: Mice of this species inhabit tropical parts of Central America and eastern Mexico (Fig. 10). They occur principally in forests and brushy situations along the forest edge, and for the most part seem to be excluded from tropical rain forests. The species' geographic range, the largest in the *mexicanus* group, extends from southern Tamaulipas, Mexico, southeastward to Honduras and possibly to Costa Rica. Some seven or eight subspecies are recognized. Though one of the common rodents in many parts of its range and abundantly represented in museum collections, the species is still poorly understood. Enough evidence has accumulated, however, to indicate that the populations now treated under the name *P. mexicanus* may not constitute a freely interbreeding chain, but instead may be comprised of two moieties of yet undetermined levels in speciation.

One moiety inhabits the coastal region of extreme southeastern Mexico and southern Guatemala. The populations are known by the name *P. m. gymnotis*. Since externally and cranially they contrast rather strongly with the more northern populations of *mexicanus,* the possibility that the southern coastal populations may not interbreed with those of *mexicanus* to the north needs to be explored. Whether *P. allophylus* belongs in the *P. mexicanus* complex is unknown. In addition, relationships of *mexicanus* with *ochraventer, yucatanicus,* and several other allopatric forms need to be assessed.

P. sloeops definitely fits in the *P. mexicanus* complex. Its describer indicates that the name should be listed as a synonym of *P. mexicanus* (Goodwin, *in litt.*).

Other species of the Mexicanus Group: Nine additional taxa are listed as "species" (see classification). These are tropical forms which inhabit mountains of Middle America, the southernmost living in Costa Rica and Panama. All of them are anatomically and ecologically more similar to *P. mexicanus* than to any temperate environment species, but all or almost all of them have differen-

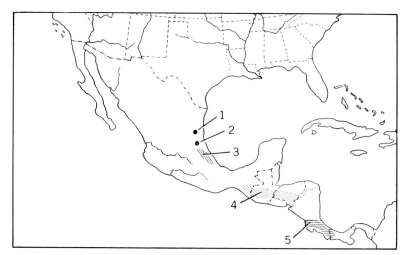

Fig. 11.   1, *P. ochraventer;* 2, *P. latirostris;* 3, *P. furvus;* 4, *P. guatemalensis;* 5, *P. nudipes.*

tiated morphologically from *mexicanus* and probably are reproductively isolated from it. Some occur sympatrically with it. Several probably are conspecific with either *P. guatemalensis* or *P. megalops.* The nine taxa are discussed briefly below, the forms arranged geographically approximately from north to southeast.

P. latirostris and *P. furvus* are inhabitants of the wet subtropical belt situated on the eastern flanks of the mountains of northeastern Mexico (Fig. 11). They are allopatric and may be conspecific.

Farther south and east in wet forests of the highlands occur the forms *zarhynchus, guatemalensis, altilaneus, grandis,* and *nudipes* (Figs. 11, 12). Like *furvus* and *latirostris* these forms are inadequately known. *P. zarhynchus* and *P. guatemalensis* are among the largest *Peromyscus,* but the size differences between those and (a) *furvus* and *latirostris* to the north and (b) *grandis* and *nudipes* to the south is of the order seen in many subspecies of the genus. On anatomical grounds, *P. nudipes* could fit either with *guatemalensis* or with *P. mexicanus.* As is true of the other forms discussed here, its precise affinities and level of speciation remain to be clarified. The form *altilaneus* shows some resemblance to *megalops,* but it is more similar to *guatemalensis* and *mexicanus* (Osgood, 1909:198). The populations designated by the name *angustirostris* apparently

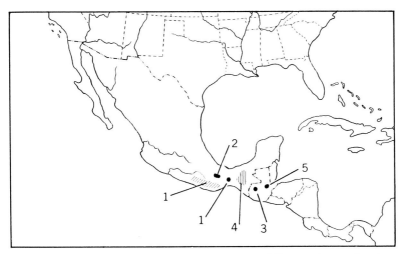

Fig. 12. 1, *P. megalops;* 2, *P. melanocarpus;* 3, *P. altilaneus;* 4, *P. zarhynchus;* 5, *P. grandis.*

are insufficiently distinct from those of *P. furvus* to warrant recognition by formal name (Musser, 1964).

The geographic ranges of *P. megalops* and *P. melanocarpus* (Fig. 12) complement those of the other forms mentioned here, and it is possible that these are two more of a single series of disjunctly distributed populations. On the other hand, in sum of anatomical characters, *megalops* is set apart somewhat from the *guatemalensis* and *furvus* complexes and may be specifically distinct from any of those forms. Its affinities, however, are not yet clear. Four geographic races of *P. megalops* are currently recognized. *P. melanocarpus* is said to be closely related to *megalops* (Osgood, 1909:216).

SUBGENUS *Osgoodomys*

Opinions regarding affinities of *P. banderanus* have changed considerably since Osgood reviewed the genus. He observed the peculiar cranial and external features which characterize its populations, and commented that in combination of characters the species is unique in the genus. In discussing geographic variation of the sampled populations, however, he indicated that three geographic races are recognizable. Two of these (*banderanus* and *vicinior*) were said to be similar. They occur in the coastal region and adjoining interior valleys of west-central Mexico, from the state of Nayarit southeastward to the state of Guerrero (Fig. 13). Samples from the coastal regions of Oaxaca (Fig. 10), however, were found to be anatomically unlike those. In Osgood's opinion the Oaxacan populations, treated under the name *angelensis*, diverged from typical *banderanus* in the direction of the *mexicanus* group. The divergence was viewed as sufficiently strong that it "would not be surprising if further material should demonstrate connection between *mexicanus* and *banderanus*" (Osgood, 1909:211). It was likely that the *angelensis* samples, which he considered to be aberrant *banderanus,* led him to underestimate the distinctness of the northwestern populations of *P. banderanus.*

It is highly probable that the *angelensis* populations, together with those named *coatlanensis,* do actually belong in the *mexicanus* complex rather than in *P. banderanus.* Externally and cranially they are similar to the population now known under the name *P. m. gymnotis,* and it may well be that they are representatives of the *gymnotis* segment of *P. mexicanus.* A detailed study of the anatomy, including the male reproductive tracts, and other facets of the biology of these forms is clearly needed.

In any event it now seems clear that the populations known by the names *banderanus* and *vicinior* are strongly differentiated from the other species of *Peromyscus,* not only in cranial and external characters (Osgood, 1909) but also in structure of the male phallus (Hooper, 1958). In the opinion of Hooper and Musser (1964*b*) the species, divorced of *angelensis* and *coatlanensis,* is unique in the genus and warrants subgeneric separation. Accordingly, those authors confined it to the newly named subgenus *Osgoodomys.* In characters of the soft and hard parts of the glans penis, *banderanus* (*sensu latu*) approaches *P. floridanus* and *Neotomodon alstoni.* To judge from the information now at hand, *banderanus* is a relict species which may well date back in its own phyletic branch to an early stage in the evolution of the genus.

SUBGENUS *Habromys*

The species *lophurus, lepturus,* and *simulatus* were grouped by Osgood (1909) on the basis of similarities in skin and skull. He indicated that though each of these taxa is distinguishable, the three seem to be close allies, distinct from the other species of

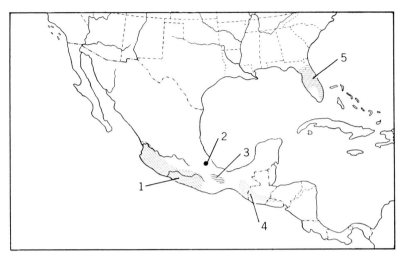

Fig. 13. 1, *P. banderanus;* 2, *P. simulatus;* 3, *P. lepturus* and *P. ixtlani;* 4, *P. lophurus;* 5, *P. floridanus.*

Peromyscus. All of the populations represented by these names inhabit the wet subtropical belt on the mountains of eastern and southern Mexico and western Central America. Each is known from few localities, *simulatus* in central Veracruz, *lepturus* in northern Oaxaca, and *lophurus* in southeastern Mexico and parts of Guatemala and El Salvador (Fig. 13).

In external and cranial morphology, *lophurus* and *lepturus* are unlike any other species in the genus save possibly the other forms which are here tentatively included in the subgenus *Habromys*. In anatomy of the glans penis *lophurus* and *lepturus* are quite different from all species with which they have been associated in the past. On the basis of the peculiar external and cranial characters and particularly of the unique glandes of these forms, Hooper and Musser (1964b) suggested that the forms be grouped in a new subgenus, *Habromys*. They appear to be relict taxa, which jointly may date to an early stage in the evolution of the genus. If Osgood's interpretation of his data is correct, *simulatus* also fits with *lophurus* and *lepturus*. The populations of these three, all inadequately known, form a geographic sequence in the subtropical wet belt on the mountains extending from Veracruz, Mexico, southward to Honduras. Their characters and distribution patterns suggest that they may be disjunctly distributed members of a single parental stock. Certainly, current evidence indicates that *lophurus* and *lepturus,* the best known forms of this series, are set apart anatomically from other species of the genus save *ixtlani* and probably *simulatus.*

P. *ixtlani* (Fig. 13) is closely similar to *P. lepturus* and probably will be found to be conspecific with it.

SUBGENUS *Podomys*

This subgenus was defined by Osgood (1909) in recognition of the peculiar set of characters of *P. floridanus,* the single species of the subgenus. Information subsequently accrued on many aspects of its biology has served to corroborate Osgood's opinion as to its distinctness. It is well differentiated from other species of the genus (Osgood, 1909; Hooper, 1958; Bader, 1959; Hooper and Musser, 1964*b*). The available information is entirely in harmony with the view that the Florida populations (Fig. 13) are relicts from a time when their progenitor was distributed widely in the southeastern United States and possibly in Middle America (in this connection see Johnson and Layne, 1961, regarding an ectoparasite of *floridanus*). Data from the male phallus suggests that some of the forms currently living in Mexico (e.g., *P. lepturus, P. lophurus, P. banderanus,* or *Neotomodon alstoni*) may be rather close relatives. These, too, may be relict forms tracing back to the same stock which gave rise to *floridanus.*

SUBGENUS *Megadontomys*

Osgood (1909) followed Merriam's recommendation that the forms *thomasi* and *nelsoni* were sufficiently differentiated from the rest of the species of *Peromyscus* to warrant their segregation in a separate subgenus. Osgood tentatively added *P. flavidus* to the subgenus, but he had reservations regarding this arrangement. He indicated that *flavidus* morphologically approaches several Neotropical genera and that it was being retained in the subgenus until various Neotropical forms were better understood. Subsequently, another form, *pirrensis,* was described. Its affinities are with *flavidus.* These forms are among the largest in size of body and skull in the genus. They are known from a few subtropical situations in the mountains of Mexico and Central America.

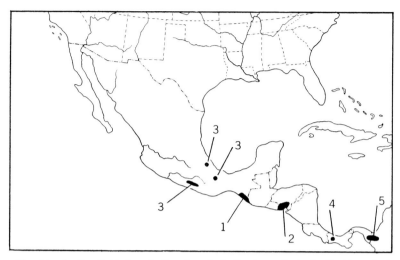

FIG. 14.   1, *P. allophylus;* 2, *P. stirtoni;* 3, *P. thomasi;* 4, *P. flavidus;* 5, *P. pirrensis.*

Recent evidence indicates that, as Osgood suggested, *P. thomasi* is well differentiated from the other species of *Peromyscus* and may warrant subgeneric separation from those species. The evidence also indicates that *P. flavidus,* together with *P. pirrensis,* should be subgenerically separated from *P. thomasi* to accord with the anatomical differences as between those sets of forms. Characters of *thomasi* are given by Osgood (1909), Hooper and Musser (1964*b*), and Musser (1964). The latter author gives information regarding habits, and presents data which indicate that the species is comprised of three geographic races, one of which is *thomasi.* The three sets of populations occur in the mountains of Guerrero, Oaxaca, and Veracruz, Mexico (Fig. 14). Each set apparently is now separated from the others by intervening areas of seemingly inhospitable terrain.

SUBGENUS *Isthmomys*

As indicated above, the inclusion by Osgood of the strongly differentiated *P. flavidus* in the subgenus *Megadontomys* was tentative. *P. flavidus* and *P. pirrensis* have now been consigned to the newly created subgeneric category *Isthmomys* (Hooper and Musser, 1964*b*) in recognition of the high level of differentiation of those two from the other species of *Peromyscus*. On anatomical grounds it makes sense to continue to include *flavidus* and *pirrensis* in *Peromyscus* because they fit in the periphery of the reticular variational pattern of that complex of species. Each is known from a few localities in Panama, *flavidus* in the western part of the Republic and *pirrensis* in the eastern part (Fig. 14). The level in speciation attained by these forms still remains to be determined. They have the aspect of geographic races of a single species.

Summary

Knowledge of the species of *Peromyscus* is highly unequal. Though there is an impressive volume of information on the genus, the amount probably exceeding that for any other native North American rodent group, most of it pertains to few species. Those few are inhabitants of the United States, the center of student interest in the group. *P. maniculatus* and *P. leucopus* have had the lion's share of attention; there probably is more information on various segments of the biology of these two than for any other kinds in the genus. But the state of knowledge of three other species, perhaps seven others, is not far behind, and when studies of some of these that are now underway are brought to completion those forms, too, may be ranked with *maniculatus* and *leucopus*. The seven species are *gossypinus, polionotus, floridanus, californicus, boylei, truei,* and *nasutus*.

Middle American populations warrant much more attention than they have received to date. They deserve study not only because they are poorly known but also because among them reside some of the major unsolved problems in the evolution of the genus. In that topographically and climatically diverse Middle American region the populations have differentiated maximally, or at least to the extent that the diversity among them exceeds that attained by the populations in the United States. Yet for none of these 30 or more Middle American taxa are there data comparable in amount and variety to those obtaining for the northern forms. They clearly warrant increased study.

This inter-taxa inequality of comparative data bears directly on classifying. A classification typically is founded on the most complete sets of data available; those spanning most populations usually constitute the prime basis for ordering the organisms into groups, while the less complete sets serve as supplemental information in the ordering process. Ideally, all sets of comparative data should be considered for possible use in the process.

A classification of *Peromyscus* which takes into account all systematic data has not yet been attempted, and a definitive one of that sort can not ensue until there is more information on some populations, particularly those in Middle America. Two of the classifications presented here are based principally on conventional data of study skin and skull. The third emerges from those two and incorporates re-arrangements of taxa to accord with a variety of new information, much of it derived from crania and skins and from the male reproductive tract. It designates six subgenera and several subgroups in representing the pattern of differentiation observed in the genus. Adjustments in the scheme will be needed as knowledge of the species is increased.

Literature Cited

ALVAREZ, TICUL. 1961. Taxonomic status of some mice of the Peromyscus boylei group in eastern Mexico, with description of a new subspecies. Univ. Kansas Publ. Mus. Nat. Hist., 14:111–20, 1 fig.

ARATA, ANDREW A. 1964. The anatomy and taxonomic significance of the male accessory reproductive glands of muroid rodents. Bull. Florida State Mus., 9:1–42, 9 figs.

BADER, ROBERT S. 1959. Dental patterns in *Peromyscus* of Florida. Jour. Mamm., 40:600–02.

BAIRD, SPENCER F. 1859. Mammals of North America. The descriptions of species based chiefly on the collections in the Museum of the Smithsonian Institution. Philadelphia: J. B. Lippincott. Pp. xxxiv + 764, 87 pls.

BAKER, ROLLIN H. 1951. Mammals from Tamaulipas, Mexico. Univ. Kans. Publ., Mus. Nat. Hist., 5:207–18.

———— 1952. Geographic range of Peromyscus melanophrys, with description of a new subspecies. *Ibid.*, 5:253–8, 1 fig.

———— 1956. Mammals of Coahuila, Mexico. *Ibid.*, 9:125–335, 75 figs.

BAKER, ROLLIN H., AND J. K. GREER. 1962. Mammals of the Mexican State of Durango. Publ. Mus. Mich. State Univ., Biol. Ser., 2:25–154, 4 pls., 6 figs.

BANKS, RICHARD C. 1967. The *Peromyscus guardia-interparietalis* complex. Jour. Mamm., 48:210–18, 3 figs.

BLAIR, W. FRANK. 1940. A contribution to the ecology and faunal relationships of the mammals of the Davis Mountain Region, southwestern Texas. Misc. Publ. Mus. Zool. Univ. Mich., 46:1–39, 3 pls.

———— 1942. Systematic relationships of Peromyscus and several related genera as shown by the baculum. Jour. Mamm., 23:196–204, 2 figs.

———— 1943a. Criteria for species and their subdivisions from the point of view of genetics. N. Y. Acad. Sci., 44:179–88.

———— 1943b. Biological and morphological distinctness of a previously undescribed species of the *Peromyscus truei* group from Texas. Contr. Lab. Vert. Biol. Univ. Mich., 24:1–8, 1 fig.

———— 1950. Ecological factors in speciation of *Peromyscus*. Evolution, 4: 253–75, 1 fig.

———— 1953. Factors affecting gene exchange between populations in the *Peromyscus maniculatus* group. Texas Jour. Sci., 5:17–33.

———— 1954. Tests for discrimination between four subspecies of deer-mice (*Peromyscus maniculatus*). *Ibid.*, 6:201–10, 1 fig.

———— 1958. Distributional patterns of vertebrates in the southern United States in relation to past and present environments. *In:* Hubbs, C. L. (ed.), Zoogeography. Washington: Amer. Assoc. Adv. Sci. Publ., 51: 433–68, 11 figs.

BLAIR, W. FRANK, AND W. E. HOWARD. 1944. Experimental evidence of sexual isolation between three forms of mice of the cenospecies *Peromyscus maniculatus*. Contr. Lab. Vert. Biol. Univ. Mich., 26:1–19.

BRADSHAW, W. N. 1965. Species discrimination in the *Peromyscus leucopus* group of mice. Texas Jour. Sci., 17:278–93.

BURT, WILLIAM H. 1932. Descriptions of heretofore unknown mammals from islands in the Gulf of California, Mexico. Trans. San Diego Soc. Nat. Hist., 7:161–82.

——— 1934. Subgeneric allocation of the white-footed mouse, Peromyscus slevini, from the Gulf of California, Mexico. Jour. Mamm., 15:159–60.

——— 1960. Bacula of North American mammals. Misc. Publ. Mus. Zool. Univ. Mich., 113:1–76, 25 pls.

BURT, WILLIAM H., AND R. A. STIRTON. 1961. The mammals of El Salvador. *Ibid.*, 117:1–69, 2 figs.

CLARK, DAVID L. 1966. Fertility of a *Peromyscus maniculatus* $\times$ *Peromyscus melanotis* cross. Jour. Mamm., 47:340.

CLARK, WILLIAM K. 1953. The baculum in the taxonomy of *Peromyscus boylei* and *P. pectoralis*. Jour. Mamm., 34:189–92, 1 fig.

COMMISSARIS, LARRY R. 1960. Morphological and ecological differentiation of *Peromyscus merriami* from southern Arizona. Jour. Mamm., 41:305–10, 2 figs.

COUES, ELLIOT, AND J. A. ALLEN. 1877. Monographs of North American Rodentia. Washington, D. C.: Govt. Printing Off., Pp. xii + x + 1091, 77 figs.

DALQUEST, WALTER W. 1953. Mammals of the Mexican state of San Luis Potosí. Louisiana State Univ. Biol. Sci. Ser., 1:1–112, 1 fig.

DICE, LEE R. 1933. Fertility relationships between some of the species and subspecies of mice in the genus Peromyscus. Jour. Mamm., 14:298–305.

——— 1937. Fertility relations in the *Peromyscus leucopus* group of mice. Contr. Lab. Vert. Genetics Univ. Mich., 4:1–3.

——— 1940a. Ecologic and genetic variability within species of Peromyscus. Amer. Nat., 74:212–21.

——— 1940b. Speciation in Peromyscus. *Ibid.*, 74:289–98.

——— 1940c. Relationships between the wood-mouse and the cotton-mouse in eastern Virginia. Jour. Mamm., 21:14–23, 1 fig.

——— 1949. Variation of *Peromyscus maniculatus* in parts of western Washington and adjacent Oregon. Contr. Lab. Vert. Biol. Univ. Mich., 44:1–34, 3 figs.

——— 1952. Variation in body dimensions and pelage color of certain laboratory-bred stocks of the Peromyscus truei group. *Ibid.*, 57:1–26, 3 figs.

DICE, LEE R., AND M. LIEBE. 1937. Partial infertility between two members of the *Peromyscus truei* group of mice. Contr. Lab. Vert. Genetics Univ. Mich., 5:1–4.

EISENBERG, JOHN F. 1962. Studies on the behavior of *Peromyscus maniculatus gambeli* and *Peromyscus californicus parasiticus*. Behaviour, 19:177–207, 4 figs.

——— 1963. The intraspecific social behavior of some cricetine rodents of the genus Peromyscus. Amer. Midland Nat., 69:240–46.

Fox, Wade. 1948. Variation in the deer-mouse (*Peromyscus maniculatus*) along the lower Columbia River. Amer. Midland Nat., 40:420–52, 2 figs.

Goodwin, George G. 1942. Mammals of Honduras. Bull. Amer. Mus. Nat. Hist., 79:107–95.

Hall, E. R., and D. F. Hoffmeister. 1942. Geographic variation in the canyon mouse, *Peromyscus crinitus*. Jour. Mamm., 23:51–65, 1 fig.

Hall, E. R., and K. R. Kelson. 1952. Comments on the taxonomy and geographic distribution of some North American rodents. Univ. Kans. Publ. Mus. Nat. Hist., 5:343–71.

———— 1959. The mammals of North America. Vol. II. New York: Ronald Press. Pp. viii + 1083 + 79, Figs. 313–553.

Hershkovitz, Philip. 1944. A systematic review of the Neotropical water rats of the genus *Nectomys* (Cricetinae). Misc. Publ. Mus. Zool. Univ. Mich., 58:1–88, 4 pls., 5 figs.

———— 1960. Mammals of northern Colombia, preliminary report No. 8: Arboreal rice rats, a systematic revision of the subgenus Oecomys, genus Oryzomys. Proc. U. S. Natl. Mus., 110:513–68, 12 pls., 6 figs.

———— 1962. Evolution of Neotropical cricetine rodents (Muridae) with special reference to the phyllotine group. Fieldiana: Zoology, 46: 1–524, 123 figs.

Hoffmeister, Donald F. 1951. A taxonomic and evolutionary study of the piñon mouse, *Peromyscus truei*. Ill. Biol. Monogr., 21:1–104, 5 pls., 24 figs.

Hoffmeister, Donald F., and W. W. Goodpaster. 1954. The mammals of the Huachuca Mountains, southeastern Arizona. Ill. Biol. Monogr., 24: 1–152, 27 figs.

Hoffmeister, Donald F., and M. R. Lee. 1963. The status of the sibling species *Peromyscus merriami* and *Peromyscus eremicus*. Jour. Mamm., 44:201–13, 4 figs.

Hoffmeister, Donald F., and L. de la Torre. 1961. Geographic variation in the mouse *Peromyscus difficilis*. Jour. Mamm., 42:1–13, 2 figs.

Hooper, Emmet T. 1952. Notes on mice of the species *Peromyscus boylei* and *P. pectoralis*. Jour. Mamm., 33:371–8, 2 figs.

———— 1955. Notes on mammals of western Mexico. Occ. Pap. Mus. Zool. Univ. Mich., 565:1–26.

———— 1957a. Records of Mexican mammals. *Ibid.*, 586:1–9.

———— 1957b. Dental patterns in mice of the genus *Peromyscus*. Misc. Publ. Mus. Zool. Univ. Mich., 99:1–59, 24 figs.

———— 1958. The male phallus in mice of the genus *Peromyscus*. *Ibid.*, 105:1–24, 14 pls., 1 fig.

———— 1961. Notes on mammals from western and southern Mexico. Jour. Mamm., 42:120–22.

Hooper, Emmet T., and G. G. Musser. 1964a. The glans penis in Neotropical cricetines (Family Muridae) with comments on classification of muroid rodents. Misc. Publ. Mus. Zool. Univ. Mich., 123:1–57, 9 figs.

———— 1964*b*.   Notes on classification of the rodent genus *Peromyscus*. Occ. Pap. Mus. Zool. Univ. Mich., 635:1–13, 2 figs.

HULL, DAVID L.   1964.   Consistency and monophyly. Syst. Zool., 13:1–11.

JOHNSON, MURRAY L., AND B. T. OSTENSON.   1959.   Comments on the nomenclature of some mammals of the Pacific northwest. Jour. Mamm., 40:571–7.

JOHNSON, PHYLLIS T., AND J. N. LAYNE.   1961.   A new species of Polygenis Jordan from Florida, with remarks on its host relationships and zoogeographic significance. Ent. Soc. Washington, 63:115–23, 10 figs.

LAWLOR, TIMOTHY E.   1965.   The Yucatan deer mouse, Peromyscus yucatanicus. Univ. Kans. Publ., Mus. Nat. Hist., 16:421–38, 2 figs.

LIU, T. T.   1954.   Hybridization between *Peromyscus maniculatus oreas* and *P. m. gracilis*. Jour. Mamm., 35:448–9.

MCCARLEY, W. H.   1954*a*.   The ecological distribution of the *Peromyscus leucopus* species group in eastern Texas. Ecology, 35:375–9, 1 fig.

———— 1954*b*.   Natural hybridization in the *Peromyscus leucopus* species group of mice. Evolution, 8:314–23, 3 figs.

———— 1959.   The mammals of eastern Texas. Texas Jour. Sci., 11:385–426, 1 fig.

———— 1963.   Distributional relationships of sympatric populations of *Peromyscus leucopus* and *P. gossypinus*. Ecology, 44:784–88, 1 fig.

MAILLAIRD, JOSEPH.   1924.   A new mouse (*Peromyscus slevini*) from the Gulf of California. Proc. Calif. Acad. Sci., ser. 4, 12:1219–1222, 3 figs.

MANVILLE, RICHARD H.   1961.   The entepicondylar foramen and *Ochrotomys*. Jour. Mamm., 42:103–4.

MAYR, ERNST.   1963.   Animal species and evolution. Cambridge: Harvard Univ. Press. Pp. xiv + 797, figs.

MEARNS, EDGAR A.   1907.   Mammals of the Mexican Boundary of the United States. Part I. Bull. U. S. Natl. Mus., 56: xv + 530, 126 figs.

MILLER, GERRIT S., JR.   1912.   List of North American land mammals in the United States National Museum, 1911. Bull. U. S. Natl. Mus., 79:xiv + 455.

MILLER, GERRIT S., JR., AND R. KELLOGG.   1955.   List of North American recent mammals. Bull. U. S. Natl. Mus., 205:xii + 954.

MUSSER, GUY G.   1964.   Notes on geographic distribution, habitat, and taxonomy of some Mexican mammals. Occ. Pap. Mus. Zool. Univ. Mich., 636: 1–22, 1 fig.

OSGOOD, WILFRED H.   1904.   Thirty new mice of the genus *Peromyscus* from Mexico and Guatemala. Proc. Biol. Soc. Washington, 17:55–77.

———— 1909.   Revision of the mice of the American genus Peromyscus. N. Amer. Fauna, 28:1–285, 8 pls., 11 figs.

———— 1945.   Two new rodents from Mexico. Jour. Mamm., 26:299–301.

PACKARD, ROBERT L.   1960.   Speciation and evolution of the pygmy mice, genus *Baiomys*. Univ. Kans. Publ., Mus. Nat. Hist., 9:579–670, 4 pls., 12 figs.

RINKER, GEORGE C.   1960.   The entepicondylar foramen in *Peromyscus*. Jour. Mamm., 41:276.

————— 1963. A comparative myological study of three subgenera of *Peromyscus.* Occ. Pap. Mus. Zool. Univ. Mich., 632:1–18.

SATTLER, ROLF. 1964. Methodological problems in taxonomy. Syst. Zool., 13: 19–27.

SCHWARTZ, ALBERT. 1954. Oldfield mice, *Peromyscus polionotus,* of South Carolina. Jour. Mamm., 35:561–69, 1 fig.

SHEPPE, WALTER, JR. 1961. Systematic and ecological relations of *Peromyscus oreas* and *P. maniculatus.* Proc. Amer. Phil. Soc., 105:421–46, 12 figs.

SIMPSON, GEORGE G. 1945. The principles of classification and a classification of mammals. Bull. Amer. Mus. Nat. Hist., 85:xvi + 350.

————— 1961. Principles of animal taxonomy. New York: Columbia Univ. Press. Pp. viii + 247, 30 figs.

————— 1964. Species density of North American Recent mammals. Syst. Zool., 13:57–73, 5 figs.

TRUE, FREDERICK W. 1894. On the relationships of Taylor's mouse, Sitomys taylori. Proc. U. S. Natl. Mus., 16:757–8.

VORONTSOV, N. N. 1959. The system of hamster (Cricetinae) in the sphere of the world fauna and their phylogenetic relations. Bull. Mosk. Obsh. Ispyt. Prirody, Biol. Sec. [Bull. Moscow Soc. Naturalists], 64:134–7.

3

SPECIATION

Lee R. Dice

Introduction

THE LARGE number of geographic races, species, and species-groups included in the genus *Peromyscus,* the diversity of habitats in which the various forms live, the abundance of the animals in many situations, and  the ease with which most forms may be kept and bred in the laboratory makes this genus of especial value for the study of speciation. Two, three, four, or as many as five species of *Peromyscus* may live in the same area without hybridizing (Dice, 1942). Even two species that evidently are closely related taxonomically may live in the same region and sometimes may be associated in the same ecologic communities without intermating. The analysis of the geographic, ecologic, behavioral, and genetic relations among the subspecies, species, and species-groups of *Peromyscus* can be expected, therefore, to furnish much information about the processes by which new species originate in nature.

Speciation can take place in either of two ways. One type of speciation occurs when over a period of geologic time a species evolves characters which differ from those which it previously exhibited (Simpson, 1960). The other kind of speciation occurs when a species divides into two or more parts, each of which is recognized as a distinct species. I shall here consider only that type of speciation which occurs when a species splits into daughter species.

A species usually is defined by vertebrate taxonomists as a population or group of populations within or among which the member individuals are potentially interfertile, but which is isolated reproductively from all other populations (Mayr, Lindsley, and Usinger, 1953; Simpson, 1960; Mayr, 1963). Genes thus may be exchanged more or less freely within a species population, but the exchange of genes between populations assigned to different species is inhibited or prevented (Dobzhansky, 1951). Any population which is not isolated reproductively from its neighbors will obviously tend through gene flow to lose whatever genetic distinctiveness it may once have had.

The process of speciation by species splitting consists, therefore, of the evolution of reproductive isolation between populations which have descended from the same ancestors. A species can be considered to be distinct only when it does not interbreed or interbreeds only rarely with any other species. The occurrence of intersterility between such separated populations may be an added criterion of their specific distinctness, but is not a necessary criterion. Populations which are intersterile but which do not differ in any evident morphologic character, however, may escape recognition as distinct species.

Isolating Mechanisms

Several kinds of isolating mechanisms may produce reproductive isolation between neighboring populations (Dobzhansky, 1951). These isolating mechanisms may be classified roughly as consisting either of geographic, ecologic, psychologic, or physiologic barriers to the exchange of genes. The existence of ecologic, psychologic, or physiologic barriers to gene interchange will nearly always be based in part upon inherited differences.

GEOGRAPHIC ISOLATION.—A geographic barrier is one of the most effective isolating mechanisms which may prevent closely related species from interbreeding. The geographic ranges of the deermouse (*Peromyscus maniculatus*) and of the oldfield mouse (*P. polionotus*), for example, are separated in the southeastern United States by a wide area in which no population of either of these species occurs (Osgood, 1909; Dice, 1940*b*; Blair, 1950). The adults of these two species are closely similar in many of their characters, but can readily be distinguished, by differences in body dimensions and in other morphologic characters. Evidently they have originated from a common ancestor, but geographic isolation now prevents their interbreeding.

Island populations are especially likely to evolve into distinct species (Wallace, 1880). Some island populations are so completely isolated that under present geographic conditions it must be assumed that no genes are being received from any other population. The course of evolution in such an isolated population is likely to be different from that in the populations of its relatives. In time, therefore, the island population may evolve distinctive characters. On the other hand, some islands are so close to the mainland or to other islands that interchange of individuals may occasionally or even frequently occur.

An island population whose characters are closely similar to those of a population on the adjacent mainland will usually be assigned to the mainland species on the assumption that the two populations are potentially interfertile. This assumption may not always be justified. For most kinds of animals, however, it will not be possible to test the interfertility of all such island and mainland populations by actual breeding experiments.

Numerous populations of *Peromyscus* are restricted to islands (Osgood, 1909; Miller and Kellogg, 1955; Cowan and Guiguet, 1956). The populations on some of these islands are indistinguishable in all their characters from populations living on neighboring islands or on the adjacent mainland (McCabe and Cowan, 1945). Further, some of these island populations have been demonstrated to be interfertile with neighboring populations by laboratory breeding tests (Dice, 1933). Many island populations of *Peromyscus,* however, possess distinctive characters and a few of them exhibit reduced fertility when they are crossed with their mainland relatives. The population of *Peromyscus polionotus leucocephalus,* which is confined to Santa Rosa Island, just off the coast of Florida in the Gulf of Mexico, for example, differs in a number of features of pelage color (Sumner, 1930) and gives some indication of being partially infertile with the mainland populations assigned to the same species (Blair, 1943*a*). This island population, therefore, may be in the process of evolving into a distinct species.

Any two related populations which become separated by a geographic barrier will usually inhabit areas which differ in certain of their environmental features. Natural selection will consequently tend to produce a somewhat different set of adaptive characters in each of the two separated populations. Differences in random genetic drift may also produce differences between the two populations in certain of their non-adaptive characters. Hereditary differences in behavior, morphology, and/or physiology, perhaps including some degree of intersterility, may evolve during a period of geographic isolation between populations which have a common ancestry. These differences may be sufficiently great to maintain their reproductive isolation should the geographic barrier between the populations ever be broken down. Unless effective reproductive barriers between such separated populations have evolved, however, they will interbreed and perhaps merge again into a single population whenever the geographic barrier which separates them is eliminated. Geographic isolation thus may provide an opportunity for the evolution of specific differences between sister populations, but geographic isolation by itself does not constitute speciation, nor does it always result in speciation.

A number of pairs or groups of related species of the genus *Peromyscus* inhabit neighboring regions, but are separated by geographic barriers of various kinds, including distance. The geographic barriers which now separate these related species, or other geographic barriers which may have been effective in the past, undoubtedly have contributed to the evolution of the specific differences which these forms exhibit (Blair, 1950). I consequently agree with Mayr (1963) that geographic isolation must have been an important factor in the origin of many or perhaps of most species of animals.

It is not necessary to conclude, however, that speciation can occur only under geographic isolation. Under certain conditions it appears to be theoretically possible for a species to split into daughter species under ecologic isolation alone, without the sister populations concerned ever being isolated geographically (Fisher, 1958; Dobzhansky, 1940).

ECOLOGIC ISOLATION.—Ecologic isolation often separates two related populations which live in the same region but are restricted to different types of habitats (Blair, 1950). In the Illinoian Biotic Province (Dice, 1943), for example, the white-footed mouse (*Peromyscus leucopus*) inhabits the groves of hardwood trees, while the prairie race of the deermouse (*Peromyscus maniculatus bairdi*) lives in the surrounding grassy fields (Dice, 1922). These two species often meet along the forest edges (Blair, 1940a). In winter they may even live together in the same nests (Howard, 1949). The ecologic barrier between these two species thus does not completely separate them. Other kinds of reproductive barriers must prevent their interbreeding. No hybrid young are produced even when males and females of these species are paired in laboratory cages (Dice, 1933).

Geographic races of the same species of *Peromyscus* also may live in the same area but be separated by an ecologic barrier (Blair, 1953a). Two races of the deermouse (*Peromyscus maniculatus*) thus live in the same part of western Montana, the race *artemisiae* inhabits the forests on the mountains, while the race *osgoodi* is restricted to the grassy plains which extend to the eastward. These two races meet along the forest edge, but so far as known they do not interbreed in nature (Murie, 1933). The differences between these two races in body dimensions and in pelage color are so great that if the intergrading populations in other regions which now connect them (Osgood, 1909) were to be eliminated, they would be considered to be distinct species.

In northern Michigan two other subspecies of the deermouse may occur together in some areas, likewise separated only by a habitat barrier. The subspecies *bairdi* inhabits open fields and lake beaches, while *gracilis* lives in the hardwood forests (Dice, 1932; Hooper, 1942). These two races must meet sometimes along the borders of their respective habitats. No natural hybrids between them have been discovered, although in the laboratory they will mate and produce fertile offspring (Foster, 1959). Still other pairs of races of deermice sometimes live together in the same areas without interbreeding (Osgood, 1909; Dice, 1931). In all such cases the two races which overlap in geographic range are presumed to differ to at least some degree in their habitats.

Where two contrasting types of habitat occur in the same region with no intermediate conditions between them, the local populations of any species which is able to survive in both types of habitats will be subject to selection in two different directions. In a region where stands of forest alternate with grassy fields, for example, the local populations of a species which lives in both types of habitat will be exposed to differential selection. Those local populations which inhabit the forest stands will be under selection pressure to evolve forest adaptations, while those populations which live in the open fields will tend to evolve a very different set of adaptations.

Laboratory experiments with fruit flies have demonstrated that populations with a common ancestry which are exposed to selection in different directions may diverge very quickly in certain characters, even when these populations continue to interbreed. The two halves of a population of *Drosophila melanogaster* exposed to selection in opposite directions for number of sternopleural chaeta, thus, diverged strongly within 20 generations in spite of 25 per cent intermating in each generation: the same amount of intermating expected under random mating (Millicent and Thoday, 1960).

In those areas where the local populations of a species have evolved adaptations to two contrasting types of habitat, any hybrids between these contrasting populations will be expected to be, in part, intermediate in their characters and consequently will not be well adapted to the habitats of either of their parents. If there are no intermediate types of habitat to which these hybrids are adapted, they will be subject to adverse selection. The loss of such hybrids constitutes a biologic waste. Natural selection, therefore, will tend to evolve types of behavior and perhaps of other characters which will inhibit cross mating (Fisher, 1958; Dobzhansky, 1940; Waddington, 1957). Sympatric populations which have evolved distinctive characters and also effective barriers to intermating will usually be classed as separate species, unless they happen to be connected through intergrading populations in other areas.

PSYCHOLOGIC ISOLATION.—Psychologic barriers may be presumed usually to constitute the mechanism by which reproductive isolation is maintained in those situations where two potentially interfertile populations meet or overlap without interbreeding. Differences in habitat preference between related populations whose morphologic characters adapt them to different kinds of habitats would be especially likely to evolve. Those individuals whose behavior fails to restrict them to the kind of habitat to which their morphologic characters are adapted would be subject to adverse selection.

Studies in the laboratory of habitat selection by a prairie race and by a forest race of the deermouse (*Peromyscus m. bairdi* and *P. m. gracilis*, respectively) have demonstrated that the members of each race do have an inherited tendency to select and to remain in their natural type of habitat (Harris, 1952). Experimental studies of *bairdi* made in the field further demonstrate that while the inherited type of habitat selection exhibited by the animals may be reinforced or perhaps modified to some degree by experience, it is not reversed by being reared in the "wrong" habitat (Wecker, 1963). The inherited differences in habitat selection between members of different sympatric populations together, perhaps, with imprinting to the natal habitat of each individual will, in considerable part, prevent intermating between those related but sympatric populations which occupy contrasting types of habitat (Thorpe, 1945).

The most important psychologic barrier between neighboring species, however, is the tendency of each form to select and to mate with members of its own species while rejecting possible matings with aliens. Many kinds of birds, reptiles, amphibians, fishes, and insects have secondary sexual characters, display movements, or characteristic calls or songs which evidently aid in the recognition of appropriate mates. Some kinds of mammals apparently identify members of their own species by special odors. Odor is thus indicated to be used for mate identification by the red-backed voles of the genus *Cleithrionomys* (Rauschert, 1963) and by some species of *Peromyscus* (Moore, 1965). Our information about the means used by wild mammals for distinguishing members of their own species from aliens, however, is very scanty.

A preference for mates of their own species definitely is exhibited by both *Peromyscus maniculatus* and *P. polionotus*. When given a choice in the laboratory, members of these species prefer to associate with their own kind, rather than with members of the other species (Blair and Howard, 1944). The number of pairs tested of the various races of these two species was few, but assortative pairing was statistically significant for the combination *P. p. leucocephalus* versus *P. m. blandus* (Fuller and Thompson, 1960). The tendency toward assortative mating which these species exhibit would serve as a partial or perhaps as a complete barrier to their interbreeding should the geographic barrier which now separates them ever become obliterated.

A small degree of assortative pairing also is exhibited in the laboratory by the two very different subspecies of deermice, *Peromyscus m. bairdi* and *P. m. gracilis,* which sometimes live in the same region but in different habitats (Harris, 1954). On the other hand, four other subspecies of deermice (*ozarkiarum, pallescens, blandus,* and *rufinus*) exhibited no assortative pairing under laboratory conditions (Blair, 1954). No two of these four races, however, are known to occur together in the same area.

Intermating between the white-footed mouse (*Peromyscus leucopus*) and the cottonmouse (*P. gossypinus*) in those situations where both occur together also is evidently prevented by a psychologic barrier of some kind. Individuals of the two species taken from an area where both species occur in nature were brought into the laboratory and there given their choice of mates (McCarley, 1964). Members of each species associated with members of their own kind more frequently than with members of the other species. Individuals from areas where the two species do not occur together in nature, however, did not associate with their own kind more frequently than with the alien species. Mating discrimination between these two species evidently is strongest, therefore, in those regions where assortative mating is operating to prevent intermating. This suggests that psychologic barriers to intermating are especially likely to evolve in those areas where two related populations occur sympatrically.

Barriers to intermating between related populations may evolve or become enhanced very quickly under experimental conditions. Two laboratory strains of *Drosophila melanogaster* selected for different morphologic characters, for example, were placed together in the same cage, but hybrids between them were removed. Within a few generations the members of each strain had evolved a preference for mating with other members of their own strain in preference to the other strain (Knight, Robertson, and Waddington, 1956). Barriers to intermating between the species *Drosophila pseudoobscura* and *D. persimilis* also rapidly became enhanced when the two species were caged together (Koopman, 1950).

It is a biologic rule that the morphologic characters of two related species differ more in those areas where both species occur together than they do where the species inhabit different areas (Brown and Wilson, 1956). The evolution of distinctive characters and of reproductive isolation between related populations thus may possibly proceed even more rapidly where the populations inhabit the same region, but occupy different habitats, than where they are separated geographically.

On the other hand, in an area where a species occupies a variety of habitats which are not sharply distinct from one another, differentiation among the local populations is unlikely to evolve. Interbreeding among the local populations will occur in the intermediate zones between the several types of habitats. Because of the frequent interchange of genes among these local populations, all of them will be expected to exhibit the same general type of adaptations. Many species of *Peromyscus* occupy a wide variety of local habitats, with no indication of specialization for any particular type. In the absence of sharp habitat differences between any of the local populations in a given area, therefore, little tendency will be expected to exist for the evolution of interpopulation differences.

Neither is speciation likely to take place in a species whose range extends over an area in which a geographic gradient occurs in a particular environmental feature, such as soil color. Natural selection usually will favor a different set of adaptive characters along this gradient. The populations of the species which occur at each end of this gradient will usually be assigned, therefore, to different geographic races. Towards the middle of such a gradient, however, the environmental conditions usually will be more or less intermediate between those prevailing at its two extremes. In this intermediate area the adaptive characters of the local populations also will usually be intermediate between those typical of the two parent races which here intergrade. Such a gradient of soil color from east to west across North Dakota, for example, is paralleled by a gradient or cline in the pelage color of the deermouse (*Peromyscus maniculatus*). The eastern, dark-colored race *bairdi* here intergrades with the paler-colored western race *osgoodi*. In this area of intergradation the pelage color of each local population is in general well adapted to the soil color in its particular habitat (Dice, 1940*c*). Neighboring local populations appear not to be separated by any kind of habitat barrier. On the contrary, interbreeding may be presumed to occur commonly between the local populations and the young individuals will usually be well adapted to the habitats into which they are born.

In an area of gradual intergradation between adjacent races, no habitat barrier or other kind of reproductive barrier will usually separate the local populations. Under such conditions, there will be no tendency for the splitting of a species into daughter species.

Geographic races usually intergrade with neighboring races of the same species. Most geographic races consequently do not constitute incipient species (Goldschmidt, 1933; Dice and Blossom, 1937; Blair, 1950; Simpson, 1960; Mayr, 1963).

Splitting of a species thus may be presumed to occur only where reproductive isolation between certain of the local populations is produced by an effective barrier to intermating. Geographic isolation undoubtedly is the most common kind of barrier which may result in speciation. It appears to be possible, however, for ecologic isolation between certain of the local populations of a species to be sufficiently effective under special conditions to permit the splitting of a species into daughter species without geographic isolation. Divergent evolution between the local populations of a species within a particular area is likely to occur, however, only where there are strong contrasts between certain of the local habitats together with an absence of intermediate habitat types. The conditions under which speciation may result from ecologic isolation may be presumed, therefore, to be rather uncommon.

PHYSIOLOGIC ISOLATION.—The evolution of intersterility between populations which have a common ancestry constitutes the final and irreversible step in speciation. Such intersterility may be presumed to be the result of a physiologic incompatibility of some kind. When complete intersterility has evolved between two related populations, it will be impossible for these populations to interbreed successfully and possibly to merge again into a single population should the reproductive barriers which now separate them ever be removed.

For speciation to be fully complete, therefore, intersterility must have evolved between two species which have descended from a common ancestral species. If two populations or groups of populations are separated only by a physical, ecologic, or psychologic barrier, the possibility will always exist for the partial or complete breakdown of this barrier and the consequent interchange of genes. Thus, where there is potential interfertility between separated populations, or even if there is partial sterility between them, some possibility will exist for the transfer of genes from one population to the other and perhaps even for their eventual merging. Only when intersterility between two populations has become sufficiently complete to prevent the interchange of genes between them will irreversible speciation have occurred.

While intersterility with other species thus constitutes the most positive measure of specific distinctness, complete intersterility with all other forms is not essential for the taxonomic recognition of a species (Mayr, 1963). Complete intersterility with all other forms would for most species be very difficult to prove. Young may occasionally be produced from matings between species which are clearly distinct and which evidently have been separated in nature for perhaps thousands of generations. The introgression of genes from one species to another resulting from such occasional cross matings may, by increasing its genetic variability, have important consequences for the further evolution of the species involved (Anderson, 1949).

The genus *Peromyscus* exhibits a considerable range of kinds and degrees of intersterility between species and their subdivisions. The members of the same local population and of the local populations of the same geographic race usually are mutually interfertile. In every population, however, many embryos are lost before birth (Liu, 1953*b*). This embryonic wastage indicates that some gene combinations, even within the same local population, may be incompatible. A small degree of intersterility between the geographic races *bairdi* and *gracilis* of the species *maniculatus* also has been reported by Harris (1954). Matings between members of different geographic races of the same species, however, often are as productive of healthy offspring as are matings between members of the same local population (Sumner, 1932; Dice, 1933). Some related species of *Peromyscus* also are interfertile to a high degree. On the other hand, crosses between related species may exhibit reduced fertility. In particular, the hybrid males produced by some species crosses are sterile. None of the crosses between species of *Peromyscus* which I have made, however, have produced any conspicuous number of sterile hybrid females. All attempted crosses between species of *Peromyscus* belonging to different species groups have failed completely to produce offspring (Dice, 1933).

No sterility barrier appears to prevent the interchange of genes between the white-footed mouse (*Peromyscus leucopus*) and the closely related cottonmouse (*P. gossypinus*). Members of various races of these two species will cross in the laboratory and their hybrid offspring are fertile (Dice, 1937). Only a few natural hybrids have been reported, however, in spite of the fact that the geographic ranges of the two species overlap in a number of places (Osgood, 1909; McCarley, 1954*a*). The ecologic preferences of the two species differ somewhat, the cottonmouse usually lives in lowland forests, while in those regions where both species occur the white-footed mouse is restricted mostly to upland forests (Howell, 1921; McCarley, 1954*b*, 1963). In some situations, however, such as in the Dismal Swamp of Virginia, both species may live in the same habitat and may be taken on the same trapline (Dice, 1940*a*).

In contrast to the potential interfertility which exists between the white-footed mouse and the cottonmouse, a small amount of intersterility separates the deermouse (*Peromyscus maniculatus*) from the old field mouse (*Peromyscus polionotus*). The geographic ranges of these two species do not overlap and they consequently have no opportunity to interbreed in nature. Some matings between them in the laboratory produce offspring and at least some of the male and female hybrids are fertile. On the other hand, some matings between these two species fail to produce offspring or exhibit reduced fertility and some of the F_1 hybrids appear to be partially or completely sterile (Sumner, 1930; Dice, 1933, 1940*b*; Blair, 1943*b*; Liu, 1953*a*, 1953*b*; Dawson, 1965). The variation in amount of interfertility exhibited by these crosses may indicate variation in the interfertility relations among the races of one or both of these two species. A particular difficulty in obtaining successful crosses between male *maniculatus* and female *polionotus* is that the large size of the hybrid fetuses makes their birth difficult and some mothers and some young die at parturition or soon afterward (Watson, 1942). *P. maniculatus* and *polionotus* have presumably diverged in fairly recent geologic time from a common ancestor, but are now completely separated geographically. Up to the present time, however, only a small degree of intersterility between them appears to have evolved.

P. oreas, another member of the *Peromyscus maniculatus* group, exhibits puzzling relationships with several of its close relatives whose geographic ranges it meets or overlaps. In the lowlands of the Puget Sound area, for example, *oreas* may live in the same forest habitats as *austerus* and may even be taken on the same traplines (Dice, 1932, 1949; Svihla, 1933; Johnson and Ostenson, 1959). In other parts of the Pacific Northwest, *oreas* may live in association with *artemisiae, macrorhinus, gambeli,* or perhaps other races of *maniculatus* (Osgood, 1909; McCabe and Cowan, 1945; Sheppe, 1961). In some of the areas where *oreas* lives sympatrically with *austerus* or with other races of *maniculatus,* no intergradation occurs and practically every adult individual can be identified as being one form or the other. In certain localities, however, many specimens cannot be certainly classified as being either *oreas* or another race associated with it. Such a relationship between two neighboring populations is called intergradation by mammalogists and usually is considered to indicate that the intergrading forms are subspecies of the same species. On the assumption that *oreas* is everywhere reproductively isolated from other members of the *maniculatus* group, however, Sheppe (1961) has elevated this form to specific rank.

The fertility relations and also the behavioral relations in nature of *oreas* with other members of the *maniculatus* group are largely unknown. In the laboratory several young have been produced from matings of *oreas* with *gracilis,* a forest-inhabiting race of *maniculatus* from northeastern North America. The only F_1 hybrid which survived, however, was a male which by breeding tests and by cytological examination was proved to be sterile (Liu, 1954). Whatever the correct taxonomic position of *oreas* may ultimately prove to be, further studies of its geographic, ecologic, behavioral, and fertility relations with other members of the *maniculatus* group will provide much valuable information about the process of speciation.

Analysis of the fertility and behavioral relations of *oreas* with its relatives is made difficult by the animals usually being infertile in the laboratory even when mated with their own kind (Svihla, 1933). My experience in breeding *Peromyscus* indicates that the grassland and desert races of *maniculatus* breed much more readily in the laboratory than most of the northern, forest-inhabiting races. This suggests that these northern forms may require a lower temperature for breeding and for normal behavior than southern forms.

The evolution of intersterility between the pinyon mouse (*Peromyscus truei*) and the juniper mouse (*P. nasutus*) has progressed a step farther than between most of the members of the *maniculatus* group. No hybrids between *truei* and *nasutus* have been discovered in nature, although in many areas the two species occur sympatrically. When males and females of these two species are paired in laboratory cages, some hybrid young are produced, although the proportion of successful matings is low. The female hybrids are at least partially fertile, but the male hybrids are sterile (Dice and Liebe, 1937; Moree, 1946). A considerable degree of intersterility, therefore, has developed between these two species.

The local populations of both *truei* and of *nasutus* often are strongly isolated on the separated desert mountain ranges which occur in the arid parts of southwestern North America. On some isolated mountains only one or the other or these species is present, but in numerous places the two species are sympatric. They usually have somewhat different habitat preferences. Frequently, however, they live in the same habitats and may even be taken on the same traplines (Hooper, 1941; Dice, 1942). When given a choice in the laboratory, members of both species were reported by Blair (1953*b*) to prefer to associate with their own kind. Tamsitt (1961), however, found no evidence for assortative pairing among the members of these two species or of the related form *comanche*.

A number of the geographic races or groups of races assigned to *Peromyscus truei* currently meet and intergrade only in a narrow mountain pass or in a similar very restricted area. Hoffmeister (1951) has pointed out that if, owing to a climatic or other environmental change, the ranges of certain of these currently intergrading forms should become geographically separated, they might then be considered to represent distinct species.

Such a step in speciation is perhaps represented by *comanche,* another member of the *truei* group. *P. comanche* inhabits rocky canyons along the eastern border of the Texas high plains. Its geographic range is separated by more than 80 miles of inhospitable habitats from the nearest population of any other member of the *truei* group (Tamsitt, 1959*a*, 1959*b*). In its superficial appearance *comanche* resembles the members of the *boylei* group, rather than either *nasutus* or *truei*. In the laboratory I attempted to cross it with various races of *boylei*, but without success. Matings of *comanche* with *nasutus* were partially successful, although less so than matings of *comanche* with *comanche*. Three F_1 hybrid males from a cross between *comanche* and *nasutus* had abnormal spermatogenesis and were evidently sterile (Blair, 1943*b*). One hybrid male, however, proved to be fertile when backcrossed to a *comanche* female and one other F_1 hybrid male was indicated by cytological examination to have normal spermatogenesis, although he failed to produce offspring. One F_1 hybrid female produced no offspring when backcrossed to a *comanche* male. On the other hand, no irregularities of spermatogenesis were detected by cytological examination of five F_1 hybrid males obtained from crosses between *comanche* males and *truei* females (Tamsitt, 1960).

The geographic isolation of *comanche* from other members of the *truei* group has presumably permitted the evolution of its distinctive characters and of its partial intersterility with its nearest taxonomic relative, *nasutus*. *P. comanche* is, nevertheless, considered to be a subspecies of *nasutus* by Hoffmeister (1951). Whatever taxonomic assignment is made of *comanche,* its distinctive characters and its geographic separation from all its near relatives illustrates certain of the conditions under which a new species may sometimes originate.

In at least two other species groups of *Peromyscus,* two closely related species may occur together in the same areas, although they usually occupy somewhat different habitats. The two related species *boylei* and *pectoralis* thus may occur sympatrically (Blair, 1940*b*) and so may the two related species *eremicus* and *merriami* (Hoffmeister and Lee, 1963). The morphologic and behavioral differences which now separate the two members of each of these species pairs may be presumed to have evolved at some time in the past when certain local populations of their parent species became reproductively isolated. Abundant opportunities for local populations to become reproductively isolated, perhaps for hundreds or thousands of generations, must have been provided by the geologic and ecologic changes which have occurred in southwestern North America in recent millennia.

The barrier of intersterility between those species of mammals which are assigned taxonomically to different species groups is usually much more complete than it is between the species within a species group. No young have been obtained from any attempted matings in the laboratory between species of *Peromyscus* which are members of different species groups (Dice, 1933). A white-footed mouse (*Peromyscus leucopus*) and a deermouse (*P. maniculatus*), for example, when paired in a laboratory cage will usually live amicably together indefinitely, but no young are ever produced. These two species are closely similar in appearance and are members of the same subgenus of *Peromyscus,* but they belong to different species groups.

Anatomical differences in their genitalia may inhibit or prevent intermating between certain species of mammals. The considerable differences in the structure of the male phallus between some of the species groups of *Peromyscus* (Hooper, 1958) thus may possibly prevent their intermating. Variation in the structure of the phallus among the species of *Peromyscus* within the same species group, however, is mostly very small.

The genetic process by which intersterility between related species of mammals originates in nature is almost entirely unknown. Theoretically, two populations descended from a common ancestor might become intersterile either by random genetic drift or by differential natural selection. Polyploidy and other kinds of chromosomal changes which are known to result in intersterility between related populations of certain plants and invertebrates have not been demonstrated to be effective causes of intersterility between species of mammals (Mayr, 1963).

Random genetic drift may be assumed to produce or to contribute to the production of intersterility between certain species which have a common ancestry. Related populations that are separated by any effective barrier to gene interchange will be subject to at least a small degree of differential genetic drift. Different mutations may occur in the separated populations and different gene loci may become homozygous through random gene fixation (Muller, 1939; Wright, 1940; Huxley, 1942). Should any of those numerous hereditary factors which control reproduction and individual growth come by chance to differ in two sister populations, these populations might become to some degree intersterile.

A serious difficulty with the hypothesis that intersterility between two related species may result from differential random genetic changes, is to explain how genetic differences producing intersterility between species can spread through the many local populations of which both species may be composed and at the same time retain interfertility among the members of each species (Dobzhansky, 1940). Genes or gene combinations which are to any important degree incompatible with those of a majority of the members of a species will be expected to be eliminated quickly by natural selection. The probability of intersterility evolving between any two separated sister species owing to random genetic processes, therefore, must be extremely small. Nevertheless, given a very large number of generations of reproductive isolation between two sister species and given the large number of gene loci which presumably may affect the process of reproduction in the higher animals, it is possible that random genetic changes may sometimes result in the production of intersterility between such species.

Another hypothesis concerning the origin of intersterility between phylogenetically related species assumes that this evolves largely as a byproduct of divergence between the species in their distinctive characters of morphology, behavior, or physiology (Stern, 1936; Muller, 1939, 1940). The underlying assumption of this hypothesis appears to be that certain of the genes or groups of genes which control the embryonic development and ultimate expression of those characters by which species differ may be pleiotropic or may in some other way affect interfertility among the individuals (Dobzhansky, 1956). The process of embryonic development is extremely complex physiologically, especially in the higher animals. Intersterility between two species presumably could arise by a genetic change which would affect any stage in this process. Even a slight difference between two species in the genetic mechanism which controls the embryonic development of an adaptive character might possibly, therefore, at the same time make these species to some degree intersterile.

This is a plausible hypothesis, but the evidence to support it is not at present available. The genes which control the adaptive characters of most species are extremely difficult to identify. Many adaptive characters are evidently controlled by polygene systems, within which the genes may be complexly interrelated. The demonstration that any gene or group of genes which controls an adaptive character in a particular population also affects the interfertility of this population with related populations, therefore, will be a most laborious task.

A new genetic modification of any kind which originates in a particular population will be unlikely to survive and to spread unless its effects on the interfertility of the members of this population with the members of other related populations is very slight.

Any genetic modification which produces an important degree of intersterility amongst the members of a population or a group of related populations will be expected to be quickly eliminated by natural selection. The evolution of intersterility between two parts of a species population which have become reproductively isolated from one another by any kind of a barrier will consequently require a series of very small steps (Stern, 1936; Dobzhansky, 1940). This will be true whether the new intersterility factor has originated by chance alone or whether it may be indirectly favored by being associated genetically with an adaptive character.

The requirement that the evolution of intersterility between reproductively isolated populations must proceed by a series of very small steps perhaps explains why some species which evidently have been separated for a very long period of time nevertheless are still interfertile or partially interfertile with certain of their relatives. Mayr (1949) estimates that a minimum of perhaps half a million years may be required for the origin of a new species of any of the higher animals.

When any appreciable degree of intersterility has developed between two neighboring sister populations, natural selection will tend to evolve effective behavioral barriers to their intermating (Fisher, 1958). Any interpopulation mating that produces no living offspring or that produces hybrids with reduced viability or lessened fertility will obviously be disadvantageous to both populations involved. Such cross matings could be especially disastrous to a local population consisting of only a few individuals. Should a considerable proportion of the matings of a small population in any generation produce defective offspring, that population might not produce enough healthy offspring to maintain itself. Natural selection will consequently tend to evolve assortative mating or other kinds of isolating mechanisms which inhibit or prevent unproductive cross matings between sympatric species.

Summary

Studies of *Peromyscus* and of other animals indicate that the development of reproductive isolation between certain of the local populations of a species is the essential first step in the splitting of that species into daughter species. This reproductive isolation will usually be produced by geographic barriers between those local populations which represent the respective incipient species. In an area where a species occupies two or more contrasting types of habitat between which there are no intermediate conditions, however, it may be possible for the resulting habitat isolation to produce speciation without the geographic separation of the populations concerned.

Those populations of a species which may have become reproductively isolated, either by geographic or by habitat barriers, will nearly always live in somewhat different types of environment. Natural selection will consequently tend to produce a different set of adaptive characters in each separated population. Random genetic drift may possibly also contribute sometimes to the divergence of such separated sister populations. Populations which are reproductively isolated, therefore, will probably in time become sufficiently distinctive in certain of their characters so that they will be described as separate species, unless they happen to be connected in other areas through an intergrading chain of intermediate populations.

The intersterility which separates many related species is presumed to originate either by random genetic drift or by being associated genetically with certain of the adaptive characters which distinguish these species from one another. The evolution of intersterility between sister species is believed to take place in a series of extremely small steps and usually to require, therefore, a very long period of time.

Our information about the interrelations of species populations in nature is unfortunately inadequate for an understanding of the steps by which any existing pair of species may have evolved from their common ancestor. The factors of behavior, anatomy, and/or physiology which actually prevent sympatric species from crossing and producing fertile offspring are still largely unknown. It is true that much has been learned in recent years about the conditions under which speciation may be expected to occur. Many further studies of the geographic, ecologic, behavioral, and physiologic interrelations in nature among the populations of such animals as *Peromyscus,* however, are needed if we are to learn the processes by which new species actually do originate.

Literature Cited

Anderson, Edgar. 1949. Introgressive hybridization. New York: Wiley and Sons.

Blair, W. Frank. 1940a. A study of prairie deer-mouse populations in southern Michigan. Amer. Midland Nat., 24:273–305.

———— 1940b. A contribution to the ecology and faunal relationships of the mammals of the Davis Mountain region, southwestern Texas. Miscl. Publ. Mus. Zool. Univ. Mich., 46:1–39.

———— 1943a. Criteria for species and their subdivisions from the point of view of genetics. Ann. N. Y. Acad. Sci., 44:179–188.

———— 1943b. Biological and morphological distinctness of a previously un-described species of the *Peromyscus truei* group from Texas. Contrib. Lab. Vert. Biol. Univ. Mich., 24:1–8.

———— 1950. Ecological factors in speciation in *Peromyscus*. Evolution, 4: 253–275.

———— 1953a. Factors affecting gene exchange between populations in the *Peromyscus maniculatus* group. Texas Jour. Sci., 5:17–33.

———— 1953b. Experimental evidence of species discrimination in the sympatric species, *Peromyscus truei* and *P. nasutus*. Amer. Nat., 87: 103–105.

———— 1954. Tests for discrimination between four subspecies of deer-mice (*Peromyscus maniculatus*). Texas Jour. Sci., 6:201–210.

Blair, W. Frank, and Walter E. Howard. 1944. Experimental evidence of sexual isolation between three forms of the cenospecies, *Peromyscus maniculatus*. Contrib. Lab. Vert. Biol. Univ. Mich., 26:1–19.

Brown, W. L., Jr., and E. O. Wilson. 1956. Character displacement. Syst. Zool., 5:49–64.

Cowan, I. McT., and C. J. Guiguet. 1956. The mammals of British Columbia. Brit. Columb. Mus. Handbook, No. 11:1–413.

Dawson, W. D. 1965. Fertility and size inheritance in a *Peromyscus* species cross. Evolution, 19:44–55.

Dice, Lee R. 1922. Some factors affecting the distribution of the prairie vole, forest deer-mouse, and prairie deer-mouse. Ecology, 3:29–47.

———— 1931. The occurrence of two subspecies in the same area. Jour. Mamm., 12:210–213.

———— 1932. Mammals collected by F. M. Gaige in 1919 at Lake Cushman and vicinity, Olympic Peninsula, Washington. Murrelet, 13:47–49.

———— 1933. Fertility relationships between some of the species and subspecies of mice in the genus *Peromyscus*. Jour. Mamm., 14:298–305.

———— 1937. Fertility relations in the *Peromyscus leucopus* group of mice. Contrib. Lab. Vert. Gen. Univ. Mich., 4:1–3.

———— 1940a. Relationships between the wood-mouse and cotton-mouse in eastern Virginia. Jour. Mamm., 21:14–23.

———— 1940b. Speciation in *Peromyscus*. Amer. Nat., 74:289–298.

———— 1940c. Intergradation between two subspecies of deer-mouse (*Pero-

myscus maniculatus) across North Dakota. Contrib. Lab. Vert. Biol. Univ. Mich., 13:1–14.

———— 1942. Ecological distribution of *Peromyscus* and *Neotoma* in parts of southern New Mexico. Ecology, 23:199–208.

———— 1943. The biotic provinces of North America. University of Michigan Press.

———— 1949. Variation of *Peromyscus maniculatus* in parts of western Washington and adjacent Oregon. Contrib. Lab. Vert. Biol. Univ. Mich., 44:1–34.

DICE, LEE R., AND PHILIP M. BLOSSOM. 1937. Studies of mammalian ecology in southwestern North America, with special attention to the colors of desert mammals. Publ. Carnegie Inst. Wash., 485:1–125.

DICE, LEE R., AND MARGARET LIEBE. 1937. Partial infertility between two members of the *Peromyscus truei* group of mice. Contrib. Lab. Vert. Gen. Univ. Mich., 5:1–4.

DOBZHANSKY, TH. 1940. Speciation as a stage in evolutionary divergence. Amer. Nat., 74:312–321.

———— 1951. Genetics and the origin of species. 3rd edition. Columbia University Press.

———— 1956. What is an adaptive trait? Amer. Nat., 90:337–347.

FISHER, R. A. 1958. The genetical theory of natural selection. 2nd edition. New York: Dover Publications.

FOSTER, DOROTHY D. 1959. Differences in behavior and temperament between two races of the deer-mouse. Jour. Mamm., 40:496–513.

FULLER, JOHN L., AND W. ROBERT THOMPSON. 1960. Behavior genetics. New York: John Wiley and Sons.

GOLDSCHMIDT, RICHARD. 1933. Some aspects of evolution. Science, 78:539–547.

HARRIS, VAN T. 1952. An experimental study of habitat selection by prairie and forest races of the deermouse, *Peromyscus maniculatus*. Contrib. Lab. Vert. Biol. Univ. Mich., 56:1–53.

———— 1954. Experimental evidence of reproductive isolation between two subspecies of Peromyscus maniculatus. *Ibid.*, 70:1–13.

HOFFMEISTER, D. F. 1951. A taxonomic and evolutionary study of the piñon mouse, *Peromyscus truei*. Illinois Biol. Monogr., 21:1–104.

HOFFMEISTER, DONALD F., AND M. RAYMOND LEE. 1963. The status of the sibling species *Peromyscus merriami* and *Peromyscus eremicus*. Jour. Mamm., 44:201–213.

HOOPER, EMMET T. 1941. Mammals of the lava fields and adjoining areas in Valencia County, New Mexico. Misc. Publ. Mus. Zool. Univ. Mich., 51:1–47.

———— 1942. An effect on the *Peromyscus maniculatus* Rassenkreis of land utilization in Michigan. Jour. Mamm., 23:193–196.

———— 1958. The male phallus in mice of the genus *Peromyscus*. Misc. Publ. Mus. Zool. Univ. Mich., 105:1–24.

HOWARD, WALTER E. 1949. Dispersal, amount of inbreeding, and longevity in a local population of prairie deermice on the George Reserve, southern Michigan. Contrib. Lab. Vert. Biol. Univ. Mich., 43:1–50.

HOWELL, ARTHUR H. 1921. A biological survey of Alabama. N. Amer. Fauna, 45:1–88.

HUXLEY, JULIAN. 1941. Evolution: The modern synthesis. New York: Harper and Bros.

JOHNSON, MURRAY L., AND B. T. OSTENSON. 1959. Comments on the nomenclature of some mammals of the Pacific Northwest. Jour. Mamm., 40:571–577.

KNIGHT, G. R., ALAN ROBERTSON, AND C. H. WADDINGTON. 1956. Selection for sexual isolation within a species. Evolution, 10:14–22.

KOOPMAN, K. F. 1950. Natural selection for reproductive isolation between *Drosophila pseudoobscura* and *Drosophila persimilis*. Evolution, 4: 135–148.

LIU, T. T. 1953a. The measurement of fertility and its use as an index of reproductive isolation among certain laboratory stocks of *Peromyscus*. Contrib. Lab. Vert. Biol. Univ. Mich., 59:1–12.

——— 1953b. Prenatal mortality in *Peromyscus* with special reference to its bearing on reduced fertility in some interspecific and intersubspecific crosses. *Ibid.*, 60:1–32.

——— 1954. Hybridization between *Peromyscus maniculatus oreas* and *P. m. gracilis*. Jour. Mamm., 35:448–449.

MCCABE, T. T., AND I. MCT. COWAN. 1945. *Peromyscus maniculatus macrorhinus* and the problem of insularity. Trans. Roy. Canadian Inst., 25:117–215.

MCCARLEY, W. H. 1954a. Natural hybridization in the *Peromyscus leucopus* group of mice. Evolution, 8:314–323.

——— 1954b. The ecological distribution of the *Peromyscus leucopus* group in eastern Texas. Ecology, 35:375–379.

——— 1963. Distributional relationships of sympatric populations of *Peromyscus leucopus* and *P. gossypinus*. Ecology, 44:784–788.

——— 1964. Ethological isolation in the cenospecies *Peromyscus leucopus*. Evolution, 18:331–332.

MAYR, ERNST. 1949. Speciation and systematics. *In*: G. L. Jepson, E. Mayr, and G. G. Simpson, eds., Genetics, Paleontology, and Evolution, Pp. 281–298. Princeton University Press.

——— 1963. Animal species and evolution. Harvard University Press.

MAYR, ERNST, E. G. LINSLEY, AND R. L. USINGER. 1953. Methods and principles of systematic zoology. New York: McGraw-Hill Book Co.

MILLER, GERRIT S., JR., AND REMINGTON KELLOGG. 1955. List of North American Recent mammals. Bull. U. S. Natl. Mus., 205:1–954.

MILLICENT, E., AND J. M. THODAY. 1960. Gene flow and divergence under disruptive selection. Science, 131:1311–1312.

MOORE, ROBERT E. 1965. Olfactory discrimination as an isolating mechanism between *Peromyscus maniculatus* and *Peromyscus polionotus*. Amer. Midland Nat., 73:85–100.

MOREE, RAY. 1946. Genic sterility in interspecific male hybrids of *Peromyscus*. Anat. Rec., 96:562.

MULLER, H. J. 1939. Reversibility in evolution considered from the standpoint of genetics. Biol. Rev., 14:261–280.

———— 1940. Bearings of the 'Drosophila' work on systematics. *In*: The new systematics, Julian Huxley, editor, Pp. 185–268. Oxford: Clarendon Press.

MURIE, ADOLPH. 1933. The ecological relationship of two subspecies of *Peromyscus* in the Glacier Park region, Montana. Occ. Pap. Mus. Zool. Univ. Mich., 270:1–17.

OSGOOD, WILFRED H. 1909. Revision of the mice of the American genus *Peromyscus*. N. Amer. Fauna, 28:1–285.

RAUSCHERT, KUNZ. 1963. Sexuelle Affinität zwischen Arten und Unterarten von Rötelmäusen (*Cleithrionomys*). Biol. Zentralbl., 82:653–664.

SHEPPE, WALTER, JR. 1961. Systematic and ecological relations of *Peromyscus oreas* and *P. maniculatus*. Proc. Amer. Philos. Soc., 105:421–446.

SIMPSON, GEORGE G. 1960. Principles of animal taxonomy. Columbia University Press.

STERN, CURT. 1936. Interspecific sterility. Amer. Nat., 70:123–142.

SUMNER, FRANCIS B. 1930. Genetic and distributional studies of three subspecies of *Peromyscus*. Jour. Genetics, 23:275–376.

———— 1932. Genetic, distributional, and evolutionary studies of the subspecies of deer-mice (*Peromyscus*). Biblio. Genetica, 9:1–106.

SVIHLA, ARTHUR. 1933. Notes on the deer-mouse, *Peromyscus maniculatus oreas* (Bangs). Murrelet, 14:13–14.

TAMSITT, J. R. 1959a. *Peromyscus nasutus* in northeastern New Mexico. Jour. Mamm., 40:611–613.

———— 1959b. Abundance of the Palo Dura mouse *Peromyscus comanche*. Southwestern Nat., 3:234–236.

———— 1960. The chromosomes of the *Peromyscus truei* group of white-footed mice. Texas Jour. Sci., 12:152–157.

———— 1961. Tests for social discrimination between three species of the *Peromyscus truei* species group of white-footed mice. Evolution, 15:555–563.

THORPE, W. H. 1945. The evolutionary significance of habitat selection. Jour. Anim. Ecol., 14:67–70.

WADDINGTON, C. H. 1957. The strategy of the genes. London: George Allen and Unwin.

WALLACE, ALFRED R. 1880. Island life. London: Macmillan Co.

WATSON, MARGARET L. 1942. Hybridization experiments between *Peromyscus polionotus* and *Peromyscus maniculatus*. Jour. Mamm., 23:315–316.

WECKER, STANLEY C. 1963. The role of early experience in habitat selection by the prairie deermouse, *Peromyscus maniculatus bairdi*. Ecol. Monogr., 33:307–325.

WRIGHT, SEWALL. 1940. Breeding structure of populations in relation to speciation. Amer. Nat., 74:232–248.

4

HABITATS AND DISTRIBUTION

ROLLIN H. BAKER

Introduction

M ICE OF THE GENUS *Peromyscus* occur widely in North America—northward to the edge of the Arctic prairie in Canada and southward to tropical undergrowth near the Colombian border of Panama. The genus is a large one containing seven subgenera; within the large subgenus *Peromyscus*, Hooper and Musser (1964) list seven species groups. By latest count there are 57 species, 25 of which are polytypic.

The living places of most of these species are poorly known; for many, all of the available published information is contained in brief "field notes" by collectors. Approximately a dozen species have provided most of our information on their home life and habitat requirements. We know most about two species: The deer mouse (*P. maniculatus*) and the white-footed mouse (*P. leucopus*). According to Blair (1956), we may know more about the former species than any other wild mammal.

Much of our knowledge of habitats of *Peromyscus* has been derived from observations made in situations where specimens were caught in traps. Locations of these captures probably reveal preferred living places or most certainly the foraging grounds of the mice. The discovery of nests provides additional information on home sites. Identification of food materials from alimentary tracts and caches yields knowledge of which parts of the environment are visited in the course of food gathering. Experiments using trap-mark-release-retrap methods provide information on environmental utilization, density, social habits, individual movements, sex and age ratios, home range, homing, and the number of species using given situations. Some of the few species studied in the above fashion have also been examined under laboratory conditions (see

TABLE 1

SIZE CLASSES OF *Peromyscus*

Size classes (Total lengths in mm)	Number of species		
	Total	Monotypic	Polytypic
126–175	6	4	2
176–225	26	14	12
226–275	20	10	10
276–325	2	2	0
326–375	3	2	1

chapters 11, 12, and 13). Consequently, statements regarding the life habits of some members of the genus may stimulate study on the ecology of others.

Peromyscus are generally active at night. In spite of their small size and host of enemies, they often move about in open situations without confining themselves to dense, overhead vegetation. Most are terrestrial, but all can climb. Movements are not usually restricted to runways, as in the case of voles (*Microtus*). All eat somewhat similar foods, namely grains, seeds, fruits, and insects (Cogshall, 1928; Hamilton, 1941; Jameson, 1952). Perhaps the bulk of their food is the mast produced by trees and shrubs. Unlike microtines that eat stems, leaves, and roots, even dense numbers of *Peromyscus* rarely alter plant cover directly. The consumption of large amounts of seeds and berries, according to Jameson (1953) and Stephenson *et al.* (1963), has no obvious effects on the habitat, even though some foresters have expressed concern for the alleged thoroughness with which these mice may collect fallen pine seeds. On the other hand, food resources may influence populations. In Plumas County, California, Jameson found that *P. boylei* and *P. maniculatus* were abundant in a year with good mast because they continued breeding until December. In a year with poor mast they were scarce because the mice ceased breeding in June, not because of excessive mortality.

Members of most species of *Peromyscus* (32 of 57) average less than 225 mm in total length. In only five species are they more than 276 mm long. Table 1 summarizes the size classes (data mostly from Hall and Kelson, 1959). Mice in the tropics show more size variation than do those in any other major area (see Table 2). Three of the five largest kinds are from restricted humid montane

TABLE 2
SIZE OF *Peromyscus* IN RELATION TO MAJOR LIVING AREAS

Size classes (Total lengths in mm)	Number of species			
	Tropical	Non-montane temperate	Mountains	Islands only
126–175	3	2	1	2
176–225	6	10	5	9
226–275	11	3	10	0
275–325	2	0	1	0
326–375	3	0	1	0

tropical situations. Mice from non-montane temperate regions are usually small with ten species in the 176–225-mm class. Montane species are generally larger than the former. The two largest occur in restricted areas south of the Tropic of Cancer. Insular species are in general small, perhaps the result of space limitation on islands (Hesse, Allee, and Schmidt, 1951).

Habitat Utilization

Although widely distributed in North America, members of the genus are inhabitants chiefly of woodlands and brushlands. They may occur in open lands of mixed weeds and grasses, characteristic of pioneer stages of grassland development, but here they usually are secondary to *Microtus* in north temperate and boreal stands of perennial grasslands and to *Oryzomys, Sigmodon,* and *Liomys* in south temperate and tropical grasslands, savannas, and swamps. Heteromyid rodents (*Dipodomys* and *Perognathus*) also usually outnumber *Peromyscus* in desert situations. In Middle America, members of the genus *Peromyscus* are also secondary to presumably better adapted Neotropical rodents in woodlands and brushlands. Nevertheless, individuals of this ubiquitous genus "may be caught in traps set in almost every conceivable situation" (Osgood, 1909: 26). Oftentimes these mice will be by far the most common mammals in the area. Among record catches is one by Hoffman (1955) who took 528 individuals of *P. maniculatus* in 1,204 trap-nights in the White Mountains of Mono County, California.

P. maniculatus or *P. leucopus* may be found in a variety of habitats; Verts (1957:57) found in strip-mined land in Illinois that "The percentage of bare ground, annuals, grasses, woody vegetation, perennials, and biennials, and the average stems per square meter show no correlation with the distribution of these mice." Others, as the insular forms, may be confined to special situations. Maps outlining the distributions of the several species (see chapter 2 and Hall and Kelson, 1959:600–658) suggest that the animals occur everywhere within geographic areas delineated. Actually, each species occupies only those parts of its geographic range which provide suitable habitats. *Peromyscus crinitus,* for example, occurs in an extensive region from Oregon south to the Gulf of California and eastward to Colorado, but within this area it is confined to rocky environments.

Each species of *Peromyscus* has a definable geographic range, the size of which for continental forms depends on the extent of the habitats to which each is morphologically, behaviorally, and physiologically adapted. The variation in size of geographic ranges is clearly illustrated by comparing the extensive distribution of *P. maniculatus* with the limited distributions of such species as *P. floridanus* (peninsular Florida) or *P. latirostris* (known only from the vicinity of Xilitla in San Luis Potosí). If only one species is present, as in much of western Canada, it may occupy a wide range of habitats; where two or more species occur together, as in many parts of the American southwest, one or more of them may be severely limited in local distribution. Ecological restriction can be predicted from the number of sympatric species in an area. However, to define the living places of individuals, information must be accumulated concerning the evolutionary history of the species, their reaction to physiographic and climatic barriers, their adaptiveness and innate responses to the environment, and their ability to cope with competitors.

Habitat utilization is a dynamic operation in which local populations attempt to maintain an equilibrium with the ever-changing environment. However, the equilibrium constantly tends to be upset by overproduction within species. If the overproduction is intraspecific, the species spreads out and may occupy even unfavorable, marginal areas (*P. maniculatus* on Beaver Island, Michigan, Ozoga and Phillips, 1964); whereas, if it is interspecific, the individuals may tend to entrench in restricted habitats in which optimum living conditions occur. The push outwardly into new habitats is ultimately blocked by barriers of one sort or another.

These may be physical (climatic), or the less obvious restraints of interspecific competition. L. Stickel (1946) and others demonstrated that the removal of all mice living within a specific natural situation was followed by an immediate immigration of members of the same species from adjacent areas. When Sheppe (1961) removed an entire isolated population of *P. oreas* (= *P. maniculatus oreas*) from a large ravine, *P. maniculatus artemisiae* from the surrounding hillsides moved in, although this subspecies had not been found in the ravine before. In short, habitable areas for mice are rarely mouseless for long.

INTRASPECIFIC HABITAT SELECTION.—That subspecies of *Peromyscus* interact with their habitats is indicated by (1) the direct relationship between the numbers of different subspecies and the size and variability of the geographic range of the species; (2) the existence of distinctive woodland and openland forms (also interior and montane and coastal forms) with possibly different morphological, behavioral, and physiological adaptations; and (3) the correlation between pelage color and soil color (Blair, 1950).

We have already seen that almost one-half (25 of 57) of the species of *Peromyscus* are polytypic; that is, divisible into two or more recognizable subspecies. Most monotypic species occupy limited geographic ranges (a notable exception is *P. melanotis*), which usually means that the animals occur in rather restricted habitats. The more widely distributed polytypic species, such as *P. leucopus* or *P. maniculatus,* are non-restrictive as far as habitats are concerned. Factors which influence spatial distribution within local populations also may have been involved in the habitat selection and the spread of the species as a whole to attain its entire present geographic range.

Relationships of subspecies of *P. maniculatus* to their environments are better known than those of any other species in the genus. Named geographic variants of this species can generally be divided into short-tailed, short-eared, and short-footed groups occupying openlands and long-tailed, long-eared, and long-footed groups occupying woodlands and brushlands. In western North America, the distribution of these two morphologic types is complicated by the irregular distribution of woodlands and openlands, as shown by A. Murie (1933) for subspecies in Glacier National

Park. Intermediate populations often occur in transitional brush-lands. Most studied are the openland subspecies of east-central United States, *P. m. bairdi,* and the woodland subspecies of north-eastern United States, *P. m. gracilis.*

When clearing of forests enabled *P. m. bairdi* to spread its range eastward, these two subspecies, formerly not in contact, came together geographically. They have not interbred probably because they remain ecologically separated (Dice, 1931; Hooper, 1942); however, Wecker (1963) finds no evidence of character displacement in areas of sympatry. Although each subspecies is restricted to a different environment, Barbehenn and New (1957) surmise that ecological barriers between them might break down under the stress of high population density. If this should occur without loss of fertility, natural interbreeding could take place between *P. m. bairdi* and *P. m. gracilis,* since it has already been demonstrated that cross-matings of these subspecies in the laboratory produce fertile offspring.

In *P. maniculatus,* the adaptation and restriction of subspecies to specialized openland or woodland habitats (Wecker, 1963) seem not to be accounted for either by food differences (Cogshall, 1928; Williams, 1959) or by temperature requirements (Stinson and Fischer, 1953). Dice (1922) found *P. m. bairdi* able to tolerate simulated differences from "normal" habitats as devised in the laboratory and concluded that the absence of this subspecies from woodlands was primarily a behavioral response. Even so, responses or traits of *P. maniculatus* have been speculatively related to the physical structure of the habitats (M. Murie, 1961:738). Andrew-artha and Birch (1954:220) point to habitat selection as resulting from a tactic or kinetic response to a particular environmental physico-chemical gradient. As Wecker (1963:308) emphasizes, psychologic factors also must be involved in habitat selection, but as yet information is scant.

Horner (1954), Foster (1959), and others have described *P. m. gracilis* as being superior to *P. m. bairdi* as a climber (the long tail and long hind foot help) and as having deliberate and steady movements, also assumed to be adaptations to semi-arboreal life; *P. m. bairdi* is described as being less able as a climber and more cautious in movements with a tendency to "freeze" in strange situations. The freezing behavior is presumably of survival value in an open environment where movement of the prey species is often an important cue for predators.

Peromyscus m. bairdi avoids wooded areas in nature even though the ground cover in them may be grassy. The catch of these animals drops off markedly as trap lines approach woodland edges or brushy field borders (Harris, 1952). In an attempt to determine the environmental factors involved in habitat selection in *P. m. bairdi* and *P. m. gracilis,* Harris presented individuals of each subspecies with a choice between an artificial "woods" and an artificial "field" in the laboratory. The individuals exhibited a preference for the artificial habitat more closely resembling the natural environment of their own subspecies. Since these experiments were performed under essentially uniform physical conditions, Harris thought that the mice were responsive to the "form" of the artificial habitats. While *P. m. bairdi* was seemingly attracted to the "field," this subspecies also might have reacted negatively toward the "woods." This hypothesis is supported by the avoidance of woodland by these mice in nature. Harris concluded also that there was a genetic basis to this behavior.

Since early experience may play an important part in the development of adult behavioral characteristics (King and Eleftheriou, 1959), Wecker (1963) suspected that early learning in *P. m. bairdi* would reinforce any hereditary response to habitat. In essence, his studies showed that the choice of openland habitat by *P. m. bairdi* is normally predetermined by heredity; that early experience in openland habitat can reinforce this innate preference but is not a necessary prerequisite for subsequent habitat selection; that early experience in contrasting environments (for example, woodland or laboratory) is insufficient to reverse the normal affinity of *P. m. bairdi* for openland habitat; that hereditary control over the habitat-selection response apparently is reduced after confinement of *P. m. bairdi* for 12–20 generations in the laboratory; and that laboratory-reared mice retained an innate capacity to utilize early experience in openland habitat by learning to respond positively to stimuli associated with this environment, whereas experience in woodland habitat or laboratory does not have a corresponding effect. Wecker concluded that *P. m. bairdi* can learn to respond to environmental cues only when the necessary stimuli are associated with the openland habitat.

Numerous dramatic characteristics, including little-studied physiologic ones, distinguish related populations of *Peromyscus*. In alpine populations, for example, the life expectancy is remarkably high (Dunmire, 1960) while the metabolism is low (Cook and Hannon, 1954) in comparison with those characteristics of animals of the same species living at lower elevations on the same mountains. Selection favoring these characteristics seemingly has to do with such alpine environmental factors as altitude, temperature, short growing season, and low productivity. Populations of *P. maniculatus* and *P. leucopus* living in xeric situations were found to survive on less water (Lindeborg, 1952) than those of the same species living naturally in more mesic situations. The ability to survive on a reduced water supply is important for the evolution of xeric-living animals.

The geographic variation of structural and certainly behavioral and physiological adaptations in different species of *Peromyscus* is usually less conspicuous than that of pelage. Coloration, which tends to conceal the animal from its enemies, often is a hereditary characteristic established as a result of environmental selection. The pelage color of *Peromyscus* often noticeably resembles the color of the background and demonstrates the possible effectiveness of selection on coat color (Dice, 1940*b*). Sumner (1926) noted striking depigmentation of hair in populations of *P. polionotus* on white sands of coastal dunes and islands of Florida as compared with darker inland populations on dark-colored soils. Dice and Blossom (1937) called attention to darker *Peromyscus* in western mountain forests as compared with paler-colored *Peromyscus* in desert foothills. Dice (1941) found paler-colored *P. maniculatus* in the sand hill country of Nebraska than on grassland prairies to the eastward or westward. The most striking cases of correlation of pelage color with that of background are on the dark lava beds in arid southwestern North America. Dark-colored individuals on such topography seemingly have been favored over pale ones in such species as *P. crinitus* (Benson, 1940), *P. eremicus* (Baker, 1960; Blair, 1947; Dice, 1939), *P. maniculatus* (Baker, *loc. cit.*), *P. melanophrys* (Baker, *loc. cit*)., *P. nasutus* [= *P. difficilis*] (Benson, 1932); *P. pectoralis* (Baker, *loc. cit.*), and *P. truei* (Baker, *loc. cit.*). Probably, as Blair (1947) points out, the existence of local populations distinguished by shade of pelage color is caused by an environmental selection pressure that exceeds the homogenizing effect of inter-breeding between the populations from adjacent dark and pale soils.

INTERSPECIFIC RELATIONSHIPS IN HABITAT SELECTION.—The geographic ranges of many of the 57 recognized species of *Peromyscus* overlap. In parts of southwestern North America it is not unusual to catch members of three and even four species of *Peromyscus* in the same or adjacent traps. These sympatric species may merely occupy a common geographic area and not necessarily identical habitats. Past geographic and/or genetic isolation may be primary while ecologic isolation may be merely secondary in the explanation of the distribution of many sympatric species. In considering the areas of overlap between the related species, *P. leucopus* and *P. gossypinus,* Dice (1940a) suggested that these species evolved when two populations of a single species became geographically separated. After differentiating and again coming in contact, the two populations had diverged genetically, so that they rarely find each other to be attractive mates (McCarley, 1954b). In the laboratory these species (both in the same species group) are completely interfertile, and both male and female hybrids are fertile. Dice also cites genetic change occurring during past geographic separation as providing isolating mechanisms which operate where *P. truei* and *P. nasutus* [= *P. difficilis*] now come together.

Competition between closely related species of *Peromyscus* for similar living places may be resolved by (1) exclusion of one species by another; (2) interdigitation of species in an area in relation to their adaptations to the different seral stages or communities present; and (3) possibly some degree of compatibility. It seems safe to assume that individuals of two or more species of *Peromyscus* do come in contact in the course of natural activity, but the methods by which the isolating mechanisms work to keep these closely related mice from crossbreeding and ultimately homogenizing the characteristics of the species involved remain generally obscure. Brown (1964) suggested that species of *Peromyscus* may occur in separate habitats because of basic differences in behavior rather than physiology. Under laboratory conditions, two species in the same species group, as classified by Osgood (1909), may be inter-fertile, but species from different species groups may be intersterile.

Species Exclusion: It was pointed out earlier that the geographic ranges of two related species, *P. leucopus* and *P. gossypinus,* overlap in parts of southeastern United States. In areas where *P. leucopus* is found alone, it occupies both upland and lowland wooded areas. Allopatric populations of *P. gossypinus* exhibit distinct preference for lowland woods and swamps and occur sparingly, at most, in upland areas. McCarley (1963) has shown that where *P. leucopus* and *P. gossypinus* are sympatric, the general restriction of the former to upland areas is the result of interspecific competition with the latter. In effect, *P. leucopus* is forced to occupy areas other than lowlands. McCarley notes further that the general scarcity of *P. leucopus* as compared with an abundance of *P. gossypinus* in parts of eastern Texas is the result of a distribution pattern which does not freely facilitate dispersal of *P. leucopus* from one upland area to another. Even so, McCarley (1954*a*) reported that populations of *P. leucopus* undergo the same seasonal fluctuations as *P. gossypinus* and are probably influenced by the same environmental conditions.

In northeastern United States, both *P. maniculatus gracilis* and *P. leucopus noveboracensis* are essentially woodland dwellers and occupy somewhat similar ecologic situations. They also resemble each other in size and other external features. In New York, Klein (1960) found that where the two lived together there was some ecologic separation—*P. l. noveboracensis* occupied dryer, warmer, more open deciduous woods and *P. m. gracilis* occupied wetter, cooler, more luxuriant coniferous woods. In captivity, Fitch (1963) found *P. leucopus,* at least initially, to be more aggressive than *P. maniculatus.* King (1957) also found *Mus* to be more aggressive than *P. maniculatus.* The latter author noted that *P. maniculatus* did not resist attacks by *Mus* and, as a result, could easily escape or avoid them. This might explain in part the positions of these two animals in Klein's study where "aggressive" *P. leucopus* dominates the temperate parts of the woodlands leaving only the boreal segments to "passive" *P. maniculatus.* A glance at maps of the geographic ranges of these two species will reveal that *P. maniculatus* thrives in many boreal situations where *P. leucopus* does not. On the other hand, where *P. m. gracilis* occurs allopatrically, it is not restricted to boreal forests alone; for example, on Beaver Island, Michigan, it occurs in most kinds of habitats (Ozoga and Phillips, 1964:327).

Seral Stage Selection: Members of various species of the genus *Peromyscus* are adapted to utilize most living places for small mammals in North America. Even within a single biotic community, species or even subspecies of *Peromyscus* may replace one another in the orderly process of ecologic succession from pioneer systems to relatively stable ones. In describing the ability of this mouse to thrive in a variety of habitats, Jameson (1955) writes that *P. maniculatus* as well as other species are among the first mammals to invade disturbed habitats and also are almost always found in old, uncut forests. Burt (1961) found *P. maniculatus* at Volcan Paricutin (in Michoacan) living in the shelter of lava that one year previously had been red hot. He found one mouse out in the middle of a lava stream where the only food available was droppings from horses and an occasional insect or seed blown in by the wind. Burt also caught *P. hylocetes* in the area of cool lava, but *P. maniculatus* seemed to be the most adventuresome pioneer in the newly developing habitat. Hoffman (1960) caught *P. maniculatus* in five of six climax plant associations in eastern Washington and northern Idaho. In Colorado, Williams (1955) found *P. maniculatus* occupying all communities in montane forest but most common in early successional stages, after fire, logging, and mining. In eastern Texas, short-term flooding (eight days) of river bottoms failed to disrupt semi-arboreal, resident *P. gossypinus,* but flood waters remaining in the river bottom for three weeks caused a 70 per cent decrease (McCarley, 1959). Chaparral fire in California, according to Lawrence (1966:288), caused a decline in chaparral-adapted *P. californicus* and *P. truei* and an increase in grassland-adapted *P. maniculatus.*

The extension of range by *Peromyscus,* according to Terman (1962), may be a diffusionlike process rather than single, long moves by individuals, if the short dispersal pattern (averaging less than 500 feet, according to Howard, 1949) of young *P. maniculatus* is indicative. Probably by such means *P. m. bairdi* invaded northeastern United States. Prior to the arrival of European man in the Great Lakes region, the range of *P. m. gracilis* conformed to hardwood forests while that of *P. m. bairdi* conformed to tall-grass prairies (Hooper, 1942). With the clearing by settlers and loggers, much habitat for *P. m. gracilis* was lost and the land became covered with communities in pioneer successional stages—among them grass, favorable to *P. m. bairdi*. In consequence, *P. m. bairdi* moved into both the Upper and Lower peninsulas of Michigan (Hooper, *loc. cit.*) and more recently into West Virginia (Wilson, 1945), New York (Moulthrop, 1938), Pennsylvania and New York (Hamilton, 1950), and Maryland (W. Stickel, 1951). This movement has brought *P. m. bairdi* into close association with *P. m. gracilis*. In the course of ecologic succession, Beckworth (1954), in southern Michigan, found *P. m. bairdi* common in the annual-biennial stage but greatly reduced in the perennial grass stage where the meadow vole (*Microtus pennsylvanicus*) abounds. In the shrub and later tree stages, *P. leucopus noveboracensis* became the principal small rodent. In New Jersey, where *P. m. bairdi* is absent, Pearson (1959) found *P. l. noveboracensis* present in both the early and late seral stages, but rare in an intermediate stage. The paucity of *Peromyscus* in the intermediate perennial grass stage points to the fact that mice of this genus do not effectively utilize this situation, in which the dominant small rodents include such genera as grass-eating *Microtus* and *Sigmodon*. No doubt grazing and fire protection in western and southwestern rangelands have increased the amount of brush and weeds and allowed for the increase of *Peromyscus* in areas formerly covered by perennial grasses.

In the Sierra Nevada of northern California, Jameson (1951:202) found *P. boylei* in brush fields at 3,500–5,000 feet but not in adjacent coniferous forests. This rodent was absent from brushy areas at 6,500 feet but abundant in a forest of oaks, maples, and bay trees at 2,200 feet. On the other hand, *P. maniculatus* occurred in coniferous forest at 3,500–5,000 feet, in sage brush at 5,000 feet, and in brush at 6,500 feet. This species also lived in the hardwood forest at 2,200 feet but was less common there than was *P. boylei*. Jameson offered two explanations for the apparent variation in the habitat preferences of the two species of *Peromyscus,* if the sampling in these areas was adequate for determining their relative abundance. Either the mice have markedly different habitat requirements in slightly different parts of their ranges or we have an incomplete concept of habitats. It is possible that at the time of Jameson's study interspecific competition was sufficiently severe to relegate each species to an area having for it optimum living conditions (Odum, 1959).

Species Compatibility: The regularity with which two and sometimes more species of *Peromyscus* are taken in the same trapline or in the same trap on subsequent nights suggests that these mice live somewhat compatibly in close association without finding each other attractive mates, notwithstanding the concept proposed by Grinnell and Swarth (1913:220) and later by Gause (Odum, 1959: 231) which proposes that no two species live in exactly the same niche. Terman (1962) found that spatial distribution in *P. maniculatus bairdi* is achieved through mutual avoidance of individuals and would appear to be of at least as much adaptive significance to the population as the practice of territoriality (Burt, 1949). Conceivably the same spatial arrangement might govern the interaction without conflict or fighting of different but related species in nature. Spatial distribution of two or more related species in the same areas may then be achieved through mutual avoidance or even perhaps passive acceptance or tolerance of one another. This concept seems to be one way to explain the apparent togetherness of such species as *P. boylei, P. difficilis* (includes *P. nasutus*), and *P. truei* (Bailey, 1931:154); the above three species and *P. melanotis* (Baker and Greer, 1962:107); *P. boylei, P. hylocetes,* and *P. melanotis* (Baker and Phillips, 1965:692); *P. eremicus* and *P. pectoralis* (Baker and Greer, *op. cit.,* 115); *P. eremicus* and *P. melanophrys* (Dalquest, 1953:155); and *P. boylei, P. leucopus,* and *P. mexicanus* (Dalquest, *op. cit.,* 143). The fact that some of these mice often differ in body size may also play a role in their alleged compatibility. However, Brown (1964), in his study of *P. maniculatus, P. leucopus,* and *P. boylei* in Missouri, thought that these species used separate habitats because of behavioral differences rather than the detectable differences in physiologic factors considered. He concluded that the aggressive competition observed between *P. leucopus* and *P. boylei* probably contributed to the "sharpness" in ecologic distribution of the two mice.

Distribution

Speciation in the genus *Peromyscus* has been greatest in temperate and tropical North America. This is seen from the numbers of species inhabiting the different political units of North America (see Table 3). Mexico has 46 species, mostly monotypic forms. Most of the geographically small political units of Central America have several species each, but temperate United States follows

TABLE 3

DISTRIBUTION OF *Peromyscus* IN VARIOUS POLITICAL UNITS OF NORTH AMERICA

Political areas (North to South)	Number of species		
	Total	Monotypic	Polytypic
Alaska	2	0	2
Canada	3	0	3
United States	14	1	13
Mexico	46	25	21
Guatemala	6	3	3
Honduras-Salvador	5	2	3
Nicaragua	1	0	1
Costa Rica	2	0	2
Panama	3	2	1

Mexico in the number of species. Numbers in Canada and Alaska dwindle sharply.

The Mexican region has been the scene of the greatest amount of speciation in the genus. The number of monotypic species and the amount of geographic variation in polytypic species emphasize the large number of environments present. In the United States more species of *Peromyscus* occur west of the Great Plains than east. The western region, especially the American southwest, presents a greater variety of environments than the eastern. Seven species are restricted to areas westward of the Great Plains and only three to the eastward. The latter three occur solely in southeastern United States in temperate, deciduous forests, shrub, and strand.

In Table 4, species of *Peromyscus* are arranged in groups according to their occurrences in major ecological areas of North America.

TABLE 4

DISTRIBUTION OF *Peromyscus* IN MAJOR ECOLOGICAL AREAS IN NORTH AMERICA

Living areas	Number of species		
	Total	Monotypic	Polytypic
Alpine	2	1	1
Northern Boreal Forest	2	0	2
Non-Montane Temperate	15	4	11
Southern Montane Woodlands	17	10	7
Tropical	23	12	11
Insular	11*	8	3

* Endemic species only.

Each of these major areas includes numerous situations to which present-day *Peromyscus* may be adapted. A brief inspection shows that tropical areas have most species and northern boreal forest and alpine areas have fewest. In the following paragraphs these areas and their resident *Peromyscus* will be discussed.

ALPINE.—Few mammals are resident in the desolate, often severe environments found above timber line. As alluded to earlier, in order for *P. maniculatus* to survive in alpine environments, according to studies by Dunmire (1960), the short breeding season with the resultant low productivity in comparison with that of relatives at lower elevations is offset in the alpine dwellers by a much greater life expectancy. Undoubtedly, species of *Peromyscus* that live in boreal areas on mountains occasionally venture upward into alpine situations; however, only *P. maniculatus* and *P. melanotis* actually reside there to any great extent. In the case of the former species, Osgood (1909:70) reports its occurrence above timberline in such western states as Washington and New Mexico while Dunmire (*op. cit.* 175) took the species as high as 14,247 feet, at the summit of White Mountain in Mono County, California. In Mexico *P. melanotis* lives in alpine areas up to 15,200 feet on Mount Orizaba, according to Goldman (1951:403). There is also evidence that *P. maniculatus* occurs at least on the edge of the Arctic prairie in Canada (Arctic Life-zone), although Harper (1961:51–53) suspects that its presence within and in the vicinity of coastal settlements on the Ungava Peninsula could be the result of introductions rather than natural spread.

NORTHERN BOREAL FOREST.—Northern woodlands, chiefly of coniferous trees, cover vast areas in the higher latitudes of North America and include, at least for purposes here, montane forests of northwestern United States. This expanse provides favorable living places for only two species of *Peromyscus*. The widespread *P. maniculatus* is the only representative of the genus in most of the area, including the large forested region of northwestern Canada (Osgood, 1909:49). In northeastern United States, *P. leucopus* also lives, at least marginally, in cool, damp boreal forests, often in near association with *P. maniculatus* (see Richmond and Rosland, 1949: 46).

TABLE 5

Peromyscus IN NON-MONTANE TEMPERATE AREAS

Area	Number of species		
	Total	Monotypic	Polytypic
Desert	7	1?	6
Grasslands	10	3	7
Brushlands	8	1	7
Woodlands	5	1	4

NON-MONTANE TEMPERATE AREAS.—Non-montane temperate areas include a variety of living places, mostly in the United States, and range from beaches, savannas, deciduous and coniferous woodlands and brushlands on coastal plains to deserts, brushlands, and grasslands on broad interior plains, foothills, basins, and elevated plateaus. Table 5 shows a breakdown of this temperate region into major living areas with the numbers of species involved.

Desert: Desert-inhabiting species are confined to southwestern North America, where they usually rank second in both diversification and density to heteromyid rodents (*Dipodomys* and *Perognathus*). At least seven species of *Peromyscus* occupy these arid openlands, where a variety of seed-bearing plants occur. One monotypic species, *P. mekisturus,* is possibly an inhabitant of desert areas in the Mexican state of Puebla (Goldman, 1951:387). Otherwise, all species are polytypic and, except for two desert endemics, occur in places other than deserts. One of these endemics, *P. merriami,* seems to be confined mostly to desert in Arizona and Sonora, while the other, *P. eremicus,* is the most typical and widespread desert-dwelling member of the genus. The ubiquitous *P. maniculatus* also lives in parts of the desert on the Mexican Plateau. Two desert dwellers, *P. crinitus* and *P. pectoralis,* are mostly restricted to rocky areas here as well as in other major habitats. *Peromyscus melanophrys* lives in rocks (including man-made rock fences) but also in desert shrub and, as described by Dalquest (1953:150), ascends joshua trees and large prickly pears.

Grasslands: Species of *Peromyscus* play a secondary role to *Microtus, Sigmodon,* and other rodents in perennial grasslands, but are conspicuous in areas of mixed annual and biennial grasses and weeds. Six of the ten species of *Peromyscus* present also reside in desert terrain. These include the rock-dwelling *P. crinitus, P.*

melanophrys, and *P. pectoralis* as well as *P. maniculatus,* possibly
P. merriami and the monotypic *P. mekisturus.* The monotypic *P. polius* occupies rocky sites in high grasslands in the Mexican state of Chihuahua. *Peromyscus leucopus* is chiefly an inhabitant of woodlands and brushlands but lives also, to some extent, in grassy areas in southwestern North America. *Peromyscus polionotus,* perhaps not strictly a grass-dwelling species, occupies some sandy, grassy areas, including old (fallow) fields and beaches in south-eastern United States (Golley, 1962:122; Howell, 1921:45–46). Sandy flats in the Mexican state of Veracruz are the homes of *P. bullatus* (Hall and Dalquest, 1963:306).

Brushlands: Since the coming of European man, land-use prac-tices (grazing by domestic livestock, lumbering, suppression of fires, clearing, and cultivation) have increased the amount of brushlands. It is likely that such species as *P. boylei, P. californicus, P. leucopus, P. maniculatus, P. melanophrys,* and *P. pectoralis* have been con-siderably favored by man's operations. Of the eight species (all but *P. floridanus* are polytypic) listed in Table 5 as living in temperate, non-montane brushy areas, several (*P. boylei, P. crinitus, P. mela-nophrys,* and *P. pectoralis*) are generally confined to rocky situa-tions. Brushlands may be occupied marginally at least by many other species, only one of which, *P. floridanus* of southeastern United States, is included in the listing.

Woodlands: Temperate woodlands, mostly deciduous, occur generally in eastern North America, where two (*P. floridanus* and *P. gossypinus*) of the five species totaled in Table 5 are more or less endemic. The arboreal habit is developed in *P. gossypinus, P. leucopus,* and *P. maniculatus* (Audubon and Bachman, 1846). *Peromyscus boylei* occurs in the western edge of this association, but as in other parts of its range shows marked preference to rocky areas (Cockrum, 1952:180; Schwartz and Schwartz, 1959:189).

SOUTHERN MONTANE WOODLANDS.—Seventeen species of *Pero-myscus* reside in non-tropical woodlands of the mountains included in the area from southwestern United States southward through Middle America. Species are grouped in Table 6 according to their distribution in three major montane areas, which are perhaps better differentiated in southwestern United States and northern Mexico than in the more southern mountains in Middle America.

TABLE 6

Peromyscus IN SOUTHERN, NON-TROPICAL MONTANE WOODLANDS

Area	Number of species		
	Total	Monotypic	Polytypic
Piñon-juniper-oak	8	1	7
Pine-oak	17	9	8
Fir	11	4	7

Piñon-juniper-oak: Mountains, often island-like in their isolation, rise out of desert terrain in southwestern North America. As one ascends from the basally located arid shrub, the first distinguishable forested belt is the piñon-juniper-oak (a part of the Upper Sonoran Life-zone of Merriam, 1898). This vegetation may be short and shrubby (chaparral) or consist of open stands of moderately tall trees with an understory of grasses and shrubs covering rocky and often steep hillsides (montane low forest, see Baker, 1956:137). Seven *Peromyscus* occur in this montane situation. Monotypic *P. polius* is probably chiefly a rock dweller in Chihuahua (Goldman, 1951:387); the same is generally true for polytypic *P. crinitus, P. difficilis* (includes *P. nasutus,* see Hoffmeister and de la Torre, 1961), *P. pectoralis, P. truei,* and probably *P. boylei. Peromyscus californicus* prefers chaparral (Vaughan, 1954:556), while the wide-spread *P. maniculatus* occurs throughout this type of environment in much of southwestern United States.

Pine-oak: Pine-oak habitats in the mountains from southwestern United States southward into Middle America vary from place to place in terms of species composition and elevation above sea level. On some mountains this vegetation is the uppermost belt; whereas on others boreal fir forest occurs above. Mesquite-grasslands and piñon-juniper-oak usually exist below the pine-oak on desert mountains and tropical deciduous forest or shrub may be directly below the pine-oak in more southern latitudes (Webb and Baker, 1962: 328). Again, rock-living species may be more restricted to rocky outcrops than to particular vegetation types. However, acorns are favored foods of such rock-dwellers as *P. boylei, P. difficilis,* and *P. truei,* so there is a strong vegetational relationship as well. Of the 17 species reported from this woodland type (see Table 6), five monotypic species in Middle America seem confined to this situation: *P. ixtlani* (Goodwin, 1964), *P. lepturus, P. lophurus, P. melanocarpus* (Goldman, 1951:210, 229, 393), and probably *P. evides* (Osgood, 1909:152). The polytypic *P. boylei, P. difficilis, P. maniculatus,* and *P. truei* also occur in piñon-juniper-oak. Two of these, *P. difficilis* and *P. maniculatus,* also live extensively in boreal forest as do the Middle American polytypic *P. guatemalensis* (Goodwin, 1934:42), *P. hylocetes* and *P. megalops* (Goldman, *loc. cit.*), and the monotypic *P. melanotis* (Osgood, 1909:109), *P. oaxacensis,* and *P. zarhynchus* (Goldman, *loc. cit.*). Polytypic *P. nudipes* of Costa Rica and Panama occurs in oak forests near timberline (Harris, 1943) as well as in humid upper tropical areas.

Fir forest: Boreal (Canadian and Hudsonian) woodlands occur on higher slopes (8,000 feet and above) in parts of Middle America and southwestern United States. Stands of fir (*Abies*) or Douglas fir (*Pseudotsuga*) with moist understory often dominate in this belt. The obscurely known monotypic *P. altilaneus* of Guatemala may be the only *Peromyscus* restricted to this montane situation. All others totaled in Table 6, the monotypic *P. melanotis, P. oaxacensis,* and *P. zarhynchus,* and the polytypic *P. difficilis, P. guatemalensis, P. hylocetes, P. maniculatus, P. megalops,* and *P. thomasi* (includes *P. nelsoni,* see Musser, 1964:18) occur also at least in pine-oak country as well. *Peromyscus maniculatus* lives in fir forest in southwestern United States, but is replaced in the higher parts of the Mexican mountains by monotypic *P. melanotis.* In numbers of species, boreal *Peromyscus* are most abundant in southern mountains, especially those in the Mexican states of Guerrero (five species), Oaxaca (six species), and Chiapas (three species). Two species are recorded from Guatemala and one from the highlands of Honduras.

TROPICAL HABITATS.—Much of the information incorporated in Table 7 has been obtained from Goldman (1951:316–362), who classified the mammals of Mexico as to their occurrences in various life zones (following Merriam, 1898). In total, 23 species of *Peromyscus* live in tropical situations, which are attractive to both monotypic and polytypic species, more occurring here than in any other major area listed in Table 4. Speciation in *Peromyscus* has flourished in the Middle American tropics, even in the presence of numerous well-adapted Neotropical rodents. Strange indeed is

TABLE 7

Peromyscus in Tropical Areas

Area	Number of species		
	Total	Monotypic	Polytypic
Humid Lower Tropical	5	1	4
Arid Lower Tropical	7	1	6
Humid Upper Tropical	17	10	7
Arid Upper Tropical	2	0	2

the fact that these seemingly aggressive and ubiquitous mice have not invaded at least tropical parts of northern South America but appear instead to stop short at the Panama-Colombia line.

Tropical habitats occur in areas which are normally frost-free throughout the year. The lower tropical division is restricted mostly to coastal plains and valleys of large river systems, while the upper tropical division generally occurs at higher elevations and ultimately is in contact with temperate areas where killing frosts are to be expected. Most workers will readily agree that numerous environments occur within each of the major areas included here.

Humid Lower Tropical: Lowland rain forest, jungle, savanna, and marshland provide homes for at least five kinds of *Peromyscus*. *Peromyscus mexicanus* is the most widespread species herein and prefers woodlands, rocky areas, and brush to openlands. A closely allied form, *P. yucatanicus,* occupies lowlands on the Yucatan Peninsula. Two temperate-acclimated species, *P. boylei* and *P. leucopus,* also thrive herein, especially in woodlands and brush. Probably the monotypic *P. stirtoni* of Honduras and El Salvador is confined to woodland situations (Goodwin, 1942:163). *Peromyscus* has not become highly diversified in this lowland area, where numerous Neotropical genera of small rodents flourish. Rarely have two species of *Peromyscus* been taken together here; Hall and Dalquest (1963:310) record *P. leucopus* and *P. mexicanus* from the same area along the forest edge in the Mexican state of Veracruz. Savanna in tropical America (for Mexico, see fig. 6 in Leopold, 1959) seems more important as a living place for rice rats (*Oryzomys*) and cotton rats (*Sigmodon*) than for *Peromyscus*.

Arid Lower Tropical: Arid thorn forest and tropical deciduous forest exist along both sea coasts and inland along streams in tropical America (for Mexico, see fig. 6 in Leopold, 1959). At least seven species of *Peromyscus* utilize these two elongated areas. The isolation of some of this environment very likely has allowed for the development of such species as *P. banderanus, P. perfulvus,* and monotypic *P. allophylus* on the westward coastal plain of Mexico, and of *P. yucatanicus* on the lowlands of the Yucatan Peninsula. *Peromyscus leucopus* also occurs in tropical shrub of the eastern Mexican coast as does *P. mexicanus,* which ranges also as far southeastward as Costa Rica. *Peromyscus boylei* reaches this tropical situation on both coasts of Mexico.

Humid Upper Tropical: Moist tropical environments occupy the slopes of seaward-facing mountains in Middle America, at least to a height of 6,500 feet. These highland jungles and cloud forests offer luxuriant habitats to numerous mammals (Baker, 1963:228). Many of these habitats are isolated or semi-isolated along river systems which have cut deep canyons in the mountains. Such conditions allow for a large assemblage of species of *Peromyscus* (18) as well as a high degree of endemism with no less than 11 alleged monotypic species, with highly restricted ranges, present. Although the ecology of these endemics is poorly known, some certainly are arboreal in tangled forest areas whereas others seek refuges in rocky substrate. Six of these monotypic species occur in Mexico with five in upper tropical situations in the Sierra Madre Oriental facing the Gulf of Mexico: *P. ochraventer* in Tamaulipas and northern San Luis Potosi, *P. latirostris* in San Luis Potosi, *P. simulatus* in Veracruz, and *P. aztecus* and *P. furvus* in Veracruz and Puebla. The drier Pacific slopes of Mexico seem less inviting to monotypic endemics; only one, *P. zarhynchus* (also in boreal fir forests, Goldman, 1951:399) in Chiapas, is reported. In other parts of Middle America, the rat-like *P. grandis* lives in Guatemala, *P. hondurensis* in Hondurus (Goodwin, 1942), and the large *P. flavidus* and *P. pirrensis* in Panama (Goldman, 1920:39). Many of the monotypic species are noted for their large size; no less than four exceed 300 mm in total length. The seven polytypic species include the ubiquitous *P. boylei, P. leucopus,* and *P. mexicanus. Peromyscus megalops* of the Mexican states of Guerrero and Oaxaca, *P. thomasi* of southern Mexico, and *P. guatemalensis* of Chiapas, Guatemala, and Honduras seem to occur both in boreal forest and humid upper tropical forest. In Costa Rica and Panama, *P. nudipes* is the only polytypic representative.

TABLE 8

Peromyscus ENDEMIC ON ISLANDS

Habitat	Number of species	
	Endemic subspecies of species with continental distribution	Species endemic on islands only
Boreal Islands	1	1
Temperate Islands	5	8
Tropical Islands	2	2

Arid Upper Tropical: Goldman (1951:357) listed *P. boylei* and *P. maniculatus* as the only mainland species occupying this dry, shrub zone that intermingles with arid temperate areas above or to the northward. This is the only tropical environment occupied at least at the edge by the widespread *P. maniculatus,* in Baja California and in southwestern Mexico.

INSULAR HABITATS.—The opportunistic behavior of mice of the genus *Peromyscus* is well illustrated by their ability to reach and inhabit islands adjacent to the coast of North and Middle America. Probably man occasionally has assisted some of them in crossing the water barrier. However, the number of insular populations sufficiently differentiated from relatives on the mainland to be classified as either different subspecies or distinctive species indicates that *Peromyscus* probably reached many areas now isolated as islands prior to the time when they were cut off from the mainland by rising sea levels following the melting back of the last glaciation. Coastal islands off New England now occupied by *Peromyscus* were formed between 5,000 and 2,000 years ago (Waters, 1963). Included among the continental species which have distinctive subspecies on islands are: *P. boylei, P. crinitus, P. eremicus, P. gossypinus, P. leucopus,* and *P. maniculatus.* In addition, 11 distinctive species of *Peromyscus* live only on islands (see Table 8).

Boreal islands: Humid forested islands in Pacific waters adjacent to the mainland of British Columbia and Alaska are occupied by distinctive subspecies of *P. maniculatus* and the endemic, polytypic *P. sitkensis.* As shown on Map 358 in Hall and Kelson (1959:617), these species seem to replace one another on the numerous islands. Distinctive subspecies of *P. maniculatus* occur on islands in the Gulf of St. Lawrence and off New Brunswick (Cameron, 1958).

Temperate islands: Continental species with subspecies on off-shore temperate islands include: *P. gossypinus* (off southeastern United States), *P. leucopus* (off New England), *P. crinitus, P. eremicus,* and *P. maniculatus* (off Sonora and Baja California). Monotypic *P. caniceps, P. collatus, P. dickeyi, P. pembertoni, P. pseudocrinitus, P. stephani,* and polytypic *P. interparietalis* and *P. guardia,* live only on islands in the Gulf of California. Most of these live in seemingly arid habitats, and in most cases no more than one occurs on any one island.

Tropical islands: Close relatives of *P. maniculatus,* monotypic *P. sejugis* and *P. slevini,* occupy arid tropical islands near the southern tip of Baja California. Aside from a subspecies of *P. boylei* on the Tres Marias Islands off the Mexican state of Nayarit and one of *P. leucopus* on Cozumel Island off the Yucatan Peninsula, there are no endemic *Peromyscus* on islands adjacent to southern parts of Middle America. This absence might be explained by the paucity, at least at the present time in Central America, of coastal lowland-dwelling *Peromyscus,* which would have had to be present in order to populate islands formed by the rising sea water.

Summary

Mice of the genus *Peromyscus* are highly successful in North and Middle America. Of the 57 species in seven subgenera currently recognized, more than one-half (32) are monotypic; that is, exhibit no definable geographic variation. Alpine areas with two species, Arctic prairie with one marginal species, and northern boreal forests with two species are the regions of the continent least used by *Peromyscus,* while tropical and subtropical regions with 24 species are the most used. In numbers of species, the genus is most common in the temperate and tropical environments of Mexico. In spite of the large amount of speciation which has taken place in the genus in tropical and subtropical areas of Mexico, the number of species dwindles rapidly southeastward of this republic in Middle America in the direction of South America.

Our knowledge of the living places of most species of *Peromyscus* is meager. We know most about two of them, *P. maniculatus* and *P. leucopus.* Except for grassland areas, swamps, and deserts, where other rodents may be dominant, species of *Peromyscus* usually are the most common mammals present. Most of them are small; more than one-half (32 of 57 species) average less than 225 mm in total length. Insular inhabitants are smallest, whereas three of the five largest (more than 276 mm long) species are restricted to tropical environments.

Species with extensive geographic ranges usually exhibit more geographic variation and are adapted for life in more kinds of habitats than species with limited geographic ranges. Localized populations of widespread species have become genetically diversified through environmental selection to the extent that distinctions may be made between different populations on the basis of morphology, behavioral characteristics, habitat preferences, physiology, and pelage color.

Where only one species of *Peromyscus* lives, it may occupy a wide range of environmental conditions; where two or more species occur together, one or more of them may be severely limited in local distribution. Such competition between species of *Peromyscus* for similar living places may be resolved by (1) exclusion of one species by another, (2) interdigitation of species in an area in relation to their preferences for different seral stages of plant communities present, and (3) perhaps some degree of compatibility.

Literature Cited

ANDREWARTHA, H. G., AND L. C. BIRCH. 1954. The distribution and abundance of animals. Chicago: Univ. Chicago Press, 782 pp.

AUDUBON, J. J., AND J. BACHMAN. 1846. The viviparous quadrupeds of North America. New York: publ. by J. J. Audubon, Vol. 1: xiv + 389.

BAILEY, VERNON. 1931. Mammals of New Mexico. N. Amer. Fauna, 53:412, 22 pls., 58 figs.

BAKER, ROLLIN H. 1956. Mammals of Coahuila, México. Univ. Kansas Publ., Mus. Nat. Hist., 9:125–335, 75 figs.

——— 1960. Mammals of the Guadiana lava field, Durango, México. Publ. Mus., Mich. State Univ., Biol. Ser., 1:303–328, 3 figs.

——— 1963. Geographical distribution of terrestrial mammals in Middle America. Amer. Midland Nat., 70:208–249, 1 fig.

BAKER, ROLLIN H., AND J. KEEVER GREER. 1962. Mammals of the Mexican state of Durango. Publ. Mus., Mich. State Univ., Biol. Ser., 2:25–154, 4 pls., 6 figs.

BAKER, ROLLIN H., AND CARLETON J. PHILLIPS. 1965. Mammals from El Nevado de Colima, Mexico. Jour. Mamm., 46:691–693.

BARBEHENN, KYLE R., AND JOHN G. NEW. 1957. Possible natural intergradation between prairie and forest deer mice. Jour. Mamm., 38:210–218, 1 pl., 2 figs.

BECKWORTH, STEPHEN L. 1954. Ecological succession on abandoned farm lands and its relationship to wildlife management. Ecol. Monogr., 24:349–376.

BENSON, SETH B. 1932. Three new rodents from lava beds of southern New Mexico. Univ. Calif. Publ. Zool., 38:335–344.

——— 1940. New subspecies of the canyon mouse (*Peromyscus crinitus*) from Sonora, Mexico. Proc. Biol. Soc. Washington, 53:1–4.

BLAIR, W. FRANK. 1947. Variation in shade of pelage of local populations of the cactus-mouse (*Peromyscus eremicus*) in the Tularosa Basin and adjacent areas of southern New Mexico. Contr. Lab. Vert. Biol., Univ. Mich., 37:1–7.

——— 1950. Ecological factors in speciation of *Peromyscus*. Evolution, 4:253–275.

——— 1956. The species as a dynamic system. Southwestern Nat., 1:1–5.

BROWN, L. N. 1964. Ecology of three species of *Peromyscus* from southern Missouri. Jour. Mamm., 45:189–202, 3 figs.

BURT, WILLIAM H. 1949. Territoriality. Jour. Mamm., 30:25–27.

——— 1961. Some effects of Volcán Parícutin on vertebrates. Occ. Papers Mus. Zool., Univ. Mich., 620:1–24, 2 pls., 1 fig.

CAMERON, AUSTIN. 1958. Mammals of the islands in the Gulf of St. Lawrence. Bull. Natl. Mus. Canada, 154:iii + 165, 8 figs., 29 maps, frontis.

COCKRUM, E. LENDELL. 1952. Mammals of Kansas. Univ. Kansas Publ., Mus. Nat. Hist., 7:1–303, 73 figs.

COGSHALL, ANNETTA STOW. 1928. Food habits of deer mice of the genus *Peromyscus* in captivity. Jour. Mamm., 9:217–221.

COOK, S. F., AND J. P. HANNON. 1954. Metabolic differences between three strains of *Peromyscus maniculatus*. Jour. Mamm., 36:553–560.

DALQUEST, WALTER E. 1953. Mammals of the Mexican state of San Luis Potosí. Louisiana State Univ. Studies, Biol. Ser., No. 1:1–233, 1 fig.

DICE, LEE R. 1922. Some factors affecting the distribution of the prairie vole, forest deer mouse, and prairie deer mouse. Ecology, 3:29–47.

———— 1931. The occurrence of two subspecies of the same species in the same area. Jour. Mamm., 12:210–213.

———— 1939. Variation in the cactus-mouse, *Peromyscus eremicus*. Contr. Lab. Vert. Genetics, Univ. Mich., 8:1–27, 1 map.

———— 1940a. Relationships between the wood-mouse and the cotton-mouse in eastern Virginia. Jour. Mamm., 21:14–23, 1 fig.

———— 1940b. Ecologic and genetic variability within species of *Peromyscus*. Amer. Nat., 74:212–221.

———— 1941. Variation of the deer-mouse (*Peromyscus maniculatus*) on the sand hills of Nebraska and adjacent areas. Contr. Lab. Vert. Genetics, Univ. Mich., 15:1–19, 1 map.

DICE, LEE R., AND P. M. BLOSSOM. 1937. Studies of mammalian ecology in southwestern North America with special attention to the colors of desert mammals. Carnegie Inst. Washington, Publ., 485:iv + 129, 8 pls., 8 figs.

DUNMIRE, WILLIAM W. 1960. An altitudinal survey of reproduction in *Peromyscus maniculatus*. Ecology, 41:174–182, 2 figs.

FITCH, JOHN H. 1963. A comparative behavioral study of *Peromyscus leucopus* and *Peromyscus maniculatus*. Trans. Kansas Acad. Sci., 66:160–164.

FOSTER, DOROTHY DICE. 1959. Differences in behavior and temperament between two races of the deer-mouse. Jour. Mamm., 40:496–513.

GOLDMAN, E. A. 1920. Mammals of Panama. Smithson. Misc. Coll., 69:1–309, 39 pls., 1 map.

———— 1951. Biological investigations in Mexico. *Ibid.*, 115:xiii + 476, frontis., 70 pls., 1 map.

GOLLEY, FRANK B. 1962. Mammals of Georgia. A study of their distribution and functional role in the ecosystem. Athens: Univ. Georgia Press, xii + 218 pp., 118 figs.

GOODWIN, GEORGE G. 1934. Mammals collected by A. W. Anthony in Guatemala, 1924–28. Bull. Amer. Mus. Nat. Hist., 68:1–60, 5 pls.

———— 1942. Mammals of Honduras. *Ibid.*, 79:107–195.

———— 1964. A new species and a new subspecies of *Peromyscus* from Oaxaca, Mexico. Amer. Mus. Novit., 2183:1–8, 4 figs.

GRINNELL, J., AND H. S. SWARTH. 1913. An account of the birds and mammals of the San Jacinto area of southern California. Univ. Calif. Publ. Zool., 10:197–406, pls. 6–10, 3 figs.

HALL, E. RAYMOND, AND WALTER W. DALQUEST. 1963. The mammals of Veracruz. Univ. Kansas Publ., Mus. Nat. Hist., 14:165–362, 2 figs.

HALL, E. RAYMOND, AND KEITH R. KELSON. 1959. The mammals of North America. New York: Ronald Press, Vol. II, viii + 547–1083 + 79 pp.

HAMILTON, W. J., JR. 1941. The food of small forest mammals in the eastern United States. Jour. Mamm., 22:250–263.

———— 1950. The prairie deer-mouse in New York and Pennsylvania. *Ibid.,* 31:100.

HARPER, FRANCIS. 1961. Land and fresh-water mammals of the Ungava Peninsula. Univ. Kansas, Mus. Nat. Hist., Misc. Publ., 27:1–178, 8 pls., 3 figs., 45 maps.

HARRIS, VAN T. 1952. An experimental study of habitat selection by prairie and forest races of the deer-mouse, *Peromyscus maniculatus.* Contr. Lab. Vert. Biol., Univ. Mich., 56:1–56, 2 pls., 1 fig.

HARRIS, WILLIAM P., JR. 1943. A list of mammals from Costa Rica. Occ. Papers Mus. Zool., Univ. Mich., 476:1–15.

HESSE, R., W. C. ALLEE, AND KARL P. SCHMIDT. 1951. Ecological animal geography. New York: 2nd ed. John Wiley & Sons, Inc., xiii + 715 pp., 142 figs.

HOFFMAN, GEORGE. 1960. The small mammal components of six climax associations in eastern Washington and northern Idaho. Ecology, 41: 571–572.

HOFFMAN, ROBERT S. 1955. A population-high for *Peromyscus maniculatus.* Jour. Mamm., 36:571–572.

HOFFMEISTER, DONALD F., AND LUIS DE LA TORRE. 1961. Geographic variation in the mouse *Peromyscus difficilis.* Jour. Mamm., 42:1–13, 2 figs.

HOOPER, EMMET T. 1942. An effect on the *Peromyscus maniculatus* Rassenkreis of land utilization in Michigan. Jour. Mamm., 23:193–196, 1 fig.

HOOPER, EMMET T., AND GUY G. MUSSER. 1964. Notes on the classification of the rodent genus *Peromyscus.* Occ. Papers Mus. Zoology, Univ. Mich., 635:1–13, 2 figs.

HORNER, B. ELIZABETH. 1954. Arboreal adaptations of *Peromyscus* with special reference to use of the tail. Contr. Lab. Vert. Biol., Univ. Mich., 61:1–85.

HOWARD, WALTER E. 1949. Dispersal, amount of inbreeding, and longevity in a local population of prairie deer-mice on the George Reserve, southern Michigan. Contr. Lab. Vert. Biol., Univ. Mich., 43:1–52, 2 pls., 24 figs.

HOWELL, ARTHUR H. 1921. A biological survey of Alabama. I. Physiography and life zones. II. The mammals. N. Amer. Fauna, 45:1–88, 11 pls., 10 figs., 1 map.

JAMESON, E. W., JR. 1951. Local distribution of white-footed mice, *Peromyscus maniculatus* and *P. boylei,* in the northern Sierra Nevada, California. Jour. Mamm., 32:197–203, 2 figs.

———— 1952. Food of deer-mice, *Peromyscus maniculatus* and *P. boylei,* in the northern Sierra Nevada, California. *Ibid.,* 33:50–60, 2 figs.

———— 1953. Reproduction of deer-mice (*Peromyscus maniculatus* and *P. boylei*) in the Sierra Nevada, California. *Ibid.,* 34:44–58, 4 figs.

———— 1955. Some factors affecting fluctuations of *Microtus* and *Peromyscus.* *Ibid.,* 36:206–209.

KING, JOHN A. 1957. Intra- and interspecific conflict of *Mus* and *Peromyscus.* Ecology, 38:355–357.

KING, JOHN A., AND B. E. ELEFTHERIOU. 1959. The effects of early handling upon adult behavior in two subspecies of deer-mice, *Peromyscus maniculatus.* Jour. Comp. and Physiol. Psychol., 52:82–88.

KLEIN, HAROLD G. 1960. Ecological relationships of *Peromyscus leucopus noveboracensis* and *P. maniculatus gracilis* in central New York. Ecol. Monogr., 30:387–407, 7 figs.

LAWRENCE, GEORGE E. 1966. Ecology of vertebrate animals in relation to chaparral fire in the Sierra Nevada foothills. Ecology, 47:278–291, 10 figs.

LEOPOLD, A. STARKER. 1959. Wildlife of Mexico, the game birds and mammals. Berkeley: Univ. Calif. Press, xii + 568 pp., 194 figs, 1 map.

LINDEBORG, ROBERT G. 1952. Water requirements of certain rodents from xeric and mesic habitats. Contr. Lab. Vert. Biol., Univ. Mich., 58:1–32, 3 figs.

McCARLEY, W. H. 1954a. Fluctations and structure of *Peromyscus gossypinus* in eastern Texas. Jour. Mamm., 35:526–532, 1 fig.

——— 1954b. Natural hybridization in the *Peromyscus leucopus* species group of mice. Evolution, 8:314–323, 3 figs.

——— 1959. The effect of flooding on a marked population of *Peromyscus.* Jour. Mamm., 40:57–63.

——— 1963. Distributional relationships of sympatric populations of *Peromyscus leucopus* and *P. gossypinus.* Ecology, 44:784–788, 1 fig.

MERRIAM, C. HART. 1898. Life zones and crop zones of the United States. Bull. U. S. Dept. Agric., Div. Biol. Surv., 10:1–79, 1 map.

MOULTHROP, PHILIP N. 1938. The prairie white-footed mouse in New York State. Jour. Mamm., 19:503.

MURIE, ADOLPH. 1933. The ecological relationship of two subspecies of *Peromyscus* in the Glacier Park region, Montana. Occ. Papers Mus. Zoology, Univ. Mich., 270:1–17, 2 figs.

MURIE, MARTIN. 1961. Metabolic characteristics of mountain, desert and coastal populations of *Peromyscus.* Ecology, 42:723–740, 6 figs.

MUSSER, GUY G. 1964. Notes on geographic distribution, habitat, and taxonomy of some Mexican mammals. Occ. Paper Mus. Zoology, Univ. Mich., 636:1–22, 1 fig.

ODUM, EUGENE P. 1959. Fundamentals of ecology. Philadelphia and London: 2nd edition. W. B. Saunders Co., xvii + 546 pp., 160 figs.

OSGOOD, WILFRED H. 1909. A revision of the mice of the American genus *Peromyscus.* N. Amer. Fauna, 28:1–285, 8 pls., 12 figs., 1 map.

OZOGA, JOHN J., AND CARLETON J. PHILLIPS. 1964. Mammals of Beaver Island, Michigan. Publ. Mus., Mich. State Univ., Biol. Ser., 2:305–348, 2 pls., 1 fig.

PEARSON, PAUL G. 1959. Small mammals and old field succession on the piedmont of New Jersey. Ecology, 40:249–255.

RICHMOND, NEIL D., AND HARRY R. ROSLAND. 1949. Mammal survey of northwestern Pennsylvania. Harrisburg: Pennsylvania Game Comm., 65 pp., 5 maps.

SCHWARTZ, CHARLES W., AND ELIZABETH R. SCHWARTZ. 1959. The wild mammals of Missouri. Columbia: Univ. Missouri Press and Missouri Conserv. Comm., xvi + 341 pp., 57 pls.

SHEPPE, WALTER, JR. 1961. Systematic and ecological relations of *Peromyscus oreas* and *P. maniculatus.* Proc. Amer. Philo. Soc., 104:421–466, 12 figs., 1 map.

STEPHENSON, G. K., P. D. GOODRUM, AND R. L. PACKARD. 1963. Small rodents as consumers of pine seed in East Texas uplands. Jour. Forestry, 61:523–526, 3 figs.

STICKEL, L. F. 1946. The source of animals moving into a depopulated area. Jour. Mamm., 27:301–307.

STICKEL, W. H. 1951. Occurrence and identification of the prairie deer-mouse in central Maryland. Proc. Biol. Soc. Washington, 64:25–32.

STINSON, R. H., AND K. C. FISCHER. 1953. Temperature selection in deer-mice. Canadian Zool., 31:404–416.

SUMNER, FRANCIS B. 1926. An analysis of geographic variation in mice of the *Peromyscus polionotus* group from Florida and Alabama. Jour. Mamm., 7:149–184, pls. 15–18, 9 figs.

TERMAN, C. RICHARD. 1962. Spatial and homing consequences of the introduction of aliens into semi-natural populations of prairie deer-mice. Ecology, 43:216–223, 4 figs.

VAUGHAN, TERRY A. 1954. Mammals of the San Gabriel Mountains of California. Univ. Kansas Publ., Mus. Nat. Hist., 7:513–532, 4 pls., 1 fig.

VERTS, B. J. 1957. The population and distribution of two species of *Peromyscus* on some Illinois strip-mined land. Jour. Mamm., 38:53–59, 3 figs.

WATERS, JOSEPH H. 1963. Biochemical relationships of the mouse *Peromyscus* in New England. Syst. Zool., 12:122–133, 5 figs.

WEBB, ROBERT G., AND ROLLIN H. BAKER. 1962. Terrestrial vertebrates of the Pueblo Nuevo area of southwestern Durango, México. Amer. Midland Nat., 68:325–333, 1 fig.

WECKER, STANLEY C. 1963. The role of early experience in habitat selection by the prairie deer-mouse, *Peromyscus maniculatus bairdi.* Ecol. Monogr., 33:307–325, 14 figs.

WILLIAMS, OLWEN. 1955. Distribution of mice and shrews in a Colorado montane forest. Jour. Mamm., 36:221–231, 1 pl., 1 fig.

———— 1959. Food habits of the deer-mouse. *Ibid.,* 40:415–420.

WILSON, L. WAYNE. 1945. The genus *Peromyscus* in West Virginia. Jour. Mamm., 26:95–96.

5

ANATOMY

DAVID KLINGENER

Introduction

FORTY YEARS ago, in the introduction to his monograph on the anatomy of the wood rat, A. B. Howell deplored the lack of interest in the common small North American mammals on the part of anatomists, who seemed preoccupied instead with a few spectacular exotic forms. Happily this situation has changed since 1926, and we now have an embryonic biological understanding of the morphology of some of our native insectivores, bats, carnivores, and rodents. The best known of our native muroid rodents is *Peromyscus*.

The following review is an attempt to summarize our present understanding of the gross adult morphology of *Peromyscus*. Certain systems and structures, including the integument and its glands, the kidney, much of the digestive system, and the endocrines, are best studied microscopically and are not reviewed here. There is not room for repetition of all available factual data; I have included only that material required for understanding the major functional and systematic problems. I have included some material on muroids other than *Peromyscus*, but only when relevant to the function of some structure or to the systematic relationships of *Peromyscus*.

The best anatomical summary is an illustration. I have included figures of the cranial foramina and carotid arteries, since these are not available elsewhere. These figures are based on specimens in the University of Massachusetts Museum of Zoology. Lack of time has prevented the preparation and rendering of more. The listed references, however, include many more illustrations.

Body Shape and Proportion

Adult individuals of *Peromyscus* range in body weight from about 15 g in some small northern species, such as *P. crinitus* and

P. polionotus, to more than 110 g in a few tropical forms, such as *P. thomasi* (Musser and Shoemaker, 1965). Length of body ranges from approximately 80 to 170 mm, and length of hind foot from 15 to 38 mm (Hall and Kelson, 1959).

Dice (1940) noted that forest-dwelling forms tend to have longer tails and larger hind feet than their prairie relatives, and he suggested a possible correlation of these features with semi-arboreal habits in the forest dwellers. Horner (1954), in her behavioral comparison of ten kinds, found that in the semi-arboreal ones (*P. maniculatus gracilis, P. oreas, P. leucopus noveboracensis, P. difficilis nasutus,* and *P. truei truei*) the long tail is used for balance when the mouse runs along a horizontal branch. The tail also serves as a tactile organ, as a prop when the animal climbs a vertical branch or trunk, and as a limited prehensile organ when a mouse slips on a branch or climbs from one branch to another. The relatively shorter tail in the terrestrial kinds (*P. maniculatus bairdi, P. m. blandus, P. m. nebrascensis,* and *P. polionotus leucocephalus*) is apparently less useful as a balancing organ, and climbing performance in the terrestrial forms is less seriously affected by experimental amputation of the tail than in those that are semi-arboreal.

Hayne (1950) found that animals from coastal populations of *P. polionotus* have longer hind feet than those from inland populations, and length of hind foot also varies less in the coastal populations. He suggested that the long hind foot might in this case be an adaptation for life on beach sands.

The pinna of the ear in some arid-land mice, particularly *P. truei,* is relatively large. On the basis of field and laboratory observations, Hoffmeister (1951) and Horner (1954) suggested that individuals of *P. truei* rely heavily on hearing for detecting the approach of predators in their natural environment, where cover is sparse.

Skeleton

The head skeleton of *Peromyscus* has received the attention of many students. In 1909 Osgood listed cranial characteristics of the genus and of the subgenera as he recognized them. Since then, numerous authors have mentioned small details of skull structure in their comparisons of different species and subspecies. Hooper and Musser (1964*b*) summarized certain major trans-specific geographic trends in skull morphology. In temperate forms the braincase tends to be rounded and somewhat inflated, the interorbital constriction has smooth outlines, the "ectopterygoid fossae" (= pterygoid fossae) are often relatively large, the "sphenopalatine vacuities" (a pair of slits between the presphenoid-basisphenoid bar and the floors of the two pterygoid fossae, not homologous to the sphenopalatine foramina) are large, and the auditory bullae tend to be inflated. Tropical species tend to have a more elongate braincase, a broader, more angular interorbital region, and relatively small pterygoid fossae, "sphenopalatine vacuities," and auditory bullae. The functional significance of these differences is unknown, but Hooper and Musser suggested several possible lines of investigation. The temperate forms referred to above include members of the *maniculatus, leucopus, eremicus, crinitus, boylei, truei,* and *melanophrys* species groups. Subtropical species of the *boylei* and *melanophrys* groups show some "tropical" characteristics, however. Tropical species include members of the *mexicanus, thomasi,* and *banderanus* species groups. *P. floridanus* and members of the *lepturus* group show some degree of intermediacy between the two morphological poles.

Hoffmeister (1951) described postnatal ontogeny of the skull in *P. truei* and compared relative growth of different regions with published data on *Neotoma* and *Citellus*. The pattern of growth of the brain may influence ontogeny of the skull in *Peromyscus*. King and Eleftheriou (1960) observed that between 10 and 20 days of age the brain grows more rapidly in *P. maniculatus gracilis* than in *P. m. bairdi*. Correspondingly, the bones of the braincase in *gracilis* grow more rapidly in relation to the bones of the facial region than they do in *bairdi* during the same period.

At first glance the cranial foramina of rodents seem to be similar to those of most other mammals. Hill (1935), after dissecting the heads of a number of different forms, showed that previous authors had been mistaken in some of their assumptions about foraminal contents, and he proposed a rather different system of homologies between rodents and other mammals. Hill included a number of muroids in his study but did not figure any of them. I have illustrated the cranial foramina of *P. maniculatus* in Figures 1 and 2, following Hill's terminology.

Within the genus *Peromyscus* and within the muroid superfamily very little variation can be expected in the foramina transmitting

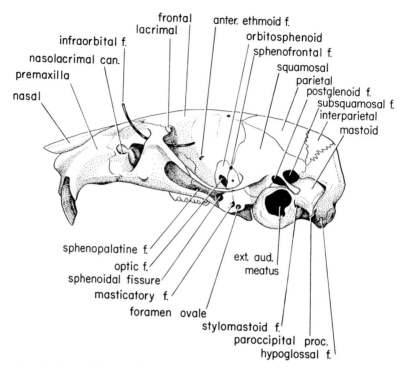

FIG. 1.   Lateral view of skull of *Peromyscus maniculatus,* Mono County, California.

cranial nerves (II through the optic foramen, III, IV, V_1, V_2, and VI through the sphenoidal fissure, the masticatory branch of V_3 through the masticatory foramen, the remainder of V_3 through foramen ovale, VII through the stylomastoid foramen, IX, X, and XI through the jugular foramen, and XII through the hypoglossal foramen). More significant variation can be found in the foramina which transmit blood vessels; that topic is considered with the carotid arteries in the present paper.

Cockerell, Miller, and Printz (1914) described and figured the auditory ossicles of *P. truei* and *P. nasutus* (= *P. difficilis*), along with those of several other muroid genera. They remarked on the absence of the orbicular apophysis of the malleus in all microtines and Old World cricetines they examined, and on its consistent presence in American muroids.

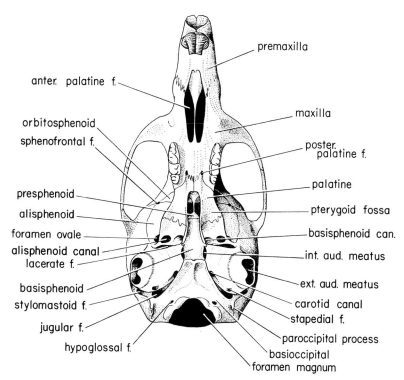

premaxilla

anter. palatine f.

orbitosphenoid
sphenofrontal f.

maxilla

poster.
palatine f.

palatine

presphenoid

alisphenoid

foramen ovale

alisphenoid canal
lacerate f.

basisphenoid

stylomastoid f.

jugular f.

hypoglossal f.

pterygoid fossa

basisphenoid can.

int. aud. meatus

ext. aud. meatus

carotid canal

stapedial f.

paroccipital process

basioccipital

foramen magnum

FIG. 2.   Ventral view of skull of *Peromyscus maniculatus,* Mono County, California.

The hyoid skeleton and its associated musculature vary significantly in muroids (Sprague, 1941, 1942; Rinker, 1954). Major differences between genera occur in the anterior horn. In some genera (*Peromyscus, Neotoma, Onychomys,* and *Reithrodontomys*) the anterior horn is composed of the ceratohyal bone and a long stylohyal cartilage which connects by a ligament with the wall of the stylomastoid foramen and forms the suspension of the hyoid apparatus. The stylohyoideus, jugulohyoideus, stylopharyngeus, and styloglossus muscles attach on the stylohyal cartilage. This condition is the most primitive one known among muroids. In certain other genera (*Sigmodon, Oryzomys, Microtus, Mesocricetus*) the hyoid suspension is formed by a sheet of fascia on the ventral surface of the bulla, and the stylohyal is reduced to a small piece of cartilage near the stylomastoid foramen. The jugulohyoideus is vestigial. The stylohyoideus, stylopharyngeus, and styloglossus originate on the fascial sheet. The anterior horn is also fascial in the gerbils, *Tatera* and *Meriones* (Sharma and Sivaram, 1959), and perhaps in *Hodomys* (= *Neotoma*) *alleni* (Sprague, 1941). The functional significance of these different hyoid suspension mechanisms is not clear, but evolutionary transformation of a bony or cartilaginous anterior horn into a ligament or fascial sheet is not uncommon among mammals.

Details of structure of the basihyal and posterior horns also vary. Sprague (1941) stated that according to these hyoid characters, *P. (Haplomylomys) californicus* and *eremicus* show affinities with *Onychomys*; *P. (Peromyscus) maniculatus, leucopus, pectoralis,* and *truei* with *Reithrodontomys;* and *Ochrotomys nuttalli* with *Oryzomys.* He also found the hyoid of *Sigmodon* to be distinct from those of all other cricetines.

Comparative morphological studies of the postcranial skeleton have been few. Rinker (1960) and Manville (1961) observed that the entepicondylar foramen is absent in *Ochrotomys nuttalli* and consistently present in all species of *Peromyscus* examined. This foramen is located above the medial epicondyle of the humerus and transmits the median nerve and sometimes the brachial artery. Selective value of presence or absence of the foramen is not understood. Rinker and Manville regarded absence of the foramen in *nuttalli* as additional evidence favoring elevation of *Ochrotomys* to generic rank.

Stains (1959) described the calcaneus in *P. leucopus* and *P. maniculatus* and noted that the two species can be distinguished locally by dimensions of the bone.

Sensenig (1943) described and illustrated embryonic development of the vertebral column in *P. maniculatus.*

Musculature

Rinker (1954) dissected the skeletal musculature of *P. maniculatus, P. leucopus,* and *P. difficilis* as part of a larger comparative myological study of *Peromyscus, Neotoma, Oryzomys,* and *Sigmodon.* In his quantitative analysis of the results he eliminated 111 of the 228 identifiable muscles because they are either similar in all four genera or else differ insignificantly. Of the remaining

117 muscles, 70 are similar in *Sigmodon* and *Oryzomys*, 71 in *Neotoma* and *Peromyscus*, 6 in *Sigmodon* and *Neotoma*, 7 in *Sigmodon* and *Peromyscus*, 8 in *Neotoma* and *Oryzomys*, and 10 in *Peromyscus* and *Oryzomys*. Furthermore, 54 of the 70 that are similar in *Sigmodon* and *Oryzomys* are also among those similar in *Neotoma* and *Peromyscus*, but differ between the pairs. *Peromyscus* differs from *Neotoma* in a total of 35 muscles, from *Oryzomys* in 96, and from *Sigmodon* in 96. *Sigmodon* differs from *Oryzomys* in 37. Rinker concluded that *Neotoma* and *Peromyscus* are more closely related to one another than to either *Oryzomys* or *Sigmodon*, and that the last two genera are closely related to one another.

In a similar study at the subgeneric level, Rinker (1963) dissected the musculature of *P. leucopus* as a representative of the subgenus *Peromyscus*, *P. nuttalli* of the subgenus *Ochrotomys* (now raised to generic rank), and *P. eremicus* of the subgenus *Haplomylomys*. He found that *O. nuttalli* differs significantly from *P. leucopus* in 17 muscles and from *P. eremicus* in 24. *P. eremicus* differs from *P. leucopus* in 17. He found no significant myological differences between *P. leucopus, P. maniculatus,* and *P. difficilis* in limited dissection of the last two species. For only eight of the muscles in which he found differences could Rinker designate primitive and advanced states. With respect to most of these eight muscles, however, *P. leucopus* and *P. eremicus* are more primitive than *O. nuttalli*.

Very few of the observed muscular differences between these muroid genera and subgenera have obvious functional significance. The selective forces operative during muscular differentiation of these rodents therefore remain obscure.

Rinker's papers (1954, 1963) are illustrated and serve as excellent guides for dissection of muroid musculature.

Digestive System

Osgood (1909) briefly described the molar patterns in a variety of species of *Peromyscus*. He regarded molar structure as important evidence in estimating relationships between these species. Hooper (1957) and Bader (1959) reported on detailed quantitative studies of molar variation in 18 species of *Peromyscus* and in *Ochrotomys*. Their results require modification of Osgood's ideas on the occurrence and constancy of certain dental structures. Hooper (1957, fig. 1) diagrammed the enamel patterns of upper and lower teeth. Most variation occurs in the "accessory" lophs and cusps: the mesoloph, mesostyle, entoloph, and entostyle in the upper molars and the mesolophid, mesostylid, ectolophid, and ectostylid in the lowers. These structures occur in the valleys between the major cusps and contribute to the complexity of the tooth. A molar without accessory elements is said to be simple. Hooper scored individual teeth for presence or absence of the accessory structures. He also scored for details of morphology of the mesolophid when present. Bader, however, reported scores only for presence or absence of accessory structures in the three species he studied.

Hooper and Bader found that the first molar tends to be more complex than the second in each toothrow. The labial sides of molars also tend to be more complex than the lingual. Species differ greatly in the nature and degree of molar variation. In some species, such as *P. eremicus,* accessory structures are almost invariably absent. In others, as exemplified by *P. yucatanicus,* the pattern is complex and almost constant. In still other species, notably *P. maniculatus, P. leucopus,* and *P. boylei,* the degree and nature of complexity may vary greatly within one local population, or throughout the geographic range of the species. Populations of *Peromyscus* living in the more humid, forested parts of North America tend to have more complex teeth than those living in the arid southwestern United States, northwestern Mexico, and the Yucatan Peninsula, where cover is sparse. Detailed illustrations of the molars of several species of *Peromyscus* have been published (Osgood, 1909; Hoffmeister, 1951; Hooper, 1957).

Van Valen (1962) measured length and width of each molar in a sample of 71 specimens of *P. leucopus.* He found that the widths of all teeth in an individual were correlated with one another, but that the lengths of the upper and lower third (last) molars were not correlated with the lengths of the first two. He attributed these results to a conflict between growth fields in the developing dentition with an unknown factor that limits the length of the toothrow.

Some species of *Peromyscus* are provided with paired internal cheek pouches used while transporting food. When present, these pouches are formed by outpocketings of the oral mucosa opposite the diastema. Cheek pouches are absent in *P. eremicus.* They are always present, but very small, in *P. leucopus*. In *Ochrotomys nuttalli* they are relatively large and extend backwards to the level of the posterior edge of the eyeball. In both of these mice the pouches are very thin-walled. Their associated musculature develops from the buccinator complex and is not extensive. The buccal part of the platysma myoides and the zygomaticolabialis pass over the pouch but do not insert on it. No pouch retractors comparable to those found in hamsters, squirrels, or geomyoids are developed (Hamilton, 1942; Goodpaster and Hoffmeister, 1954; Rinker, 1963). We would expect to find differences in food gathering behavior between species without pouches and those with them. To my knowledge no behavioral evidence on this point has been published.

In muroid rodents the stomach is "bilocular," consisting of two morphological zones. The left, or cardiac, side is lined with cornified epithelium. The right, or pyloric, side is wholly or partly glandular. Externally visible constriction between cardiac and pyloric parts is weak in most New World forms, but is so strong in some Old World hamsters that the stomach is truly two-chambered (Tullberg, 1899; Vorontsov, 1957). Tullberg described and illustrated the stomach of *Hesperomys* (= *Peromyscus*) *leucopus*. Constriction between cardiac and pyloric parts is weak, and invasion of the pyloric region by cornified epithelium is extensive. The glandular area, or fundus, is a small thick region of the pyloric wall along the greater curvature and is recognizable in gross dissection. Vorontsov reported that distribution of cornified and glandular epithelia in the stomach of *P. californicus* resembles that in *P. leucopus*, but that the pyloric region of the former species differs in being weakly divided into two parts by a second constriction, so that the stomach is rather poorly divided into three successive chambers. Tullberg and others have suggested that cornification protects most of the stomach wall from abrasion by harsh food. In *Onychomys* and *Oxymycterus*, which feed mostly on terrestrial arthropods, the glands of the fundus open into a small pocket and are thereby protected from abrasion by chitin fragments in the lumen (Horner, Taylor, and Padykula, 1965).

Vorontsov (1957) found that the large intestine in *P. californicus* is relatively short in comparison with the small intestine. In most Old World cricetines the large intestine is relatively long.

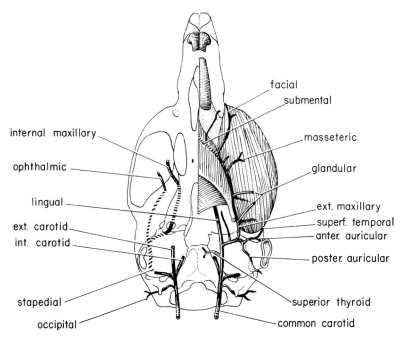

facial
submental
internal maxillary
masseteric
ophthalmic
glandular
lingual
ext. maxillary
ext. carotid
superf. temporal
int. carotid
anter. auricular
poster. auricular
stapedial
superior thyroid
occipital
common carotid

FIG. 3.   Ventral view of the major branches of the carotid arteries in *Peromyscus maniculatus*, Crawford County, Pennsylvania. Superficial structures are shown on the right, deeper structures on the left.

Circulatory System

The carotid arteries are better known than most other parts of the mammalian circulatory system. Since they are reasonably complicated, are fairly constant at specific and generic levels, and leave indications of their routes in the bony structure of the skull, they are popular among systematists as sources of taxonomic information. As part of his comprehensive study of the carotid arteries of rodents, Guthrie (1963) described the pattern found in *Peromyscus leucopus*. I have dissected these arteries in uninjected specimens of *P. maniculatus* and *P. truei* and have illustrated the major ones as they appear in the former species (Fig. 3).

As in most other myomorphs, the common carotid in *Peromyscus* divides into external and internal carotids at the level of the posterior edge of the auditory bulla. The external carotid gives off superior thyroid and lingual branches and then turns laterad, passing between the anterior and posterior horns of the hyoid apparatus and their musculature. It then divides into the external maxillary artery and the common trunk of the superficial temporal and anterior and posterior auricular arteries. The external maxillary passes forward along the ventral edge of the masseter, supplying that and other muscles and salivary glands, and eventually branches into submental and facial parts.

Immediately after branching from the common carotid, the internal carotid gives off the stapedial artery. The internal carotid enters the skull via the carotid canal at the medial edge of the bulla and supplies the brain. The stapedial enters the bulla through the stapedial foramen, emerges into the bullar cavity from the bony stapedial canal, and passes through the stapes. At the anterior edge of the bulla the stapedial divides into ophthalmic and internal maxillary arteries. The ophthalmic leaves the bulla and passes forward in a groove on the inner surface of the squamosal bone, emerging into the orbit through the sphenofrontal foramen. At the point where the internal groove of the ophthalmic artery crosses the external groove left by the masticatory nerve a small foramen is found in many skulls. This foramen transmits nothing. The internal maxillary artery passes out of the skull through the lacerate foramen and enters the posterior opening of the alisphenoid canal (which is not a true canal in *Peromyscus*, since it is not roofed over internally). The orbital part of the internal maxillary then emerges into the orbit through the sphenoidal fissure. My dissections left me unsatisfied as to where the inferior alveolar and pterygoid branches of the internal maxillary leave the main artery. Presumably they do so within the bulla as in *Microtus* (Guthrie, 1963). Guthrie prefers not to apply the name internal maxillary to the artery in the alisphenoid canal and refers to it as the continuation of the stapedial.

The pattern seen in *Peromyscus* probably occurs, with minor variations, in most other muroids. Guthrie found that skulls of "most members of the tribe Hesperomyini and all the members of the tribe Cricetini" show stapedial canals and the groove on the inner surface of the squamosal bone left by the ophthalmic artery. In the skulls of the Myospalacini and *Neotoma,* on the other hand, the stapedial canal is occluded and the groove on the squamosal is absent. Presumably the carotid pattern in these animals differs significantly from that in *Peromyscus*.

Guthrie also found cases of absence of the posterior part of the ophthalmic; in *Rattus* the ophthalmic originates from the orbital artery within the orbit, and in *Microtus* and *Clethrionomys* it originates from the orbital within the braincase just before both arteries emerge into the orbit.

There also seems to be some variation in the structure of the alisphenoid canal. In some species of *Peromyscus,* including *P. maniculatus* and *P. leucopus,* the canal is complete externally. In others, including *P. californicus* and to a greater extent *P. truei* and the Pliocene *P.* (= *Copemys*) *russelli,* the posterior part of the alisphenoid canal is incomplete externally, and part of the internal maxillary artery is uncovered (James, 1963). This condition is similar to those figured for *Rattus* and *Microtus* by Guthrie (1963, figs. 2 and 3). This situation leads to some uncertainty in naming the posterior opening of the alisphenoid canal. The opening given that name by Guthrie (1963, figs. 2 and 3) and James (1963, figs. 49a and b) does not correspond to the opening so labeled in my Figure 3.

Other parts of the circulatory system in *Peromyscus* have been neglected. As in most mammals, the veins of rodents are so variable individually (Guthrie, 1963) that few authors attempt to describe them.

Reproductive System

The male reproductive tract in *Peromyscus* is fairly well known anatomically, owing mainly to its usefulness in classification. The baculum has been described and figured in many species (Blair, 1942; Clark, 1953; Tamsitt, 1958; Burt, 1960; Lawlor, 1965). Hooper (1958) and Hooper and Musser (1964*b*) compared and illustrated the gross structure of the entire distal tract (the part of the penis beyond the flexure) in most of the known species of the genus.

In comparison with those of most other rodents, the glans penis in *Peromyscus* is structurally rather simple (Hooper, 1958, fig. 1). The baculum is a symmetrical shaft of bone, slightly expanded basally where it is attached to the corpora cavernosa penis. Distally it is capped by a single cartilaginous element of variable length and form. The urethra runs distad along the ventral surface of the corpora cavernosa and of the baculum. Its opening, the meatus urinarius, may be terminal or subterminal. The external surface of at least the proximal part of the glans is often covered with spines.

In most species of the subgenus *Peromyscus* (as constituted by Hooper and Musser) the glans penis is elongate and rod-shaped and has a protrusible tip. The baculum tends to be long and relatively slender. The meatus urinarius is subterminal. The external surface may be smooth or fluted, but always bears small spines or tubercles. Species and species groups of the subgenus differ in absolute and relative size of the glans and of its component parts.

In species of the subgenus *Haplomylomys* (P. *eremicus* and *P. californicus*) the glans is relatively broader and is noticeably expanded and flared distally. It has no protrusible tip. The baculum is relatively thick. As in the subgenus *Peromyscus* the meatus urinarius opens subterminally.

In *P.* (*Isthmomys*) *flavidus* and *pirrensis* the glans broadens greatly distally and is studded with long, sharp spines. A bulbous protractile mound covers the tip of the baculum and contains the meatus urinarius. The cartilaginous cap of the baculum is a diffuse mass and is not rod- or spine-shaped.

In *P.* (*Megadontomys*) *thomasi* the glans is large and awl-shaped. Proximally it is deeply fluted and armed with heavy spines. Internally it has a large crater and a urethral flap, structures characteristic also of *Neotoma* and certain other genera. Position of the meatus urinarius varies with state of engorgement of the urethra. The baculum and its cartilaginous cap are relatively long.

The glans in *P.* (*Osgoodomys*) *banderanus* is smooth, awl-shaped, and very small. The baculum is also very small. The meatus urinarius is terminal. Glans and baculum are also small in *P.* (*Podomys*) *floridanus*. The glans is smooth distally; proximally it is covered by small spines. The meatus urinarius is terminal. *P.* (*Habromys*) *lepturus* and *lophurus* are also characterized by small, smooth-surfaced glandes, but the baculum is large relative to length of glans. Structure of the meatus urinarius differs in detail from that in *Podomys* and in *Osgoodomys*.

Ochrotomys nuttalli has a very short, urn-shaped phallus with structural peculiarities that differentiate it from those of all other species of *Peromyscus*.

In several New World muroid genera (*Reithrodontomys, Ochrotomys, Neotomodon, Onychomys, Baiomys, Scotinomys, Nelsonia, Neotoma, Xenomys, Ototylomys, Tylomys*) phallic structure is simple and basically similar to that in *Peromyscus*, though in the wood rats and allied genera the form of the glans is somewhat complicated by the presence of a terminal crater developed in the distal urethral lumen. In most other muroids, notably in Old World hamsters, murines, and gerbils, and in microtines and New World cricetines other than those listed above, the glans is typically much more complex (Hooper and Musser, 1964a and included references). The baculum in these forms usually consists of a basal proximal bone and three distal digits which may be cartilage, calcified cartilage, or bone, and which may articulate with the basal member by a true synovial joint (Arata, Negus, and Downs, 1965). A large terminal crater is typically sunk into the end of the glans. Numerous soft structures may be present along its rim and on its floor and sides. Internally, large paired blood sinuses are often developed alongside the baculum. The phalli in a few genera, such as *Nyctomys* and *Ellobius,* are seemingly morphologically intermediate between the simple and complex types.

Arata (1964) studied the anatomy of the male accessory reproductive glands of 24 muroid genera. He found that in *Peromyscus gossypinus* the preputial glands are absent, the vesicular glands are large and recurved, and ampullary glands, bulbo-urethral glands, and four pairs of prostate glands are present. Among the species he studied, the forms characterized by the complex type of phallus possess also a complete or nearly complete set of accessory glands. The forms characterized by the simple sort of phallus are generally missing one or more of the accessory glands.

Known anatomical features of the male reproductive tract in muroid rodents have never been convincingly correlated with structure of the female tract or with copulatory behavior of the different species.

Nervous System and Sense Organs

Pilleri (1960) briefly described, but did not figure, the brain of one specimen of an unidentified species of *Peromyscus*. From his study of gross brain structure in eight muroid genera he concluded that these rodents generally have a poorly developed neocortex and cerebellum in comparison with some other rodents.

Eleftheriou and Zolovick (1965) prepared an atlas of the fore-brain of *P. maniculatus bairdi* in stereotaxic coordinates.

Young individuals of some strains of *Peromyscus* respond to certain noises with waltzing behavior and epileptic seizures. Older mice of these strains seem to become deaf and no longer respond to the noises with waltzing behavior, seizures, or even twitching of the ears. Ross (1962, 1965) studied in detail the anatomy of the auditory pathway in the brain of an epileptic waltzer strain of *P. maniculatus.* She compared young, partially deaf epileptic waltzers with older, completely deaf individuals of the same strain, and with control mice of a non-waltzing, non-epileptic strain of the same species. In the older mice she found almost complete atrophy of the primary acoustic nuclei and partial or complete degeneration of the remainder of the auditory centers and pathway. The degenerative process appears to be gradual.

King (1965) studied the relationship between adult brain and body weights in seven forms of *Peromyscus.* Body weights ranged from approximately 13 to 55 g, and brain weights from about 487 to 1029 mg. As body weight increases in the series, absolute brain weight also increases, but the ratio of brain weight to body weight decreases. King used an analysis of covariance to remove the variance in brain weight attributable to body weight and found that brain weight in the larger animals (*P. californicus* and *P. floridanus*) is still greater than that in the small ones (*P. maniculatus bairdi* and *P. polionotus*). He concluded that other factors not directly associated with increase in body size are responsible for the greater brain weight in the large animals.

King (1965) also reported data on the relationship between body weight and weight of the crystalline lens of the eye. Lens weight, unlike brain weight, is not closely related to body weight. The largest individual does have the relatively smallest lens, and vice versa, but there is considerable scatter on King's plot of lens weight vs. body weight for the seven forms. King suggested that differential operation of natural selection in the diverse environments occupied by the various species could account for his observations, but he could provide no concrete explanation of the mechanics in any specific case.

Other aspects of the anatomy of the eye and other sense organs have not been described.

Summary and Suggestions

Our understanding of the anatomy of *Peromyscus* is poor and fragmentary in comparison with that of the dog and laboratory rat, but rich and extensive in comparison with the dozens of other genera of muroid rodents of the world. In the musculature, digestive system, and skeleton we are reasonably aware of basic structure and major variation within the genus. In other areas, most particularly the central nervous system and sense organs, we are advanced but little beyond total ignorance. Perhaps the greatest flaw in our understanding is our inability to correlate most of the structural differences we have observed with differences in function and behavior. Without this correlation, anatomical facts have little evolutionary meaning.

Recent morphological investigations have caused some to question earlier ideas on the place of *Peromyscus* among the muroids and on the relationships of species of *Peromyscus* to one another. Hooper and Musser (1964*a* and included references) divided the muroids they studied into six groups on the basis of male phallic anatomy. Their "peromyscine-neotomine" group includes *Peromyscus* and 11 other New World genera (listed earlier in the present paper) characterized by a simple glans penis. The other five groups (murines, gerbillines, microtines, hamsters, and the "South American cricetines") are characterized for the most part by possession of a complex glans penis. The simple phallus was probably derived from the complex, and certain genera among the groups with complex phalli, notably *Ellobius* and *Nyctomys*, show partial simplification.

Evidence from some other quarters supports the suggested separation of the peromyscine-neotomine group from the others. Rinker's (1954) quantitative myological study demonstrated the distinctness of *Peromyscus* and *Neotoma* from *Sigmodon* and *Oryzomys*. The structure of the anterior horn of the hyoid also leads to the same sort of generic grouping; except for *Neotoma alleni,* all of the peromyscine-neotomines studied have a primitive, cartilaginous anterior horn, and in the other muroids studied it has been functionally replaced by a fascial sheet. Partial support also derives from Arata's (1964) study of the male accessory reproductive glands. Arata reported on only one species per genus, unfortunately, and we lack a basis for estimating the variation to be expected within a genus like *Peromyscus* or *Neotoma*. Arata's results also show that while genera having a complex phallus agree in having also a rather complete set of accessory glands, the peromyscine-neotomine genera differ from one another in their patterns of glandular loss.

Evidence from still other sources either does not support, or else flatly contradicts the hypothetical distinction of the peromyscine-neotomine group from the others. Structural details of the basihyal and posterior horn of the hyoid (Sprague, 1941), the accessory cusps and lophs of the molars (Hershkovitz, 1962), morphology of the stomach (Tullberg, 1899; Vorontsov, 1957), morphology of the auditory ossicles (Cockerell, *et al.*, 1914), and morphology of certain parts of the carotid arterial system (Guthrie, 1963) are cases in point. Structure of the basihyal and posterior horn of the hyoid, however, does not seem to be a reliable indicator of relation-ships of genera, judging from Sprague's results on the subgenera of *Peromyscus*. Morphology of the stomach and remainder of the digestive tract seems to be correlated to some extent with diet. Independent evolution of similar protective fundic structures in the stomachs of insectivorous muroids (Horner, *et al.,* 1965) is an apt illustration of parallel evolution. Molar structures in muroids vary greatly in populations of some living species, and to judge from the fossil record, have varied greatly within species since the Oligocene. Their morphology is probably also correlated with diet and subject to rapid parallel evolution. It is more difficult to assess the possible values of auditory ossicles and carotid arteries in estimating intergeneric relationships, especially because we do not understand the selective value of certain conditions over others.

Whether the peromyscine-neotomine group is a natural, endemic American group or not cannot be answered definitely on the basis of present anatomical evidence. While anatomical data on the phallus cover a wide range of genera, the same cannot be said for data on any of the other systems or parts. It is possible that simplification of the phallus occurred more than once among the

12 genera in question, and that the simple condition is not homologus in all. Unfortunately, it is difficult to detect parallelism or convergence in a structure or organ system which is undergoing evolutionary simplification.

The most recent subgeneric grouping of species of *Peromyscus* (Hooper and Musser, 1964*b*) is based heavily on morphology of the skin, skull, and phallus. Dental characters are not given as much weight as in some previous classifications. Available morphological information from other systems is sparse. Rinker's (1963) myological study is not in conflict with the Hooper and Musser classification. Studies of the carotid arteries, the male accessory reproductive glands, and the digestive tract, among other systems and structures, have thus far not covered many species.

In conclusion, I should like to suggest a few of many possible lines of future morphological investigation:

1) Functional morphological analysis in practically all systems, especially in the nervous system and sense organs where the results might have immediate relevance to behavioral studies.

2) Further analysis of the carotid arteries, and perhaps other aspects of the arterial system, both among the subgenera of *Peromyscus,* and among different genera of muroids.

3) A survey of the anatomy of the hyoid apparatus in the Muroidea, with particular reference to the anterior horn and its musculature.

4) Further studies of the male reproductive tract, with particular reference to the male accessory reproductive glands in several different species of a genus, and to possible functional relationships of anatomical features of male and female tracts in the same species.

5) Further quantitative studies on the skeletal musculature in different muroid genera. Further investigation at the subgeneric level in *Peromyscus* is not likely to be very productive (Rinker, 1963).

6) Further studies of the digestive tract, especially variation of the stomach in different species of *Peromyscus.*

Literature Cited

ARATA, A. A. 1964. The anatomy and taxonomic significance of the male accessory reproductive glands of muroid rodents. Bull. Florida State Mus., Biol. Sci., 9:1–42.

ARATA, A. A., N. C. NEGUS, AND M. S. DOWNS. 1965. Histology, development, and individual variation of complex muroid bacula. Tulane Studies in Zool., 12:51–64.

BADER, R. S. 1959. Dental patterns in *Peromyscus* of Florida. Jour. Mamm., 40:600–602.

BLAIR, W. F. 1942. Systematic relationships of *Peromyscus* and several related genera as shown by the baculum. Jour. Mamm., 23:196–204.

BURT, W. H. 1960. Bacula of North American mammals. Misc. Publ. Mus. Zool. Univ. Mich., 113:1–76, Pls. I–XXV.

CLARK, W. K. 1953. The baculum in the taxonomy of *Peromyscus boylei* and *P. pectoralis*. Jour. Mamm., 34:189–192.

COCKERELL, T. D. A., L. I. MILLER, AND M. PRINTZ. 1914. The auditory ossicles of American rodents. Bull. Amer. Mus. Nat. Hist., 33:347–380.

DICE, L. R. 1940. Ecologic and genetic variability within species of *Peromyscus*. Amer. Nat., 74:212–221.

ELEFTHERIOU, B. E., AND A. J. ZOLOVICK. 1965. The forebrain of the deermouse in stereotaxic coordinates. Kansas State Univ. Agric. Experiment Station, Tech. Bull., 146:1–4, Figs. 1–29.

GOODPASTER, W. W., AND D. F. HOFFMEISTER. 1954. Life history of the golden mouse, *Peromyscus nuttalli*, in Kentucky. Jour. Mamm., 35:16–27.

GUTHRIE, D. A. 1963. The carotid circulation in the rodentia. Bull. Mus. Comp. Zool., 128:455–481.

HALL, E. R., AND K. R. KELSON. 1959. The mammals of North America. New York: Ronald Press, 1162 pp.

HAMILTON, W. J., JR. 1942. The buccal pouch of *Peromyscus*. Jour. Mamm., 23:449–450.

HAYNE, D. W. 1950. Reliability of laboratory-bred stocks as samples of wild populations, as shown in a study of the variation of *Peromyscus polionotus* in parts of Florida and Alabama. Contrib. Lab. Vert. Biol. Univ. Mich., 46:1–56.

HERSHKOVITZ, P. 1962. Evolution of Neotropical cricetine rodents (Muridae) with special reference to the phyllotine group. Fieldiana: Zool., 46: 1–524.

HILL, J. E. 1935. The cranial foramina in rodents. Jour. Mamm., 16:121–129.

HOFFMEISTER, D. F. 1951. A taxonomic and evolutionary study of the piñon mouse, *Peromyscus truei*. Illinois Biol. Monogr., 21(4):i–x + 1–104.

HOOPER, E. T. 1957. Dental patterns in mice of the genus *Peromyscus*. Misc. Publ. Mus. Zool. Univ. Mich., 99:1–59.

———— 1958. The male phallus in mice of the genus *Peromyscus*. *Ibid.*, 105:1–24, Pls. I–XIV.

HOOPER, E. T., AND G. G. MUSSER. 1964a. The glans penis in Neotropical cricetines (Family Muridae) with comments on classification of muroid rodents. Misc. Publ. Mus. Zool. Univ. Mich., 123:1–57.

———— 1964b. Notes on classification of the rodent genus *Peromyscus*. Occ. Papers Mus. Zool. Univ. Mich., 635:1–13.

HORNER, B. E. 1954. Arboreal adaptations of *Peromyscus*, with special reference to use of the tail. Contrib. Lab. Vert. Biol. Univ. Mich., 61:1–85.

146   Klingener

HORNER, B. E., J. M. TAYLOR, AND H. A. PADYKULA. 1965. Food habits and gastric morphology of the grasshopper mouse. Jour. Mamm., 45: 513–535.

HOWELL, A. B. 1926. Anatomy of the wood rat. Monogr. Amer. Soc. Mamm., No. 1. Baltimore: Williams and Wilkins, 225 pp.

JAMES, G. T. 1963. Paleontology and nonmarine stratigraphy of the Cuyama Valley Badlands, California. Part I. Geology, faunal interpretations, and systematic descriptions of Chiroptera, Insectivora, and Rodentia. Univ. Calif. Publ. Geol. Sci., 45:i–iv + 1–171.

KING, J. A. 1965. Body, brain, and lens weights of *Peromyscus*. Zool. Jahrb. Anat., 82:177–188.

KING, J. A., AND B. E. ELEFTHERIOU. 1960. Differential growth in the skulls of two subspecies of deermice. Growth, 24:179–192.

LAWLOR, T. E. 1965. The Yucatan deer mouse, *Peromyscus yucatanicus*. Univ. Kans. Publ., Mus. Nat. Hist., 16:421–438.

MANVILLE, R. H. 1961. The entepicondylar foramen and *Ochrotomys*. Jour. Mamm., 42:103–104.

MUSSER, G. G., AND V. H. SHOEMAKER. 1965. Oxygen consumption and body temperature in relation to ambient temperature in the Mexican deer mice, *Peromyscus thomasi* and *P. megalops*. Occ. Papers Mus. Zool. Univ. Mich., 643:1–15.

OSGOOD, W. H. 1909. Revision of the mice of the American genus *Peromyscus*. N. Amer. Fauna, 28:1–285.

PILLERI, G. 1960. Materialen zur vergleichenden Anatomie des Gehirns der Myomorpha. Acta Anat., Suppl. 40 = 1 ad Vol. 42:69–88.

RINKER, G. C. 1954. The comparative myology of the mammalian genera *Sigmodon, Oryzomys, Neotoma,* and *Peromyscus* (Cricetinae), with remarks on their intergeneric relationships. Misc. Publ. Mus. Zool. Univ. Mich., 83:1–124, Figs. 1–18.

——— 1960. The entepicondylar foramen in *Peromyscus*. Jour. Mamm., 41:276.

——— 1963. A comparative myological study of three subgenera of *Peromyscus*. Occ. Papers Mus. Zool. Univ. Mich., 632:1–18.

ROSS, M. D. 1962. Auditory pathway of the epileptic waltzing mouse I. A comparison of the acoustic pathways of the normal mouse with those of the totally deaf epileptic waltzer. Jour. Comp. Neurol., 119:317–339.

——— 1965. The auditory pathway of the epileptic waltzing mouse II. Partially deaf mice. *Ibid.*, 125:141–163.

SENSENIG, E. C. 1943. The origin of the vertebral column in the deer mouse, *Peromyscus maniculatus rufinus*. Anat. Rec., 86:123–141.

SHARMA, D. R., AND S. SIVARAM. 1959. On the hyoid region of the Indian gerbils. Mammalia, 23:149–167.

SPRAGUE, J. M. 1941. A study of the hyoid apparatus of the Cricetinae. Jour. Mamm., 22:296–310.

——— 1942. The hyoid apparatus of *Neotoma*. *Ibid.*, 23:405–411.

STAINS, H. J. 1959. Use of the calcaneum in studies of taxonomy and food habits. Jour. Mamm., 40:392–401.

TAMSITT, J. R. 1958. The baculum of the *Peromyscus truei* species group. Jour. Mamm., 39:598–599.

TULLBERG, T. 1899. Ueber das System der Nagethiere, eine phylogenetische Studie. Nova Acta Reg. Soc. Sci. Upsaliensis, ser. 3, 18:i–v + 1–514 + A1–A18.

VAN VALEN, L. 1962. Growth fields in the dentition of *Peromyscus*. Evolution, 16:272–277.

VORONTSOV, N. N. 1957. Structure of the stomach and relative development of parts of the intestine in Cricetinae (Rodentia, Mammalia) of the Palaearctic and the New World. Doklady Akad. Nauk SSSR, 117: 526–529. (In Russian).

6

ONTOGENY

JAMES N. LAYNE

Introduction

A S IS TRUE OF ANY GROUP of organisms, knowledge of the ontogenetic patterns of mice of the genus *Peromyscus* is essential for a thorough understanding of many other aspects of their biology. Growth and development in *Peromyscus* have been investigated by numerous authors; but knowledge of this aspect of the life history of the genus, although presently exceeding that of many other smaller North American mammals, is still far from complete. Data pertaining to various aspects of growth and development are presently available for only 10 of the 58 species in the genus. The information for several of these species is superficial, and even in the case of those that have been better studied many important details are still lacking. Furthermore, only species whose distribution includes the United States and Canada have thus far been investigated, although the majority of the members of this genus are restricted to Mexico or Central America.

The literature on peromyscan growth and development is widely scattered, and there have been few previous attempts at a comparative analysis of ontogenetic patterns of the genus. Svihla (1932) carried out the first extensive study of reproductive patterns of *Peromyscus* and presented comparative data on neonatal size and condition and several early postnatal developmental events in five species representing the subgenera *Peromyscus* and *Haplomylomys*. McCabe and Blanchard (1950) analyzed in considerable detail the growth and development of *P. californicus, P. truei,* and *P. maniculatus*. King (1961*b*) reviewed some aspects of development in connection with behavioral evolution, and Layne (1966) presented a tabulation of selected developmental and growth characteristics of nine species in three subgenera. Growth and development of *Baiomys* and *Ochrotomys*, which in earlier classifications were included in *Peromyscus*, have been described by Blair (1941), Goodpaster and Hoffmeister (1954), and Layne (1960).

148

This paper summarizes the literature on morphological and behavioral development and growth of *Peromyscus* and provides additional information for several species, including *P. megalops* and *P. thomasi*, which have not been studied previously. These are the first exclusively Mexican species for which growth and developmental data have been obtained and are further noteworthy in being among the largest members of the genus. A primary objective of this review is to obtain some indication of the extent to which growth and developmental patterns and associated aspects of reproduction reflect ecological and phylogenetic relationships within this group of mammals. The species treated are listed below according to the classification of Hooper and Musser (1964). Subspecies represented, as designated in original papers or determined on the basis of locality from which stocks were collected, are given in parentheses following the specific name. Subspecific terminology follows Hall and Kelson (1959).

SUBGENUS *PEROMYSCUS*
MANICULATUS GROUP:
   P. polionotus (lucubrans, rhoadsi, subgriseus).
   P. maniculatus (artemisiae, austerus, bairdi, blandus, gambeli, gracilis, nebrascensis, nebrascensis × mutant yellow *maniculatus, osgoodi, rubidus, rufinus, sonoriensis)*.
LEUCOPUS GROUP:
   P. leucopus (leucopus, noveboracensis).
   P. gossypinus (gossypinus).
CRINITUS GROUP:
   P. crinitus (pergracilis).
BOYLEI GROUP:
   P. boylei.
TRUEI GROUP:
   P. truei (truei, gilberti).
MEXICANUS GROUP:
   P. megalops (auritus).

SUBGENUS *HAPLOMYLOMYS*
   P. eremicus (eremicus).
   P. californicus (californicus, parasiticus).

SUBGENUS *PODOMYS*
   P. floridanus.

SUBGENUS *MEGADONTOMYS*
P. *thomasi* (*thomasi*).

Prenatal Development and Growth

Two brief notes by Ryder (1887*a*, 1887*b*) and unpublished studies by Smith (1939) and Laffoday (1957) appear to comprise the sum total of information on the prenatal phase of development and growth of *Peromyscus*.

Ryder (1887*a*) reported in *Mus, Rattus,* and *Peromyscus* the presence of a thickened band of decidua extending beyond the placenta and encircling the uterine horn. This he believed represented a vestige of a zonary placenta. He (1887*b*) also confirmed the occurrence of "inversion" of the ectoderm and entoderm in *Peromyscus,* as in other more advanced rodents, and described the blastocyst and its relationships to the uterus.

Smith (1939) and Laffoday (1957) studied the embryology of *P. polionotus.* The former described a series of early embryonic stages ranging from 18 to 116 hours estimated post-copulation age, while the latter presented a comparatively detailed account of the first 18 days of prenatal development.

Both authors used laboratory stocks derived from a population referable to *P. p. subgriseus.* The method of obtaining known-age embryos was essentially the same in both studies, thus the data may be considered at least grossly comparable. The occurrence of copulation in females paired with males was determined by the presence of spermatozoa in vaginal smears taken in early morning. Females in which sperm were detected were isolated and subsequently sacrificed at appropriate intervals. Ages of the embryos thus obtained were, with one exception, calculated from the time of finding sperm in the vaginal smear. As most copulations apparently occurred during the night, there may have been an interval of as much as 12 hours or more between actual copulation and the taking of the vaginal smear. One set of embryos obtained by Smith was aged from an observed copulation rather than a vaginal smear containing sperm. Smith also experimented unsuccessfully with artificial insemination in an effort to obtain embryos more precisely aged.

TABLE 1

SYNOPSIS OF PRENATAL GROWTH AND DEVELOPMENTAL STAGES OF *Peromyscus polionotus* BASED ON SMITH (1939) AND LAFFODAY (1957)

Age	No. of specimens	Developmental status
0	5	Recently ovulated ova in Fallopian tube; wide zona pellucida; corona radiata conspicuous; fertilization status not known, spermatozoa present at same level of Fallopian tube; maximum dimensions of ova (in microns) ; 62.0 × 46.5, 69.0 × 54.2, 62.0 × 54.2, 62.0 × 46.5, 62.0 × 50.0.
15 hours	4	Pronuclear stages in ampulla of Fallopian tube; pronuclei showing various degrees of fusion; polar body in perivitelline space; corona radiata disorganized.
18 hours*	2	Pronuclear stages in distal portion of ampulla; pronuclei close together near center of egg; cytoplasm granular; no zona pellucida visible; corona radiata dispersed, scattered coronal cells in lumen of Fallopian tube.
24 hours	3	Two-celled embryos in isthmus of Fallopian tube; cytoplasm granular; nuclei round, centrally located; zona pellucida still present; no evidence of corona radiata.
29 hours	3	Two-celled stages in isthmus; cytoplasm finely granular; zona pellucida persists but incomplete.
47 hours	3	Four-celled embryos in proximal segment of isthmus.
48 hours	3	Four-celled stages in isthmus; zona pellucida still evident.
60 hours	3	Four-celled stages but slightly advanced over 48-hour examples; maximum diameter, 77.5 microns.
64 hours	4	Six to eight-cell embryos; zona pellucida evident but broken in one.
72 hours	5	Eight-cell stages in lower end of Fallopian tube in one specimen; blastocysts in uterus of another; implantation begun in latter case, embryos in pits on antimesometrial side of uterine horn; trophoblast of 14–16 cells; inner cell mass of 8–10 cells; mean length and width of blastocysts, 77.5 × 62.0 microns, respectively.
82 hours	3	Morula stages, 12–20 cells; in intramural portion of Fallopian tube near juncture with uterine horn.
96 hours	4	Two-cell stage, 12–16-cell stage, morula with beginning of segmentation cavity, and late blastocyst (93.0 × 46.5 microns) ; all free in uterine lumen.
116 hours	3	Blastocysts, two in uterine horn and one in proximal end of Fallopian tube; embryos in uterine horn free in lumen, one in shallow pit in mucosa.

TABLE 1 (Continued)

Age	No. of specimens	Developmental status
5 days	2	Implanting embryos; implantation sites grossly visible as slightly swollen areas with increased vascularity (length, 3.0 mm; width, 2.5 mm); inner cell mass differentiated into egg cylinder extending about half way to floor of yolk cavity; ectoplacental cone, embryonic and extra-embryonic ectoderm differentiated; proamniotic cavity developing; total length of egg cylinder exclusive of ecto-placental cone, 108.5 microns.
6 days	3	Embryos deeply imbedded in uterine mucosa, lumen of latter reduced to narrow slit; uterine swelling 3.5 mm long, 3.0 mm wide; proamniotic cavity present as a narrow slit confined to embryonic ectoderm.
7 days	3	Egg-cylinders completely invaginated into yolk cavity; proamniotic cavity extends into extra-embryonic ectoderm as far as ectoplacental cone and more prominent than earlier; posterior amniotic fold developing; no definite mesodermal layer though scattered cells present; uterine swellings average 3.5×3.2 mm in length and width, respectively; egg cylinder 294.0×93.0 microns.
8 days	3	Ectoplacental, amniotic, and exocoelomic cavities present; distinct layer of mesoderm developed; posterior amniotic fold nearly in contact with anterior fold; foregut invagi-nation apparent; mean measurements of embryos, 620.0×201.5 microns.
9 days	4	Variable stages of development represented in series avail-able; earliest with amnion formed, exocoelom developing, no allantois evident; one with fully formed exocoelom, allantois present, notochord differentiated; one with larger allantois and paraxial mesoderm developed but no somites; size range of first three embryos, $775.0–861.0 \times 310.0–589.0$ microns.
10 days	4	Uterine swelling 5.5 mm in length and width; body of embryos U-shaped, length in fixative 2.86 mm; heart swelling evident beneath head process; mandibular arch formed; first pair of aortic arches developed; midgut open; notochord prominent; neural groove present; optic vesicles differentiated; allantoic stalk joins chorion.

TABLE 1 (Continued)

Age	No. of specimens	Developmental status
12 days	4	Uterine swellings average 5.5 × 6.0 mm; body of embryos strongly flexed into C-shape; limb buds evident; mesencephalon bulges prominently, brain in five-vesicle stage; auditory pit evident; mandibular and hyoid arches prominent; nasal pits present; third aortic arch formed; somites differentiating; giant cells numerous in decidua adjacent to embryo; mean weight of embryo, 0.0057 g.
13 days	4	Mean dimensions of uterine swellings, 7.6 × 7.5 mm; head at least one-half total length; tail evident; limb buds fan-shaped; eyes prominent, lens distinct, no lids; no development of pinna; nasolateral and nasomedial processes in contact with maxillary process but not completely fused; third, fourth, and sixth aortic arches intact; mean weight, 0.0281 g.
14 days	3	Uterine swellings average 8.5 × 7.5 mm; torsion of embryo less pronounced than earlier and body slightly less strongly flexed; slight umbilical hernia; small genital tubercle evident; metanephros represented by aggregation of mesenchymal cells posterior to mesonephros; no differentiation of skeletal elements; mean weight, 0.0458 g.
15 days	3	Uterine swellings average 9.2 × 8.2 mm; trunk relatively straight, head axis at right angles to trunk axis; pinna developed as short triangular projection; vibrissal follicles evident; slight evidence of formation of digits; Müllerian duct evident as short, solid cord; nasal septum and Meckel's cartilage being laid down as dense concentration of pre-cartilaginous mesenchyme; vertebrae, ribs, shoulder girdles outlined; mean weight, 0.0801 g.
16 days	4	Uterine swellings average 10.0 × 9.0 mm; vibrissal follicles more conspicuous than earlier, those of mystacial, supraorbital, and mental series evident; slight development of teats; digits deeply outlined though still fully fused; umbilical herniation more pronounced; cartilage present in cranial region; mean weight, 0.1335 g.
17 days	3	Uterine swellings average 12.5 × 9.0 mm; head less flexed than earlier; plantar tubercles evident; fore digits separate at tips, hind still fully webbed; hair follicles evident on various parts of body; eyelids not developed; pinnae turned down anteriorly; incisor tooth germs present; Müllerian ducts still unfused, genital system in indifferent stage; ossification centers present in cranial region; mean weight, 0.2115 g.

TABLE 1 (Continued)

Age	No. of specimens	Developmental status
18 days	4	Uterine swellings average 14.5 × 10.0 mm; appearance essentially similar to full term fetus; pinnae folded over but unsealed; eyelids undergoing formation, eyes half closed; umbilical hernia reduced; hind feet notched; enucleated erythrocytes present; Müllerian tubes fused; increased ossification of skull; tooth germs of molars, as well as of incisors, present; mean weight, 0.3851 g.

* Timed from observed copulation.

A brief summary of prenatal development and growth of *P. polionotus* based on the combined data of Smith and Laffoday is presented in Table 1. The descriptions of stages at 18, 29, 47, 64, 82, and 116 hours estimated age are from Smith, while the data for the remaining stages are based on Laffoday's study. The information in Table 1 on the earlier embryos is relatively complete, whereas external morphology is emphasized in the older stages and no attempt is made to more completely summarize the detailed description of organogenesis and other features of development presented in the original paper.

All embryos 64 hours or less in estimated post-copulation age were contained in the Fallopian tubes, and tubal embryos were found up to 116 hours. Embryos first occurred in the uterus at 72 hours, the data suggesting that most enter the uterine portion of the tract at 82–96 hours post copulation. The most advanced embryonic stage occurring in the Fallopian tubes was the blastocyst, but the majority of the embryos located here represented earlier developmental stages. Implantation takes place apparently between about 116 hours and 6 days, by which time uterine swellings are evident.

Laffoday compared the embryology of *polionotus* with other muroid rodents that have been studied, including *Mus, Rattus,* and *Mesocricetus,* and concluded that there were no fundamental differences. *P. polionotus* does, however, appear to depart from the other genera in certain relatively minor developmental features. According to Laffoday, the zona pellucida of the species of *Pero-*

TABLE 2

MEAN LITTER SIZES AND GESTATION PERIODS OF *Peromyscus*

Species and subspecies	Approx. adult wt. (g)	Mean litter size	Gestation Period Normal Mean	Normal Range	Lactating Mean	Lactating Range	Source
P. poliomotus lucubrans		2.8*	—	—	—	—	Carmon et al., 1963
P. p. rhoadsi	13	4.0*	—	—	—	—	Rand and Host, 1942
P. p. subgriseus	15	3.35* (a)	23.80 (a, b)	23–24 (a, b)	28.00 (a)	25–31 (a)	a. Laffoday, 1957; b. Smith, 1939
P. maniculatus artemisiae	21	4.43*	23.39	22–26	30.60	27–34	Svihla, 1932, 1934
P. m. bairdi	15	3.05*	—	23–27	—	—	Svihla, 1932
P. m. blandus		3.85*	23.57	22–25	25.33	23–29	Svihla, 1932
P. m. gambeli	19	3.20 (a) 5.00*+ (b)	—	—	—	—	a. Svihla, 1932; b. McCabe and Blanchard, 1950
P. m. osgoodi		4.59*	23.61	22–27	27.00	22–35	Svihla, 1932
P. m. rufinus		4.26*	23.55	23–24	26.57	23–32	Svihla, 1932
P. l. leucopus	22	3.43*	23.33	22–24	28.62	23–34	Svihla, 1932
P. l. noveboracensis	21	4.36*	23.00	22–25	30.00	23–37	Svihla, 1932
P. g. gossypinus	29	3.7*	23.10++	—	30.2++	(1 record)	Pournelle, 1952
P. crinitus pergracilis	14	3.0*	—	24–25	—	(1 record)	Egoscue, 1964
P. t. truei		2.84*	26.20	25–27	—	40 (1 record)	Svihla, 1932
P. t. gilberti	27	3.43*+	—	—	—	—	McCabe and Blanchard, 1950
P. megalops auritus	71	1.60*	—	21 (1 record)	—	—	Present study
P. e. eremicus	20	2.60*	—	—	—	—	Svihla, 1932
P. c. californicus	38	1.87*	23.60	21–24	—	—	Svihla, 1932
P. c. parasiticus		1.91*+	—	—	—	—	McCabe and Blanchard, 1950
P. floridanus	27	3.1+	—	—	—	—	Layne, 1966
P. t. thomasi	77	3.5*+	—	—	—	—	Musser, in litt.; present study

* Laboratory conceived.
+ Wild conceived.
++ Calculated from average maximum and minimum gestation periods based on absence and presence of sperm in two successive daily vaginal smears.

myscus could not be detected beyond the eight-celled stage, whereas in *Mus* it reportedly persists to the blastocyst. The ectoplacental cone also appears to begin to form somewhat earlier in *Peromyscus* than in *Mus, Rattus,* and *Mesocricetus.* The extent to which these differences are real or simply reflect the variation in preparative techniques or methods of aging employed in different studies is open to question. Perhaps of more significance than the other differences noted by Laffoday is the variation in shape of the ectoplacental cone among the several muroid rodents compared. *Peromyscus, Mus,* and *Rattus* have a tall, conical-shaped ectoplacental cone, whereas in *Mesocricetus* the structure is much flattened.

Length of Gestation

The normal gestation period in various species of *Peromyscus* ranges from 21 to 27 days, with an overall average, based on reported mean values for eight species and ten subspecies, of 23.47 days (Table 2). A noteworthy feature of these data is the absence of correlation between mean length of gestation and adult body size at either the specific or subspecific level. Thus, over a range of body size from 13 to 40 g the majority of the forms have a normal gestation period approximating 23 days. The only species that appear to depart from this value are *truei, eremicus,* and *crinitus,* the difference being at most about three days and bearing no relationship to body size. The ranges of normal gestation periods given in Table 2 indicate that, as in the case of means, there is no correlation between absolute or relative variability in length of gestation and body size.

The data suggest that length of gestation tends to be relatively constant in different subspecies within a species. For example, Svihla (1932) found no significant difference in mean length of gestation in several subspecies of *maniculatus,* or between *P. l. leucopus* and *P. l. noveboracensis.* Some of the relatively slight differences between species also appear to be correlated with categories at the subgeneric or species-group level. Thus, *polionotus* and *maniculatus* show rather close agreement as do *gossypinus* and *leucopus.* Though the differences are admittedly slight, the fact that the smaller species tend to have longer gestation periods when the reverse would be expected adds some weight to the possibility that the similarities do in fact reflect phyletic relationships. The subgenus *Haplomylomys* appears to be characterized by a relatively short gestation period. If the single value for *eremicus* is representative, this species has a distinctly shorter gestation than species of equivalent size in other subgenera, while the much larger *californicus* has a mean gestation equal to or less than that of many smaller species and a minimum value (21 days) matched only by *eremicus*.

Svihla (1932) studied individual variation in the gestation period of non-lactating females of *maniculatus* and *leucopus*. Successive gestation periods in a given individual often varied by one or two days. For example, six consecutive gestation periods in each of two females of *maniculatus* were 22, 24, 23, 22, 24, 23 and 24, 23, 24, 23, 24, 23 days. It may be significant that each of these series shows a pattern of alternating shorter and longer gestation periods that seems to have greater regularity than would be expected on the basis of chance alone.

Those species of *Peromyscus* experiencing post partum estrus exhibit prolongation of gestation, as a result of delayed implantation, when nursing a previous litter. The extent to which the gestation period is extended under such conditions averages two to seven days in most species, with a single, probably extreme, case of approximately two weeks in *truei* (Table 2). Svihla (1932) found variation in lengths of successive gestation periods in individual females of *leucopus* and *maniculatus* to be greater in lactating than in non-lactating mice.

Ectopic pregnancy is known to occur in the genus (Moore, 1929).

Litter Size

Mean litter size of species and subspecies of *Peromyscus* for which developmental and growth data are available is approximately 3.4, with a mean range of 1.6 in *megalops* to between 4 and 5 in some subspecies of *maniculatus* (Table 2). Data on litter sizes of other species in the genus are given by Asdell (1964). It should be emphasized that estimates of litter size in at least some instances must be regarded as only broadly representative of the particular taxon. Average litter size for a given species or subspecies is an approximation because much variation between and within natural populations may be expected as a result of the combined effects of genetic differences and the influence of numerous environmental factors. Furthermore, estimates of litter size based on laboratory stocks may not be comparable to those calculated from field data (Layne, 1966; McCabe and Blanchard, 1950). In this connection, the majority of the data on litter size given in Table 2 are derived from laboratory populations.

Despite the limitations noted, the present data on litter size in different stocks of *Peromyscus* may be assumed to be broadly representative and satisfactory for at least general comparisons. At the specific level, *polionotus, maniculatus, leucopus, gossypinus,* and *thomasi* are characterized by relatively large litters, while *megalops, eremicus,* and *californicus* have small litters. *P. crinitus, truei,* and *floridanus* occupy a more or less intermediate position in the series. These groupings do not closely reflect presumed taxonomic relationships. Litter size is not obviously correlated with adult body size, but there does appear to be an inverse correlation between litter size and neonatal weight, as will be discussed in a later section.

Judging from the present data, variation in litter sizes between subspecies of the same species may be as great as that between some species. In addition, there appears to be no obvious correlation between litter size and either adult or neonatal weight at the subspecific level.

Parturition

Observations on birth of young and associated aspects of parturition have been published for three species: *P. maniculatus* (Clark, 1937; Svihla, 1932), *P. gossypinus* (Pournelle, 1952), and *P. polionotus* (Laffoday, 1957). I have also obtained some fragmentary information relating to parturition in *P. floridanus, P. leucopus, P. megalops,* and *P. thomasi.* Though the information on this phase of the reproductive biology of the genus is meager indeed, it nevertheless exceeds that available for most other comparable taxa of wild mammals.

PREPARTURITIONAL BEHAVIOR.—Behavior of three female *gossypinus* preceding the onset of parturition was described by Pournelle (1952). One individual confined to an aquarium for observation was active and restless during the night preceding birth. She frequently stretched her body full length along the bottom of the container or vertically against the wall. In some cases fetal movements were detected while the female was stretching. The following morning the mouse became quiet and appeared to sleep continuously in a corner of the aquarium for about 45 minutes before the onset of parturition. A second female was less active than the first and slept more prior to the beginning of parturition. This mouse also stretched along the floor of the container at intervals, but was not seen to stretch vertically in the manner of the first individual. The third female was restless on the morning of parturition, stretching the body frequently and rolling from side to side on the abdomen. Marked fetal movements were noted about two hours prior to the birth of the first young.

A single *polionotus* observed shortly before parturition by Laffoday (1957) rested on her back, at which time movements of the fetuses could be seen. At intervals this female also stretched her body by keeping the hind feet in place and pulling with the fore feet. During the stretching, the sides of the abdomen were depressed toward the midline.

Several near-term pregnant *floridanus* that I have observed appeared to be less active and more sluggish than nonpregnant or early pregnant animals. When placed in a small container for observation they tended to huddle quietly in one place and were reticent to move even when gently prodded with a pencil or finger. One individual observed almost continuously for four or five hours before birth sat quietly hunched up in the jar until a short time before partum, when she became restless. She then began to move around, groom frequently, and at intervals stretched her body vertically against the sides of the jar.

PARTURITION.—During actual delivery of the young, female *Peromyscus* typically seem to assume a quadrupedal or bipedal, hunched over, crouching position. Such a posture has been observed in *P. m. artemisiae* (Svihla, 1932), *P. m. nebrascensis* × mutant yellow *P. maniculatus* (Clark, 1937), *P. m. nebrascensis* (Clark, 1937), *P. gossypinus* (Pournelle, 1952), and *P. polionotus* (Laffoday, 1957). In some instances, as described by Clark for a female *P. m. nebrascensis* and Pournelle for a *P. gossypinus,* the mouse may rise up on the hind feet to a more erect position while the young is being expulsed.

Abdominal contractions often precede the appearance of the young, and individuals of *P. m. artemisiae, P. gossypinus,* and *P. polionotus* have been observed to pull the skin around the vulva to the side, press the lower abdomen, or scratch the vulvar area with the forepaws prior to the beginning of delivery.

In a number of the deliveries observed the female used her teeth or fore feet to aid the passage of the young from the birth canal. The female *P. m. artemisiae* observed by Svihla (1932) assisted the entire delivery of the young by gently pulling on them with her teeth. Similar behavior in the birth of several litters of *gossypinus* was noted by Pournelle. Clark (1937) witnessed a *P. m. nebrascensis* gently pulling on the head of a young, that had emerged from the vulva, with her front feet. After several attempts she succeeded in freeing it from the birth canal. The female *polionotus* observed by Laffoday (1957) assisted the birth of the three young in the litter in a similar manner. A third type of maternal assistance in the birth process was noted by Clark (1937) in hybrid female, *P. m. nebrascensis* × mutant yellow *maniculatus.* When the head of the young appeared, the female placed her front feet on each side of the vulva and began to lick and prod the young with her snout. Another *P. m. nebrascensis* did not aid the birth of the young until the latter was almost free, at which time she appeared to reach down and seize the umbilical cord with her teeth, pulling both young and placenta clear.

Under abnormal circumstances birth may proceed rapidly without being accompanied by the typical postural responses or maternal aid described above. Svihla (1932), for example, disturbed a *P. m. blandus* in her nest while she was in the process of giving birth. Two young and associated placentas were expelled almost simultaneously in the time taken for the female to leave the nest and move to a corner of the cage about 5 inches away.

In keeping with the generally similar anterior and posterior size and proportions of the neonate, both breech and head presentations may occur in the births of individuals of the same litter. It appears, however, that breech delivery tends to be more prevalent. In 23 recorded cases involving three species, *P. maniculatus, P. gossypinus,* and *P. polionotous,* 14 births involved breech presentation and nine cephalic (Svihla, 1932; Laffoday, 1957; Pournelle, 1952). In *gossypinus* alone, 11 (64.7 per cent) of 17 births involved breech delivery.

The placenta and fetal membranes may be discharged almost immediately after delivery of the young or not until a few minutes afterwards. The placentas and umbilical cords of two young *gossypinus* were 10 and 11 mm in diameter and 14 and 15 mm in length, respectively (Pournelle, 1952).

All published descriptions of normal births in the genus indicate that the placenta is consumed soon after it has been discharged. Females of *P. gossypinus* observed invariably consumed the collapsed fetal membranes before the placenta proper (Pournelle, 1952).

A single observation on *floridanus* shows that this species also consumes the afterbirth. A female examined shortly after birth of a young was found eating the placenta while standing on the hind feet in a hunched over position. It is likely that the female had given birth in this position. Female *P. gossypinus* and *polionotus* have been observed aiding the passage of the placenta by pulling on it with teeth and forepaws, respectively, in the same manner as during birth of the young. The umbilical cord is either broken by stretching when the placenta is being eaten or at the time of birth (*P. m. artemisiae, P. m. nebrascensis, P. gossypinus*) or is bitten off while the placenta is being consumed (*P. m. nebrascensis* hybrid stock, *P. gossypinus*). Laffoday (1957) was unable to determine the manner in which the umbilical cord was disposed of in *polionotus*. A female *P. gossypinus* observed by Pournelle nibbled at the stalk of the umbilical cord remaining attached to the neonate after the placenta was eaten, and I have observed a similar instance in *floridanus*. These few observations do not appear to be indicative of any consistent difference between species or subspecies in the manner of disposal of the umbilical cord.

BEHAVIOR BETWEEN DELIVERIES.—In the intervals between births a female *P. m. artemisiae* rested and stretched her body in the manner noted above under preparturitional behavior (Svihla, 1932). One female *P. m. nebrascensis* observed by Clark (1937) groomed herself and rested with her front feet on her mate; her eyes were closed between births. Another female of the same stock ran about her cage as if in pain, and chewed on the bottom following birth of her second young. She did not give birth to a third individual until the following day. A *polionotus* groomed and stretched frequently in the interim between births (Laffoday, 1957). A *floridanus* that I observed cleaned her vulva and groomed or rested quietly in the corner of the cage. Despite only slight to moderate activity, her breathing rate was distinctly more rapid than normal.

RESPONSE TO NEONATES.—Available accounts indicate variability in the responses of female *Peromyscus* to their newly born young. A female *P. m. artemisiae* began to clean the young while delivery was still in progress (Svihla, 1932). Two of the three females of *maniculatus* studied by Clark (1937) cleaned the young during or soon after birth, while the third ignored her litter for an extended period, perhaps as a result of being placed into a new cage. Females of *P. gossypinus* observed by Pournelle (1952) cleaned each young following birth, while a *polionotus* did not wash the young until the last in the litter had been born (Laffoday, 1957). The conditions under which the mice were housed for observation may have influenced their behavior toward the neonates.

As in the case of other mammals, female *Peromyscus* sometimes destroy, and often consume, recently born young. This behavior often follows disturbance of the animal during parturition or shortly afterwards.

The sparse data on parturitional behavior in the genus *Peromyscus* suggest the possibility of specific and subspecific differences in such aspects as position at birth, use of teeth or forepaws to assist birth, and tendency to clean young. The data are far too few, however, to allow conclusions as to whether such variation is individual or indicative of actual differences in populations.

TIMING OF PARTURITION.—Birth in *Peromyscus* appears to occur typically during the daylight hours, the adaptive advantage of such timing is obvious for a nocturnally active rodent. The births of *P. m. artemisiae* and *P. m. nebrascensis* reported by Svihla and Clark occurred in the morning and at 4:00 P. M., respectively. Of 119 births of *P. m. bairdi* recorded by C. R. Terman, 98 were born during the period of light and 21 during the dark (King, 1963). Sixty-one per cent of 52 litters were born between 8:00 A. M. and noon, 22 per cent between noon and 4:30 P. M. and 17 per cent after 4:30 P. M. Seventeen (65.4 per cent) of 26 litters of *gossypinus* were born between 6:00 A. M. and noon, and 4 (15.4 per cent) between noon and 6:00 P. M. (Pournelle, 1952). No birth was definitely known to have occurred during the nocturnal period. Nine of 14 litters of *polionotus* were born between 8:00 A. M. and noon (Laffoday, 1957).

In my own files are several records of births in *P. floridanus, P. megalops, P. thomasi,* and *P. leucopus* in which the time of parturition was established within general limits. Two litters of *floridanus* were born between noon and 2:10 P. M. and between 5:30 and 9:00 P. M., respectively. The latter case may not be normal, as the female was handled shortly before parturition and the disturbance might have induced premature delivery. Births of single litters each of *megalops, thomasi,* and *leucopus* occurred between 6:00 and 10:00 A. M., 7:30 A. M. and 6:00 P. M., and 9:00 A. M. and noon, respectively.

DURATION OF PARTURITION.—The length of time taken for the birth of a litter varies widely, although the factors responsible for such variation are not clear. A litter of three *P. m. artemisiae* was born in an hour, while in other cases births of members of a litter extended over several hours (Svihla, 1932). Seven young in a litter of *P. m. nebrascensis* × mutant yellow *P. maniculatus* were born at intervals of 5 to over 25 minutes, while in a litter of three *P. m. osgoodi* one young was born at least 15 hours after the other two (Clark, 1937). The latter case is of interest in connection with the possibility of a tendency toward diurnal parturition in the genus, as it suggests that if birth of a litter begins late in the day its completion may be delayed until the following morning. Total parturition times recorded for litters of *P. gossypinus* by Pournelle (1952) were 36, 45, 64, 67, and 55 minutes. In the single parturition observed in *P. polionotus,* three young were born over a 45-minute period, with intervals of 13 and 32 minutes between the birth of the first and second and the second and third, respectively (Laffoday, 1957).

AGGRESSIVENESS OF FEMALES.—Observations on aggression of female *Peromyscus* with litters or in advanced pregnancy toward mates or other conspecifics are available for several species and subspecies. Although individual females of the same population may vary appreciably in their aggressiveness, depending perhaps on genetic as well as experimental and environmental factors such as length of time housed with other individuals, etc.; the data provide some evidence of differences between species and perhaps subspecies as well.

Clark (1937) noted that females of *P. m. nebrascensis* and of hybrid *nebrascensis* showed no antagonism toward males housed with them at the time of parturition. Eisenberg (1962) and Svihla (1932), however, found that females of other subspecies of *maniculatus* excluded mates or older young from the nest before birth and for a day or two afterwards. Horner (1947) described cases of paternal care in *P. m. bairdi* and *P. m. gracilis* in which males were in some cases observed washing young only one day of age. However, she also noted a male *bairdi* interfering with the birth of young on one occasion. Howard (1949) found male *P. m. bairdi* in nest boxes with females and young in natural populations.

Although Horner (1947) observed parental care in *P. leucopus*, there is evidence that in the wild, females with recent litters are antagonistic to males (Nicholson, 1941). A male *P. gossypinus* placed in an aquarium with a female with a new litter was promptly attacked (Pournelle, 1952), and female *floridanus* in advanced pregnancy or with recently born young are also aggressive toward other individuals (Layne, 1966). According to Eisenberg (1963), females of *P. crinitus* are more intolerant of males than are females of *maniculatus* during parturition.

Females of the subgenus *Haplomylomys* may be less aggressive than those of other groups mentioned. Eisenberg (1962, 1963) and McCabe and Blanchard (1950) found females of *californicus* to be highly tolerant of young of a previous litter and the male during and after birth of a new litter, and females of *eremicus* show similar tendencies (Eisenberg, 1963). In other aspects of its behavior, *californicus* also gives evidence of a more highly developed social system than other *Peromyscus* thus far studied (Eisenberg, 1962, 1963).

MORTALITY AT PARTUM.—Death of females apparently resulting from parturition has been recorded in *P. leucopus* (Seton, 1920) and *P. m. bairdi* (Svihla, 1932). Watson (1942) and Liu (1953) cited difficulty in parturition caused by large hybrid fetuses as one of the factors causing reduced fertility in female *P. polionotus* and male *P. maniculatus* crosses. Dawson (1965), however, found no conclusive evidence of this.

TABLE 3

ABSOLUTE AND RELATIVE BODY WEIGHTS (GRAMS) OF NEWBORN *Peromyscus*

Species and subspecies	N	Mean	Range	Per cent of adult weight	Source
P. polionotus rhoadsi	23	±1.3	—	10	Rand and Host, 1942 (estimated from graph)
P. p. subgriseus	57	1.61	1.1–2.2	11	Laffoday, 1957
P. maniculatus artemisiae	70 (a)	1.71 (a) 1.80 (b)	1.3–2.2 (a)	8	a. Svihla, 1932; b. Svihla, 1934
P. m. bairdi	39	1.67 (a) 1.62 (b)	1.1–2.3 (a, b)	10	a. Svihla, 1932; b. Svihla, 1935
P. m. blandus	22	1.80	1.5–2.1	—	Svihla, 1932
P. m. gambeli	62	1.40	—	7	McCabe and Blanchard, 1950
P. m. osgoodi	121	1.72	0.8–2.0	—	Svihla, 1932
P. m. rufinus	44	1.67	1.3–2.0	—	Svihla, 1932
P. oreas	25+	1.63	1.1–2.3	—	Svihla, 1936
P. l. leucopus	16	1.85	1.5–2.4	8	Svihla, 1932
P. l. noveboracensis	98	1.87	1.4–2.4	9	Svihla, 1932
P. g. gossypinus	45	2.19	—	8	Pournelle, 1952
P. g. palmarius	—	—	1.5–1.7	5	Rand and Host, 1942
P. crinitus pergracilis	7	2.2	1.8–2.6	16	Egoscue, 1964
P. t. truei	21	2.31	1.7–3.0	—	Svihla, 1932
P. t. gilberti	41	2.34	—		McCabe and Blanchard, 1950
P. megalops auritus	3	3.9	2.9–5.0	5	Present study
P. e. eremicus	9	2.54	2.1–2.9	13	Svihla, 1932
P. c. californicus	19	4.92	3.6–5.8	12	Svihla, 1932
P. c. parasiticus	46	4.31	—		McCabe and Blanchard, 1950
P. floridanus	15	2.4	1.9–2.9	9	Layne, 1966
P. t. thomasi	9	4.5	3.0–5.9	6	Present study*

* Estimate based on weights of preserved litter of 4 at 2 days of age and two captive conceived litters of 2 and 3 young.

Condition of the Neonate

SIZE.—Mean weights of newborn young of various species and subspecies range from approximately 1.3 to 4.9 g, with an overall mean for the genus of about 2.2 g (Table 3). In those species and subspecies for which adult weights are given in Table 2, the neonate averages approximately 9 per cent of adult weight, with extremes of 4 and 16 per cent. Those species (*P. crinitus, P. eremicus,* and *P. californicus*) having neonates of largest relative weight are characteristic of xeric environments, and those (*P. thomasi* and *P. megalops*) with the smallest are the largest members of the genus for which information on the neonate is available.

Where ranges or other measures of variability have been given, considerable variability in neonatal weight within a population is evident. Sex differences in weight of newborn *Peromyscus* have thus far been reported only in *P. polionotus* (Carmon *et al.,* 1963). In this study, female neonates were slightly heavier than males. Individual variation in weight within litters is often pronounced, in some cases nearly approximating the total sample range. For example, Svihla (1932) recorded ranges from 1.4 to 2.0 g in a litter of five *P. m. rufinus* and from 1.1 to 1.6 g in a litter of four *P. m. bairdi.* He found no consistent trends in the average weights of individual young in small and large litters within a given subspecies or in the young produced by a single female. Maximum differences in weights of young in six litters of *P. floridanus* ranged from 0.1 to 0.3 g (mean, 0.18 g), but showed no correlation with litter sizes of 2 or 3 (unpublished observations). However, weights of neonates in three litters of *P. thomasi* were correlated with litter size, averaging 3.4 (at two days of age), 4.9, and 5.8 in litters of 4, 3, and 2, respectively. Carmon *et al.* (1963) calculated the relative contribution of family and individual components of variation in birth weight of *polionotus*. Intra-litter variation accounted for 33.23 per cent and familial variation for 66.76 per cent of the total variation.

Although subspecies within a species may vary in mean birth weight, such differences have not been shown to be statistically significant in the cases studied (Svihla, 1932).

The relationships between neonatal and adult weight and neonatal weight and mean litter size are shown in Figures 1 and 2. The scatter diagram of birth weight plotted against adult weight in-

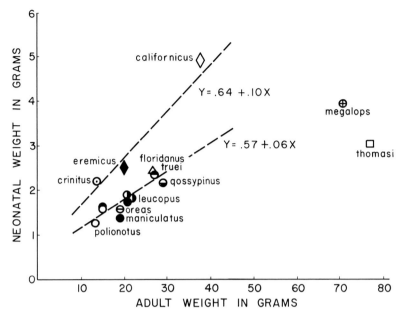

FIG. 1. Relationship between adult weights and neonatal weights of species and subspecies of *Peromyscus*. All subspecies of a given species are denoted by the same symbol but are not individually identified.

dicates that these two variates are positively correlated (r = +0.69, P < 0.01). The equation of the regression line fitted to these points by the method of least squares is: Y = 1.4 + 0.03X. Inspection of the distribution of points suggests that they tend to be segregated into three groups: (1) a main, rather narrowly linear cluster representing the species *P. polionotus, P. oreas, P. maniculatus, P. leucopus, P. truei, P. floridanus,* and *P. gossypinus;* (2) a series of three points representing *P. crinitus, P. eremicus,* and *P. californicus* displaced considerably above the *polionotus-gossypinus* sequence; and (3) a pair of values for the largest species, *P. megalops* and *P. thomasi* lying well to the right of the others. One interpretation of this pattern of points is that birth weight and adult size are related by different functions in the *polionotus-gossypinus* and *crinitus-californicus* series. Based on this assumption, separate regression lines are fitted to the two sets of points in Figure 1. The equation for the *crinitus-californicus* regression is:

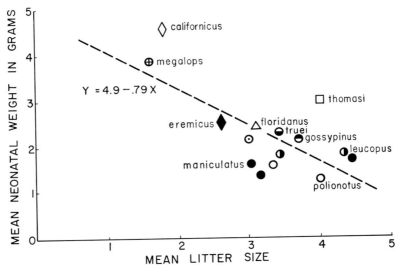

FIG. 2. Relationship between neonatal weight and litter size in *Peromyscus*. Different subspecies of a species are denoted by the same symbol but are not individually identified.

$Y = 0.64 + 0.10X$ and for that of the *polionotus-gossypinus* group: $Y = 0.57 + 0.06X$. The respective correlation coefficients are $+0.98$ ($P < 0.05$) and $+0.90$ ($P < 0.01$).

It remains to account for the position of the points for *megalops* and *thomasi*. They obviously do not lie on the regression line extrapolated from either cluster of points for smaller species, and it might thus be inferred that while the relationship between neonatal weight and adult weight over a range of approximately 13 to 40 g is essentially rectilinear, it is curvilinear over the full size range represented. It is clear that further data on neonatal size, particularly of species in the 40- to 70-g weight range, are needed for a more satisfactory analysis of the relationships between neonatal and adult size in this genus.

Neonatal weight and litter size appear to be inversely related, the correlation coefficient being -0.70 ($P < 0.01$). The equation of the regression line fitted to these points is $Y = 4.9 - 0.79X$. The inverse relationship between birth weight and litter size has the effect of dampening variation in the absolute biomass of young

TABLE 4

MEASUREMENTS (IN MM) OF NEWBORN *Peromyscus*

Species and subspecies	Total length			Body length			Tail length			Hind foot			Source
	N	Mean	Range	N	Mean	Range	N	Mean	Range	N	Mean	Range	
P. maniculatus													
artemisiae	14+	45.57	38–57	14+	34.27	—	14+	11.30	9–13	14+	6.44	5.8–7.3	Svihla, 1934
P. m. bairdi	39	43.62	40–53		33.13	—	39	10.49	9–12	39	6.15	5.6–6.6	Svihla, 1935
P. m. gambeli	—	—	—		—	—	43	8.56	—	43	6.77	—	McCabe and Blanchard, 1950
P. oreas	25+	46.09	39–52	25+	33.82	—	25+	12.27	9–15	25+	6.50	5.6–7.1	Svihla, 1936
P. leucopus													
noveboracensis	1	47	—	1	35	—	1	12	—	1	7	—	Present study
P. g. gossypinus	45	47.16	—	45	36.18	—	45	10.98	—	45	6.53	—	Pournelle, 1952
P. truei gilberti	1 (a)	±41	—	1 (a)	±31	—	36 (b)	11.45	—	36 (b)	7.15	—	a. Hoffmeister, 1951 (full term fetus); b. McCabe and Blanchard, 1950
P. megalops													
auritus	2	55.0	55, 55	2	39.5	38.0, 41.5	2	15.5	13.5, 17.0	2	7.8	75, 80	Present study
P. crinitus													
pergracilis	—	—	—	—	—	—		12.75	12–14	—	7.5	7–8	Egoscue, 1964
P. californicus													
parasitcus	—	—	—	—	—	—	44	19.85	—	44	10.63	—	McCabe and Blanchard, 1950
P. floridanus	2	44.0	43, 45	2	31.5	31, 32	2	12.5	12, 13	2	8.2	8.0, 8.5	Layne, 1966
P. t. thomasi	3	62.6	61–65	3	45.3	43–48	3	17.3	14–18	3	8.7	8.5–9.0	Present study

produced by different species and subspecies of the genus. Based on the birth weights and litter sizes given in Table 2, the average litter weight of all species and subspecies combined is roughly 7 g, with extremes of about 5 g for *P. polionotus* and 12 g for *P. thomasi*. Litter weights of 9 of the 15 stocks available for comparison fall between 6 and 8 g. In terms of average adult weight, litter weight varies from about 9 per cent in *P. megalops* to about 47 per cent in *P. crinitus,* the mean for all species and subspecies being about 30 per cent. In view of the considerable variability in litter size encountered in different populations, the above values should be regarded only as approximations useful for general rather than detailed comparative purposes.

Data on measurements of newborn *Peromyscus* are fewer than those for weight. Information on standard body measurements presently available for ten species are summarized in Table 4. Where not given in the original publication, body length has been calculated by subtracting tail length from total length. Overall means for the genus calculated from the means of each species or subspecies included in Table 4 are as follows: total length, 48.2 mm; body length, 35.6 mm; tail length, 13.0 mm; and hind foot, 7.5 mm.

Neonatal length of tail and size of hind foot are most strongly correlated with birth weight, the respective correlation coefficients being $+0.74$ $(0.05 > P > 0.02)$ and $+0.75$ $(0.02 > P > 0.01)$. The respective correlation coefficients between total length and weight and between length of body and weight are $+0.65$ $(0.1 > P > 0.05)$ and $+0.60$ $(0.1 > P > 0.05)$. These results are somewhat unexpected, as it seems reasonable to suspect that size of hind foot and tail, being perhaps more likely to be influenced by selective factors, would not show as close a correlation to neonatal weight as total length or length of body, particularly the latter. However, total length is a more difficult measurement to take than that of tail or hind foot on newborn young; thus the relatively poor correlation of total length and length of body with weight suggested by the present analysis may result from inaccuracies in the available data.

Neonatal body proportions of nine species are compared in Table 5. The proportions are based on neonatal measurements given in Table 4 and mean adult measurements for the same taxon

TABLE 5

BODY PROPORTIONS OF NEONATAL AND ADULT *Peromyscus*

Species and subspecies	Neonatal size as percentage of adult measurement				Size as percentage of respective total length					
					Body		Tail		Hind foot	
	Total length	Body	Tail	Hind foot	Adult	Young	Adult	Young	Adult	Young
P. maniculatus artemisiae	27	38	14	33	52	75	48	25	11	14
P. m. bairdi	30	38	18	35	59	76	41	24	12	14
P. oreas	26	41	13	31	47	73	53	27	12	14
P. l. noveboracensis	27	36	15	33	55	74	45	26	12	15
P. g. gossypinus	26	35	15	28	58	77	42	23	13	14
P. t. gilberti	27	30	10	29	51	73	49	27	11	17
P. megalops	20	30	11	26	48	72	52	28	11	14
P. floridanus	25	32	16	34	55	72	45	28	13	19
P. californicus	—	—	16	39	45	—	55	—	11	—
P. thomasi	20	30	10	28	47	72	53	27	10	14

and specific population as far as possible. The sources of adult measurements for each species or subspecies are as follows: *P. m. artemisiae* (Svihla, 1934) ; *P. m. bairdi* (Svihla, 1935) ; *P. oreas* (Svihla, 1936) ; *P. l. noveboracensis* (Osgood, 1909; Massachusetts sample) ; *P. gossypinus* (Pournelle, 1952) ; *P. t. gilberti* (Hoffmeister, 1951) ; *P. megalops auritus* (Osgood, 1909) ; *P. floridanus* (Layne, 1966) ; *P. californicus* (Osgood, 1909; McCabe and Blanchard, 1950) ; *P. t. thomasi* (Musser, 1964).

Of the four linear dimensions considered, body and hind foot have the largest relative size as compared with corresponding adult measurements. Tail is relatively the smallest of the four body components and slightly more variable than the others. As in the case of weight, the neonates of the largest species are relatively small in linear dimensions. Variability in relative size of total length and other dimensions in the remaining species seems to show no obvious trends correlating with adult size, presumed taxonomic relationships, or ecological orientation. The apparent absence of such trends may reflect the inadequacy of the present data. There is a suggestion of some ecological correlation in relative size in total length and other dimensions in the case of the two subspecies of *P. maniculatus* for which data are available and the related *P. oreas*. *P. m. bairdi* is an open country ecotype, often nesting in fairly accessible situations such as beneath driftwood on beaches or under isolated piles of dry vegetation in bare fields. In keeping with the supposed advantages of relatively precocial development and growth in a primarily grassland form, neonatal *P. m. bairdi* appear to be relatively larger than those of forest stocks.

The data in the second part of Table 5 show that body proportions of the newborn young differ markedly from those of adults. The body is relatively larger and the hind foot slightly so, while the tail is proportionately much smaller. Relative size of hind foot at birth and in the adult are not strongly correlated. Whereas in the majority of the taxa represented, neonatal hind foot is from 1 to 3 per cent larger in relative size than in the adult, the difference is 6 per cent in *truei* and *floridanus*. Both of these species are inhabitants of dry, relatively open habitats; *floridanus*, at least, is highly terrestrial and occurs on very sandy soils (Layne, 1963). Possibly the relatively large neonatal hind foot in these species is adaptive for such environmental conditions.

In contrast to hind foot, tail and body proportions of neonates and adults are more clearly correlated. Thus, species or subspecies with relatively long tails as adults tend to have longer tails at birth. The correlation is, however, not exact, as species with similar adult body-tail proportions may have slightly different neonatal proportions. *P. floridanus* and *P. leucopus* exemplify this. Whether these slight discrepancies represent actual differences with consequent implications in connection with postnatal growth patterns or are simply a result of sampling error is open to question.

There are almost no other meristic data besides external measurements presently available for newborn *Peromyscus*. King and Eleftheriou (1960) give femur lengths, skull measurements, and brain weights of newborn young of two subspecies of *P. maniculatus,* and Hoffmeister (1951) gives the length (13.5 mm) of the head of a neonatal *P. truei.*

MORPHOLOGY.—Morphological development of the young at birth has been described in the following nine species of *Peromyscus*: *P. polionotus* (Laffoday, 1957; Rand and Host, 1942), *P. maniculatus* (McCabe and Blanchard, 1950; Svihla, 1932), *P. oreas* (Svihla, 1936), *P. leucopus* (Svihla, 1932), *P. truei* (McCabe and Blanchard, 1950; Svihla, 1932), *P. gossypinus* (Pournelle, 1952), *P. eremicus* (Svihla, 1932), *P. californicus* (McCabe and Blanchard, 1950; Svihla, 1932), and *P. floridanus* (Layne, 1966). In addition, I have observed neonates of *P. megalops* and *P. thomasi.*

The newborn young of all species thus far studied, with the exception of *crinitus, eremicus,* and *californicus,* are pinkish or flesh-colored above and below. *P. crinitus* has been described as being "lightly pigmented" on the dorsum (Egoscue, 1964), while the neonates of *eremicus* and *californicus* are heavily pigmented dorsally (McCabe and Blanchard, 1950; Svihla, 1932). Other species do not attain a similar stage of pigmentation, caused by a concentration of pigment in developing hair follicles, until a day or more following birth.

Two young in each of two litters of *P. megalops* that I observed agreed with the majority of the species in the genus in having no pigmentation on the dorsum, whereas a single, much larger neonate of a third litter was distinctly dusky above. Young of *P. thomasi* are reddish above and below and are not appreciably darker on the dorsum.

Mystacial vibrissae have been reported present at birth in all species studied. The longest vibrissae of newborn *maniculatus, truei,* and *californicus* averaged 2.70, 3.26, and 4.25 mm, respectively (McCabe and Blanchard, 1950) ; and in *floridanus* they range from 3 to 4 mm, extending to the middle of the eye region (Layne, 1966). The vibrissae of neonates of *leucopus* are relatively short, reaching only to about the anterior margin of the eye region; those of *megalops* extend to the middle of the eye, as is true also of a two-day-old litter of *thomasi* examined. I have noted carpal vibrissae on neonates of *megalops* and assume that these and the vibrissae other than the mystacials found in this genus are present at birth in other species as well, although they have not been specifically noted in published descriptions. The prominent development of vibrissae in newborn *Peromyscus* as well as other mammals suggests that these specialized tactile hairs may have an important function in the nursing behavior of the young.

With regard to general development of pelage, newborn *Peromyscus* appear hairless to the naked eye. In at least some species, however, hairs other than vibrissae may be present at birth. Tiny, sparsely distributed hairs are present on the dorsum of the body and more numerous short, bristly hairs occur on the chin and snout of newborn *floridanus* (Layne, 1966). Pournelle (1952) states that *gossypinus* is hairless except for vibrissae, and McCabe and Blanchard (1950) record a similar condition for *maniculatus, truei,* and *californicus,* with the exception of white hairs up to 0.5 mm on the chin of the latter. Svihla (1932), however, found that although *maniculatus* appears to be naked at birth, microscopic examination of skin sections shows that body hairs have already pushed above the skin surface. Bristles were present on the snout of newborn *megalops* examined, but no hairs were detected on the remainder of the body, even under magnification. I observed a similar condition in two litters of *leucopus* and one litter of *thomasi.*

Although the extent to which hairs are developed at birth probably varies within a species as a result of normal variation in gestation period or rates of prenatal development, the rather limited data presently available on this point suggest that there may also be species differences in this characteristic.

The eyes of newborn *Peromyscus* are covered with skin, although the dark iris ring and lens vesicle are discernible. The future line of separation of the lids may be indicated by a faint line of pigment, but there is no thickening of the lids.

In all species thus far observed, including *P. megalops* and *P. thomasi,* the pinna is folded over the site of the external auditory meatus and fused to the skin of the head.

The digits are fused throughout their length, and claws are apparently present at birth in all species. The latter are 0.1–0.2 mm long in *maniculatus* (McCabe and Blanchard, 1950) and are blunt and soft-appearing in *floridanus, megalops, thomasi, leucopus,* and presumably in other species as well.

BEHAVIOR.—Motor responses of newborn *Peromyscus* are poorly developed. According to McCabe and Blanchard (1950), *P. maniculatus gambeli* and *P. truei* make continuous sucking-like movements of the jaws and weak, spasmodic movements of the body, mostly of the hind limbs. They are unable to raise the head, drag along, or roll over. Observations made by the same authors on the newborn young of *P. californicus* suggested that young of this species are perhaps slightly better coordinated at birth than are those of the preceding species. The sucking jaw movements were less persistent and automatic. Neonates of *P. floridanus* give uncoordinated torsal twists and involuntarily roll from side to side or completely over. When resting quietly, they frequently twitch spasmodically (Layne, 1966). Three newborn young of *P. megalops* that I observed usually rested on their sides and were capable of only feeble, uncoordinated movements. Occasionally the body would twitch spasmodically when the mouse was inactive. Neonates of seven *P. l. noveboracensis* studied were helpless. They were quiet much of the time and moved weakly. Neonates of *P. thomasi* twist about in an uncoordinated manner and appear to be incapable of moving from a given spot. They lie on their sides when undisturbed. The individuals I have observed did not give spasmodic twitches, as noted in some species.

Vocalizations have been recorded in the newborn young of all species studied to date. Pournelle (1952) described the vocalizations of *P. gossypinus* as being very faint high-pitched twitters. He recorded vocalizations of young during birth. Neonates of *P. l.*

noveboracensis also utter barely audible high-pitched squeaks. *P. m. gambeli* and *P. truei* give persistent sucking sounds but no squeaks, while newborn *P. californicus* are more versatile in their vocal repertoire, which includes "minute shrieks, chirpings, chucklings, made with convulsive efforts" (McCabe and Blanchard, 1950). Neonates of *P. floridanus* also emit several kinds of sounds, including "smacking" or "sucking" notes, and high-pitched squeaks with accompanying jaw movements (Layne, 1966). I have recorded the occurrence of repetitious, sucking squeaks in newborn *P. megalops* and *P. thomasi.* The sounds of the former species were distinctly louder and lower-pitched than those of *leucopus.* Egoscue (1964) noted that most newly born *P. crinitus* squeak when handled.

On the basis of the present admittedly fragmentary and generalized descriptive evidence, there appears to be relativly little variation in the behavioral capacities of different species of *Peromyscus* at birth. It is not obvious, in contrast to what might be expected, that the neonates of larger species have significantly less well-developed behavioral capacities than those of the smaller forms. In fact, one gains the impression that *californicus* and *megalops,* particularly the former, might actually be somewhat more advanced at birth than such smaller species as *maniculatus* and *leucopus.*

Postnatal Morphological Development

FORM.—The neonatal appearance characterized by relatively large, blunt head; fleshy pinnae; large hind feet; and short tail is retained by young *Peromyscus* for the first 10 to 14 days. Species differences exist in the age at which the young begin to lose the infantile form. For example, at two weeks of age, *P. maniculatus* is more advanced in appearance than either *P. truei* or *P. californicus* (McCabe and Blanchard, 1950).

The pinnae become erect within 24 hours after birth in members of the subgenus *Haplomylomys* (*P. californicus* and *P. eremicus*), and may become unsealed in *californicus* as early as 6 hours post partum (McCabe and Blanchard, 1950). The pinnae do not become elevated in other species until the second to eighth day of age (Table 6). The mean age at which the pinnae become erect in most species is three or four days, with the largest species, *megalops* and *thomasi,* being slowest in this developmental feature. For some

TABLE 6

CHRONOLOGY OF SELECTED POSTNATAL DEVELOPMENTAL STAGES IN *Peromyscus*

Species and subspecies	Elevation of pinnae (days)		Eruption of lower incisors (days)		Opening of eyes (days)		Source
	Mean	Range	Mean	Range	Mean	Range	
P. p. rhoadsi	—	4–5	6	—	14.0	13–16	Rand and Host, 1942
P. p. subgriseus	3.7 (b)	3–5 (b)	usually 6 (b)	6–7 (b)	13.7 (a) 13.6 (b)	10–16 (a, b)	a. King, *in litt.*; b. Laffoday, 1952
P. maniculatus artemisiae	—	3–4 (a, b)	—	—	14.50 (a) 15 (b)	13–19	a. Svihla, 1932; b. 1935
P. m. bairdi	3* (a)	2–4 (a, b)	5.2 (c)	—	13.67 (a) 12.1 (c)	12–16 (a)	a. Svihla, 1932; b. 1935; c. King, 1958
P. m. blandus	3*	—	—	—	14.36	13–16	Svihla, 1932
P. m. gracilis	—	—	5.7	—	16.8	—	King, 1958
P. m. gambeli	3* (a)	—	—	—	13.33 (a) 90% by 14 (b)	12–15 (a)	a. Svihla, 1932; b. McCabe and Blanchard, 1950
P. m. nebrascensis	3*	—	—	—	13.69	12–17	Svihla, 1932
P. m. rubidus	3*	—	—	—	15.00	—	Svihla, 1932
P. m. rufinus	3*	—	—	—	15.14	14–17	Svihla, 1932
P. oreas	by 4	—	—	—	—	—	Svihla, 1936
P. leucopus ssp.	—	—	—	—	12.5	10–15	King, *in litt.*
P. l. leucopus	—	—	—	—	13.13	12–15	Svihla, 1932

TABLE 6 (Continued)

Species and subspecies	Elevation of pinnae (days)		Eruption of lower incisors (days)		Opening of eyes (days)		
	Mean	Range	Mean	Range	Mean	Range	Source
P. l. noveboracensis	2.8 (a)	2–4 (a)	5.7 (a)	4–7 (a)	13.0 (a) / 13.45 (b)	10–15 (a, b)	a. Present study; b. Svihla, 1932
P. g. gossypinus	usually 4	—	80%, 5–7	—	76% 12–14	10–18	Pournelle, 1952
P. t. truei	—	—	—	—	17.50	15–20	Svihla, 1932
P. t. gilberti	—	3–6 (a)	—	—	—	14–21 (b)	a. Hoffmeister, 1951; b. McCabe and Blanchard, 1950
P. megalops auritus	5.8	5–8	9.3	8–10	22.8	20–28	Present study
P. crinitus pergracilis	—	—	—	—	16 (median)	15–17	Egoscue, 1964
P. californicus ssp.	—	—	—	—	14.4	12–20	King, *in litt.*
P. c. californicus	less than 1	—	—	—	13.67	13–14	Svihla, 1932
P. eremicus ssp.	—	—	—	—	12.8	10–15	King, *in litt.*
P. e. eremicus	less than 1	—	—	—	15.50	15–17	Svihla, 1932
P. floridanus	3.9 (a)	2–6 (a)	6.8 (a)	4–9 (a)	16.5 (a), 14.2 (b)	12–20 (a, b)	a. Layne, 1966; b. King, *in litt.*
P. thomasi	3.5 (b) / 5 or 6 (a)	3–6? (a, b)	—	—	19.5 (median) (a)	17–22 (a)	a. Musser, *in litt.*; b. Present study

* General statement for species *P. maniculatus, P. leucopus,* and *P. truei.*

time after elevation, the pinnae are thick and fleshy, unlike the adult condition.

McCabe and Blanchard (1950) recorded the rate of development of the feet and claws of *P. maniculatus gambeli, P. truei,* and *P. californicus;* and I have obtained similar data for *P. leucopus, P. megalops,* and *P. thomasi.* The digits of *P. m. gambeli* are separated 1–1.5 mm more at one week of age than at birth, and the feet are nearly of adult form by the second week. At one week of age the claws are from 0.2 to 0.5 mm longer than at birth. In contrast, the digits of *truei* are scarcely more separate at one week than at birth, although they are fully formed by the second week. The claws of one-week-old young of this species are from 0.5 to 0.9 mm long. In keeping with its generally advanced condition with respect to *maniculatus* and *truei,* at one week *californicus* has the front digits almost fully separated and the hind partially so. The claws are from 0.9 to 1.2 mm long. Although this species is not so distinctly advanced as the others at two weeks, the feet are fully developed. The fore digits are fully formed at one week in *P. leucopus,* and those of the hind are no longer fused by the eighth or ninth day of age. Formation of the digits of *megalops* is complete by the thirteenth or fourteenth day. The feet of *thomasi* may be slower to develop than other members of the genus, as the fore digits of one twelve-day-old litter examined were still fused for one-half to three-fourths of their length and the hind were separated at the very tips only.

The anus of the neonatal *leucopus* is closed, and a similar condition probably obtains in other members of the genus. In one litter of *leucopus* whose development I observed, the anal orifice appeared to be distinctly patent by the eighth day.

PIGMENTATION AND PELAGE.—In those species in which newborn are unpigmented, pigmentation begins to develop on the dorsum within 24 hours after birth and by the second day the upper parts, except the pinnae, are a darker gray. General pelage hairs develop first on the dorsum, exhibiting a cephalo-caudal gradient in density. Anteriorly, the hairs may be dense enough to conceal the skin of the head and nape while the posterior region is glabrous or only sparsely clothed. In the early stages of growth of the dorsal pelage, epithelial scales may be evident. New pelage then extends onto the venter, limbs, and tail. The juvenal pelage when first completed is usually fluffy and luxurious, but with further growth of the body it may appear sparser in general aspect.

The pinnae are essentially unpigmented at birth and do not begin to acquire pigmentation for several days; it first develops marginally. Hairs do not appear on the pinnae until pelage growth on the remainder of the dorsum is fairly advanced. The soles of the hind feet are unpigmented or only faintly pigmented at birth, and in those species for which data are available the hind feet begin to show pigmentation during the first week.

As pelage development is a continuous process and has not been described in detail in most papers relating to ontogeny of *Peromyscus,* it is difficult to treat this phase of development in quantitative terms or to make adequate comparisons at the specific or subspecific levels.

The dorsa of young *P. polionotus rhoadsi* are pigmented at two days, and fine hair is macroscopically visible at four days; it shows on the venter at eight or nine (Rand and Host, 1942). *P. p. subgriseus* shows sparse hair on the dorsum at three days, and hairs are visible on the venter by the fifth day. By the eighth or ninth day hairs have appeared on the pinnae (Laffoday, 1957). Golley *et al.* (1966) state that the juvenal pelage of the old-field mouse is "complete within about one week." This statement presumably refers to covering of the body rather than full development of the pelage.

P. maniculatus artemisiae and *P. m. bairdi* exhibit pigmentation of the dorsum within 24 hours of birth and dark hair is visible on the upper parts by the second to fourth day (Svihla, 1932, 1934, 1935). Young of *P. oreas* also become pigmented on the dorsum within a day of birth (Svihla, 1936). At one week of age *P. m. gambeli* have hairs over the entire dorsum, although the covering is thickest on the crown and nape and sparse on the venter. The body is often scurfy. The young at this age are still naked on the venter except for the chin. By the end of the second week the coat is full but the hairs are still short, imparting a sleek appearance to the animal. The eyes are furred over, and the ventral hairs have grown to the point where the gray basal portions are exposed. The juvenal pelage is complete at three weeks, but the density of the hair decreases because of the increase in surface area of the body

(McCabe and Blanchard, 1950). Pigmentation of the skin is fully developed in *P. m. bairdi* at the mean age of 3.6 days, 5.4 days in *P. m. gracilis* (King, 1958).

According to Pournelle (1950), young *P. gossypinus* are well pigmented above by the second day, and at five days fine hair is visible on the dorsum, whereas the abdomen is still hairless. Rapid growth of hair on the back and its appearance on the venter occurs between the fifth and tenth days.

P. t. truei becomes pigmented within 24 hours of birth (Svihla, 1932), and *P. t. gilberti* is reported to become darkly pigmented on the dorsal region between the second and third day and on the legs by the fourth (Hoffmeister, 1951). Pelage development in the second subspecies has been described by McCabe and Blanchard (1950) and Hoffmeister (1951). The former found that *truei* lagged behind *maniculatus* in pelage development at one week. Although the crown and nape were relatively well covered, the remainder of the dorsum appeared naked, although sparsely dis- tributed, mostly tiny, hairs were present. Hoffmeister (1951) noted that differentiation of underhair, overhair, and guard hair had occurred by 11 days of age. The juvenal pelage is not completed until four weeks of age, at which time the tail pencil has developed also.

The dorsum of *P. floridanus* becomes noticeably darker within a day or two of birth, and the margin of the pinnae show developing pigmentation by three days. The pelage of the upper parts is evident to the unaided eye by the third day and is densest on the head and nape. Scattered hairs appear on the tail, and the longest (6–7 mm) mystacial vibrissae reach to the rear edge of the eye. Epithelial scales occur abundantly on the backs of some young at this age. Hairs are first evident on the venter at two or three days and appear on the upper surfaces of the feet by the fifth day. By one week, the soles and plantar tubercles of the hind feet are pigmented, and scattered short hairs are found on the edge of the pinna. The dorsum shows evidence of the agouti pattern at ten days and by the sixteenth the gray bases of the ventral hairs have been exposed. The juvenal pelage is apparently fully developed by the third week (Layne, 1966).

At one week *californicus* is more advanced in pelage development than other species (*maniculatus* and *truei*) with which it has been compared (McCabe and Blanchard, 1950). The young are completely haired above and below, except for the posterior abdomen, which remains bare or nearly so. This advance seems to persist through the second week, the hair being longer and denser than in the smaller species; but at three weeks the pelage of *californicus* is further from completion than that of the other species. Its juvenal pelage is fully developed by the fourth week, and is less dense at five weeks.

Young *P. leucopus* that I observed developed pigment the first day, and tiny widely scattered hairs were visible on the head and upper back region. Sparsely distributed hairs were present on the venter by the second day and on the legs and tail by the fourth or fifth. By the fifth or sixth days the dorsal pelage was clearly visible to the naked eye. Epidermal scales may be present. The soles of the hind feet and the pinnae are pigmented by the sixth day. The gray bases of the ventral hairs are evident between ten and 12 days, and the juvenal pelage is complete by the third week of age.

The dorsum of *P. megalops* is darker at one day than at birth and very dark by the third, at which time the pinnae are slightly pigmented, and the longest mystacial vibrissae extend to the rear edge of the eye. Although probably present earlier, hairs on the dorsum are first clearly visible to the naked eye at four or five days of age; none is visible on the venter, tail, or limbs at this time. Tiny hairs appear on the tail and venter at six days and on the upper surfaces of the feet by the seventh or eighth. Hairs are present on the pinnae and entire length of the tail between the eighth and eleventh day. The under- and overhair components of the pelage are differentiated by the tenth day and by the thirteenth or fourteenth the agouti pattern of the dorsal pelage and gray bases of the ventral hairs are evident. The juvenal pelage appears fully grown in the young at four weeks of age.

In three litters of *P. thomasi* observed by Musser (*in litt.*), the young were darkly pigmented on the dorsum at three days of age but hair was not distinctly visible on the dorsum until the eighth. Formalin-preserved individuals of one of these litters, which died at 12 days, had well-developed dorsal pelage, with the longest guard hairs being about 2 mm. Hairs were present to the tip of the tail and on the feet. The ventral pelage was sparser than that on the upper parts, the longest hairs being less than 1 mm. The longest mystacial vibrissae reached to the base of the ears. The pinnae were faintly pigmented along the margins and possessed short hairs. The soles and plantar tubercles of the hind feet were pigmented, and there were patches of dark hairs at the ankles and wrists. The two young of a litter I observed had a grayish dorsum on the second day, but hairs were not visible to the naked eye. Fine hairs were noted on the dorsum at three days and on the lower sides at five days. The juvenal pelage was completed sometime before seven weeks of age in the young observed by Musser.

DENTITION.—Although the exact age at which the incisors erupt is difficult to determine accurately, such data have been presented for several species of *Peromyscus* (Table 6). The eruption of the lower incisors, which often slightly precedes that of the uppers, occurs from the fourth to tenth day of age, averaging the fifth or sixth in most species studied. The incisors appear to develop somewhat later in *P. floridanus* and *P. megalops* than in other species. In the former, the upper incisors erupted between the seventh and eleventh days, mean 9.0. The incisors begin to exhibit wear shortly after eruption, presumably as the result of occlusion.

The incisors of *P. floridanus* are white and somewhat translucent when they first appear. Faint yellowish coloration is evident in the lowers of some individuals at two weeks, but the typical yellowish color of the adult does not develop until about the time of weaning, at three or four weeks (Layne, 1966). I have noted the first suggestion of coloration of the lower incisors of young *megalops* at about 22 days.

The cheek teeth of *truei* erupt before the 30th day of age, and wear appearing at an early period is probably attributable to occlusion (Hoffmeister, 1951).

AUDITORY CANAL.—Although the opening of the external auditory meatus is an important developmental event, especially from the standpoint of behavioral ontogeny, data on this stage have rarely been included in accounts of development in the genus.

In 17 young *floridanus* studied, the external auditory meatus became patent between 10 and 15 days of age, with a mean of 12.2 days (Layne, 1966). The young were routinely tested for response to sound by giving a loud squeak within several inches of the animal; such a stimulus almost invariably elicits a startle response in adults. The mean age at first response to the sound in the 17 young thus tested was 13.2 days. The meatus of young *P. polionotus* was not patent at 9 days of age, but at 11 days a response to sound was recorded (Laffoday, 1957).

I have also made observations on the opening of the auditory meatus and reaction to sound in three litters (11 individuals) of *P. leucopus noveboracensis* and three (5 individuals) of *P. megalops auritus*. In the first species, the meatus opened between 9 and 13 days of age (mean, 10.4), and the mean age at which a response to the squeak was shown was 10.5 days (9–13). The mean age at which the meatus of *P. megalops* was patent was 17.7 days, with a range of from 14 (possibly 13) to 23 days. The first response to a sound was recorded between 14 and 23 days (mean 18.0). The auditory meatus opened between the twelfth and sixteenth day of age in a litter of *P. thomasi* studied by Musser (*in litt.*).

Eyes.—Data available on the age at which the eyes open in various species and subspecies of *Peromyscus* are summarized in Table 6. For the genus as a whole, opening of the eyes occurs from the tenth to 28th day of age, with a mean (calculated from individual means of all species and subspecies) of approximately 14.8 days. Age of eye opening does not appear to be strongly correlated with either adult size or neonatal weight. Although most of the smaller species, such as *polionotus, maniculatus,* and *leucopus,* are characterized by early opening of the eyes, 13 or 14 days usually, and the largest species available for comparison, *megalops* and *thomasi,* are the slowest in this regard, the species in the intermediate range do not exhibit much consistency in connection with this ontogenetic feature. Thus, *truei* and *floridanus* are distinctly less advanced than *gossypinus,* although all three are roughly equivalent in size. Likewise, the eyes of *crinitus* are slower to open than those of species of similar size, whereas *californicus* and *eremicus* appear to be relatively (in the case of the first) or actually (in the case of the second) precocial in the opening of the eyes.

In some cases, the chronology of eye opening in different species agrees with presumed phylogenetic relatedness. Thus, *polionotus* and *maniculatus,* which are members of the same species group, tend to average about 14 days in age when the eyes open, while the two members of another species group, *leucopus* and *gossypinus,* open the eyes at about 13 days. That this is not a spurious correlation is supported by the fact that the age differences between these species groups in average eye opening are the reverse of what would be expected on the assumption that larger body size means more prolonged development. The relative precocity of eye opening in *californicus* and *eremicus* also agrees with their placement in the same subgenus, *Haplomylomys.* On the other hand, the tendency toward delayed opening of the eyes in *truei* and *floridanus,* which also exhibit a number of other developmental similarities, is apparently the result of convergence, as these species are quite unrelated according to current interpretation of various morphological data.

As shown in *maniculatus* and *leucopus,* subspecies within a species tend to be generally similar in regard to the mean and extreme ages of eye opening. However, significant differences may occur between subspecies in some cases. King (1958) found a difference of about five days in the mean ages and no overlap in the ranges at which the eyes of *P. m. bairdi* and *P. m. gracilis* opened. Such differences in physical ontogeny at the subspecific and specific levels undoubtedly have important implications in relation to the development of behavior.

Differences in mean values of ages of eye opening or other developmental events reported by different authors for the same species or subspecies (e.g., *P. m. bairdi* and *P. floridanus* in Table 6) may reflect actual genetic differences as the result of the use of different breeding stocks or environmental effects stemming from different maintenance procedures, such as animal room temperatures, feeding regimens, amount of handling, caging arrangements, and so forth.

Weaning

Weaning is a gradual process and thus difficult to quantify for comparative purposes. In most *Peromyscus* studies, age of weaning has been estimated on the basis of observations of young taking solid food, appearance of solid food traces in feces, or evidence of independence of the young from the mother. Obviously, for the most part these methods provide only rough approximations as to the age of weaning. The only systematic attempt thus far to obtain

quantitative data on weaning in the genus is that of King *et al.*
(1963). These authors defined weaning "as the age at which the
young mice maintain or gain weight during a 24-hour period of
isolation." Although this technique provides a clearly definable
standard for comparative purposes and does indicate a certain level
of physiological development of the young, it does not provide a
measure of the time young of different stocks are normally permitted to nurse by the mother, a datum that is of considerable
importance from the standpoint of behavioral development.

The majority of species and subspecies of *Peromyscus* are apparently weaned between three and four weeks of age. Rand and Host
(1942) stated that *P. polionotus* are weaned between 20 and 25
days of age. Svihla (1932, 1934) gave 22 to 37 days as the period
of weaning in a number of subspecies of *P. maniculatus,* and
McCabe and Blanchard (1950) and King *et al.* (1963) have also
reported values for some of the same and additional subspecies.
McCabe and Blanchard found the earliest age at weaning to be
19 days in *P. m. gambeli,* and King *et al.,* using the criterion of
weaning noted above, determined the age of weaning in *P. m.
bairdi* and *P. m. gracilis* to be 18 and 24 days, respectively.

Svihla (1932) included *P. leucopus* in his general statement of
22–37 days as the period of weaning in *Peromyscus,* while Nicholson
(1941) indicates weaning usually occurs between 25 and 30 days
in this species. My own observations on several litters of *P. l. noveboracensis* indicate that weaning may begin as early as 19 days.
Young *P. gossypinus* are normally weaned between 20 and 25 days
of age, although individuals 13 or 14 days old can survive on solid
food (Pournelle, 1952). Ages of weaning in *P. truei* and *P. floridanus* have been given as 21 to 28 days (Layne, 1966; McCabe and
Blanchard, 1950). Svihla (1932), however, indicates that weaning
in *P. truei* occurs later than in *P. leucopus* and *P. maniculatus.*
According to him, young removed from the mother earlier than a
month of age rarely survive, and one litter was not weaned until
40 days of age. Egoscue (1964) stated that young *P. crinitus* are
weaned by 28 days.

Svihla (1932) indicated that the period of lactation in *P. californicus* and *P. eremicus* is prolonged, although, with the exception
of one litter of *P. californicus* that was not weaned until 44 days of
age, he presents no actual values. McCabe and Blanchard (1950)

note that female *californicus* still have an abundant milk supply at three weeks and that the young nurse continuously. Judging from their notes on behavioral and morphological development, the majority of the young are probably weaned by five weeks of age.

Despite the large size, the young of *P. megalops* and *P. thomasi* do not appear to nurse for an excessively long period as compared with smaller forms. Musser (*in litt.*) observed young *thomasi* gnawing on food stuffs as early as 19 days of age, and individuals separated from the mother as early as 22 days showed no apparent ill effects. My observations on *megalops* indicate that weaning may begin as early as 21 to 23 days in this species. This is based upon ages at which young first began to nibble on solid food outside the nest.

In species that experience post partum estrus, nursing of a litter may be terminated rather abruptly by the birth of a new litter. Svihla (1932) noted that female *maniculatus* and *leucopus* usually cease to suckle the older litter and force them out of the nest when a new litter is born. In some instances, however, young of both the old and new litters may be nursed for a few days. Howard (1949) was able to prolong lactation in a captive *leucopus* to approximately 56 days by continuously supplying immature young to replace those maturing. The production of milk in this female apparently began to diminish, despite the continued presence of infants, after the 50th day, suggesting that an endogenous component is involved in the lactation cycle. Rand and Host (1942) note that female *polionotus* isolated with their litters so that they do not become pregnant again may nurse the young up to 35 days, whereas normally lactation is of shorter duration.

Young *Peromyscus* may continue to associate with the mother past the age of weaning. In the wild this tendency appears to vary seasonally in at least some species. Nicholson (1941) found that in *P. leucopus* the longest time individuals of a given litter remained with a female during the summer was 37 days, while in the late fall and winter litters sometimes remained with the mother until 45 to 60 days of age. Howard's (1949) data indicated a similar trend in *P. m. bairdi*. The persistence of families in the fall and winter months was associated with the cessation of reproduction in adults, hence probably a reduction in aggressive tendencies of the adult female, and delay in sexual maturity of the late-season young. Females of *P. californicus* are reported to be highly tolerant of grown young in the nest even during and after birth of a litter (McCabe and Blanchard, 1950). Evidence of persistence of the mother-young bond after weaning has also been obtained for captive *P. floridanus* and is suggested by field data as well (Layne, 1966). The maintenance of the family group for a longer period in the fall or winter may be adaptive in cold regions in terms of the metabolic saving resulting from huddling. The laboratory findings of Rand and Host (1942) suggest that, in addition to other factors noted, prolongation of lactation in the last-born litters of the season may contribute to the continued cohesion of the family unit.

Sexual Development

Sexes of newborn *Peromyscus* can ordinarily be distinguished by the position and relative development of the urogenital papilla. The teats of females become visible about a week after birth and remain relatively prominent until the ventral pelage is fairly well developed.

Age of onset of reproductive activity has been determined for a number of species and subspecies of *Peromyscus* raised under laboratory conditions. Criteria used to assess reproductive status of females include occurrence of estrous cycles as determined by vaginal smears, perforation of the vulva, and conception. Clark (1938) noted that the correlation of age at vaginal estrus and earliest age of conception indicated the former was a valid indicator of puberty in this genus. Where data on the vaginal cycle, earliest conception, and opening of the vulva are available for a given species or subspecies, they give comparable estimates of age at puberty. Appearance of spermatozoa in the cauda epididymis has been commonly used as an indication of male fertility. Data from natural populations of *Peromyscus* and other small mammals give abundant evidence of effects of various environmental factors such as season, food supply, population density, etc., on age at puberty. Although their effect may be less pronounced, environmental variables such as season of birth, diet, social factors, caging arrangements, temperature, photoperiod, and others may also influence sexual development in mice under laboratory conditions. Such effects may be responsible for some of the apparent differences in age at puberty of various *Peromyscus* stocks for which laboratory data are presently available, and this possible source of variation should be taken into account in interpretations of these data.

Following is a summary, by species, of information on sexual development in *Peromyscus*.

P. polionotus: The mean age of first estrus, as determined by vaginal smears, in four subspecies of this species was 29.6 ± 0.5 days (Clark, 1938). The earliest age at attainment of puberty was 23 days. In a series of 12 matings, one female conceived a litter at 102 days of age (Clark, 1938). Rand and Host (1942) recorded opening of the vulva in *P. p. rhoadsi* at 26 days. One captive female was estimated to have conceived a litter at 35 days of age, and a field-caught specimen with small embryos was estimated to be approximately 40 days of age.

P. maniculatus: The mean age of first estrus in four subspecies studied by Clark (1938) was 48.7 ± 1.2 days, the earliest being 23 days. The youngest female to participate in a fertile mating was 34 days. He found abundant sperm in the epididymis of 16.5 per cent of 79 mutant hybrid males of this species at 40 days of age. Seventy-nine per cent of 43 individuals about 60 days of age were fertile. Fertile matings were recorded for males of mutant stocks at about 40 days.

Wild females of *P. m. bairdi* are known to have produced litters at 55, 56, and 59 days of age (Howard, 1949). Assuming a 23-day gestation period, their estimated ages at mating were 32, 33, and 36 days. In the wild population studied by Howard (1949), females born after September or October did not breed until the following spring at an age of 24 to 27 weeks or older. Males of this subspecies may be capable of fertile matings as early as 33 days of age. Blair (1940) estimated the youngest breeding males in the population he studied to be five to six weeks old.

In the subspecies *gambeli*, the vulva is infrequently perforate at five weeks and usually open, with frequent evidence of coitus, at seven weeks (McCabe and Blanchard, 1950). Some males show slight development of the cremaster at five weeks. The cremaster is usually well developed at seven weeks, and a copulation patch is often present on the abdomen by the ninth week.

P. leucopus: Clark (1938) determined the average age of first estrus in *P. l. noveboracensis* as 46.2 ± 3.2 days, with the earliest at 28 days. This author also records a case of conception in a female 39 days old.

P. gossypinus: Of 36 females studied by Pournelle (1952), six failed to show a perforate vulva by the 65th day of age. Of the remainder, the mean age of opening of the vulva was 43 days (earliest 36 days). The youngest female to conceive was 73 days old.

Sperm were absent from the seminiferous tubules in specimens under 40 days and were not present in the epididymis in animals less than 45 days of age. Sperm were abundant in the seminiferous tubules and epididymis of young by 70 days, and known fertile matings occurred prior to this age.

P. crinitus: Egoscue (1964) states that breeding in this species begins at about 70 days of age.

P. boylei: Mean age of females at puberty was 50.9 ± 1.9 days, the earliest estrus occurring at 37 days (Clark, 1938). The youngest females involved in fertile matings were over four months of age.

P. truei: The earliest age at first estrus recorded by Clark (1938) was 28 days, with a mean of 50.1 ± 2.5 days. The youngest females to conceive in 24 matings were all over three months old. The study of McCabe and Blanchard (1950) also provides evidence for relatively delayed puberty in this species. The vulva was sometimes partly open in females at nine weeks of age, although there was no sign of coitus. Males showed slight scrotal development by the ninth week.

P. megalops: In a single litter of two females, the vulva of one individual opened between 46 and 48 days and that of the other at 49 (personal observations).

P. eremicus: The mean age at first estrus was 39.2 ± 1.5 days, with the earliest being at 28 days of age (Clark, 1938). The earliest age of conception in a series of eight successful matings was at 58 days. One of eight males had mature spermatozoa in the epididymis at 40 days of age.

P. californicus: McCabe and Blanchard (1950) found no indication of sexual activity in females at nine weeks of age, whereas the vulva was usually slightly open at 11 weeks.

P. floridanus: The mean age of opening of the vulva (as determined from weekly examinations) was 35 days with a range of from 18 to 48 (Layne, 1966). Slight cremaster development was noted in males 44, 49, and 52 days of age. No sperm were found in the epididymides of specimens seven and eight weeks old, while one 11-week-old individual was fertile.

P. thomasi: A preserved male 75 days of age had enlarged testes and abundant sperm in the cauda epididymis (personal observation).

The foregoing data, although they do not lend themselves to critical comparisons, do suggest that the age at onset of reproductive activity in *Peromyscus* is, as in the case of other developmental features, not strongly dependent upon body size. The fragmentary information available for such large species as *thomasi* and *megalops* seem to indicate that these forms do not lag as far behind smaller members of the genus as might be expected. There is also some evidence that in at least some species (e.g., *P. leucopus, P. maniculatus, P. gossypinus, P. floridanus*) males may mature somewhat later than females. This trend, which may prove to be more widespread when further data have been accumulated, would seem to constitute an adaptation tending to reduce breeding among litter mates.

Developmental Molts

Two types of molts can be recognized in *Peromyscus*: (1) those associated with seasonal changes (seasonal or annual molts) and (2) those associated with ontogeny (developmental molts). Three pelage phases, juvenal, post-juvenal or subadult, and adult are apparently typical of the life cycle in the genus (Collins, 1923). These pelages are generally distinguishable by coloration, structural differences of individual types of hair, or differences in the relative abundance of different hair types. For the purposes of the following descriptions, the developmental molts leading to the acquisition of the post-juvenal or subadult and first adult pelages are designated as *post-juvenal* and *post-subadult* molts, respectively. Hoffmeister (1951) suggested that an additional pelage and molt might occur in *P. truei* between the juvenal and subadult phases, and Lawlor (1965) found possible evidence of a similar condition in *P. yucatanicus*.

Osgood (1909) believed that a single annual molt was typical of the genus *Peromyscus,* although he found evidence of two annual molts in *P. melanotis*. In actuality, the occurrence of two seasonal molts in adults may be of more common and perhaps even regular occurrence than is presently believed. Hoffmeister (1951) pointed out the possibility of two annual molts in *P. truei,* and Brown (1963) and Lawlor (1965) have demonstrated the existence of this pattern in *P. boylei* and *P. yucatanicus,* respectively. Extensive data from live-trapping studies and snap-trapped samples of *P. floridanus* that I have obtained also provide clear evidence of two molts a year in this species as well.

POST-JUVENAL MOLT.—*Chronology*: The juvenal pelage of all species presently included in the genus *Peromyscus* is of some shade of gray. Compared with the second developmental molt and subsequent seasonal molts, the post-juvenal molt appears to be less variable in the age of onset and completion within a given species or subspecies and more regular in its pattern.

In most species for which data are available this molt begins at 30 to 45 days of age, although *P. polionotus* appears to be somewhat precocial in this respect. Golley *et al.* (1966) found the onset of molt in the latter species to occur at about 20 days of age in females and 25 days in males, whereas Laffoday (1957) observed first evidence of the molt in *P. p. subgriseus* between 32 and 36 days of age.

Collins (1923) found that in *P. maniculatus gambeli* the transition from juvenal to subadult pelage usually commenced at four weeks of age and was completed by about the twelfth week. Working with the same subspecies, McCabe and Blanchard (1950) recorded a single individual in the post-juvenal molt at four weeks, with first indications of the molt appearing more commonly early in the fifth week. In most individuals, the ventral phase of the molt was complete by six weeks, and by the ninth week the dorsal molt was finished except for the crown and vestiges of juvenal pelage at the base of the tail. The crown may retain gray juvenal pelage for sometime thereafter. Sumner (1917) noted the occurrence of the post-juvenal molt between six and ten weeks in young of hybrid *maniculatus* parents. The duration of the molt (from the first appearance of new pelage) in *maniculatus* studied under field conditions ranged from 8 to 35 days, averaging about 25 (Storer *et al.,* 1944).

Data on the timing of the post-juvenal molt of *P. leucopus* have been given by several authors. Gottschang (1956) found that most (95 per cent) individuals began the molt between 40 and 50 days of age. The earliest age of onset of the molt was 38 days. There was no correlation between or within litters between weight and the beginning of the post-juvenal molt. However, males tended to begin the molt earlier than females. Most individuals completed the molt in approximately three and one-half weeks (12 to 29 days). These data pertain to captive mice housed outside under natural conditions. Five mice in juvenal pelage when first caught in the course of live-trapping studies completed the molt in from 10 to 53 days (mean = 27). Nicholson (1941) recorded retention of the juvenal pelage until 40 to 50 days of age and nearly complete subadult pelage by 60 to 75 days. The earliest molt in laboratory populations studied by Bendell (1959) was at four weeks. The first mice with completed molts were nine weeks of age, and all had subadult pelage by the eleventh week. His data suggest that field populations may differ in the chronology of the post-juvenal molt. In one experimental island population, the first records of molt occurred at an estimated age of five weeks, and the first animals to have completed the molt were estimated to be nine weeks. Some mice estimated to be 12 weeks old were still undergoing molt. In a second natural population, mice from 3 to 12 weeks were molting. The chronological variation in molts suggested by these data may actually have been somewhat less, as ages were estimated from a weight-age curve derived from captive-raised young. Considering the extent of variation in weight often occurring in young of litters of the same chronological age, it is likely that not all young were assigned to the correct age class.

Pournelle (1952) stated that young *P. gossypinus* usually begin the post-juvenal molt between 34 and 40 days of age. In a sample of 90 individuals, 76.7 per cent (32–53) began the molt in this interval.

The post-juvenal molt in *P. boylei* begins at about five weeks (mode = 35 days), and the subadult pelage is nearly complete at 12 to 15 weeks (Brown, 1963).

Hoffmeister (1951) recorded the beginning of the post-juvenal molt in *P. truei* between five and seven weeks of age, and, as noted earlier, he also observed what might have been a prior molt in which the gray juvenal pelage was replaced by a slightly lighter gray coat. McCabe and Blanchard (1950) found individuals in early stages of the molt at six weeks. The molt was roughly half completed at eight weeks, and one individual had nearly completed the molt at 11 weeks. By the fourteenth week, the majority of the mice had completed the molt, except for retention of juvenal pelage on the crown.

The two young of a litter of *P. megalops* that I studied showed first evidence of the post-juvenal molt at 39 days of age. One of these had an essentially complete subadult coat by 91 days, while the other was still undergoing the molt. One young of a second litter of two began the molt between 35 and 42 days, while the other was not found in an early stage of the molt until 49 days of age. These observations refer to the dorsal phase of the molt.

The earliest age at which post-juvenal molt (on venter) was observed in *P. californicus* was five weeks (McCabe and Blanchard, 1950). By the sixth week about 30 per cent of the mice showed incipient molt. Although one individual had completed the molt as early as the ninth week, the majority did not have full subadult pelage until 17 weeks of age.

The mean age of the beginning of the post-juvenal molt in *P. floridanus* was between 36 and 42 days, 35 days being the earliest age that molting was observed, and 70 days the latest (Layne, 1966). The maximum duration of the molt averaged about 55 days, with a range of from approximately 28 to 84 days.

Musser (*in litt.*) observed changes in the pelage coloration, presumably indicative of the post-juvenal molt, in three litters of *P. thomasi* between 47 and 51 days of age. Most of the young were in fresh subadult pelage 24 days later, although the molt was still in progress in some.

Pattern: The course of the post-juvenal molt has been described, in various degrees of detail, in ten species: *P. polionotus, P. maniculatus, P. leucopus, P. gossypinus, P. boylei, P. truei, P. californicus, P. eremicus, P. yucatanicus,* and *P. floridanus* (Brown, 1963; Collins, 1918, 1923; Golley *et al.,* 1966; Gottschang, 1956; Hoffmeister, 1951; Laffoday, 1957; Lawlor, 1965; Layne, 1966; McCabe and Blanchard, 1950; Pournelle, 1952; Storer *et al.,* 1944). To this list can also be added *P. megalops* (personal observations) and *P. thomasi* (Musser, *in litt.*).

Although the pattern of the molt is generally similar throughout the genus, differences in certain details appear to exist at both the specific and subspecific levels. The extent of individual variation in this molt has not been systematically studied in most cases, and further data may result in a breakdown of the apparent distinctions between taxa.

In most instances stages of the post-juvenal molt are characterized by contrasting coloration of the gray juvenal pelage and the browner post-juvenal or subadult pelage. Rarely, however, are individuals encountered with juvenal pelage indistinguishable from the subadult phase (McCabe and Blanchard, 1950). Although seasonal differences occur in the juvenal pelage of some rodents (Layne, 1954), there is presently no evidence that such is the case in *Peromyscus*. Gottschang (1956) found no differences in the juvenal pelage nor in any aspects of the post-juvenal molt in spring-, summer-, and fall-born litters of *P. l. noveboracensis* raised under natural conditions.

The post-juvenal molt of the venter tends to precede, and is sometimes completed before, the dorsal phase. Replacement of the pelage on the underparts also appears to be less regular than that of the upper parts of the body. As the ventral phase of the molt is a more subtle process because of the absence of contrasting coloration between the old and new hairs, its details have received less attention, and different descriptions of the pattern and chronology in the same species or subspecies are in some cases somewhat contradictory. Collins (1923) reported differences in the points of origin of the post-juvenal molt among subspecies of *P. maniculatus* and between that species and *P. eremicus* and *P. californicus*. In *P. m. gambeli* the molt begins either on the throat in the region of the angle of the jaws or on the anterior margin of the forelimbs and proceeds medially and posteriorly. The subspecies *sonoriensis* and *rubidus* agree with *gambeli,* whereas in *austerus* initial molt centers appear to develop on the posterior ventral regions. The earliest appearance of molt in *P. eremicus* is in the form of paired patches on the midventral part of the posterior thoracic region, whereas *P. californicus* shows simultaneous appearance of new pelage on the anterior aspect of the forelimbs and axillae.

The observations of McCabe and Blanchard (1950) indicate greater variability in the point of origin of the ventral phase of the post-juvenal molt than implied by Collins (1923). They found that *P. m. gambeli* and *P. truei* commonly, but not invariably, showed the molt first on the sides of the chest below the line between the upper and lower parts. In contrast, about three-fourths of *P. californicus* had the molt first appearing in the axillary region. In some cases, *maniculatus* and *truei* showed the *californicus* pattern and vice versa. New pelage commonly appeared first on the throat and chest of *truei* and *californicus*.

The earliest stage of ventral molt observed in *P. floridanus* involved narrow rings of new pelage around the bases of the forelimbs (Layne, 1966). In one litter, secondary centers appeared in the inguinal region; at about the same time new ventral pelage appeared anteriorly. The post-juvenal molt began in two areas lateral to the midline on the chest in *P. boylei* (Brown, 1963). The earliest stage of the molt that I observed in *P. megalops* involved new pelage on the throat and thoracic region of the venter. According to Musser (*in litt.*) the first post-juvenal pelage in *P. thomasi* appeared along the sides of the abdomen and neck.

Following its inception, the molt of the venter moves generally in a posterior direction. In species in which a chest band or spot occurs, there appears to be some variation in the time of appearance of the marking. The patch is present in the juvenal pelage of *truei*, but in *californicus* it appears as a consequence of the post-juvenal molt, the first indication of its development being at seven weeks (McCabe and Blanchard, 1950). *P. floridanus* also typically possesses a pectoral patch in adult pelage, although the intensity and size of the mark is geographically variable. In one population studied, some individuals showed evidence of the patch as soon as the ventral molt reached the breast region, whereas in others the patch did not appear over a long period of observation (Layne, 1966).

McCabe and Blanchard (1950) stated that, although it may appear in some species that the dorsal and ventral molts proceed from a common center, the centers do actually originate separately. The initial center of the dorsal phase of the post-juvenal molt has been reported as being the anterior aspect of the forelimb or shoulder in *P. m. gambeli, P. eremicus, P. californicus,* and *P. boylei* (Brown, 1963; Collins, 1923). This is also the usual case in *P. floridanus,* although in some individuals a second center on the anterior margin of the upper thigh appears shortly after the first (Layne, 1966). McCabe and Blanchard (1950) state that the dorsal molt of *P. m. gambeli, P. truei,* and *P. californicus* begins at two centers, one on the forelimbs and one on the flank color line at the rear of the trunk, although the anterior one tends to appear somewhat earlier. According to Gottschang (1956) new dorsal pelage in *P. leucopus* first appears as a patch or narrow line in front of the hind leg, with a second center developing later on the shoulders.

My limited observations on *P. megalops* indicate that the post-juvenal molt of the dorsum of this species begins on the middle and posterior region of the shoulder.

Although Pournelle's (1952) description of the molt in *P. gossypinus* is not detailed enough to accurately fix the center of origin of the dorsal phase, his statement that the band of adult pelage along the sides of the body representing the first indication of the post-juvenal molt "lengthens anteriorly over the shoulders and sides of the face" suggests that this species agrees with *leucopus* in having a posterior center of origin. In the earliest stage of the dorsal molt of *P. polionotus* recorded by Laffoday (1957), patches of new pelage occurred below the eye, on the hind leg, and as a fine line along the side of the body. This condition might be indicative of progression from anterior and posterior centers which develop nearly simultaneously.

The above data suggest that there are two principal foci from which the dorsal molt in *Peromyscus* may proceed and that in some species the anterior center is most active, in some the posterior center has precedence, and in still others both centers are equally "potent." It remains, however, to determine just how consistent the point of origin of the molt may be in different species and subspecies.

Since the growing hairs of the subadult pelage are concealed for a time by the longer juvenal pelage, the post-juvenal molt may actually be rather well advanced before any actual color change is discernible. Following its initial appearance on the dorsum, the new pelage typically spreads along the lower sides then extends upwards on each side of the body to coalesce along the middorsal line of the shoulder or upper back region, forming a saddle. Collins (1923) indicates that subspecific differences may exist in the position of this saddle. From the saddle stage, the molt proceeds anteriorly and posteriorly, sometimes at different rates. The juvenal pelage of the crown and rump is normally the last to be replaced, and frequently these patches of old pelage persist for an extended period after the molt has been completed on the remainder of the body (Brown, 1963; Collins, 1923; Hoffmeister, 1951; Layne, 1966; McCabe and Blanchard, 1950; Storer *et al.*, 1944). In some species the old pelage persists longer on the posterior dorsum than on the crown, as is true of *P. boylei* (Brown, 1963), whereas in others the reverse obtains.

In *P. m. gambeli* the molt of the crown and cheeks occurs toward the end of the molt of the rest of the dorsum or soon after it has been completed (McCabe and Blanchard, 1950), the pelage on the crown being replaced by the posterior extension of new hair originating from a center on the tip of the snout and forward extension from the shoulder region (Collins, 1923). A similar mode of replacement may occur in other species. In *truei* and *californicus*, the juvenal pelage on the crown may persist for such a long time after the post-juvenal molt has been completed elsewhere that its replacement might more properly be regarded as part of the post-subadult molt (McCabe and Blanchard, 1950). Hoffmeister (1951) suggested that juvenal, or perhaps post-juvenal, pelage persisted on the top of the snout of *truei* throughout the life of the animal. Collins (1923) also demonstrated the persistence of juvenal hairs in the post-juvenal pelage of *P. m. gambeli, P. eremicus* and *P. californicus,* thus indicating that the post-juvenal molt does not involve total replacement of the first coat.

The various phases of the post-juvenal molt vary between themselves and interspecifically in their timing. Of the three species studied by McCabe and Blanchard (1950), the molt of *maniculatus* generally proceeds relatively smoothly without prolonged arrests or lags. In *truei* and *californicus*, however, the molt frequently pauses on one part of the body before proceeding, the arrest lasting up to three weeks. Irregular patches of juvenal pelage also frequently persist for some time in the subadult coat.

Post-subadult Molt.—The post-juvenal or subadult pelage is typically more brownish than the juvenal coat but duller than that of the full adult. These differences are not invariable (Hoffmeister,

1951; Storer *et al.*, 1944). Knowledge of the second developmental molt leading to the first adult pelage in *Peromyscus* is less complete than in the case of the post-juvenal molt. That the subadult molt is properly regarded as a true developmental rather than a seasonal molt is evidenced by the fact that it occurs as a regular age-related phenomenon independent of season (McCabe and Blanchard, 1950). This molt may occasionally be omitted; this occurred in less than five per cent of the *P. truei* and *P. californicus* studied by McCabe and Blanchard (1950). The post-subadult molt also appears to be generally less regular in timing and pattern than the post-juvenal molt. The most detailed accounts of this molt in *Peromyscus* are those of Collins (1923) and McCabe and Blanchard (1950) for *P. maniculatus, P. truei,* and *P. californicus.*

Chronology: In those species for which data are available, the adult molt begins at an average age of about 15 (10–19) weeks. Collins (1923) stated that the molt in *P. maniculatus gambeli* began at 10 to 18 weeks of age, with an average of slightly over 12 weeks. McCabe and Blanchard (1950) indicate that the post-subadult molt in this subspecies does not normally begin until the sixteenth week, although abnormal molting may be in progress between the twelfth to fifteenth weeks. The molt is essentially complete in all individuals by the twenty-first week of age.

The post-subadult molt of *P. boylei* begins between 15 and 19 weeks of age (mode = 16) and is complete by 21 to 28 days (mode = 22) (Brown, 1963). Although *P. truei* may undergo abortive, partial molts earlier, the definitive molt does not begin until 17 to 21 weeks of age. It has been completed in most individuals by the latter age (McCabe and Blanchard, 1950). Young of *P. floridanus* show evidence of the post-subadult molt between the ages of 13 and 21 weeks (Layne, 1966). Despite its large size, *P. thomasi* does not appear to lag behind smaller members of the genus in the acquisition of the first adult pelage. In one individual observed by Musser (*in litt.*), the molt was in an early stage at 15 weeks and complete by 17 weeks. Other young were in various stages of the molt between 17 and 22 weeks of age.

P. californicus appears to commence this molt somewhat later than species mentioned earlier. The youngest molting individuals of this species noted by McCabe and Blanchard (1950) were 19 weeks of age. Most 20-week-old young were in initial stages of the molt, although the molt had not begun in one mouse at 26 weeks of age.

Golley *et al.* (1966) stated that female *P. polionotus* older than 35 days and males 65 to 70 days of age exhibited adult molt (assumed to be the post-subadult molt as defined here).

Pattern: In its general features the post-subadult molt resembles the post-juvenal molt, although there appear to be some significant differences. An intermediate partial molt occurring in the interim between the completion of the post-juvenal and initiation of the post-subadult molt has been noted in *P. maniculatus gambeli* and *P. truei* by McCabe and Blanchard (1950).

This incomplete molt involves the appearance of small patches of adult pelage behind the shoulder above, and rarely below, the color line. Infrequently, this intermediate stage may be continued into the actual post-subadult molt; in most instances, however, it is suspended in advance of the beginning of the latter molt. Collins (1923) has reported an apparently different pattern in *P. m. gambeli*. He stated that the molt proceeds in two distinct waves, with a pause of about two weeks between the successive phases. The major part of the replacement is completed during the first wave, while the remainder of the adult pelage is acquired in a series of irregular molts which constitute the second phase. It is difficult to reconcile these two descriptions, as the second wave of Collins would seem to correspond to the intermediate molt occurring earlier as described by McCabe and Blanchard (1950). If not simply the result of different interpretations of the same events, this apparent difference may be genetic or environmental (associated with different conditions under which the laboratory stocks were maintained in the two studies).

Collins (1923) and McCabe and Blanchard (1950) state that the origin and course of the post-subadult molt of *P. m. gambeli* is generally similar to the post-juvenal molt. Collins (1923) indicates that it differs in being less regular and always marked by a molt line. There is also no alteration of the general color pattern resulting from the first molt wave in the course of the second phase. McCabe and Blanchard observed that the post-subadult molt of *P. maniculatus* was somewhat less diffuse on the dorsum than in the case of *P. truei* and *P. californicus*.

Brown (1963) found the pattern of the second developmental molt of *P. boylei* to agree with that of the post-juvenal. New pelage on the venter originated on the pectoral region and moved anteriorly and posteriorly. A second center was established in the inguinal area and spread forward. A distinct molt line was sometimes evident on the dorsum, and there was a tendency toward retardation of the molt on the postdorsum.

The molt of the venter in *P. t. truei* begins first or quickly follows the initiation of the replacement on the upper parts (McCabe and Blanchard, 1950). Its course is difficult to follow, as it progresses erratically and rapidly. The original centers may be the same as those of the post-juvenal molt. According to McCabe and Blanchard (1950), the centers of origin and general progression of the molt on the upper parts are the same as for the post-juvenal molt, although the progress of the molt is more continuous, the moving front being characteristically marked by a molt line, and the replacement on the dorsum being more diffuse. In contrast, Hoffmeister (1951) states that the post-subadult molt of *P. t. gilberti* begins between the ears, where the post-juvenal molt terminates, and on the lower sides, moving posteriorly and upwards, respectively, from the two centers.

The post-subadult molt of *P. californicus* is also generally similar to the post-juvenal molt and, as in the case of *truei,* proceeds with less hesitation and tends to be more extensive over the dorsum (McCabe and Blanchard, 1950). The specific site or sites of origin of the molt in this species have not as yet been satisfactorily demonstrated (McCabe and Blanchard, 1950).

P. floridanus appears to agree with other members of the genus in having a post-subadult molt pattern corresponding to the post-juvenal one, although it is apparently less regular (Layne, 1966).

According to Musser (*in litt.*) the molt of *P. thomasi,* although difficult to trace, also follows the post-juvenal pattern, beginning along the lower sides, moving dorsad to fuse along the midline, then proceeding anteriorly and posteriorly.

Behavioral Development

Behavioral ontogeny of *Peromyscus* has received less attention than morphological aspects of development. With the exception of studies by King (1958, 1961a, 1963), King and Eleftheriou (1959),

King and Shea (1958, 1959), and Clark (1936), the data on this aspect of the biology of *Peromyscus* are descriptive and mostly of such a general nature as to be of limited value for critical comparisons between species or subspecies. Admittedly, even purely descriptive behavioral information is more difficult to obtain and interpret than other types of ontogenetic data.

Handling of mothers or young or other manipulations associated with behavioral observations may, in themselves, alter the behavior being studied. Evidence for this is found in studies of laboratory rodents (e.g., Denenberg and Morton, 1962; Denenberg *et al.*, 1962), and such effects may be even more pronounced in wild species. The possibility that the behavioral responses of young of different species or subspecies to a given environmental factor may differ either quantitatively or qualitatively, or both, further complicates the problem of making valid comparisons of behavioral ontogeny. For example, the young of a species or subspecies of more nervous temperament may be more strongly motivated to seek cover when removed from the nest for examination and thus make greater efforts to move than young of a more docile species or subspecies. As a consequence, the young of the first kind might give the impression of faster neuromuscular maturation. Differential effects of handling on subsequent behavior of young have been demonstrated in two subspecies of *P. maniculatus* (King and Eleftheriou, 1959). Although more emphasis on systematic testing of various aspects of behavioral development should be given in further work on this phase of the biology of *Peromyscus,* the difficulty of determining whether differences in certain behaviors are indicative of maturation level or simply the result of differences in responses to the testing situation will remain.

Another problem that arises in an attempt to develop complete schedules of behavioral development in different species or subspecies is the difference in probability of observing certain behavioral acts in a given unit of time. Thus, certain behaviors of adults, such as drumming or tail shaking, that occur relatively rarely and under rather specific circumstances cannot be expected to be recorded as readily during development as a more generally occurring act such as grooming. Further, if a given behavior is commoner in one species than another, it might be recorded at an earlier age in the young of the first species even though there was no basic difference in the chronology of its development in the two species.

In the following account, an attempt is made to summarize what is presently known concerning behavioral development in the genus *Peromyscus*. The categories under which various types of behavior are discussed have been selected largely as a matter of convenience for descriptive purposes. The behaviors included within each are neither equivalent in general complexity nor necessarily related to a specific function. The categories are also not mutually exclusive. For example, activity and temperament are obviously interrelated as are activity and locomotion. In fact locomotor capacities underlie most of the behavioral characteristics considered here.

LOCOMOTION.—Although the young of *Peromyscus* are essentially helpless at birth, development of locomotor ability begins within the first few days of postnatal life. *P. polionotus* have slight coordination and crawl by the fifth day of age and by the seventh or eighth walk with good coordination, run with the venter dragging, and attempt to climb (Laffoday, 1957; Rand and Host, 1942). At two weeks of age they run and jump like adults and land on their feet when dropped. Digging, although ineffective because the forefeet are not employed, is first noted at ten days, and by the fifteenth day the young dig efficiently in the manner of adults.

Various aspects of the development of locomotor capacity have been studied in three subspecies of *P. maniculatus*. Young of *P. m. gambeli* can raise the head and roll or drag the body a few inches, but are unable to stand, at one week (McCabe and Blanchard, 1950). At two weeks they can run slowly to a distance of 15 or 20 feet and show a slight suggestion of development of the escape jump. They are able to run and dodge well by the third week. King (1958) studied the chronological development of clinging and climbing ability of *P. m. bairdi* and *P. m. gracilis*. The young were able to cling to a vertical screen at five to six days of age, climb vertically at nine to ten days, cling upside down at about ten days, and climb from the lower to upper surface of the screen as it was slowly turned at 13 to 14 days. *P. m. bairdi* tended to be somewhat more advanced than *gracilis* in these capacities, but the differences were not statistically significant. King (1961*a*) also tested the swimming abilities of the same subspecies. Young of both taxa swam with coordinated movements between ten and 14 days of age. Again *bairdi* gave evidence of more rapid maturation than *gracilis* in that it began to reach a platform in the test apparatus earlier.

P. leucopus begin to exhibit clinging tendencies as early as five days. Some individuals may attempt to walk and climb at this age (personal observations). By the tenth day the young climb, cling, and walk well, and at 15 days their locomotor abilities are very good. The only published information on locomotor development of the closely related *P. gossypinus* is the statement that movements are more directed in young about ten days old (Pournelle, 1952).

McCabe and Blanchard (1950) note that *P. truei* at one week are distinctly less advanced in locomotor ability than *P. maniculatus* of the same age and the lag continues into the second week, at which time they run poorly for less than two feet and can execute only abortive escape jumps. By the third week, they run fast and jump and have good endurance. At four weeks they are faster and stronger than *maniculatus,* although their escape jump is less developed.

Young *P. floridanus* can elevate the forepart of the body at one day of age and crawl and climb clumsily at two (Layne, 1966). By the fifth day they can move with a normal walking gait, although the venter drags. By the eighth day they walk with the body better supported and the tail curved stiffly upward. At 13 or 14 days their locomotor patterns are much like the adult. Young 20 days old were able to climb out of a four-inch box and gave abrupt escape leaps.

Young *P. californicus* exhibit an early precocity in their ability to move, but appear to lose this lead in comparison with *maniculatus* and *truei* later in development (McCabe and Blanchard, 1950). At one week they can almost stand and can scramble many feet before tiring. By the second week, their locomotor capacities are about equivalent to those of *truei.* At three weeks they are behind *truei* and *maniculatus,* running slowly and awkwardly and showing only incipient development of the escape jump. The species still appears to be less competent in its movements than *truei* or *maniculatus* at the fourth week. On the basis of these observations, it is tempting to assume that *californicus* matures more slowly than the smaller species. On the other hand, differences in temperament and activity between these species complicate comparisons. *P. californicus* is calm and gentle and young are not inclined to seek cover or escape. Thus, the apparent lag in development of locomotor capacities of this species may be at least partly attributable to a lesser tendency to move around as compared with the more active young of *maniculatus* and *truei*.

Musser (*in litt.*) made some observations on the development of locomotion in *P. thomasi*, and I have obtained details of this aspect of behavioral ontogeny for *P. megalops*. Young of the first species crawl awkwardly by the sixth day and their coordination is noticeably improved by the eleventh. By the twenty-sixth day they are running and crawling well.

P. megalops attempt to crawl at five days, and one young of this age being dragged outside the nest on the teats of the female made coordinated walking movements with the limbs, although it was unable to keep on its feet. By ten days of age, the young can cling momentarily to a finger before falling off, also, they crawl with fairly well coordinated limb movements. Young begin to sit up, though unsteadily, and walk with the typical quadrupedal gait at 12 days. One individual placed on the edge of the nest cup promptly crawled into the nest, although the eyes were closed and the slope of the rim was the same toward and away from the nest. The individual may have been guided into the nest by vocal cues from the young and adult still in the nest. Oriented movements such as these have been noted in young of other species while the eyes were still sealed. Fourteen-day-old young walk steadier and with the tail curved upward.

Young 17 days of age climbed a three-inch wall and balanced themselves on the narrow edge. They also made jumps to the ground from a height of six inches, and ran off as soon as they landed. Although the eyes were closed, the movements of the young were directionally well oriented. One individual was observed digging vigorously in the cotton nesting material in typical adult fashion. Standing on the hind feet and extending the body was first noted at 20 days. Young at this age clung tenaciously to objects they were climbing on and were difficult to pull off. By 23 days the young showed more caution about jumping from an elevated surface than earlier. When placed on a balance pan about six inches above the table top, they would move to a point on the perimeter, peer over the edge, then back up and move to another point to repeat the performance. The movements of these young were generally deliberate and rather slow, as is typical of the adults.

Some of the data presently available suggest that differences in the development of locomotor patterns in the young may have an ecological basis. For example, the early appearance of digging behavior in *polionotus* correlates with the semi-fossorial habits of this species. Although the evidence is highly subjective, I have gained the impression that young of *floridanus*, another burrow-dwelling, highly terrestrial form, are less inclined to climb than are those of semi-arboreal species. Furthermore, strongly arboreal types may show a greater inclination to freeze and move more deliberately, as such behavior would appear to be more advantageous to a form not highly specialized for rapid movement above ground (Layne, 1960). Clark (1936) studied climbing behavior in ground-dwelling and semi-arboreal stocks of *Peromyscus* and obtained results which indicate the existence of locomotor differences correlated with habits in young at an early age. The tests were based on the fact that when young are placed on an inclined plane in the absence of light they tend to crawl upwards at an angle. Young of non-arboreal subspecies and semi-arboreal subspecies of *P. maniculatus* and the semi-arboreal *P. californicus* and *P. boylei* were tested on slopes of 20, 40, and 60 degrees. The arboreal forms all had higher mean orientation angles at each slope than the non-arboreal stocks. They were especially sensitive to the 20-degree slope as compared with the non-arboreal mice. Differences in locomotor performances between young *P. m. gracilis* and *P. m. bairdi* also appear to have an ecological basis (King, 1958).

RIGHTING.—Although neonates and young a few days of age may attempt to right themselves and eventually do so, the ability to immediately assume the normal dorsum up position when rolled or pushed over represents a significant advance in neuromotor development. Data on this aspect of behavioral development are available for only a few species of *Peromyscus*, and in only one, *P. maniculatus*, has righting behavior been systematically investigated.

King (1958) tested young of *P. m. bairdi* and *P. m. gracilis* for righting ability and found a fully developed righting response at mean ages of 7.2 and 8.9 days, respectively, although in both sub-species the capacity to right appeared earlier. In comparison, *P. floridanus* exhibits a good righting response at five days (Layne, 1966), and *P. leucopus* at seven days (personal observations). Young of the large *P. megalops* resist being pushed over at seven days and begin to show good righting ability between eight and 12 days. Individuals at the latter ages are able to right imme-diately (personal observations). Laffoday (1957) noted that young *P. polionotus* were unable to right at three days but could even-tually do so at five, and Rand and Host (1942) record righting attempts in this species at four days of age. On the basis of these observations, *polionotus* agrees fairly closely with *maniculatus* and *leucopus* in the age of full development of the righting response.

Vibrissae Response.—The age at which young give a recognizable response to stimulation of the mystacial vibrissae has been recorded for three species of *Peromyscus*. *P. floridanus* at two days of age jerk the head away from firm displacement of the vibrissae and make pawing movements toward the source of the stimulus (Layne, 1966). A similar response was given to a lighter touch of the hairs at one week, and at 12 days one young recoiled so violently from such a stimulus that it performed a backward somersault.

The earliest response to movement of vibrissae noted in *P. leu-copus* was at five days (personal observations), the young moved the head but showed no clear orientation of the movement with regard to the direction of the stimulus. By nine days of age all individuals in three litters tested for this behavior showed a definite response. All young at this and later ages moved the head toward the stimulus, although one individual at eight days gave a weak response by withdrawing the head. By the thirteenth day, the young were more variable in their responses, some ignored the stimulus and others turned toward it. There was increasingly less overt vibrissae response from this age on, although a mouse 20 days old reacted slightly to a touch.

The earliest age a vibrissal response was recorded in two litters of *P. megalops* I studied was eight days, the mouse attempted to bring its head into contact with the probe with which the vibrissae were stroked. Two nine-day-old individuals also gave similar positive responses. A pronounced reaction to a touch of the vibrissae was elicited from one young at 13 days. It raised its head and turned toward the side stimulated, giving every indication of attempting to bring the rostrum into contact with the source of the stimulus.

These few data suggest that the chronological age of appearance of a visible response to stimulation of the vibrissae is independent of size over a relatively wide range of body weight. An additional point of interest is the tenuous possibility of species differences in the type of response elicited by vibrissae stimulation, *floridanus* appears to turn away from and *megalops* and *leucopus* to turn toward the stimulus. If such a difference should prove to be real when further data have been accumulated, it may prove to be somehow related to the ecology of the species concerned. Of the three species noted above, *floridanus* is semi-fossorial while the others tend to be more arboreal.

Grooming.—Grooming is an important element of adult behavior in *Peromyscus,* as in other mammals. While it apparently evolved in connection with care of integument, it has secondarily acquired other functions, including social ones. The chronological age of appearance of this functionally important and complex behavior should provide a rather good index of neuromuscular maturation. Unfortunately the ontogeny of grooming has not been studied in detail in any species of *Peromyscus,* and general information is now available for only a few species.

Young *P. polionotus* have been observed washing the face in response to ether fumes at seven days (Laffoday, 1957). They can sit up and groom in the adult manner by the tenth day (Rand and Host, 1942). The first indication of grooming efforts in *P. floridanus* was noted at 11 days (Layne, 1966). The young of this species could sit up unsteadily to groom at 13 days and two days later groomed with well-coordinated movements of both fore and hind feet.

I observed incipient grooming at ten days in two litters of *P. leucopus,* scratching with the hind foot at 11, and fully developed grooming by 15. One individual in two litters of *P. megalops* was seen to make an abortive grooming attempt (brushing at the snout with forefeet) at one week, and in both litters 12-day-old young washed their faces while lying on their sides. They sat partly erect to groom at 13 days, and at 14 were observed scratching the ears with a hind foot. Mutual grooming was noted in a litter of 26 days (personal observations). Musser (*in litt.*) notes scratching of the body in young *P. thomasi* at 11 days.

ROOTING.—As in other mammals, young *Peromyscus* tend to force the snout into cracks or crevices, a behavior probably functionally associated with nursing. I have specifically noted this behavior in *P. floridanus* by the second day of age and in *P. megalops* by the fifth day. The tendency of the young to push beneath each other and press against objects often exhibited to a fairly advanced age may be a further extension of this type of behavior. I have noted tendencies of the young to aggregate, in *P. floridanus* and *P. megalops* up to three weeks of age, and Rand and Host (1942) state that *P. polionotus* at 20 days huddle if removed from the mother. Thigmotactic responses, however, seem to reach their peak development prior to the opening of the eyes.

SUCKLING.—During the first two to three weeks of postnatal life, young *Peromyscus* are usually found on the teats of the mother when she is in the nest. The amount of time the young actually spend in nursing is not known. As the time of weaning approaches, the young are found off the teats with increasing frequency.

Nursing females generally crouch or lie in a normal dorsum-up position, and the young burrow beneath the body to reach the teats. My own observations indicate that the posterior teats tend to be utilized more than the anterior ones. This correlates with the fact that in those species (members of the subgenus *Haplomylomys* and at least *P. crinitus* and *P. megalops* of the subgenus *Peromyscus*) with a reduced number of mammae, it is the anterior-most pair that has been lost.

When she wishes to leave the nest under normal conditions, the female apparently disengages the suckling young by stretching her body and moving off slowly. When the young are disturbed, however, they may cling tightly and the mother must remove them forcibly, often with considerable difficulty (King, 1963).

Under certain circumstances the nestlings may be dragged from the nest while clinging to the teats of the mother. This behavior is usually observed when the mother is startled and makes a precipitous exit from the nest. The weight of the young does not seem to interfere with the female's movements nor slow her down appreciably, and at such times the mother appears unmindful of the presence of the attached young. When very young, the nestlings may be roughly dragged and bounced on their backs or sides, but older young generally manage to keep the dorsum uppermost while being pulled along and may even attempt to keep up with the parent by walking or running. Even at an age of only a few days the young are often able to retain their hold despite the most frantic running, leaping, or climbing movements of the mother. Observations by several authors attest to the tenacity with which the young can cling to the teats. Pournelle (1952) noted that attempts to pull ten-day-old *P. gossypinus* off the mother could result in tearing the teats from the body. Rand and Host (1942) demonstrated that at six days *P. polionotus* could grip the teats firmly enough to suspend the weight of the mother and other young (total weight 23.4 g) for three or four seconds. An eight-day-old young kept the female (17 g) suspended for 30 seconds. I have found also that young *P. floridanus* may grip the teats so tightly that it is extremely difficult to remove them without injury to the mother.

Present data indicate that the tendency of young to cling to the teats varies with age. Young *P. floridanus* do not attach as tightly to the teats for the first few days after birth as they do later (Layne, 1966). The grip seems to improve by the fourth day and by one week is well developed. At 17 days, young may still cling tenaciously, but they tend to come off more easily after this age. I have observed a generally similar pattern in clinging tendencies in two other species. Young *P. megalops* drop off the teats readily the first and second day after birth and again begin to drop off more readily from about three weeks of age on. *P. leucopus* can cling tightly by the seventh day, but the decline in the response appears to begin earlier than in *floridanus* or *megalops*. Young have been recorded as coming off the teats easily at ten to 13 days. Pournelle (1952) notes that the young of *P. gossypinus* cling most tenaciously at about ten days of age, and Rand and Host (1942) state that the clinging ability of young *P. polionotus* reaches its highest development at about eight days. McCabe and Blanchard (1950) found that young *P. truei* and *P. maniculatus* were almost constantly attached to the teats until three weeks of age. At four weeks *P. truei* tended to drop off more readily.

The results of an experimental investigation (King, 1963) of nipple-clinging tendencies in *P. maniculatus gracilis* and *P. m. bairdi,* between three and 20 days of age, indicate a general increase in clinging tendency in the second subspecies to a peak at seven to ten days followed by a general decline. There was no such obvious age trend in *P. m. gracilis,* and the two subspecies differed in other respects in regard to this behavior. It is likely that additional testing of nipple clinging in young of this genus would reveal further differences at both the specific and subspecific levels.

Variation in nipple-clinging tendencies of young within a given population may result from various causes. King (1963) noted that hungry or malnourished infants seem to be more strongly inclined to cling tenaciously than better nourished individuals, and my own observations on several species suggest that young of nervous or wild females are more likely to cling than are those of more docile mothers. As young of wild females also appear to be in poorer condition than those of tamer individuals, there is some question as to the primary cause of this intensification of nipple clinging.

Nipple-clinging tendency appears to increase within a few days of birth and becomes essentially fully developed at about a week to ten days, which roughly corresponds to the time of appearance and early growth of the incisors. This suggests that the incisors may aid the young in grasping the nipple. Rand and Host (1942) were of the opinion that incisor structure or development did not contribute to clinging ability because the clinging response was already highly developed at about the time the incisors were barely visible above the gums. This observation does not necessarily rule out the hypothesis, as the incisor teeth often appear just below or at the gum several days before they extend through the soft tissues and could probably function in holding the teat even at this stage. The incisor teeth of *Peromyscus* do not appear to be specialized in connection with nipple clinging as has been observed in some rodents (e.g., Birkenholz and Wirtz, 1965; Hamilton, 1953) .

The principal adaptive value of nipple clinging is probably reduction of litter losses through predation, although this assumption cannot presently be supported with good observational or experimental evidence. The fact that females sometimes flee the nest without the young or lose some of the attached young in their flight does not necessarily contradict this interpretation of the adaptiveness of the behavior. Even if only a slightly higher proportion of litters were saved in this manner than would otherwise be the case, the behavior would benefit the population. The loss of some young along the escape route might even contribute to a higher probability of survival of the mother and those young that remained attached, as fewer young might give the female a somewhat greater advantage in speed and agility and the young that fall off might serve to divert the attention of the predator. One might also speculate that the weaker members of the litter would perhaps have a greater likelihood of dropping off the teats during an escape than the stronger and more vigorous individuals, which would also have implications for the betterment of the population as a whole.

Variation in nipple-clinging tendencies between species and subspecies are probably also of an adaptive nature. King (1963) related differences in clinging behavior between two subspecies of *P. maniculatus* to the divergence in their habits and ecology.

Nipple clinging is occasionally observed under conditions that suggest it may function in ways other than escape behavior. I have seen *P. floridanus* three weeks old grasp the teats of the mother and walk around the cage with her, neither individual showing any sign of disturbance. Such behavior might tend to keep the young and female together during early forays from the nest and in event of danger aid the young in quickly gaining refuge. Although all species of *Peromyscus* commonly carry young in the mouth, Rand and Host (1942) suggested that young of *P. polionotus* may often be transported to a new burrow by clinging to the teats of the mother.

Although the mechanisms of nipple-clinging behavior of *Peromyscus* have not been subjected to critical analysis, somewhat casual observation gives the impression that the infant and not the mother plays the active role. When a female in the nest is startled some stimulus must reach the nursing young causing reflex clamping of the jaws so that the female drags them involuntarily from the nest when she hastily departs. Evidence for this conjecture is that the characteristic differences in nipple-clinging tendencies between young *P. m. bairdi* and *P. m. gracilis* persist in young raised by foster mothers of the other subspecies (King, 1963). I have also noted that if females with young are disturbed, but not immediately driven from the nest, they will frequently divest themselves of the attached young before leaving.

VOCALIZATIONS.—Published data on sounds produced by young *Peromyscus* pertain primarily to those within the human audible range, although higher frequency sounds are also given. In view of the fact that vocal behavior of the young may play an important role in integrating adult-infant relationships and may be involved in interactions between the members of a litter, this aspect of the developmental biology of the genus has received surprisingly little attention. Much of the data presently available are insufficiently detailed to provide a satisfactory basis for interpretations of the functions of particular sounds in a given species or make possible critical comparisons of vocal patterns at the specific or subspecific level. King (1963) and Hart and King (1966) have presented sonagrams of squeaks of young of different species and subspecies, and a definitive analysis of vocal behavior and its ontogeny in *Peromyscus* will probably depend upon further utilization of this technique (see Chapter 12).

The basic vocal repertory of the young of most species studied to date appears to consist of squeaks and softer, lower pitched sucking, smacking, or clicking sounds. Both types have previously been noted in *P. truei* and *P. floridanus* (Layne, 1966; McCabe and Blanchard, 1950), and I have also recorded them in *P. leucopus* and *P. megalops*. Although only squeaks have been reported for young *P. polionotus* (Laffoday, 1957; Rand and Host, 1942) and *P. maniculatus* (Eisenberg, 1963; King, 1963; McCabe and Blanchard, 1950), it is probable that the other sounds are produced by these species as well. I have noted an additional sound, a very faint grating or grinding note, in young *P. megalops*. This sound was heard once from a 15-day-old young and twice when young 24 and 26 days of age were in the nest with the female. In the latter instances the source of the sound, whether the young or adult, was not determined. I have also heard a similar sound from three-week-old *floridanus* on a single occasion. The characteristics of this sound suggest that it might be produced by grinding of the teeth. Although chattering of incisors is not frequent in *Peromyscus*, it is occasionally done by adults of at least some species (Eisenberg, 1962). *P. californicus* is much more vocal than other species thus far studied. Its young utter a variety of chirps, chuckles, mewing notes, and squeaks (McCabe and Blanchard, 1950), and the adults are highly vocal as well. The more complex vocal pattern of this species correlates with its relatively highly developed social organization (Eisenberg, 1962).

Although vagueness of descriptions make close comparison impossible, there appear to be differences in the frequency, intensity, and other characteristics of the squeaks of different species and subspecies. Rand and Host (1942) state that *polionotus* give "tiny" squeaks, and I have observed that the squeaks of recently born *leucopus* are also very faint and high-pitched. Based on personal impressions, the squeaks of *floridanus* at the same age are decidedly louder than those of the latter species, while day-old *megalops* utter especially loud, powerful, pulsed squeaks reminiscent of the "peeping" of a newly hatched bird. These limited observations suggest a correlation between body size and intensity and probably other of the characteristics of the sounds produced. Recorded squeaks of three-day-old *P. maniculatus bairdi* and *P. m. gracilis* differed in several respects when analyzed, and those of *P. floridanus* were even more divergent (King, 1963). The two subspecies of *P. maniculatus* also differed in the number of squeaks uttered per unit of time when measured under similar conditions (Hart and King, 1966).

As in adults, variations in the types of squeaks produced by a given species are apparent, and particular types may tend to be relatively consistent in the age at which they appear. The squeaks given at the early ages seem to be associated with discomfort and may be regarded as distress signals which function to attract the attention of the mother. For example, on the basis of rather casual observations on several species, I have the impression that females tend to retrieve squeaking individuals more rapidly than silent ones. As development proceeds, the young may give louder, shorter, and sharper squeaks when disturbed. This type may correspond to the sharp, metallic "chit" uttered by adults during aggressive encounters. The occurrence of these two types of squeaks in young *truei* is suggested by the descriptions of vocalizations in McCabe and

Blanchard (1950), and I have noted them in *megalops*. In the latter, louder, sharper squeaks were given by 7- and 8-day young when they were handled, and a typical aggressive squeak was given by one 26-day-old individual when it was being pushed into a box.

In addition to age differences in the appearance of particular vocalizations, the overall frequency of vocalization, particularly squeaks, appears to decline with increasing age in the young of most species that have been studied. Squeaking appears to diminish sooner than the sucking notes. From birth through the first few days of age the young of all species tend to be highly vociferous. McCabe and Blanchard (1950) report that *truei* are less vocal at one week than earlier, and at two weeks, although sucking notes were given during handling, the young rarely uttered squeaks. By the fifth week the young are like adults in being normally silent, squeaking only when hurt. *P. maniculatus* are usually silent at one week and squeak only when hurt after the second week. *P. floridanus* is also less vociferous at one week than during the first few days after birth and seldom make sounds after the seventeenth day of age (Layne, 1966). Older young, like the adults of this species, are almost invariably silent even when handled roughly.

A few observations on young of *P. leucopus, P. megalops,* and *P. thomasi* indicate similar trends. By the sixth or seventh day young of the first species squeak only infrequently, although sucking notes may be given almost continuously. From about the tenth day on, they squeak only when handled. By the eighth day, young *megalops* squeak noticeably less than earlier and tend to squeak only when provoked from about the fourteenth day on. Two young of a litter of *thomasi* appeared quieter on the fifth day than earlier. *P. californicus* differs from other species in continuing to be quite vocal beyond five weeks of age, although the young of this species also apparently squeak less at one week and only when frightened or injured by the fourth week (McCabe and Blanchard, 1950).

It might be concluded that the greater frequency of vocalization, particularly squeaks, at an early age functions in establishing the mother-infant bond and increasing the probability that young will be retrieved if accidentally scattered or lost from the nest. In this connection, Noirot (1964) has presented evidence that the distress calls of young laboratory mice (*Mus*) exert a strong influence on the retrieving and nest-building behavior of females. Once the mother-infant bond is established and the young are less helpless and can stand exposure for longer periods without ill effects, vocalization diminishes, as any advantages of its continuation beyond the period of maximum usefulness would be counteracted by the detrimental effects of attracting predators to the nest.

DRUMMING AND TAIL SHAKING.—Species differences in the chronological age at which drumming appears and the tendency of the young to perform this behavior apparently exist in the genus, and further study will probably reveal that such differences are reflected in adult behavior as well. McCabe and Blanchard (1950) recorded no drumming in *maniculatus* at two weeks. The frequency of drumming when the young were isolated was highest at three weeks and began to diminish by four weeks. In contrast, *truei* drummed only occasionally at two weeks and not at all at three or four. Young *P. californicus* did no drumming, although this behavior has been noted in adults of this species (Eisenberg, 1963). *P. floridanus* was first recorded drumming at 35 days (Layne, 1966). I never observed drumming in the young of several litters of *P. leucopus* and *P. megalops*.

A violent, convulsive quivering of the tail is often given by *Peromyscus,* in certain stressful situations. This behavior is apparently widespread among rodents. I have recorded this behavior for the first time in *leucopus* at 18 days of age, and there appear to be no published observations of its occurrence in the young of other members of the genus.

DEFENSE AND AVOIDANCE BEHAVIOR.—Newborn *Peromyscus* do not exhibit any behavior that can be clearly classified as defensive or avoidance, and there appear to be specific and subspecific differences in the chronology of development of such behavior. King (1958) has made the only systematic study of the ontogeny of some aspects of such behavior in the genus.

Young *P. floridanus* first exhibit resistance to being picked up and attempt to squirm out of the hand at five days of age; at ten days their escape tendencies are more pronounced. During early postnatal life, the avoidance responses of the young are rather generalized, but by the nineteenth day they are clearly aware of the observer and their escape attempts are more oriented. One young at 20 days flipped onto its back and kicked with the fore and hind feet when a finger was extended toward it. By the third week of age the defense and avoidance patterns of the young are generally similar to those of adults (Layne, 1966).

McCabe and Blanchard (1950) provided some data on the development of aggressive behavior in *P. maniculatus, P. truei,* and *P. californicus.* Young of the first species attempt to move away when a finger is placed near the nose when a week old. By the second week they bite feebly, and a week later more forcibly. At three weeks, they show a tendency to freeze when disturbed. In contrast, young *truei* struggle to escape and bite hard and readily at two weeks. *P. californicus* will not bite at two weeks; they show no inclination to hide or freeze when disturbed. At three weeks they may bite gently when sufficiently provoked, and at this age exhibit a tendency to freeze.

King (1958) determined the age of appearance of two types of agonistic behavior in two subspecies of *P. maniculatus. P. m. bairdi* responded to the approach of an object by falling back and pushing with the feet at a median age of nine days, as compared with 11 days for *P. m. gracilis.* The median ages at which *bairdi* and *gracilis* young fought forceps when picked up by the tail was 12 and 14 days, respectively.

I recorded the first overt response to handling at seven days of age in *megalops.* In this case, a young gave the impression of "lunging" toward the hand when touched and simultaneously uttered a sharp squeak. Individuals eight days old made escape attempts when held, and by the seventeenth day young actively sought to escape being picked up; they struggled violently when held. Gentle bites were given at 25 days. The earliest record of biting obtained for several litters of *P. leucopus* was 15 days (personal observations). Laffoday (1957) recorded biting in *P. polionotus* at 11 days of age.

ACTIVITY.—Young *Peromyscus* tend to increase their general activity along with neuromotor improvement, but there is evidence in at least some species of periods of especially heightened activity which correspond chronologically with the opening of the auditory canal and eyes. Such a phase of hyperactivity or jumpiness has been described by King (1958) as reaching a peak at about 15 days in *P. maniculatus bairdi* and at about 20 days in *P. m. gracilis*. A quantum jump in alertness and exploratory tendency was noted in *P. floridanus* at 13 days coincident with the development of hearing, and a second increase occurred about the seventeenth day when the eyes opened (Layne, 1966). Young *P. gossypinus,* which develop more rapidly than *P. floridanus,* exhibit the alert, nervous manner characteristic of adults at about ten days and just prior to eye opening show an increased responsiveness (Pournelle, 1952).

Musser *(in litt.)* observed 19-day-old young *P. thomasi* exploring their cage and sniffing and gnawing at objects encountered. Young of several litters of *P. leucopus* that I observed became very active between about the fourteenth and nineteenth day of age. The shift in activity level around the age of eye opening was especially pronounced in several young *P. megalops* studied. Up to the eleventh day these young were easy to handle and rather sluggish when removed from the nest for examination, tending to remain in one place. Between the fourteenth and sixteenth days they began to increase their activity, and from about the seventeenth day until a short time after the eyes opened, they became extremely active and difficult to handle. Following this period they again became calm and less active.

TEMPERAMENT.—Inter- and intraspecific differences in temperament are expressed relatively early in ontogenetic development. A particularly clear illustration of this is provided by McCabe and Blanchard's (1950) study of *P. californicus, P. truei,* and *P. maniculatus.* The pronounced differences in the temperament of the adults of these species are clearly evident in their young by the second week of age. The same is true of other species. The adults of *P. megalops* are docile and easily tamed, and the young exhibit similar characteristics early in their development. In contrast, the adults of *P. leucopus* and *P. floridanus* are more nervous and irritable than *megalops,* and this is evident in the general behavior of the young of these species from an early age (personal observations). King (1958) demonstrated early differences in the temperaments of young of two subspecies of *P. maniculatus.*

Postnatal Growth

Various aspects of postnatal growth in *Peromyscus* have been dealt with in a number of papers. Growth curves of weight and linear measurements of *P. polionotus rhoadsi* were presented by Rand and Host (1942), and Laffoday (1957) gave weights and measurements of young *P. p. subgriseus* at different ages from birth to 60 days. Carmon *et al.* (1963) also studied growth and sources of variability in weight of *P. polionotus.* Growth data for *P. gossypinus* and *P. floridanus* were given by Pournelle (1952) and Layne (1966), respectively. Svihla (1934, 1935, 1936) described growth in weight and linear dimensions of *P. m. artemisiae, P. m. bairdi,* and *P. oreas* from parturition through 58 to 60 days of age, while Dice and Bradley (1942) studied growth in weight of seven subspecies of *P. maniculatus* from 15 days to approximately six months of age, and McCabe and Blanchard (1950) presented a detailed analysis of growth in *P. m. gambeli, P. truei,* and *P. californicus.* King (1958) and King and Eleftheriou (1960) gave data on weights of *P. m. gracilis* and *P. m. bairdi* from birth to about one month of age. Dice (1932, 1933, 1936) analyzed age changes beyond one year in body and skeletal size in several subspecies of *P. maniculatus.* Hoffmeister (1951) described ontogenetic changes in size and proportions of the skull of *P. truei,* and King and Eleftheriou (1960) studied differential growth in skulls and brain weight of *P. maniculatus bairdi* and *P. m. gracilis.* McIntosh (1955) also analyzed differences in relative growth of body and skeletal elements in the latter two subspecies.

In addition to the preceding species, I have obtained growth data for litters of *P. leucopus noveboracensis, P. megalops auritus,* and *P. t. thomasi,* and Guy G. Musser has provided me with preserved specimens of *P. t. thomasi* at 2, 12, and 75 days of age from which weights and measurements have been obtained.

It should be noted that all growth data presently available represent laboratory-reared specimens. In some cases the young have been wild-conceived (*P. floridanus, P. leucopus*), but for the most part the mice involved have been laboratory-bred and the stocks frequently have been maintained in the laboratory for many generations. Thus, the extent to which the growth patterns of these mice correspond to those in natural populations is open to question. Also of importance from the standpoint of comparison of growth patterns of species and subspecies of *Peromyscus* studied under laboratory conditions is the extent to which differences in the conditions under which the mice in different studies were maintained have influenced growth characteristics. Ample evidence of the effects of physical and psychological environmental factors on growth parameters in laboratory rodents has been obtained, and wild species are likely to be even more sensitive to such influences. Finally, it must be admitted that determination of what constitutes the "normal" growth pattern in a given species, subspecies, or lesser population unit is probably more difficult than in the case of developmental characteristics. This is because growth rates of a given stock appear to vary over a wider range in response to environmental influences than do developmental events. The apparently greater lability of growth rates as compared with developmental rates may be adaptive in the sense of allowing the population to better adjust its energy budget to the resources of the environment without sacrificing the advantages of a given developmental schedule.

WEIGHT.—Body weight is probably the best single criterion of growth of the whole organism, and knowledge of growth in mammalian species is largely based on this parameter. An additional advantage of weight as an indicator of over-all growth is the relative ease with which it can be obtained as compared with linear or volumetric measurements. At the same time, weight also has certain drawbacks as an indicator of "true" growth.

Postnatal growth curves of body weight of nine species of *Peromyscus* for periods ranging from birth to four to ten weeks of age are shown in Figure 3. In addition, three widely spaced age/weight values for a tenth species, *P. thomasi,* are also plotted. For the most part the curves are based upon weekly means, either as reported in the original sources or estimated from graphs or time intervals other than weekly. Carmon *et al.* (1963) showed that growth curves based on weights at six-day intervals were as good as those based on weights taken every two days.

Figure 3 shows that in all species weight increase is rapid for the first few weeks following birth and then begins to diminish. The change in rate is gradual in all cases, and it is difficult to identify a specific point of inflection of the curve. The period of most rapid weight increase in most species extends from birth through about the second or third week, and growth begins to level off by the fourth to sixth weeks, roughly the time of puberty in a number

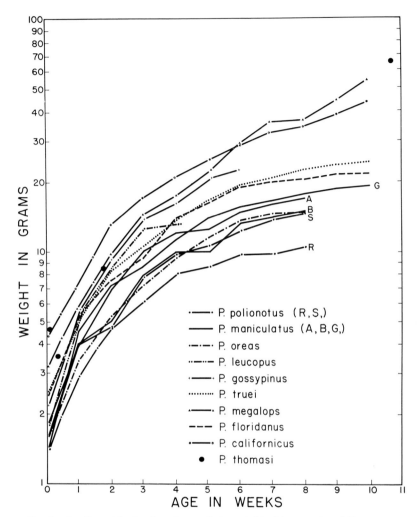

Fig. 3. Semilogarithmic plot of growth in weight of ten species of *Peromyscus*. Curves of two subspecies, *rhoadsi* (R) and *subgriseus* (S) of *P. polionotus* and three subspecies, *artemisiae* (A) , *bairdi* (B) , and *gracilis* (G) of *P. maniculatus* are shown.

of the species. The two largest species for which weight data are presently available, *P. megalops* and *P. thomasi,* are obvious exceptions. The curve of the former is still relatively steep and shows no indication of leveling off at ten weeks, and the three weight values plotted for the latter indicate a closely similar trend. Within the remaining species, there is some indication of an earlier approach to asymptotic levels in smaller species, but the differences in the curves are less impressive than the similarities and the correlation is by no means perfect. Some species of closely similar body size grow at distinctly different rates, while others differing widely in adult size show no clear difference in the shapes of their weight curves. As an example of the first case, *P. gossypinus,* which differs only about 2 g from *P. truei* and *P. floridanus* in mean mature body weight, grows more rapidly than the latter species during the first four weeks of postnatal life and appears to be approaching its asymptotic level at an earlier age. *P. leucopus,* which agrees rather closely in weight with *P. oreas* and the subspecies of *P. maniculatus* shown, has a steeper growth curve for the first three weeks and appears to level out earlier. On the other hand, the curves for *P. californicus, P. floridanus,* and *P. truei* are essentially similar in shape, yet *californicus* is approximately 40 per cent larger than the others.

Comparison of the growth curves of subspecies of *P. polionotus* and *P. maniculatus* suggests that differences at the subspecific level may in some cases be as great or greater than those between species. The data indicate that *P. p. subgriseus* grows faster and levels off more gradually than *P. p. rhoadsi.* Although Svihla (1935) concluded that no marked difference existed in the weight curves of *P. m. bairdi* and *P. m. artemisiae* from birth to 58 days of age, the curves as plotted differ as much or more than those of *P. floridanus* and *P. truei.* King (1958) found no significant weight differences between mean individual growth rates of *P. m. bairdi* and *P. m. gracilis* between birth and 31 days of age, although relative differences in weight changes between the two subspecies were apparent. Dice and Bradley (1942) found that smaller subspecies of *P. maniculatus* tended to slacken their rate of growth at slightly earlier ages than larger subspecies. The weight curves of the seven subspecies figured by these authors are generally similar in shape for the first three months or so, with the exception of *P. m. sonoriensis.* The latter appears to grow more slowly than subspecies of comparable size over the period from about 40 to 100 days.

Comparison of growth rates of various species and subspecies of *Peromyscus* in terms of adult weight is made in Table 7. Values

TABLE 7

POSTNATAL GROWTH IN WEIGHT EXPRESSED AS A PERCENTAGE OF MATURE WEIGHT AND APPROXIMATE AGE AT ONE-HALF GROWTH IN *Peromyscus*

Species and subspecies	Age in weeks										One-half growth, days
	1	2	3	4	5	6	7	8	9	10	
P. polionotus rhoadsi	22	35	46	60	64	70	72	78	—	—	23
P. p. subgriseus	22	35	49	60	79	85	91	—	—	—	22
P. maniculatus artemisiae	19	33	47	56	59	70	75	80	—	—	27
P. m. bairdi	26	32	52	65	84	91	91	97	—	—	20
P. m. gambeli	24	38	45	60	74	81	86	92	98	99	23
P. oreas	18	28	39	52	63	71	77	80	—	—	23
P. leucopus noveboracensis	25	40	58	63	—	—	—	—	—	—	18
P. g. gossypinus	20	33	46	60	69	77	81	85	87	89	23
P. truei	19	30	39	51	62	72	77	—	—	—	27
P. megalops	9	14	20	27	32	41	52	51	64	75	48
P. floridanus	19	27	35	51	60	68	73	74	79	78	28
P. californicus	23	34	44	55	66	73	78	85	90	90	25
*P. thomasi**	7	12	18	24	32	40	49	58	69	79	ca. 49

* Values based on estimated growth curve.

for adult weights utilized in these calculations are those given in Table 2. The values for *P. thomasi* are approximations calculated from a growth curve fitted by eye to the four points plotted in Figure 3. Ages at half growth estimated from the growth curves in Figure 3 are also included in Table 7. Age at half growth is a convenient and less variable point for comparative purposes than age at full growth. Half-growth values obtained for *P. m. gambeli, P. truei,* and *P. californicus* by the present method differ by no more than a day from those calculated by McCabe and Blanchard (1950) according to the equation developed by Robertson (1908).

In general, proportional growth data support the previous interpretations based on absolute weight curves. With the exception of the two largest species, which agree closely in their growth patterns, the species and subspecies included in Table 7 do not show any strong correlation between adult size and age of half growth, and those trends suggested by the absolute curves are also evident. *P. maniculatus, P. oreas,* and *P. polionotus,* despite their generally smaller size, appear to grow somewhat slower than *P. leucopus* and *P. gossypinus.* The relatively protracted growth of *P. truei* and *P. floridanus* is also clearly evident in the proportional data, while the large *P. californicus* is not obviously behind smaller species in its rate of approach to mature size up to the age of ten weeks.

Geometric growth rates provide an especially sensitive basis of comparison of growth patterns between species or subspecies, and such values, calculated by the method of Simpson, Roe, and Lewontin (1960), for weights of *Peromyscus* at weekly intervals are presented in Table 8. All species show rapidly declining geometric growth rates with increasing age, those species approaching mature weight most rapidly experience the faster decay of rates. As in the previous comparisons, *P. megalops* and *P. thomasi* are unique in sustaining relatively high geometric growth rates for the longest period of time. The greatest differences in geometric growth rates are found during the first postnatal week, and it will be noted that there is little obvious correlation between initial growth rate and adult size, even in the case of species of such extremely divergent size as *P. polionotus* and *P. californicus.* This shows that the relatively precocial growth of the latter species does not result from differences in geometric growth rates but rather from its larger neonatal size. In other cases, for example, *P. flori-*

TABLE 8

GEOMETRIC GROWTH (PER CENT PER DAY) IN WEIGHT OF SPECIES AND SUBSPECIES OF *Peromyscus*

Species and subspecies	Age intervals, weeks									
	0–1	1–2	2–3	3–4	4–5	5–6	6–7	7–8	8–9	9–10
P. polionotus rhoadsi	10.40	6.89	3.96	3.82	0.86	1.27	0.30	1.28	—	—
P. p. subgriseus	9.81	6.93	3.71	3.92	3.91	1.06	0.99	—	—	—
P. maniculatus artemisiae	11.41	7.99	5.95	2.60	0.58	0.26	—	—	—	—
P. m. bairdi	12.91	3.19	6.71	3.19	0.0	3.75	—	—	—	—
P. m. gambeli	16.87	6.36	2.73	3.99	2.96	1.31	0.86	0.99	—	—
P. oreas	10.77	6.34	4.57	4.06	2.81	1.59	1.43	0.49	—	—
P. leucopus noveboracensis	15.64	6.68	5.50	1.12	—	—	—	—	—	—
P. gossypinus	13.66	7.30	4.91	3.81	1.98	0.12	—	—	—	0.36
P. truei	11.19	6.81	3.43	4.01	2.63	2.17	0.99	1.34	0.40	0.36
P. megalops	11.62	2.25	4.38	4.51	7.12	3.76	3.31	0.11	3.24	2.23
P. floridanus	10.76	5.17	3.84	5.21	2.44	1.76	0.90	0.28	0.93	0.0
P. californicus	10.06	5.76	3.61	3.07	2.59	1.51	1.03	1.14	0.83	0.06
P. thomasi*	9.41	7.04	5.02	3.07	5.73	3.07	2.90	2.41	2.33	2.00

* Values based on estimated growth curve.

danus and *P. gossypinus,* differences in absolute postnatal growth appear to be attributable primarily to differences in geometric growth rates rather than size at birth.

In most species, geometric rates decline steadily with age. In some, however, a marked decrease in rate at one week followed by an increase the following week is seen in the early period of growth. Such a trend is evident in *P. m. bairdi* and *P. megalops* between the second and third weeks, between the third and fourth weeks in *P. m. gambeli, P. truei,* and *P. floridanus,* and between the fourth and fifth weeks in *P. thomasi.* The inflection of the curve in *P. megalops* is apparently a result of abnormal growth of a litter, and its existence in *P. thomasi* is questionable. The phenomenon appears to be real in *P. m. gambeli* and *P. floridanus* (McCabe and Blanchard, 1950; Layne, 1966) and may reflect a waning milk supply as weaning is approached followed by the beginning of utilization of new food sources by the young. This explanation may also pertain to *P. m. bairdi.* A similar growth trend has been shown for *P. polionotous* (Carmon *et al.,* 1963), and in such other rodents as *Reithrodontomys* (Layne, 1959) and *Citellus* (Johnson, 1931). At later ages growth curves tend to become more irregular. Carmon *et al.* (1963) found that the family component was largely responsible for variation in litter growth rates prior to weaning, whereas individual differences contributed more to total variation in the post-weaning growth phase.

The curves in Figure 3 and data in Tables 7 and 8 indicate that growth in weight of at least some species continues beyond the tenth week of age. *P. floridanus* at 15 weeks are still below the average adult weight for the corresponding wild population, while individuals between 6½ and 9 months of age are close to the adult mean (Layne, 1966). Dice and Bradley (1942) found that slow growth in weight of subspecies of *P. maniculatus* continued from about the sixth week to an age of six months or more. The trends in weight in older mice tended to be somewhat less regular than those of linear measurements. McCabe and Blanchard (1950) observed a tendency in *P. californicus* and *P. truei* for weight to attain a plateau and then after an appreciable delay begin to rise again. This trend was much less evident in the smaller *P. m. gambeli.* Animals a year or two old did not show evidence of continued weight increase.

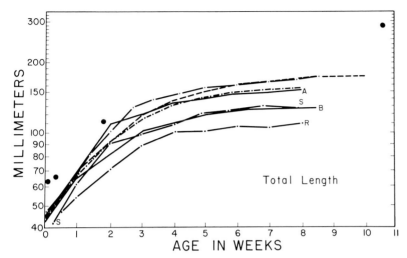

FIG. 4.   Semilogarithmic plot of growth in total length of *P. polionotus, maniculatus, oreas, gossypinus, floridanus,* and *thomasi.* Symbols are the same as those in Fig. 3.

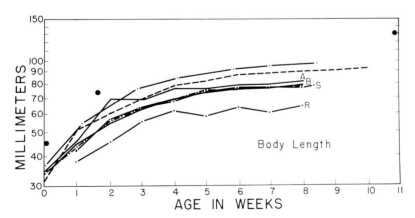

FIG. 5.   Semilogarithmic plot of growth of body (total-tail length) in *P. polionotus, maniculatus, oreas, gossypinus, floridanus,* and *thomasi.* Symbols are the same as those in Fig. 3.

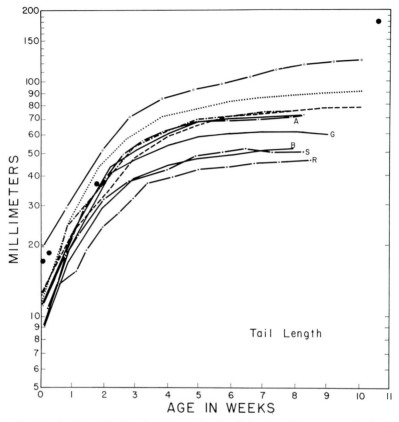

FIG. 6. Semilogarithmic plot of growth of tail in *P. polionotus, maniculatus, oreas, gossypinus, truei, floridanus, californicus,* and *thomasi.* Symbols are the same as those in Fig. 3.

It is doubtful if the slow increase in weight shown by older *Peromyscus* can be equated with "true" growth, as it is probably largely attributable to the laying on of fat; accumulation of secretions, intercellular or bony matrix; development of the reproductive system; and similar accretionary processes. Although, as Laird *et al.* (1965) point out, this aspect of growth as distinguished from self replicative processes probably extends back to the prenatal period, it becomes relatively more apparent with the cessation of exponential growth as the animal approaches maturity.

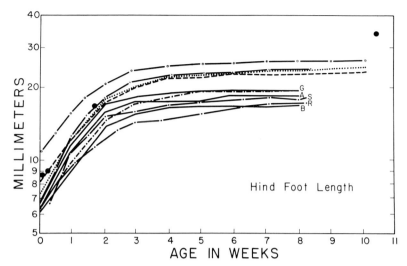

Fig. 7. Semilogarithmic plot of growth of hind foot in *P. polionotus, maniculatus, oreas, gossypinus, truei, floridanus, californicus,* and *thomasi.* Symbols are the same as those in Fig. 3.

Excessive fat production may be an especially important factor to consider in assessing limits of "true" growth in laboratory stocks. McCabe and Blanchard (1950) noted a prominent build-up of fat in *P. californicus* and *P. truei* for three or four weeks before the age of first pregnancy, and my observations on several species in the laboratory indicate rather substantial increases in weight of captives to the point where they greatly exceed the mean weight of adults in the wild populations from which they were collected. An example of the magnitude of such increase is provided by *P. thomasi*. Mean weight of six adults kept in captivity for a prolonged period was 124.7 g as compared with a mean of 77.0 g reported by Musser (1964) for wild adults from the same locality.

BODY MEASUREMENTS.—Absolute growth curves of total, tail, hind foot, and ear length in various species and subspecies of *Peromyscus* are shown in Figures 4 to 7. Growth data for body dimensions are also presented as percentages of adult size in Tables 9 to 13. The sources of the adult measurements used in these calculations have been cited in the section on neonatal size.

TABLE 9

PROPORTIONAL GROWTH (PERCENTAGE OF ADULT MEASUREMENT) AND APPROXIMATE AGE OF ONE-HALF GROWTH OF TOTAL LENGTH IN *Peromyscus*

Species and subspecies	Age in weeks*										One-half growth, days
	1	2	3	4	5	6	7	8	9	10	
P. polionotus rhoadsi	43	56	71	81	81	86	84	89	—	—	11
P. p. subgriseus	46	68	76	84	93	96	100	99	—	—	8
			(20)		(34)	(40)	(50)				
P. maniculatus artemisiae	40	66	69	78	82	86	87	87	—	—	10
P. m. bairdi	42	57	69	77	82	86	88	89	—	—	10
P. oreas	39	53	66	74	81	84	87	89	—	—	13
P. gossypinus	46	—	71	83	—	89	92	93	—	—	11
	(9)		(19)	(29)		(41)		(59)			
P. floridanus	40	52	66	76	82	88	91	92	94	95	13
P. thomasi	—	35	—	—	—	—	—	—	—	95	26+
		(12)								(75)	

* Percentages not representing weekly intervals are entered under the nearest week, with actual age in days given below in parentheses.
+ Based on estimated growth curve.

TABLE 10

PROPORTIONAL GROWTH (PERCENTAGE OF ADULT MEASUREMENT) AND APPROXIMATE AGE AT ONE-HALF GROWTH OF BODY LENGTH IN *Peromyscus*

Species and subspecies	Age in weeks*										One-half growth, days
	1	2	3	4	5	6	7	8	9	10	
P. polionotus rhoadsi	51	63	76	82	78	85	81	87	—	—	7
P. p. subgriseus	54	73	80	86	93	96	99	99	—	—	6
P. maniculatus artemisiae	53	77	76	84	85	88	88	92			6
			(20)		(34)	(40)	(48)				
P. m. bairdi	52	62	73	80	84	89	89	91	—	—	6
P. oreas	53	66	76	83	89	91	93	96	—	—	5
P. gossypinus	52	—	73	83	88	88	91	92	—	—	8
	(9)		(19)	(29)		(41)		(59)			
P. floridamus	50	61	70	79	84	88	90	91	93	94	7
P. thomasi	—	50	—	—	—	—	—	—	—	86	12+
		(12)								(75)	

* Percentages not representing weekly intervals are entered under the nearest week, with actual age in days given below in parentheses.

+ Based on estimated growth curve.

TABLE 11

PROPORTIONAL GROWTH (PERCENTAGE OF ADULT SIZE) AND APPROXIMATE AGE
AT ONE-HALF GROWTH OF TAIL LENGTH IN *Peromyscus*

Species and subspecies	Age in weeks*										One-half growth, days
	1	2	3	4	5	6	7	8	9	10	
P. polionotus rhoadsi	30	47	64	78	84	88	90	91	—	—	15
P. p. subgriseus	35	62	72	81	92	96	102	98	—	—	11
			(20)		(34)	(40)	(48)				
P. maniculatus artemisiae	27	48	62	72	79	82	85	85	—	—	15
P. m. bairdi	29	49	64	72	79	81	86	86	—	—	15
P. m. gambeli	34	63	78	89	96	99	—	—	—	—	12
P. oreas	26	40	56	66	74	78	82	83	—	—	18
P. gossypinus	37	—	68	83	—	91	93	94	—	—	14
	(9)		(19)	(29)		(41)		(59)			
P. truei	29	48	64	78	86	90	93	95	97	98	15
P. floridanus	28	41	60	73	81	89	93	94	95	96	18
P. californicus	27	42	57	70	76	80	89	94	97	99	18
P. thomasi	—	22	—	—	—	—	—	—	—	98	34+
		(12)								(75)	

* Percentages not representing weekly intervals are entered under the nearest week, with actual age in days given below in parentheses.
+ Based on estimated growth curve.

TABLE 12

PROPORTIONAL GROWTH (PERCENTAGE OF ADULT SIZE) AND APPROXIMATE AGE AT ONE-HALF GROWTH OF HIND FOOT LENGTH IN *Peromyscus*

Species and subspecies	Age in weeks*										One-half growth, days
	1	2	3	4	5	6	7	8	9	10	
P. poliomotus rhoadsi	54	76	88	90	95	97	100	—	—	—	6
P. p. subgriseus	58	84	88	94	98	98	102	97	—	—	6
			(20)		(34)	(40)	(48)				
P. m. artemisiae	57	82	93	93	93	98	98	98	—	—	5
P. m. bairdi	51	79	90	96	96	96	96	96	—	—	7
P. m. gambeli	59	88	94	99	—	—	—	—	—	—	5
P. oreas	48	71	86	91	95	95	95	95	—	—	8
P. gossypinus	67	—	90	99	—	100	—	—	—	—	6
	(9)		(19)	(29)							
P. truei	51	74	85	92	94	96	98	96	101	—	7
P. floridanus	50	74	86	93	95	98	98	98	99	99	7
P. californicus	60	78	89	93	97	98	99	99	98	100	4
P. thomasi	—	53	—	—	—	—	—	—	—	106	11+
		(12)								(75)	

* Percentages not representing weekly intervals are entered under the nearest week, with actual age in days given below in parentheses.

+ Based on estimated growth curve.

All body measurements considered here approach mature size at a more rapid rate than body weight. As the slopes of the growth curves of measurements are generally less steep than those for weight during the initial phase of postnatal life, it follows that the earlier approach of linear measurements to asymptotic levels is a consequence of their proportionately larger size at birth. Thus, while weights of neonatal *Peromyscus* are only about 5 to 16 per cent of the adult, body measurements range from 10 to 35 per cent of corresponding adult values. The growth curves of body measurements also tend to be more similar in shape than those of weight. This difference is particularly apparent in the segments of the curves representing later stages (roughly 4 to 10 weeks) of the postnatal growth period and indicates a generally greater uniformity of growth of body dimensions as compared to weight. As in the case of weight, however, there is some tendency for the curves of smaller species to approach the asymptotic level at an earlier age than those of larger forms.

The hind foot approaches adult size most rapidly of the four body parts, measurements of which are considered here. Half-adult size in eight species is attained between approximately 4 and 11 days of age (Table 12). The curve of length of hind foot shows a relatively well-defined point of inflection at about two weeks in smaller species (*maniculatus* and *polionotus*), whereas in larger forms it levels off more gradually.

Ear length follows hind foot in the velocity of its approach to adult proportions. The age at half growth of the pinna varies from 10 to 18 days in the seven species for which data are available (Table 13), with all but *californicus* showing a pronounced reduction of growth rate between three and four weeks of age (Figure 7). Because of the fleshiness and difference in shape of the pinna for a time after its erection, early measurements of this structure may not be fully comparable to later ones.

Total length approaches adult size more rapidly than length of tail for the first four or five weeks of postnatal life, by which time length of tail tends to be of approximately the same percentage of adult size. The estimated age at half growth of these two measurements also reflects this difference, ranging from 8 to 26 days in the case of total length and 11 to 34 days for length of tail. The growth curves of total and tail length show that early

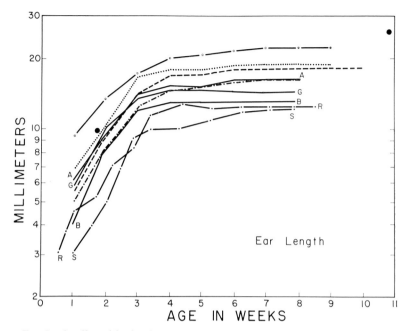

Fig. 8. Semilogarithmic plot of growth of the pinna in *P. polionotus, maniculatus, oreas, truei, floridanus, californicus,* and *thomasi.* Symbols are the same as those in Fig. 3.

postnatal geometric rates of the latter are much higher than those of total length, which means that the other component of total length, body length, is increasing more slowly during this period. Geometric growth rates of length of body and tail calculated for *floridanus* from birth through ten weeks of age suggest that in the latter part of this period growth of the body contributes more than tail to increase in total length (Layne, 1966).

A further generality concerning growth characteristics of body dimensions is that, as in weight, the geometric growth rate of a given dimension appears to be inversely correlated with its relative size at birth.

In none of the growth curves of body measurements is there any clear indication of a decline and reacceleration of growth at about the time of weaning as is apparent in the weight curves of several species.

Although limitations of the present data prevent critical comparisons, there is ample indication of specific and subspecific differences in growth patterns of body structures in addition to the general trends for the genus as a whole outlined above. At the species level, smaller forms tend to approach asymptotic values somewhat earlier than larger ones but variation in this respect is considerable. For example, *truei, floridanus, gossypinus,* and *californicus* do not slacken growth of the hind foot as early as the smaller *maniculatus, oreas,* and *polionotus.* However, the slightly larger *gossypinus* and much larger *californicus* do not exhibit any significant retardation of hind foot growth as compared with *floridanus* and *truei. P. gossypinus* also grows faster than *P. floridanus* and most other smaller species in total and tail length, although in the latter case it is exceeded by *P. truei.* A further exception is the case of *polionotus* and *maniculatus,* where in growth of all linear dimensions the former tends to lag behind the latter. The relatively huge *P. thomasi* clearly has the most retarded growth rates of any member of the species thus far studied.

In some cases, differences in body proportions apparently contribute to differences in growth patterns of specific body dimensions in species of grossly similar body weight. In general, for example, those species with relatively long tails tend to approach asymptotic levels somewhat more slowly than shorter-tailed forms, but this is not always true.

Subspecific differences in the growth of linear dimensions are also suggested by the curves for *P. polionotus rhoadsi* and *P. p. subgriseus,* and the several subspecies of *P. maniculatus.* Dice and Bradley (1942) demonstrated differences in growth patterns of seven subspecies of *P. maniculatus.* In general, those stocks with smallest adult size tended to decrease their growth rates at somewhat earlier ages. But there were obvious exceptions, such as the case of *P. m. sonoriensis* which grew more slowly than a number of subspecies of considerably larger size. As is the case at the species level, hind foot of all subspecies tended to approach definitive size earlier than other dimensions. Growth curves of body measurements tended to fluctuate less than those of weight at later ages.

Growth in at least some linear dimensions of *Peromyscus* may continue into relatively advanced age, but the data are few. Such increase probably more closely reflects "true" growth than the

TABLE 13

PROPORTIONAL GROWTH (PERCENTAGE OF ADULT SIZE) AND APPROXIMATE AGE
OF ONE-HALF GROWTH OF EAR LENGTH IN *Peromyscus*

Species and subspecies	Age in weeks*										One-half growth, days
	1	2	3	4	5	6	7	8	9	10	
P. polionotus rhoadsi	30	35	77	—	87	—	—	—	—	—	18
	(8)	(12)	(24)		(32)						
P. p. subgriseus	25	42	77	—	—	100	—	—	—	—	15
			(20)			(44)					
P. m. artemisiae	—	—	—	—	—	—	—	—	—	—	12
P. m. bairdi	30	61	91	98	98	98	98	98	—	—	12
P. m. gambeli	41	72	90	96	100	—	—	—	—	—	10
P. oreas	31	50	75	88	94	97	100	—	—	—	14
P. truei	37	53	87	95	97	100	—	—	—	—	13
P. floridanus	28	45	72	86	86	91	93	93	93	93	16
P. californicus	42	59	79	93	94	98	100	—	—	—	10

* Percentages not representing weekly intervals are entered under the nearest week, with actual age in days given below in parentheses.

similar changes observed in the case of weight. *P. floridanus* between 18 and 25 months of age averaged larger than young at 15 weeks of age in all body measurements (total, body, tail, and ear length) except hind foot (Layne, 1966). Dice and Bradley (1942) found continued slow growth in certain body dimensions of some subspecies of *P. maniculatus* beyond six months of age. In earlier studies, Dice (1932, 1933, 1936) demonstrated growth in external measurements of *P. m. bairdi* and *P. m. rufinus* between one and two years of age and in *P. m. gracilis* beyond the second year. These data are of course from laboratory stocks; few individuals in natural populations would survive to such ages.

VIBRISSAE.—McCabe and Blanchard (1950) recorded lengths of the longest mystacial vibrissae in *P. maniculatus, P. truei,* and *P. californicus* at weekly intervals from birth to eight or nine weeks of age. The initial geometric growth rates of these specialized hairs exceeded both weight and body measurements, those of *P. maniculatus* grew more rapidly than those of the other species during the first two weeks. In all species relative growth of the vibrissae declined precipitously at about the termination of weaning.

SKELETAL GROWTH.—In view of the extensive work on ecology, behavior, genetics, and taxonomy that has been carried out with mice of the genus *Peromyscus* there are surprisingly few data on development and growth of the skeletal system.

Hoffmeister (1951) studied age-related changes in the skull of *P. truei* and found that the greatest change of size in the various bones occurred between birth and five weeks. By the twenty-fifth to thirty-fifth day, the greatest length of skull is about 88 per cent of adult size; and in mice estimated to be from 35 to 100 days of age the skull averages approximately 99 per cent of maximum average size. Ontogenetic changes in cranial proportions of this species resembled those described for other rodents (Allen, 1894; Hall, 1926).

King and Eleftheriou (1960) analyzed differential growth in the cranial and facial bones of the skulls of *P. maniculatus bairdi* and *P. m. gracilis* from birth through 30 days of age and demonstrated that each subspecies was characterized by different growth characteristics. Of particular significance was the finding that the cranial portion of the skull grew at a faster rate in relation to the facial region in *gracilis* than in *bairdi*. This difference was in turn correlated with faster growth of the brain in *gracilis,* a relationship that has important implications with regard to the neural basis of differences in the ontogeny of behavioral patterns.

A more detailed study of body and skeletal measurements of the same pair of subspecies by means of covariance analysis was carried out by McIntosh (1955). The subspecies differed significantly in each regression. The evidence suggested that the genetic mechanisms responsible for the subspecific differences included factors controlling growth rates of particular parts of the body as well as those for general size. Although not specifically demonstrated, the findings did not exclude the possibility of some proportional differences being directly connected with differences in adult size. Clark (1941) and King and Eleftheriou (1960) also commented on the genetic basis of size and proportional differences between species and subspecies of *Peromyscus.*

Continued growth of certain skeletal elements to the second year of age or beyond was demonstrated in several subspecies of *maniculatus* by Dice (1932, 1933, 1936).

Discussion

This review suggests that much of the variation in growth and developmental patterns and associated aspects of reproduction in the genus *Peromyscus* can be interpreted in terms of body size and environmental relationships.

The data for species that have thus far been studied exhibit the expected trend of slower development and growth with increased body size (Fig. 9). However, considering the relatively wide range of sizes represented in this series, the relationship between size and growth rate appears to be less pronounced than might be predicted. While the two largest species, *thomasi* and *megalops,* are clearly more retarded in their growth and development than smaller species, the trend is less evident among the remaining species which range from approximately 13 to 40 g in weight. While it is true that the larger members of the latter group do tend to be generally less precocial than the smaller species, the differences in chronology of specific developmental or growth events are relatively minor. Moreover, the trend is not consistent, as some larger species exceed smaller ones in their growth and developmental rates.

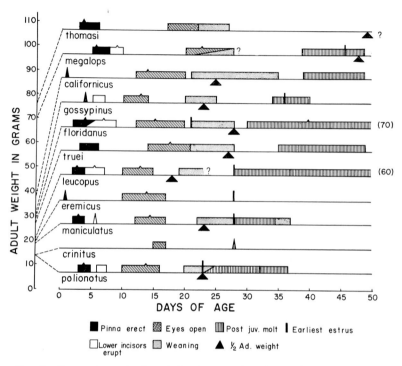

Fig. 9. Comparison of selected developmental and growth characteristics of 11 species of *Peromyscus*. Horizontal bars denote range of ages over which a given developmental event is known to occur. Small projections on the bars indicate the mean, either as given in the original source or estimated from other data. Tall triangles signify that the characteristic in question was stated in the original paper as occurring usually or typically at that age. In some cases data for two or more subspecies of the same species are combined. The data upon which this compilation is based are given in greater detail in appropriate sections of the text and in Tables 6 and 7.

Even in the case of the very large *thomasi* and *megalops,* the slowing of development relative to smaller species does not appear to be directly proportional to the marked size disparity. For example, although *thomasi* is approximately six times larger than the smallest species of the series, *polionotus,* the two do not differ appreciably in the age at which the pinnae become erect and weaning begins. The difference of six or seven days in mean age of eye opening is not much greater than that between some species of more nearly equivalent body size. The seemingly modest reduction in developmental rates of *thomasi* and *megalops* compared with those of smaller *Peromyscus* stands in sharp contrast to the wide disparity in growth rates between the two groups. In effect, the young of the large species reach a developmental state adequate for independent existence at about the same age as those of smaller *Peromyscus*, but undergo a greater proportion of their growth after weaning. As a result, lactating females are subjected to lower energy demands than would be the case if growth rates comparable with smaller species were maintained, while at the same time advantages of early developmental maturity are retained.

The present evidence of a relatively weak correlation between postnatal ontogenetic rates and body size in *Peromyscus* suggests that selection favoring larger body size has generally been accompanied by selection favoring acceleration of growth and development. If this were not the case and increased size was achieved through processes leading to an appreciable extension of the ontogenetic period, the resultant delay in age at maturity would presumably negate some of the advantages of larger size. McCabe and Blanchard (1950) have pointed out that although the growth phase of large and small mammals with seasonal breeding cycles comprises about the same proportion of the total life span, the small species actually live a much shorter time under conditions favorable for reproduction. It follows, therefore, that especially strong selection pressure on growth and developmental patterns leading to early puberty is to be expected in small mammals such as *Peromyscus*. It should also be noted, however, that this is only one of numerous selective forces operating on mode of growth and thus, although it may be assumed that there is a general tendency for all species to maximize their developmental rates, the actual extent to which this is accomplished may differ greatly even among closely related forms.

The apparent tendency toward relative acceleration of growth and development in larger *Peromyscus* may also be related to degree of specialization. The larger members of the genus may generally be more highly specialized in their ecological relationships than smaller ones, hence more rapid development might contribute to better survival of their young during the critical period between weaning and establishment in the population.

In addition to trends presumably associated with body size, growth and developmental patterns of *Peromyscus* appear to reflect strongly differences in habitat relationships. Hershkovitz (1962) described two major adaptive patterns in Neotropical cricetine rodents. These include forest, or sylvan, species and pastoral forms inhabiting "pampas, Punas, llanos, savannas, tundras, scrublands, rocklands, deserts, and other types of grazing and grazing-browsing habitats." Pastoral lines are derived from sylvan stocks and show greater specialization in tooth patterns, pelage characteristics, and other aspects of their biology. These adaptive trends are also evident in the genus *Peromyscus* at both the specific and subspecific level. Of the species included in this review, *leucopus, gossypinus, oreas, thomasi,* and *megalops* are essentially forest dwellers. *P. maniculatus,* though it contains both forest and pastoral subspecies, is probably also properly regarded as essentially a sylvan species. The pastoral group consists of *californicus, eremicus, crinitus, floridanus, truei,* and *polionotus.* With the exception of *polionotus,* these species are all associated with arid habitats, including scrub, pine-oak woodlands, chaparral, and deserts. Most of these habitat types have probably been derived from the xeric Madro-Tertiary Geoflora, which originated in scattered dry sites in southwestern North America within the region generally occupied by the mesic Neotropical-Tertiary flora (Axelrod, 1958).

Sylvan species of *Peromyscus* appear to share a number of common reproductive characteristics, which may attest to an over-all uniformity in forest environmental conditions. The gestation period is close to the mean value for the genus (approximately 23 days) and shows relatively little variation. With the exception of *P. megalops,* litter size is comparatively large, averaging about four. *P. megalops* with its very small litter size is aberrant in this respect, although it agrees with other forest species in the remainder of its reproductive features. The adaptive significance of this difference is unknown. The neonates of sylvan species are relatively small to medium in size, averaging about eight per cent of adult weight; and rates of postnatal growth and development are moderately high.

In contrast, pastoral species exhibit a tendency toward increased neonatal size, averaging about 12 per cent of adult weight, and a reduction in litter size to a mean of approximately three. This trend is most evident in the species inhabiting arid habitats, reaching its extreme in *californicus* and *eremicus* of the subgenus *Haplomylomys*. This tendency toward an inverse relationship between litter size and neonatal size is suggestive of some basic limitation in the amount of energy that species of *Peromyscus* can expend in the production of new individuals. If this interpretation is correct, then the relatively narrow range of total biomass of young produced by the members of this genus may reflect a fundamental similarity in their niche relationships, with the particular manner in which a given species or subspecies divides this biomass depending upon its environmental relationships.

In other aspects of reproduction, pastoral species exhibit greater variability than the sylvan group, which correlates with the greater diversity of pastoral habitats. The gestation period is more variable, the species with the extreme maximum and minimum gestation periods in the genus being pastoral types. Pastoral species also exhibit two principal trends in postnatal growth and development. One group, including *californicus, eremicus,* and *polionotus,* is characterized by precocial tendencies. The other, presently including *crinitus, truei,* and *floridanus,* exhibits prolongation of postnatal development and growth. *P. californicus* and *eremicus* of the first group are distinctly more precocial relative to their size than is *polionotus.* The primary reason for assigning the latter species to the precocial group is its somewhat faster development than in *maniculatus,* to which it is most closely related and generally similar in size. The comparatively fast-growing *californicus* and *eremicus* also appear to have absolutely or relatively (in terms of body size) shorter gestation periods than the average for the genus. Limited data point to a contrasting situation in the slow-growing species. *P. truei* has a somewhat longer gestation period compared with other *Peromyscus,* and the same may be true of *crinitus.* The gestation period of *floridanus* is not known.

The slight evidence for a correlation of gestation length with postnatal growth and developmental rates together with other data indicates that differences in the postnatal phase of ontogeny reflect actual changes in over-all growth and developmental schedules rather than simply a shift in the point of birth relative to the total period of ontogeny from zygote to maturity.

A further aspect of reproductive biology in which xeric-adapted pastoral species appear to differ from sylvan species is in breeding success under laboratory conditions. Reports in the literature indicate that *californicus, eremicus, truei,* and *floridanus* do not breed as readily in captivity as *maniculatus, leucopus,* and *gossypinus* (Svihla, 1932; Dice, 1954; Layne, 1966, pers. obs.; McCabe and Blanchard, 1950). *P. polionotus* (several subspecies) reproduces well in the laboratory and thus is more like sylvan species in this respect (pers. obs.).

The trend toward large neonates and small litter size in xeric pastoral species may be an adaptation for life under conditions of water or food scarcity or both, as water loss should be less and efficiency of energy utilization greater in a mother-litter unit consisting of fewer and larger young than one comprised of a greater number of smaller young. Lack (1963) has interpreted similar trends in clutch and egg size relationships in populations of certain lizards as responses to food or water shortage in the environment.

The tendency toward precocial growth and development in *californicus* and *eremicus* may be a further adaptation to conserve water and possibly energy, as has been suggested for certain relatively fast-growing heteromyid rodents living in xeric regions (Eisenberg and Isaac, 1963). In addition, such an ontogenetic pattern would seem to be advantageous to species living in some types of arid habitats, particularly deserts, where conditions favorable for reproduction are largely contingent upon rainfall and associated food abundance and thus tend to be of very limited duration and of cyclic or sporadic occurrence. Under such circumstances accelerated growth and development would contribute to a higher breeding rate and perhaps better chances for survival of young.

The type of growth and development exhibited by *truei, floridanus,* and *crinitus* may represent an adaptive response to environments characterized by chronically limited or widely dispersed food resources. The prolongation of development and growth in these species has the effect of prorating the energy cost of raising a litter over a longer interval, thereby reducing the daily food requirements of the female. Similar modifications of growth patterns as an adaptation to poor quality or quantity of food have been recorded in other mammalian species (e.g., Ealey, 1963; Hamilton,

1962). Although there are presently no good comparative data on relative food levels in habitats of slow-growing pastoral species and forest forms, observations indicate that in one of the typical habitats (pine-oak woodland) of *P. floridanus* food is less abundant than in the characteristic forest habitat types of *P. gossypinus* in the same geographic region (Layne, 1966).

Factors of the microhabitat as well as major environmental parameters have probably also been involved in evolution of growth and developmental patterns in *Peromyscus*. For example, differences in nest sites have been suggested as a possible influence on the evolution of the contrasting growth and developmental rates of *P. floridanus* and *P. gossypinus* (Layne, 1966). The former species apparently nests exclusively in burrows, whereas the latter typically places its nest in more accessible places such as holes or crevices in logs and stumps or beneath objects. The more secure sites of *floridanus* may thus have permitted selection favoring prolongation of development and growth to proceed more readily, whereas the more exposed nest sites of *gossypinus* may have increased selective pressure favoring acceleration of ontogeny.

The apparent tendency of xeric pastoral species to be poorer breeders than forest species in the laboratory may indicate that their reproductive physiology is more sensitive to the influences of food level or quality, temperature, population density, social interactions, or other environmental factors than in the case of sylvan species. This would seem to be adaptive to species living under more severe conditions by insuring closer restriction of breeding activity to periods when environmental conditions are most favorable for survival of young, thus reducing needless expenditure of energy in mate-finding, copulation, gestation, and care of young when chances of successful reproduction are low.

P. polionotus appears to be less specialized in its reproductive pattern than the other pastoral species considered here. This may be owing to the fact that its habitats tend to be less arid than those of the other forms, that it may have diverged more recently from a forest-inhabiting ancestor, or a combination of both factors.

The present and admittedly limited data for different subspecies of *P. maniculatus* suggest that radiation of populations within a species from forest to pastoral habitats is accompanied by the same pattern of reproductive adaptations found at the species level. The more mesic grassland subspecies *bairdi* is more precocial than the forest race *gracilis*. On the other hand, pastoral subspecies from more arid regions in western North America may be characterized by slower growth as in the case of certain xeric pastoral species, and presumably for the same adaptive reasons. *P. m. sonoriensis* appears to be an example of this trend, its growth is somewhat more prolonged than that of *bairdi*. Limited data on eye opening of other subspecies of *maniculatus* from open, relatively dry habitats also seem to give some indication of slower development than in *bairdi*.

Regarding the taxonomic significance of growth and developmental patterns, and other reproductive characteristics of *Peromyscus*, it appears that these features are too adaptively labile to be of much use as criteria of phyletic affinity. For the most part, resemblances between species appear to result from convergence or parallelism and show little correspondence to degree of taxonomic relationship as presently understood (Hooper and Musser, 1964). Only where closely related species have retained a generally similar environmental orientation, as in the subgenus *Haplomylomys* and in the *leucopus* species group of the subgenus *Peromyscus,* do reproductive resemblances clearly reflect taxonomic affinity.

Morphological characteristics linked to reproductive patterns are also of questionable reliability as taxonomic characters. The number of teats is a case in point. Although the number of teats has been used as a diagnostic character at the subgeneric level, loss of the anterior pair in connection with reduction of litter size has apparently occurred independently in at least three lines (subgenus *Haplomylomys* and *P. crinitus* and *P. megalops* of the subgenus *Peromyscus*).

Although it is possible that more detailed study and sophisticated methods of analysis would provide a basis for distinguishing between homologous and convergent similarities in reproductive patterns of *Peromyscus,* it is doubtful that the present data are adequate for this purpose.

Summary

This paper reviews present knowledge of pre- and postnatal development and growth and certain related aspects of the reproductive biology of *Peromyscus*. Knowledge of embryological development in the genus is limited to a single species, *P. polionotus,* whereas data on at least some aspects of postnatal development and growth are available for 12 species in four (*Peromyscus, Haplomylomys, Podomys,* and *Megadontomys*) of the seven subgenera recognized in the most recent classification of the genus.

Inter- and intraspecific variation in the reproductive parameters considered appear to be related primarily to factors associated with body size and environmental orientation. Of the two, environmental relationships seem to be most strongly correlated with trends in reproductive characteristics.

Although body sizes of the species that have been studied range from approximately 13 to 80 g, there is less indication of prolongation of growth and development with increasing body size than might be predicted. Larger species do tend to be slower in their growth and development than the smaller forms, but within the 13- to 40-g range of body size that includes most species this trend is not marked and even the very large species appear to develop relatively fast compared with much smaller ones. In distinct contrast to their developmental pattern, the largest species have much slower growth rates than smaller forms. In adaptive terms, this reduces energy requirements of lactating females without sacrificing advantages of relatively rapid development. The evidence of a comparatively weak correlation between body size and developmental rate suggests that selection favoring large body size in this genus has also involved acceleration of developmental processes to avoid prolongation of the age of puberty. The possibly greater degree of specialization of larger species may also be a factor involved in their tendency to retain relatively fast developmental rates.

Peromyscus typically inhabiting forests tend to be characterized by a gestation period of about 23 days, relatively large litter size, neonates of small to medium size, and comparatively rapid postnatal growth and development in relation to body size. In comparison, pastoral species tend to have smaller litter size and larger neonates. This tendency is most pronounced in species characteristic of arid habitats such as scrub, pine-oak woodland, chaparral, and desert. Pastoral species are more variable in their gestation periods and postnatal growth and developmental patterns than sylvan types. Some species tend toward precocity, whereas others have somewhat prolonged growth and development. Both trends are again most pronounced in species from xeric habitats. Xeric-adapted pastoral species may also have a reproductive physiology that is more sensitive to biotic and abiotic environmental factors than is the case in the sylvan group. These differences in the reproductive patterns of sylvan and pastoral species of *Peromyscus* appear to relate principally to availability of water or food or both. Growth and developmental characteristics of sylvan and pastoral subspecies of *P. maniculatus* appear to show the same correlations with type of habitat as at the specific level.

Present evidence indicates that the reproductive features of *Peromyscus* are too readily modified in response to environmental conditions to provide a useful criterion of taxonomic relationships. This applies also to morphological characters, such as number of teats, that are clearly linked to reproductive processes.

Acknowledgments

I am indebted to Guy G. Musser, University of Michigan,[1] for live specimens of *P. megalops* and *P. thomasi* and for notes on the development of the latter and to John A. King, Michigan State University, for data on age of eye opening in several species. Katherine Carter, Margaret Moore, and Melzar Richards aided in making some of the observations on *P. leucopus, megalops, thomasi,* and *floridanus* reported here. These studies were aided by Grant G-3072 from the National Science Foundation.

Literature Cited

ALLEN, J. A. 1894. Cranial variations in *Neotoma micropus* due to growth and individual differentation. Bull. Amer. Mus. Nat. Hist., 6:233–246.

ASDELL, S. A. 1964. Patterns of mammalian reproduction. 2nd ed. Ithaca, New York: Cornell University Press, 670 pp.

AXELROD, D. I. 1958. Evolution of the Madro-tertiary geoflora. Bot. Rev., 24:433–509.

BENDELL, J. F. 1959. Food as a control of a population of white-footed mice, *Peromyscus leucopus noveboracensis* (Fischer). Can. Jour. Zool., 37:173–209.

BIRKENHOLZ, D. E., AND W. O. WIRTZ, II. 1965. Laboratory observations on the vesper rat. Jour. Mamm., 46:181–189.

BLAIR, W. F. 1940. A study of prairie deer-mouse populations in southern Michigan. Amer. Midland Nat., 24:273–305.

---

[1] Presently at The American Museum of Natural History.

———— 1941. Observations on the life history of *Baiomys taylori subater.* Jour. Mamm., 22:378–383.

BROWN, L. N. 1963. Maturational molts and seasonal molts in *Peromyscus boylei.* Amer. Midland Nat., 70:466–469.

CARMON, J. L., F. B. GOLLEY, AND R. G. WILLIAMS. 1963. An analysis of the growth and variability in *Peromyscus polionotus.* Growth., 27:247–254.

CLARK, F. H. 1936. Geotropic behavior on a sloping plane of arboreal and non-arboreal races of mice of the genus *Peromyscus.* Jour. Mamm., 17:44–47.

———— 1937. Parturition in the deer-mouse. *Ibid.,* 18:85–87.

———— 1938. Age of sexual maturity in mice of the genus *Peromyscus. Ibid.,* 19:230–234.

———— 1941. Correlation in body proportions in mature mice of the genus *Peromyscus.* Genetics, 26:283–300.

COLLINS, H. H. 1918. Studies of normal moult and of artifically induced regeneration of pelage in *Peromyscus.* Jour. Exp. Zool., 27:73–95.

———— 1923. Studies of the pelage phases and the nature of color variations in mice of the genus *Peromyscus. Ibid.,* 38:45–107.

DAWSON, W. D. 1965. Fertility and size inheritance in a *Peromyscus* species cross. Evolution, 19:44–55.

DENENBERG, V. H., AND J. R. C. MORTON. 1962. Effects of environmental complexity and social groupings upon modification of emotional behavior. Jour. Comp. and Physiol. Psych., 55:242–246.

DENENBERG, V. H., D. R. OTTINGER, AND M. W. STEPHENS. 1962. Effects of maternal factors upon growth and behavior of the rat. Child Devel., 33:65–71.

DICE, L. R. 1932. Variation in a geographic race of the deer-mouse, *Peromyscus maniculatus bairdii.* Occ. Pap. Mus. Zool. Univ. Mich., 239, 26 pp.

———— 1933. Variation in *Peromyscus maniculatus rufinus* from Colorado and New Mexico. *Ibid.,* 271, 32 pp.

———— 1936. Age variation in *Peromyscus maniculatus gracilis.* Jour. Mamm., 17:55–57.

———— 1954. Breeding of *Peromyscus floridanus* in captivity. *Ibid.,* 35:260.

DICE, L. R., AND R. M. BRADLEY. 1942. Growth in the deer-mouse, *Peromyscus maniculatus.* Jour. Mamm., 23:416–427.

EALEY, E. H. M. 1963. The ecological significance of delayed implantation in a population of the hill kangaroo (*Macropus robustus*), Pp. 33–48. *In:* A. C. Enders (ed.), Delayed implantation. Rice University Semicentennial Publ. Chicago: Univ. of Chicago Press.

EGOSCUE, H. J. 1964. Ecological notes and laboratory life history of the canyon mouse. Jour. Mamm., 45:387–396.

EISENBERG, J. F. 1962. Studies on the behavior of *Peromyscus maniculatus gambelii* and *Peromyscus californicus parasiticus.* Behaviour, 19:177–207.

———— 1963. The intraspecific social behavior of some cricetine rodents of the genus *Peromyscus.* Amer. Midland Nat., 69:240–246.

250   *Layne*

Eisenberg, J. F., and D. E. Isaac. 1963. The reproduction of heteromyid rodents in captivity. Jour. Mamm., 44:61–67.

Golley, F. B., E. L. Morgan, and J. L. Carmon. 1966. Progression of molt in *Peromyscus polionotus*. Jour. Mamm., 47:145–148.

Goodpaster, W. W., and D. F. Hoffmeister. 1954. Life history of the golden mouse, *Peromyscus nuttalli*, in Kentucky. Jour. Mamm., 35:16–27.

Gottschang, J. L. 1956. Juvenile molt in *Peromyscus leucopus noveboracensis*. Jour. Mamm., 37:516–520.

Hall, E. R. 1926. Changes during growth in the skull of the rodent *Otospermophilus grammurus beecheyi*. Univ. Calif. Publ. Zool., 21:355–404.

Hall, E. R., and K. R. Kelson. 1959. The mammals of North America. New York: Ronald Press, 2 v., 1083 pp.

Hamilton, W. J., Jr. 1953. Reproduction and young of the Florida wood rat, *Neotoma f. floridana* (Ord). Jour. Mamm., 34:180–189.

Hamilton, W. J., III. 1962. Reproductive adaptations of the red tree mouse. Jour. Mamm., 43:486–504.

Hart, F. M., and J. A. King. 1966. Distress vocalization of young *Peromyscus maniculatus*. Jour. Mamm., 47:287–293.

Hershkovitz, P. 1962. Evolution of Neotropical cricetine rodents (Muridae) with special reference to the phyllotine group. Fieldiana: Zool., Chicago Nat. Hist. Mus., 46:1–524.

Hoffmeister, D. F. 1951. A taxonomic and evolutionary study of the Piñon mouse, *Peromyscus truei*. Ill. Biol. Monogr., 21:1–104.

Hooper, E. T., and G. G. Musser. 1964. Notes on classification of the rodent genus *Peromyscus*. Occ. Pap. Mus. Zool. Univ. Mich., 635, 13 pp.

Horner, B. Elizabeth. 1947. Paternal care of young mice of the genus *Peromyscus*. Jour. Mamm., 28:31–36.

Howard, W. E. 1949. Dispersal, amount of inbreeding, and longevity in a local population of prairie deer mice on the George Reserve, southern Michigan. Contrib. Lab. Vert. Biol., Univ. Mich., 43, 50 pp.

Johnson, G. E. 1931. Early life of the thirteen-lined ground squirrel. Trans. Kansas Acad. Sci., 34:282–290.

King, J. A. 1958. Maternal behavior and behavioral development in two subspecies of *Peromyscus maniculatus*. Jour. Mamm., 39:177–190.

——— 1961a. Swimming and reaction to electric shock in two subspecies of deer mice (*Peromyscus maniculatus*) during development. Anim. Behav., 9:142–150.

——— 1961b. Development and behavioral evolution in *Peromyscus*, Pp. 122–147. *In*: Blair, W. F. (ed.). Vertebrate speciation. Austin: Univ. Texas Press.

——— 1963. Maternal behavior in *Peromyscus*, Pp. 58–93. *In*: Rheingold, H. L. (ed.), Maternal behavior in mammals. New York: John Wiley and Sons, Inc.

King, J. A., and B. E. Eleftheriou. 1959. Effects of early handling upon adult behavior in two subspecies of deer mice, *Peromyscus maniculatus*. Jour. Comp. and Physiol. Psychol., 52:81–88.

———— 1960. Differential growth in the skulls of two subspecies of deer mice. Growth, 24:179–192.

KING, J. A., J. C. DESHAIES, AND R. WEBSTER. 1963. Age of weaning in two subspecies of deer mice. Science, 139:483–484.

KING, J. A., AND NANCY SHEA. 1958. Behavioral development in two subspecies of *Peromyscus maniculatus*. Anat. Rec., 131:571–572.

———— 1959. Subspecific differences in the responses of young deer mice on an elevated maze. Jour. Hered., 50:14–18.

LACK, D. 1963. The evolution of reproductive rates, Pp. 172–187. *In*: Huxley, J., *et al.* (eds.), Evolution as a process. New York: Collier Books.

LAFFODAY, S. K. 1957. A study of prenatal and postnatal development in the oldfield mouse, *Peromyscus polionotus*. Doctoral dissertation, Univ. of Fla., 124 pp.

LAIRD, ANNA K., SYLVANUS A. TYLER, AND A. D. BARTON. 1965. Dynamics of normal growth. Growth, 29:233–248.

LAWLOR, T. E. 1965. The Yucatan deer mouse, *Peromyscus yucatanicus*. Univ. Kans. Pub., Mus. Nat. Hist., 16:421–438.

LAYNE, J. N. 1954. The biology of the red squirrel, *Tamiasciurus hudsonicus loquax* (Bangs), in central New York. Ecol. Monogr., 24:227–267.

———— 1959. Growth and development of the eastern harvest mouse, *Reithrodontomys humulis*. Bull. Fla. State Mus. Biol. Sci., 4:61–82.

———— 1960. The growth and development of young golden mice, *Ochrotomys nuttalli*. Quart. Jour. Fla. Acad. Sci., 23:36–58.

———— 1963. A study of the parasites of the Florida mouse, *Peromyscus floridanus*, in relation to host and environmental factors. Tulane Studies in Zool., 11:1–27.

———— 1966. Postnatal development and growth of *Peromyscus floridanus*. Growth, 30:23–45.

LIU, T. T. 1953. Prenatal mortality in *Peromyscus* with special reference to its bearing on reduced fertility in some interspecific and intersubspecific crosses. Contrib. Lab. Vert. Biol. Univ. Mich., 60:1–32.

McCABE, T. T., AND BARBARA D. BLANCHARD. 1950. Three species of *Peromyscus*. Santa Barbara, Calif.: Rood Associates, Publ., 136 pp.

McINTOSH, W. B. 1955. The applicability of covariance analysis for comparison of body and skeletal measurements between two races of the deer mouse, *Peromyscus maniculatus*. Contr. Lab. Vert. Biol. Univ. Mich., 72:1–54.

MOORE, A. W. 1929. Extra-uterine pregnancy in *Peromyscus*. Jour. Mamm., 10:81.

MUSSER, G. G. 1964. Notes on geographic distribution, habitat, and taxonomy of some Mexican mammals. Occ. Pap. Mus. Zool. Univ. Mich., 636, 22 pp.

NICHOLSON, A. J. 1941. The homes and social habits of the wood mouse (*Peromyscus leucopus noveboracensis*) in southern Michigan. Amer. Midland Nat., 25:196–223.

NOIROT, E. 1964. Changes in responsiveness to young in the adult mouse: The effect of external stimuli. Jour. Comp. Physiol. Psychol., 57:97–99.

Osgood, W. H. 1909. Revision of the mice of the American genus *Peromyscus*. N. Amer. Fauna, 28:1–285.

Pournelle, G. H. 1952. Reproduction and early postnatal development of the cotton mouse, *Peromyscus gossypinus gossypinus*. Jour. Mamm., 33: 1–20.

Rand, A. L., and P. Host. 1942. Mammal notes from Highlands County, Florida. Results of the Archbold Exped., no. 45, Bull. Amer. Mus. Nat. Hist., 80:1–21.

Robertson, T. B. 1908. On the normal growth rate of an individual, and its biochemical significance. Archiv für Entwicklungsmechanik der Organismen, 25:581–614. (cited by McCabe and Blanchard, 1950)

Ryder, J. A. 1887a. The vestiges of a zonary decidua in the mouse. Amer. Nat., 21:780–784.

———— 1887b. The inversion of the germinal layers in *Hesperomys*. *Ibid.*, 21:863–864.

Seton, E. T. 1920. Breeding habits of captive deer mice. Jour. Mamm., 1:135.

Simpson, G. G., Anne Roe, and R. C. Lewontin. 1960. Quantitative zoology. New York: Harcourt, Brace and Co., 440 pp.

Smith, W. K. 1939. An investigation into the early embryology and associated phenomena of *Peromyscus polionotus*. MS thesis, Univ. of Fla., 35 pp.

Storer, T. I., F. C. Evans, and F. G. Palmer. 1944. Some rodent populations in the Sierra Nevada of California. Ecol. Monog., 14:165–192.

Sumner, F. B. 1917. Several color "mutations" in mice of the genus *Peromyscus*. Genetics, 2:291–300.

Svihla, A. 1932. A comparative life history study of the mice of the genus *Peromyscus*. Misc. Publ. Mus. Zool. Univ. Mich., 24:1–39.

———— 1934. Development and growth of deer mice (*Peromyscus maniculatus artemisiae*). Jour. Mamm., 15:99–104.

———— 1935. Development and growth of the prairie deer mouse, *Peromyscus maniculatus bairdii*. *Ibid.*, 16:109–115.

———— 1936. Development and growth of *Peromyscus maniculatus oreas*. *Ibid.*, 17:132–137.

Watson, M. L. 1942. Hybridization experiments between *Peromyscus polionotus* and *P. maniculatus*. Jour. Mamm., 23:315–316.

ADDENDUM

Brand, L. R., and R. E. Ryckman. In press. Laboratory life histories of *Peromyscus eremicus* and *Peromyscus interparietalis*. Jour. Mamm.

King, J. A. 1965. Body, brain, and lens weights of *Peromyscus*. Zool. Jb. Anat., 82:177–188.

Linzey, D. W., and A. V. Linzey. 1967. Growth and development of the golden mouse, *Ochrotomys nuttalli*. Jour. Mamm., 48:445–458.

Rood, J. P. 1966. Observations on the reproduction of Peromyscus in captivity. Amer. Midland Nat., 76:496–503.

Scudder, C. L., A. G. Karczmar, and L. Lockett. 1967. Behavioural developmental studies of four genera and several strains of mice. Anim. Behav., 15:353–363.

SICHER, E. A., AND W. N. BRADSHAW. 1966. The reproductive pattern of the wood mouse (*Peromyscus leucopus*) in northern West Virginia. Proc. W. Va. Acad. Sci., 38:23–31.

VESTAL, B. M., AND J. A. KING. 1967. Relation of age of eye opening to first optokinetic response in four taxa of *Peromyscus*. Amer. Zool., 7:216–217.

7

PARASITES

JOHN O. WHITAKER, JR.

Introduction

BECAUSE *Peromyscus* is one of the best known genera of mice in North America it is surprising that more information is not available concerning its parasites. The major purpose of this chapter will be to summarize the available information on its larger external and internal parasites, and to point out areas for further research.

It will become evident that there is much to be learned concerning the parasites of *Peromyscus,* and that much of the information available concerns relatively few species. Those for which the most information on parasites is available are *Peromyscus maniculatus, P. leucopus, P. truei, P. boylei, P. crinitus, P. eremicus, P. californicus, P. gossypinus,* and *P. floridanus,* listed in order by decreasing number of reports.

A paper on the latter species by Layne (1963) might be used as a model for a mammalogist who might wish to study parasitism in a particular species.

Many specimens of *Peromyscus* are discarded before examinations have been made for parasites. This is indeed a waste. It is unfortunate that more mammalogists do not take the time to examine their specimens for at least the larger internal and external parasites.

There are several reasons why more information of this type is not accumulated. First, most biologists are interested either in the parasites or the mice, seldom both. Most parasitologists are concerned primarily with the taxonomy and life history of the parasites, not with the parasites of a particular species of mammal. For this reason, most literature on parasites is arranged by taxonomic grouping of the parasites, not the mice, although many papers include host lists.

The mammalogist who is interested in parasites cannot simply collect and identify them. There are usually too many closely related species and the literature is scattered; there is no single reference.

Likewise, the mammalogist cannot simply send his materials to somebody for identification. He usually has to deal with a different individual for each group of parasites. These individuals usually do the identifications as a favor. Often a donation of some of the materials will be an added incentive for the parasitologist to make identifications.

Some of the major types of information which can be accumulated concerning the parasites of *Peromyscus* are listed below.

1. Percentage of hosts infested. How many of the animals being examined harbored the various species of parasites?
2. Mean number of parasites per host, both overall and of those infested.
3. Interrelations of species of parasites. Are particular mice apt to harbor several species or do parasites occur at random?
4. Are there differences in infestations of parasites between habitats?
5. Are there seasonal differences in infestations? (Many of the fleas are present on mice only at one season.)
6. Are there differences in infestation rates with age or sex of the host?
7. Are there geographic differences in infestation rates?
8. Host specificity of parasites. Do the parasites occur only on one or a few closely related hosts, or over a wide range of hosts?

Individuals more interested in parasitism than in the mammals might study some of the following:

1. Life history of the parasites.
2. Mode of transmission from host to host.
3. Adaptations of the parasites to the morphology and life history of the host.

Methods of Collecting Parasites

There are numerous ways to collect parasites of *Peromyscus*. At Indiana State University, all mammals are routinely processed as follows. Most are snap-trapped and immediately placed in individual plastic bags which are then securely closed. This keeps all parasites with the particular host. In the laboratory the mice are examined with a 10- to 30-power zoom dissecting microscope. Needles are used to probe carefully through the fur for external parasites. Fleas are searched for first because they are the most likely to escape. Special care is used to be sure that the bases of the hairs and the skin itself are examined.

This method works especially well for finding some of the mites which cling tightly to the hairs or skin such as *Labidophorus* and some myobiids. A further advantage of this method is that the distribution of mites on the mice can be determined. I have yet to compare this method with the immersion method described below by first examining a series of mice by hand and then treating the same series with detergent.

Many workers remove mites from small mammals by immersing them in various liquids, such as detergent and water (McDaniel, 1965) or ether, detergent and water (Howell and Elzinga, 1962). The latter authors then filter the liquid through a 200-mesh screen. The screen is placed in a dish of water and the mites washed free and collected in a micropipette with the aid of a dissecting microscope.

Fleas and all other external parasites are immediately preserved in 70% alcohol (although mites can be mounted live; see methods of preservation). Ticks will be found crawling on the fur or imbedded in the skin. In the past most parasitologists have dug the ticks from the skin, preserved them, and then have had the problem of cleaning the flesh from the proboscis, since it is necessary to see that organ for identification. Purely by accident I discovered a much easier method. On one occasion after extracting the tick from the flesh I forgot to add alcohol. In half an hour the tick had removed himself from the flesh. I now use this method with consistently good results.

After the examination for external parasites the animals are opened, and superficial examination is made of the underside of the skin, the lungs, the heart, the liver, and the bladder. If there is indication of disturbance these organs are examined in more detail.

The digestive tract, especially the forepart of the intestine, is a prime location for internal parasites. In our examinations the intestines are removed and two pairs of forceps are used to squeeze the contents into a watch glass of water. This material is then washed if necessary by pouring off the water carefully and slowly so that the heavier items, including the parasites, will remain on the bottom. Most of the internal parasites are white or red and show up rather well either with reflected light or with a white background and direct lighting. Usually I use both. This method works because the intestines of mice are relatively smooth inside and most forms will slide out easily. It has the disadvantage that cestodes are often broken. In collecting cestodes it is imperative that the scoleces be preserved to facilitate identification. In our work, internal parasites are usually preserved in FAA, but many special solutions are available. A parasitology manual should be consulted for these.

The methods described above often are satisfactory for a mammalogist, but they have some major limitations and I have been criticized for using some of them. Some of the problems are as follows:

METHODS OF TRAPPING.—Live trapping has the advantage that presumably none of the parasites leave the host. When snap traps are used at least some of the parasites may leave before the traps are checked. This is probably true of fleas. It is best to check the traps early in the morning (for *Peromyscus* at least, since they are essentially nocturnal) to help compensate for this. Even then the problem can be severe when the mammals were taken early in the night rather than just before dawn. The advantage of snap traps is that the worker can use many more traps in the field because of the ease of handling. Probably forms other than fleas are not too greatly affected over relatively short periods of time by the death of the host, but the worker must keep this limitation in mind.

SEARCHING THE FUR WITH NEEDLES.—In using this method one cannot be sure that he has found all the parasites, but with reasonable care, and by spending 5 to 10 minutes (longer if parasites are numerous) per specimen, the worker is usually convinced that he has a good sample of what is present. There are other methods used whereby the mouse is immersed in various liquids in which the parasites float to the surface. They are then skimmed off and preserved. The major disadvantage of this method is that it is messy and not desirable for making up study skins. I am not convinced that this type method is best for collecting all types of mites. For example, we have found that mites of the genus *Labidophorus* are very common on some species of small mammals, but there are relatively few reports of this genus from North America. It appears to us that the method of looking through the fur with the needles turns up some forms not otherwise easily taken.

SUPERFICIAL EXAMINATION OF INTERNAL ORGANS.—Many parasites occur in organs other than the intestines and, depending on the objectives of a particular study, concentrated examination of these organs deserves much more attention than given in the Indiana studies.

STRIPPING THE INTESTINE.—Another method of making the examination for internal parasites is to slit open the intestine throughout its entire length. When many cestodes are present, this is a necessity. For nematodes and trematodes, I am not convinced that it is worth the extra bother. Using *Zapus*, I once slit open the intestines of 50 mice which had yielded intestinal parasites by stripping. No additional parasites were found.

Other parts of the body should be examined, such as the nasal passages, esophagus, kidneys, and blood. When working on specific problems many workers will be interested in looking for smaller species of parasites such as bacteria and protozoa which often occur in the blood, in the intestines (especially the cecum), and in the stomach. Usually these take relatively specialized techniques and often are not easily incorporated into the routine examination.

Methods of Preservation

KILLING.—The killing of the parasites is one of the important steps in preparation for permanent mounting or storage. Parasites taken from frozen mammals will be dead and can be preserved directly in 70% alcohol or in FAA solution. Mites, fleas, lice, ticks, and *Cuterebra* can all be preserved directly in 70% alcohol whether dead or alive.

A simple method I have used for killing worm parasites is by adding a tiny amount of alcohol (to make about a three per cent solution) to water and then gently rocking the worms in this mixture. This method is not recommended by parasitologists. Worm parasites are usually killed in various hot liquids. Nematodes, in particular, should not be treated in this way, nor should they be immersed in water. They will often burst. They can be killed in hot 70% ethyl alcohol, and then preserved in 70% alcohol to which 5% glycerine is later added.

For more details, consult a general reference on parasitology or one on microscopic methods.

PREPARATION OF PARASITES FOR STUDY.—The methods outlined below are not necessarily the best for the various forms of parasites, nor are they necessarily in general use by parasitologists. However, they are methods which have proved useful to me and should be useful to mammalogists in general.

Trematodes: The following rather simple method has yielded good results.
1. Remove from 70% alcohol and place in water overnight.
2. Put in Ehrlich's stain for a few minutes (Preparation of Ehrlich's stain is outlined in Corrington, 1941.) (5 to 30 minutes).
3. 50% alcohol for 10 minutes.
4. 70% alcohol for 10 minutes.
5. Destain in acid alcohol until staining is correct. (This is critical; don't go too far.) Blue in alkaline alcohol.
6. Two changes of fresh 70% alcohol.
7. 80% alcohol for 15 minutes.
8. 95% alcohol for 15 minutes.

9. 100% alcohol for 15 minutes.
10. Clear in beechwood creosote and later add balsam to the creosote.
11. Mount in balsam.

Cestodes and Acanthocephalans: Individuals working with Cestodes and Acanthocephalans often like to prepare their own material. For this reason, I simply preserve them in 70% alcohol.

Nematodes: Nematodes are fixed in 100% acetic acid for five minutes, rinsed in tap water, then put into a 5% solution of glycerine in 70% alcohol and stored for 10 days. They are then placed in open containers so that the alcohol evaporates, leaving pure glycerine, in which they are permanently stored.

Mites:
1. Transfer live mites from 70% alcohol to Nesbitts solution for clearing. Leave 12–14 hours.
2. Mount mites on slides in Hoyer's solution.
3. Center specimen and ease coverslip into position, so that no air bubbles are captured. Mount mite ventral side up, head toward top of slide.
4. Let slide set for 24 hours or longer. Fill in any air spaces that may have developed around edges with Hoyer's solution. Let stand again. Remove excess Hoyer's solution with razor blade or water.
5. Ring slides with quick drying Asphaltum, for permanent storage.

 Nesbitts Solution
  80 gms. chloral hydrate
  50 ml H_2O
  5 ml HCE
  (Dissolve ingredients and add tiny amount of acid fuchsin to color solution red.)

 Hoyer's Solution
  50 gms distilled H_2O
  30 gms gum arabic
  200 gms chloral hydrate
  20 gms glycerine
  (Mix and dissolve *in above sequence* at room temperature. If wrong sequence is used, the materials will not dissolve.)

Fleas:
1. Specimens in 70% alcohol. Arrange with camel hair brush or spread with piece of glass as weight.
2. 95% alcohol, 10 to 30 minutes.
3. Oil of cloves or Carbol-xylol, about 15 minutes.
4. Xylol, about 15 minutes.
5. Mount in balsam.

Lice:
1. Specimens in 70% alcohol.
2. Pipette off alcohol and add 82% alcohol, 10 minutes.
3. Pipette off and add 95% alcohol, 10 minutes.

4. Pipette off and add a mixture of 95% alcohol and terpineol, 50–50. Puncture specimen with a minuten nadeln under microscope. Allow to evaporate for 20 to 30 minutes under lamp. Add a drop or two of pure terpineol. Let stand 10 minutes.
5. Transfer specimen to terpineol or to xylol for 10 minutes.
6. Mount in balsam.

Ticks and dipterous larvae (Cuterebra) : Ticks and *Cuterebra* larvae are stored permanently in 70% alcohol.

Parasites of Various Species of Peromyscus

Most of the species of *Peromyscus* have not been studied in terms of their parasites. Of the 56 currently recognized species of *Peromyscus* I have been unable to find any reports at all concerning parasitism of 35 species.

Only isolated records of parasites are available for *P. difficilis, P. guatemalensis, P. hylocetes, P. melanophrys, P. melanotis, P. mexicanus, P. nudipes, P. oaxacensis, P. pectoralis,* and *P. sitkensis.* There are fewer than 25 records for *P. polionotus* and *P. nasutus.*

This leaves only the nine species mentioned in the introduction for which we have extensive information concerning internal and external parasites.

PENTASTOMIDS.—Adult Pentastomids of the genus *Porocephalus* are worm-like Arachnoids with two pairs of ventral rectractile hooks beside the mouth. Adults are parasites of Crotalid snakes, but the larvae are parasites of rodents and are found in the internal organs. One of ten specimens of *Peromyscus leucopus* collected in Oklahoma harbored ten larvae of *P. crotali* Humboldt in the lungs, pancreas, and mesenteries (Self and McMurray, 1948). Layne (1963) found larvae in or on the liver, mesovarium, bladder, mesenteries, intestine, lungs, body wall, kidney, testes, and epidimymis of *Peromyscus floridanus* from Florida, and Komarek (1939) found "pentastomid larvae" in *Peromyscus gossypinus* in Florida.

ACANTHOCEPHALA.—Acanthocephalans, or "spiny headed worms," occur as adults only in the intestine of vertebrate animals, but are not at all common in *Peromyscus.* I have found only two reports of them in this genus. Adults attach to the intestinal wall by a protrusible proboscis which is normally covered with hooklets. There is no true segmentation, but acanthocephalans can be confused with cestodes because they sometimes have pronounced pseudosegmentation. The proboscis (sometimes difficult to see) and the extending of the internal organs through the length of the body (rather than being separate in each segment) should serve to identify them.

Relatively little is known of the life cycles of acanthocephalans but an alternation of hosts is probably present in all. An insect often serves as an intermediate host.

The major source of information concerning the acanthocephalans of North American vertebrates is that of Van Cleave (1953). *Moniliformis clarki* (Ward, 1917) has been found in *Peromyscus maniculatus* in Utah by Grundmann and Frandsen (1960), and in Minnesota by Erickson (1938; identified by Van Cleave, 1953).

TREMATODES.—Trematodes have been reported for only five species of *Peromyscus*. During the examination of digestive tracts of 286 specimens of *Peromyscus leucopus* and 457 of *P. maniculatus bairdi,* from Vigo County, Indiana, only two trematodes were found, one from each species.

Trematodes have a rather complicated life cycle, all require a snail as a secondary host. The predominantly vegetarian diet of *Peromyscus* may help explain the relatively small number of parasitic trematodes.

Some good general references on trematodes are Dawes (1946) and Yamaguti (Vol. I).

Records of trematodes from *Peromyscus* follow:

Peromyscus floridanus
   Ova of dicrocoelid fluke (172*)
Peromyscus gossypinus
   Scaphiostomum pancreaticum McIntosh, 1934 (187)
   Zonorchis komareki McIntosh, 1939 (189)
Peromyscus guatemalensis
   Brachylaema chiapensis Ubelaker and Dailey, 1966 (256)
   * Numbers in parentheses refer to references in Literature Cited section.
Peromyscus leucopus
   Alaria mustelae, Bosma, 1931 (33)
   Brachylaema peromysci Reynolds, 1938 (120, 214)
   B. virginianum Rausch and Tiner, 1949 (211)
   Entosiphonus thompsoni Sinitsin, 1931 (124, 169, 260)
   Postharmostomum helicis (Leidy, 1847) (215) (experimental infection)
Peromyscus maniculatus
   Brachylaema microti Kruidenier and Gallicchio, 1959 (102, 117)
   Euryhelmis pacificus Senger and Macy, 1952 (223)
   Postharmostomum helicis (Leidy, 1847) (257) (experimental infection)
   P. sexconvolutum Miller, 1936 (281)

CESTODES.—Adult cestodes, or tapeworms, are nearly always parasitic in the alimentary tract of vertebrates. Most are composed of a number of segments, or proglottids. There is no digestive system. Food is taken in through the body wall. Most cestodes require an intermediate host. Larval cestodes may be found in various tissues of vertebrates or arthropods. They are variable in structure, but many possess characteristic bladder-like structures and are called bladderworms.

Tapeworms can harm the host by utilizing its food and sometimes by causing a blockage of the lumen of the intestine. However, in *Peromyscus*, usually only one or a few cestodes, if any, are present, and harm to the individual is probably not great, although the points where the scoleces are attached to the intestinal wall may be damaged.

The scoleces, or heads, of cestodes are needed for identification, hence it is most important that this organ is found and preserved.

Good general references on tapeworms are those of Wardle and McLeod (1952) and Yamaguti (Vol. II).

Records of cestodes in *Peromyscus* are listed below:

Peromyscus boylei
  Hymenolepis horrida (von Linstow, 1900)   (261, 262)
  Mesocestoides kirbyi Chandler, 1944, or
  M. variabilis Mueller, 1927 (266)
  Mesocestoides larvae (262)
Peromyscus californicus
  Hymenolepis horrida (von Linstow, 1900)   (261, 262)
Peromyscus crinitus
  Mesocestoides carnivoricolus Grundmann, 1956 (117)
Peromyscus floridanus
  Cysticerci- of *Cladotaenia* or *Paruterina* (172)
  Hydatigera lyncis (Skinker, 1935) (172)
  Spirometra (mansonoides?) (Mueller, 1935) (188)
  Vampirolepis nana (Siebold, 1852) (172)
Peromyscus gossypinus
  Spirometra mansonoides (Mueller, 1935) (188)
  Zonorchis komareki (McIntosh, 1939) (189)
Peromyscus leucopus
  Cladotaenia sp. (200)
  (*Cysticercus talpae*) (120)
  Paruterina sp. (cysticercus) (120)
  Taenia sp. (84)
  Undetermined larvae (lung) (84)

Peromyscus maniculatus
  Catenotaenia dendritica (Goeze, 1782) (174)
  C. linsdalei McIntosh, 1941 (116, 117)
  C. peromysci Smith, 1954 (228)
  Choanotaenia peromysci (Erickson, 1938) (84, 123, 185, 262)
  Cladotaenia circi Yamaguti, 1935 (larv.) (117)
  C. globifera (Batsch, 1786) (104)
  Cladotaenia sp. (84)
  Cysticercus sp. (227)
  Hydatigera lyncis (Skinker, 1935) (227)
  Hymenolepis citelli McLeod, 1933 (117)
  H. diminuta (Rudolphi, 1819) (208)
  Hymenolepis sp. (84, 123, 174)
  Mesocestoides carnivoricolus Grundmann, 1956 (116, 117)
  Paruterina candelabraria (Goeze, 1872) (117)
  P. rauschi Freeman, 1957 (104)
  Prochoanotaenia sp. (84)
  Taenia mustelae (Gmelin, 1790) (84, 104) (larvae in liver)
  T. ovis (Gobbold, 1869) (218)
Peromyscus truei
  Hymenolepis horrida (von Linstow, 1900) (261, 262)
Peromyscus sp.
  Andrya communis (Douthitt, 1915 (73)
  Anoplocephala sp. (118)

NEMATODA.—Hall (1916), in the only work in which an attempt has been made to summarize the information specifically on rodents, lists only one record of a nematode in any of the members of the genus *Peromyscus*. In the work of Yorke and Maplestone (1962) the genus *Peromyscus* is not mentioned, although there are references to nematodes in "rodents" or "mice." The latest summary of information on nematode parasites of vertebrates is by Yamaguti (Vol. III).

It would appear from these and the paucity of later records that relatively little is known of the nematode parasite fauna of *Peromyscus*. Records for only ten species of *Peromyscus* have been found, and only *P. floridanus, P. leucopus,* and *P. maniculatus* have been studied to any extent.

Most parasitic nematodes have only one host and a relatively direct life cycle. There is normally a free-living stage, often the egg. Infection usually occurs by ingestion of the free-living stage, or by the free-living stage actively burrowing into the tissues

The majority of the reported nematodes of *Peromyscus* are from the intestinal tract, but they are occasionally found in the stomach. Wilson (1945) found 13 of 130 stomachs infected with *Rictularia* sp. Records of nematodes in rodents of the genus *Peromyscus* are listed below:

Peromyscus boylei
  Gongylonema peromysci Kruidenier and Peebles, 1958 (168)
  Heligmosomum sp. (263)
Peromyscus californicus
  Trichurus peromysci Chandler, 1946 (49, 263)
Peromyscus crinitus
  Protospirura numidica Seurat, 1914 (117)
Peromyscus eremicus
  Gongylonema peromysci Kruidenier and Peebles, 1958 (168)
Peromyscus floridanus
  Aspicularis americana Erickson, 1938 (172)
  Capillaria hepatica (Bancroft, 1896) (172, 173)
  Rictularia (*coloradensis?*) (Hall, 1916) (172)
  Syphacia peromysci Harkema, 1936 (172)
  Trichostrongylus ransomi Dikmans (172)
Peromyscus gossypinus
  Protospirura numidica Seurat, 1914 (71)
Peromyscus leucopus
  Ascaris larvae (Raccoon type) (249)
  Aspicularis americana Erickson, 1938 (84)
  A. tetraptera (Nitsch, 1821) (120)
  Capillaria americana Read, 1949 (212)
  Capillaria sp. (120)
  Larvae of nematodes (120, 124)
  Mastophorus muris (Hall, 1916) (71)
  "microfilariae" (79)
  Nematospiroides dubia Baylis, 1926 (120)
  Rictularia coloradensis Hall, 1916 (120, 124, 247, 248)
  R. onychomis Cuckler, 1939 (50, 213)
  Rictularia sp. (211, 275)
  Syphacia peromysci Harkema, 1936 (120, 124)
  S. samorodini Erickson, 1938 (84)
  Trichurid sp. (120)
Peromyscus maniculatus
  Aspicularis americana Erickson, 1938 (84)
  A. tetraptera Schultz, 1924 (115) (experimental infection)
  Capillaria americana Read, 1949 (212)
  C. hepatica (Bancroft, 1896) (104)
  Gongylonema peromysci Kruidenier and Peebles, 1958 (168)
  Longistriata carolinensis Dikmans, 1935 (68)

Nematospiroides dubia Baylis, 1926 (281)
Nippostrongylus brasiliensis (Travassos, 1914) (117)
Nippostrongylus sp. (174)
Protospirura numidica Seurat, 1914 (174)
Rictularia coloradensis Hall, 1916 (117, 174, 208, 247)
Rictularia sp. (115)
Syphacia obvelata (Rudolphi, 1802) (208)
S. peromysci Harkema, 1936 (117, 174, 218, 246)
S. samorodini Erickson, 1938 (84, 115)
Trichuris perognatha Chandler, 1945 (117)
Peromyscus nasutus
Syphacia obvelata (Rudolphi, 1802) (117)
Peromyscus truei
Protospirura numidica Seurat, 1914 (117)
Gongylonema peromysci Kruidenier and Peebles, 1958 (168)

MITES, EXCLUDING CHIGGERS.—Mites are generally the most abundant parasites on mice of the genus *Peromyscus* as well as on other small mammals. They can be found on a high proportion of the animals examined, and on all parts of the animals, sometimes in great numbers. They can often be seen crawling on, or in, the fur, especially in the nape region.

There are few papers dealing strictly with parasites of *Peromyscus*. Those available are by Layne (1963), Wilson (1945), and a series of papers by Allred (1956–1958) on five species in Utah, *P. boylei, P. crinitus, P. eremicus, P. maniculatus,* and *P. truei.*

To examine host specificity within the genus, data of Allred for the five species mentioned above are used (Table 1). If no host specificity were involved one would expect the average number of mites per individual to be the same from species to species. This was not the case. *Eubrachylaelaps circularis* and *Androlaelaps fahrenholzi* were most abundant on *P. boylei* and *E. circularis* reached its second greatest abundance on *P. crinitus,* but was rare on *P. maniculatus,* while *A. fahrenholzi* was second in abundance on *P. maniculatus* but occurred only once on *P. crinitus. E. hollisteri* was most abundant on *P. crinitus,* next most abundant on *P. eremicus* but was rare on the other three species. *Hirstionyssus occidentalis* and *Ornithonyssus bacoti* were most abundant on *P. maniculatus* and *P. truei,* respectively.

It is evident that the mite faunas are very different on the different species of mice of the genus *Peromyscus* in Utah. It seems likely that this is at least partly a result of host specificity, but it

TABLE 1

COMPARISON OF DENSITY OF FIVE SPECIES OF MITES
ON FIVE SPECIES OF *Peromyscus* FROM UTAH
(data from Allred, 1957)

	P. boylei	P. crinitus	P. eremicus	P. maniculatus	P. truei
No. mice	37	67	201	2907	59
Eubrachylaelaps circularis					
No. mites	145	35	65	79	68
Average No. per mouse	3.92	0.52	0.32	0.03	1.15
Eubrachylaelaps debilis					
No. mites	0	2	2	623	10
Average No. per mouse	0.00	0.03	0.01	0.21	0.17
Eubrachylaelaps hollisteri					
No. mites	0	57	70	17	3
Average No. per mouse	0.00	0.85	0.35	0.01	0.05
Androlaelaps fahrenholzi					
No. mites	48	1	36	1158	10
Average No. per mouse	1.30	0.01	0.18	0.40	0.17
Hirstionyssus occidentalis					
No. mites	0	6	1	409	0
Average No. per mouse	0.00	0.09	0.01	0.14	0.00
Ornithonyssus bacoti					
No. mites	1	1	2	61	11
Average No. per mouse	0.02	0.01	0.01	0.02	0.19

could be attributed to differing habitat requirements of the mites. If habitat is a major factor, a species of mite should be most abundant on the species of mice most often in the correct habitat and least often on mice seldom in that habitat. To assess the relative habitat differences versus host specificity the breakdown of average number of mites per host in the same habitats is needed. Also seasonal differences in collections of mice, if any, could be involved, since mites are often seasonal in occurrence.

Few workers have tried to evaluate the role that the habitat plays in determining the distribution and abundance of a particular parasite. Elzinga and Rees (1964) examined the parasite infestations of *Peromyscus maniculatus* and *Reithrodontomys megalotis* in three different habitats. They found different rates of occurrence of the parasites in different habitats.

It would seem that if habitat and climatic factors did not play a role in determining the distribution and abundance of parasites, that the parasite should generally occur throughout the range of the host species. This is not true. There are many cases where parasites occur in only part of the range of the host species.

Parasite records from the nests of *Peromyscus* were excluded from this report. However, this is an interesting area which deserves much further work. Drummond (1957) has studied the Acarine fauna of the nests of *Peromyscus leucopus* in Maryland. During his study 141 species belonging to 37 different families were taken. Some of these were forms often occurring on *Peromyscus leucopus*, such as *Haemogamasus liponyssiides*, *Ornithonyssus bacoti, Eulaelaps stabularis, Androlaelaps fahrenholzi, A. megaventralis*, and *Dermacentor variabilis* (the last is a tick). Other species have not been reported on *Peromyscus* and many are probably strictly nest inhabitants.

Records of mites from the various species of *Peromyscus* are listed below, but a few taxonomic notes are in order, especially concerning the genera *Haemolaelaps* and *Haemogamasus,* since there have been several recent name changes for common and widespread species. The genus *Haemolaelaps* is considered synonymous with *Androlaelaps,* which has priority, and *Haemolaelaps glasgowi,* one of the most common species on North American small mammals, is now considered to be a synonym of the earlier described Old World species *Androlaelaps fahrenholzi* Berlese (Till, 1963).

The other changes are in the genus *Haemogamasus.* According to Evans and Till (1962), *H. alaskensis* Ewing should now be referred to as *H. ambulans* (Thorell), while the form that has been called *H. ambulans* (Thorell) by American authors (to 1966) should be known as *H. nidi* Michael.

Some of the major references on mites are those of Baker and Wharton (1952) and Strandtmann and Wharton (1958).

Mite records from *Peromyscus* are listed below:
Peromyscus boylei
  Androlaelaps casalis (Berlese, 1887) (8)
  A. fahrenholzi (Berlese, 1911) (6, 8)
  Brevisterna utahensis Strandtmann and Allred, 1956 (8)
  Bryobia praetiosa Koch, 1836 (6)
  Dermanyssus becki Allred, 1957a (4, 8)
  Eubrachylaelaps circularis (Ewing, 1933) (6, 8, 106, 143)
  E. debilis Jameson, 1950 (8, 143)
  Eubrachylaelaps sp. (268)

Haemogamasus nidi Michael, 1892　(143)
H. pontiger (Berlese, 1904)　(8)
Hermannia sp.　(6)
Hirstionyssus affinis (Jameson, 1950)　(242)
H. occidentalis (Ewing, 1923)　(241)
Ornithonyssus bacoti Hirst, 1913　(6, 8)
Ornithonyssus sp.　(6)
Peromyscus californicus
　Androlaelaps fahrenholzi (Berlese, 1911)　(242)
　Eubrachylaelaps circularis (Ewing, 1933)　(106)
　E. hollisteri (Ewing, 1925)　(106)
Peromyscus crinitus
　Androlaelaps casalis (Berlese, 1887)　(8)
　A. fahrenholzi (Berlese, 1911)　(6, 8, 160)
　A. leviculus (Eads, 1951)　(8, 160)
　Brevisterna utahensis Strandtmann and Allred, 1956　(6, 8, 10, 240)
　Dermanyssus becki Allred, 1957　(6, 8, 10)
　D. gallinae (DeGeer, 1778)　(8)
　Dermanyssus sp.　(160)
　Eubrachylaelaps circularis (Ewing, 1933)　(6, 8)
　E. debilis Jameson, 1950　(6, 8, 9, 160, 161)
　E. hollisteri (Ewing, 1925)　(6, 8, 9, 106)
　Haemogamasus pontiger (Berlese, 1904)　(9)
　Hirstionyssus carnifex (Koch, 1839)　(9)
　H. incomptus (Eads and Hightower, 1952)　(6, 8)
　H. occidentalis (Ewing, 1923)　(6, 8)
　H. triacanthus (Jameson, 1950)　(8)
　H. utahensis Allred and Beck, 1966　(8)
　Hypoaspis sp.　(6)
　Ischyropoda armatus Keegan, 1951　(9)
　Kleemania sp.　(9)
　Ornithonyssus bacoti Hirst, 1913　(6, 8)
　Ornithonyssus sp.　(6)
　Parasitid sp.　(6)
Peromyscus eremicus
　Androlaelaps fahrenholzi (Berlese, 1911)　(6, 8)
　Brevisterna utahensis Strandtmann and Allred, 1956　(6, 8)
　Dermanyssus becki Allred, 1957a　(6, 8)
　Dermanyssus sp.　(6)
　Dermannysid sp.　(6)
　Eubrachylaelaps circularis (Ewing, 1933)　(6, 8)
　E. debilis Jameson, 1950　(6, 8)
　E. hollisteri (Ewing, 1925)　(6, 8)
　Haemogamasus pontiger (Berlese, 1904)　(6, 8)
　Hirstionyssus femuralis Allred, 1957a　(6, 8)
　H. hilli (Jameson, 1950)　(6, 8)
　H. incomptus (Eads and Hightower, 1952)　(8)

H. occidentalis (Ewing, 1923)   (6)
H. triacanthus (Jameson, 1950)   (8)
Ornithonyssus bacoti Hirst, 1913   (6, 8)
Peromyscus floridanus
   Androlaelaps fahrenholzi (Berlese, 1911)   (172)
   Eulaelaps stabularis Koch, 1836   (172)
   Haemogamasus liponyssoides Ewing, 1925   (172)
   Ornithonyssus bacoti Hirst, 1913   (172)
Peromyscus gossypinus
   Androlaelaps fahrenholzi (Berlese, 1911)   (126, 192, 279, 280)
   Dermatophagoides scheremetewskyi   (279, 280)
   Eulaelaps stabularis Koch, 1836   (126, 192)
   Haemogamasus liponyssoides Ewing, 1925   (126, 159, 192)
   Listrophorus sp.   (279, 280)
   Ornithonyssus bacoti Hirst, 1913   (192, 279, 280)
Peromyscus hylocetes
   Eubrachylaelaps circularis (Ewing, 1933)   (160)
Peromyscus leucopus
   Androlaelaps fahrenholzi (Berlese, 1911)   (74, 98, 126, 130, 245, 269)
   Blarinobia simplex (Ewing, 1938)   (245)
   Blarinobia sp.   (197)
   Brevisterna morlani Strandtmann and Allred, 1956   (240)
   Dermacarus hypudaei (Koch, 1841)   (271)
   Dermanyssus sp.   (74)
   Haemogamasus liponyssoides Ewing, 1925   (74, 159)
   H. nidi Michael, 1892   (197)
   Hirstionyssus carnifex (Koch, 1839)   (245)
   H. cynomys (Radford, 1941)   (209)
   H. occidentalis (Ewing, 1923)   (241)
   H. talpae Zemskaya, 1955   (269)
   Laelaps alaskensis Grant, 1947   (98)
   L. kochi Oudemans, 1936   (74, 269)
   Listrophorus sp.   (74, 98)
   Myobia musculi   (269)
   Myocoptes musculinus Koch, 1844   (269)
   Myocoptes sp.   (74)
   Notoedres sp.   (197)
   Ornithonyssus bacoti Hirst, 1913   (74, 130, 269)
   O. bursa (Berlese, 1888)   (74)
   Radfordia subuliger Ewing, 1938   (74, 269)
   Resinacarus sp.   (245, 269)
Peromyscus maniculatus
   Androlaelaps casalis (Berlese, 1887)   (6, 8)
   A. fahrenholzi (Berlese, 1911)   (6, 8, 10, 83, 122, 160, 170, 209, 219, 269)
   A. leviculus (Eads, 1951)   (6, 8, 160, 161)
   Antennophorid sp.   (6)
   Ascaid sp.   (6)

Brevisterna morlani Strandtmann and Allred, 1956 (240)
B. utahensis Strandtmann and Allred, 1956 (6, 8, 240)
Camisiid sp. (6)
Cheyletid sp. (6)
Dermacarus hypudaei (Koch, 1841) (271)
Dermanyssus becki Allred, 1957*a* (6, 8)
Eremaeid sp. (6)
Eubrachylaelaps circularis (Ewing, 1933) (6, 8, 106, 143)
E. debilis Jameson, 1950 (6, 8, 143, 160)
E. hollisteri (Ewing, 1925) (6, 8)
Eulaelaps stabularis Koch, 1836 (8, 122, 170, 269)
Eulaelaps sp. (6)
Gamasolaelaptid sp. (6)
Garmania sp. (6)
Glycyphagus sp. (6)
Haemogamasus ambulans (Thorell, 1872) (6, 8, 159)
H. liponyssoides Ewing, 1925 (159, 170)
H. longitarsus (Banks, 1910) (6, 8)
H. nidi Michael, 1892 (6, 8, 122)
H. pontiger (Berlese, 1904) (6, 8)
Haemogamasus sp. (6)
*Hirstionyssus carnifex (Koch, 1839) (160)
H. cynomys (Radford, 1941) (209)
*H. geomydis Keegan, 1946 (6)
H. hilli (Jameson, 1950) (6, 8, 160)
H. incomptus (Eads and Hightower, 1952) (6, 8, 83)

* Allred and Beck (1966) feel these are misidentifications.

H. isabellinus Oudemans, 1913 (6, 8)
H. neotomae Eads and Hightower, 1951 (8)
H. obsoletus (Jameson, 1950) (6, 141)
H. occidentalis (Ewing, 1923) (83, 122, 241)
H. punctatus Allred and Beck, 1966 (8)
H. talpae Zemskaya, 1955 (122, 269)
H. tarsalis Allred and Beck, 1966 (8)
H. triacanthus (Jameson, 1950) (8)
H. utahensis Allred and Beck, 1966 (8)
Hirstionyssus sp. (219)
Hirstionyssus sp. "P" (143)
Hypoaspis lubrica Oudemans and Voigts, 1904 (6, 8)
H. miles (Berlese, 1892) (6, 8)
Hypoaspis sp. (6)
Ischyropoda armatus Keegan, 1951 (6, 8, 159)
I. furmani Keegan, 1956 (8, 161)
Kleemania sp. (8, 10, 269)
Labidophorid sp. (219)
Laelaps incilis Allred and Beck, 1966 (8)

L. kochi Oudemans, 1936 (269)
L. nuttalli Hirst, 1915 (6, 8)
Listrophorus sp. (6, 8)
Macrocheles sp. (8, 122)
Macrochelid sp. (269)
Myobiid sp. (219)
Myocoptes sp. (8, 83)
Neoparasitid sp. (6)
Ondatralaelaps multispinosus (Banks, 1909) (6, 8)
Ornithonyssus bacoti Hirst, 1913 (6, 8, 269)
Ornithonyssus sp. (6)
Pachylaelaptid sp. (6)
Parasitid sp. (6)
Phytoseiid sp. (6)
Poecilochirid sp. (269)
Poecilochirus sp. (6, 122)
Pyemotid sp. (6)
Pygmephorus sp. (122)
Radfordia affinis (Poppe, 1896) (269)
R. bachai Howell and Elzinga, 1962 (136)
R. lemnina (Koch, 1841) (6, 8)
R. subuliger Ewing, 1938 (6, 8, 83, 232, 269)
Resinacarus sp. (269)
Rhizoglyphus echinopus (Fumouze and Robin, 1868) (6)
Tetranychid sp. (6)
Typhlodromus mariposus (Fox, 1946) (6)
Uropodid sp. (6)
Peromyscus melanotis
   Haemogamasus keegani (Jameson, 1952) (142)
Peromyscus mexicanus
   Eubrachylaelaps jamesoni Furman, 1955 (106)
Peromyscus oaxacensis
   Eubrachylaelaps circularis (Ewing, 1933) (106)
Peromyscus polionotus
   Androlaelaps casalis (Berlese, 1887) (192)
   A. fahrenholzi (Berlese, 1911) (126, 192)
   Cheyletus eruditus (Schrank, 1781) (192)
   Eulaelaps stabularis Koch, 1836 (192)
   Hypoaspis lubrica Oudemans and Voigts, 1904 (192)
   Ornithonyssus bacoti Hirst, 1913 (192)
   Radfordia ensifera (Poppe, 1896) (192)
Peromyscus truei
   Androlaelaps casalis (Berlese, 1887) (6, 8)
   A. fahrenholzi (Berlese, 1911) (6, 8, 10, 130, 160)
   Brevisterna morlani Strandtmann and Allred, 1956 (240)
   B. utahensis Strandtmann and Allred, 1956 (8)
   Dermanyssus becki Allred, 1957a (8)

Eubrachylaelaps circularis (Ewing, 1933) (6, 8, 10, 89, 160)
E. debilis Jameson, 1950 (6, 8, 10)
E. hollisteri (Ewing, 1925) (6, 8)
Haemogamasus nidi Michael, 1892 (8)
Hirstionyssus affinis (Jameson, 1950) (8)
H. neotomae Eads and Hightower, 1951 (8)
H. utahensis Allred and Beck, 1966 (8)
Ischyropoda armatus Keegan, 1951 (6, 8)
Kleemania sp. (10)
Ornithonyssus bacoti Hirst, 1913 (6, 8, 130)
Ornithonyssus sp. (6)
Peromyscus sp.
Androlaelaps fahrenholzi (Berlese, 1911) (154)
A. leviculus (Eads, 1951) (10)
Eubrachylaelaps circularis (Ewing, 1933) (106)
E. debilis Jameson, 1950 (106)
E. hollisteri (Ewing, 1925) (106)
E. spinosus Furman, 1955 (106)
Eulaelaps stabularis Koch, 1836 (154)
Haemogamasus ambulans (Thorell, 1872) (8)
Hirstionyssus carnifex (Koch, 1839) (241)
H. obsoletus (Jameson, 1950) (241)
Ornithonyssus bacoti Hirst, 1913 (82)

CHIGGERS.—Chiggers are the larval forms of mites of the family Trombiculidae and are usually parasitic on vertebrates. Adult and nymphal trombiculids are free living, feeding on eggs and immature forms of small arthropods. They are usually treated separately from the mites proper because they are so distinct from them morphologically and ecologically, because they form a relatively homogeneous assemblage, and because they can be easily recognized by the mammalogist. The latest key to the chiggers is that of Brennan and Jones (1959).

Chiggers have three pairs of legs and often are found around the ears of the hosts. They may have their mouthparts embedded when found; for this reason, one must exercise extreme care when removing them to avoid damage.

An exception to the generalization stated above concerning the place of attachment on the host is the finding by Loomis (1963). *Microtrombicula nasalis,* and *M. wrenni* from *Peromyscus* in California, and possibly also *M. ornata* from rodents in Kansas and Colorado, are parasites of the anterior portions of the nasal passages. To find these, one can open the nasal passages and examine them with a dissecting microscope or use the nasal flushing technique described by Yunker (1961). Intra-nasal mites should be looked for in other species of *Peromyscus* (and other mammals) in other areas.

Chiggers are relatively non-host specific, although some forms are rather restricted to one major vertebrate group.

A number of records of chiggers are available, and are summarized below.

Peromyscus boylei
  Chatia ochotona (Radford, 1942) (110, 143, 252)
  Euschoengastia criceticola Brennan, 1948 (6, 8, 40, 41, 110, 143)
  E. lacerta Brennan, 1948 (41, 110, 143)
  E. peromysci (Ewing, 1929) (41, 110, 143)
  E. pomerantzi Brennan and Jones, 1954 (41, 110)
  E. radfordi Brennan and Jones, 1954 (41, 110, 143)
  Neotrombicula dinehartae (Brennan and Wharton, 1950) (41, 110)
  N. jewetti Brennan and Wharton, 1950 (41, 44, 110)
  Odontacarus hirsutus (Ewing, 1931) (143)
  Pseudoschoengastia occidentalis Brennan, 1952 (36, 110, 143)

Peromyscus californicus
  Euschoengastia californica (Ewing, 1925) (182)
  E. criceticola Brennan, 1948 (41, 110, 182)
  E. frondifera Gould, 1956 (182)
  E. otophila Loomis and Bunnell, 1962 (182)
  E. peromysci (Ewing, 1929) (41, 110)
  E. pomerantzi Brennan and Jones, 1954 (41, 110)
  E. radfordi Brennan and Jones, 1954 (41, 110)
  Leeuwenhoekia dolosa Gould, 1956 (110)
  Miyatrombicula scottae (Brennan, 1952) (182)
  Neotrombicula californica (Ewing, 1942) (41, 110, 182)
  N. dinehartae (Brennan and Wharton, 1950) (110, 182)
  N. jewetti (Brennan and Wharton, 1950) (41, 110)
  N. scottae (Brennan, 1952) (41, 110)
  Pseudoschoengastia occidentalis Brennan, 1952 (36, 41, 110, 182)

Peromyscus crinitus
  Euschoengastia decipiens Gould, 1956 (81)
  E. radfordi Brennan and Jones, 1954 (81)
  Microtrombicula nasalis Loomis, 1963 (181)
  M. wrenni Loomis, 1963 (181)
  Trombicula montanensis Brennan, 1946 (8)

Peromyscus eremicus
  Euschoengastia criceticola Brennan, 1948 (8, 40, 110)
  E. decipiens Gould, 1956 (8)
  Whartonia whartoni Hoffman, 1951 (110)
  E. lanceolata Brennan and Beck, 1955 (8)

E. radfordi Brennan and Jones, 1954 (8)
Microtrombicula nasalis Loomis, 1963 (181)
M. wrenni Loomis, 1963 (181)
Peromyscus floridanus
    Euschoengastia peromysci (Ewing, 1929) (172)
    Gahrliepia americana (Ewing, 1942) (172)
    Microtrombicula crossleyi Loomis, 1954 (172)
Peromyscus gossypinus
    Euschoengastia setosa (Ewing, 1937) (90)
    Gahrliepia americana Ewing, 1942 (91)
Peromyscus leucopus
    Cheladonta micheneri Gould, 1956 (180)
    Euschoengastia blarinae Ewing, 1931 (93)
    E. carolinensis Farrell, 1956 (93)
    E. crateris Farrell, 1956 (93)
    E. criceticola Brennan, 1948 (180)
    E. diversa Farrell, 1956 (180)
    E. jonesi Lipovsky and Loomis, 1954 (177, 180)
    E. peromysci (Ewing, 1929) (74, 93, 180, 245)
    E. rubra Farrell, 1956 (93)
    E. setosa (Ewing, 1937) (74, 93, 180)
    Euschoengastoides hoplai (Loomis, 1954) (178, 180)
    E. loomisi (Crossley and Lipovsky, 1954) (180, 183)
    Eutrombicula alfreddugesi (Oudemans, 1910) (178, 180, 183, 278)
    Fonsecia gurneyi (Ewing, 1937) (179, 180, 183)
    F. kansasensis Loomis, 1955 (179, 180)
    Gahrliepia americana (Ewing, 1942) (74, 180)
    Leptotrombidium myotis (Ewing, 1929) (46)
    L. panamensis (Ewing, 1925) (183)
    Microtrombicula crossleyi (Loomis, 1954) (178, 180, 183)
    M. jonesae Brennan, 1952 (74)
    M. ornata (Loomis and Lipovsky, 1954) (178, 180, 183)
    Neotrombicula harperi (Ewing, 1928) (44)
    N. lipovskyi Brennan and Wharton, 1950 (158, 180)
    N. richmondi Brennan and Wharton, 1950 (158)
    N. sylvilagi Brennan and Wharton, 1950 (158, 180)
    N. whartoni (Ewing, 1929) (158)
    Odontacarus cayolargoensis Brennan, 1958 (183)
    O. morlani Brennan, 1958 (37)
    O. plumosus Greenberg, 1951 (180, 183)
    O. polychaetus Greenberg, 1952 (180, 183)
    Pseudoschoengastia audyi Brennan and Jones, 1959 (183)
    P. farneri Lipovsky, 1951a (180, 183)
    Trombicula jessiemae Gould, 1956 (39)
Peromyscus maniculatus
    Chatia ochotona (Radford, 1942) (6, 8, 40, 110, 252)
    C. setosa Brennan, 1946 (8, 34, 40, 110, 143)

Cheladonta micheneri Lipovsky, Crossley and Loomis, 1955 (180)
Euschoengastia californica (Ewing, 1925) (110, 182)
E. cordiremus Brennan, 1948 (8, 10, 35, 40, 110)
E. crateris Farrell, 1956 (93)
E. criceticola Brennan, 1948 (6, 8, 10, 35, 40, 41, 46, 83, 110, 143)
E. decipiens Gould, 1956 (8, 40)
E. diverse Farrell, 1956 (180)
E. frondifera Gould, 1956 (110)
E. jonesi Lipovsky and Loomis, 1954 (177, 180)
E. lacerta Brennan, 1948 (110)
E. lanei Brennan and Beck, 1955 (8, 10)
E. luteodema Brennan, 1948 (8, 40)
E. obesa Brennan and Beck, 1955 (8, 40)
E. oregonensis (Ewing, 1929) (8, 40)
E. peromysci (Ewing, 1929) (41, 93, 110, 143, 180)
E. pomarantzi Brennan and Jones, 1954 (8, 110)
E. radfordi Brennan and Jones, 1954 (8, 41, 83, 110, 143)
E. rotunda Brennan and Beck, 1956 (8)
E. sciuricola (Ewing, 1925) (6, 8, 110, 143)
E. setosa (Ewing, 1937) (170, 180)
E. trigenuala Farrell, 1956 (180)
Euschoengastoides hoplai (Loomis, 1954) (41, 110)
E. loomisi Crossley and Lipovsky, 1954 (61, 180)
Eutrombicula alfreddugesi (Oudemans, 1910) (180, 278)
E. lipovskyana (Wolfenbarger, 1952) (180)
Fonsecia gurneyi (Ewing, 1937) (179, 180)
F. kansasensis Loomis, 1955 (179, 180)
Leeuenhoekia americana (Ewing, 1942) (8, 110, 180)
L. dolosa Gould, 1956 (110, 143)
Leptotrombidium myotis Ewing, 1929 (6, 8, 170)
L. panamensis (Ewing, 1925) (8)
Microtrombicula crossleyi (Loomis, 1954) (178)
M. ornata Loomis and Lipovsky, 1954 (180, 184)
Miyatrombicula esoensis (Sasa and Ogata, 1953) (6, 8)
M. scottae Brennan, 1952 (41, 110)
Miyatrombicula sp. (170)
Neotrombicula browni (Brennan and Wharton, 1950) (44, 158)
N. californica (Ewing, 1942) (6, 8, 40, 44, 110)
N. cavicola (Ewing, 1931) (44, 110)
N. dinehartae (Brennan and Wharton, 1950) (41, 44, 110)
N. harperi (Ewing, 1928) (6, 8, 44, 158, 170)
N. jewetti (Brennan and Wharton, 1950) (6, 8, 41, 44, 110, 143)
N. lipovskyi (Brennan and Wharton, 1950) (158, 180)
N. loomisi (Kardos, 1954) (158)
N. microti (Ewing, 1928) (6, 44, 46, 143)
N. sylvilagi (Brennan and Wharton, 1950) (158, 180)
N. whartoni (Ewing, 1929) (158)

Odontacarus galli (Ewing, 1946) (113, 180)
O. hirsutus (Ewing, 1931) (143)
O. linsdalei Brennan and Jones, 1954 (8, 10)
O. plumosus Greenberg, 1961 (112, 180)
Pseudoschoengastia farneri Lipovsky, 1951 (175, 180)
P. hungerfordi Lipovsky, 1951a (175, 176, 180)
P. occidentalis Brennan, 1952 (182)
Trombicula arenicolor Loomis, 1954 (8, 40, 83)
T. bakeri Ewing, 1946 (8)
T. montanensis Brennan, 1946 (6, 8)
Peromyscus mexicanus
Pseudoschoengastia extrinseca Brennan, 1960 (38)
Peromyscus nudipes
Euschoengastia spissa Brennan and Jones, 1961 (43)
Hoffmannina handleyi Brennan and Jones, 1961 (43)
Trombicula caccabulus Brennan and Jones, 1961 (43)
T. dicrura Brennan and Jones, 1961 (43)
T. keenani Brennan and Jones, 1961 (43)
T. tiptoni Brennan and Jones, 1961 (43)
Peromyscus truei
Euschoengastia criceticola Brennan, 1948 (8, 41, 110)
E. decipiens Gould, 1956 (8)
E. peromysci (Ewing, 1929) (41, 110)
E. pomerantzi Brennan and Jones, 1954 (8, 41, 110)
E. radfordi Brennan and Jones, 1954 (41, 110)
Euschoengastoides hoplai (Loomis, 1954) (8, 40)
Eutrombicula belkini (Gould, 1950) (6, 8, 40, 41, 110, 111)
Microtrombicula nasalis Loomis, 1963 (181)
M. trisetica Loomis and Crossley, 1953 (41, 110)
Miyatrombicula scottae Brennan, 1952 (41, 110)
Neotrombicula jewetti Brennan and Wharton, 1950 (41, 110)
Odontacarus bakeri Hoffman, 1951 (128)
Trombicula arenicolor Loomis, 1954 (8)
T. bakeri Ewing, 1946 (8)
T. montanensis Brennan, 1946 (6, 8)
Peromyscus sp.
Euschoengastia criceticola Farrell, 1956 (93)
E. radfordi Brennan and Jones, 1954 (41, 110)
Hoffmannina handleyi Brennan and Jones, 1954 (43)
Miyatrombicula jonesae (Brennan, 1952) (206)
Neotrombicula carterae Brennan and Wharton, 1950 (44)
N. lipovskyi Brennan and Wharton, 1950 (44)
Polylopadium kramisi Brennan and Jones, 1961 (43)
Pseudoschoengastia whartoni Brennan, 1960 (38)
Sasacarus sp. "W" (10)
Trombicula caccabulus Brennan and Jones, 1961 (43)
T. chiriguensis Brennan and Jones, 1961 (43)

T. dicrura Brennan and Jones, 1961 (43)
T. keenani Brennan and Jones, 1961 (43)
Whartonia whartoni Hoffman, 1951 (183)

TICKS (IXODOIDEA).—There are a number of reports of ticks from *Peromyscus.* Ticks are essentially large, specialized mites, but can be readily distinguished from mites by their very large size and prominent, toothed hypostome. With the exception of the argasid tick, *Ornithodorus,* all ticks reported from *Peromyscus* are members of the family Ixodidae, or "hard ticks." In this family there is sexual dimorphism. The hard dorsal scutum covers the entire dorsum in males, but only the anterior part in females. Smaller larval ticks, at first glance, might be confused with large mites, but can be readily identified as ticks, because at this stage they have six rather than the eight legs characteristic of adult mites. Larval mites also have six legs, but larval mites are much too small to be confused with ticks at any stage.

Ixodid ticks exist by sucking blood. The larvae and nymphs feed once during each stage of their development but remain attached for days or even weeks. The adult female takes one blood meal before dropping from the host and laying her eggs. Ticks in search of a blood meal crawl upon vegetation and may wait for exceptionally long periods (months) until a warm-blooded host passes near them, at which time the tick drops. If the tick misses the host, or if the host for some reason is not suitable, the tick climbs back up the vegetation to wait for another host.

Many of the ticks found on mice of the genus *Peromyscus* are immature forms and often several are found on the same mouse. Earlier it was very difficult to identify larval ticks, but Clifford, Anastos, and Elbl (1961) have presented a very good key to eastern larval ticks which the mammalogist can use with a minimum of effort and preparation.

Ticks reported from *Peromyscus* are listed below.

Peromyscus boylei
  Ixodes angustus Neumann, 1899 (59)
Peromyscus californicus
  Haemaphysalis leporispalustris (Packard, 1869) (31, 129)
  Ixodes pacificus Cooley and Kohls, 1943 (31)
  Ixodes sp. (129)
Peromyscus crinitus
  Ixodes angustus Neumann, 1899 (9)

I. kingi Bishopp, 1911 (9)
I. ochotonae Gregson, 1941 (9)
Ixodes sp. (9)
Peromyscus eremicus
  Ixodes kingi Bishopp, 1911 (9)
  I. pacificus Cooley and Kohls, 1943 (9)
  I. spinipalpis Hadwen and Nuttall, 1916 (9, 21)
  Ixodes sp. (9)
Peromyscus floridanus
  Amblyomma americanum (Linnaeus, 1758) (172)
  A. maculatum Koch, 1844 (172)
  Dermacentor variabilis (Say, 1821) (172)
  Ixodes minor Neumann, 1902 (172)
Peromyscus gossypinus
  Dermacentor variabilis (Say, 1821) (31, 52, 192, 255, 279, 280)
  Ixodes scapularis Say, 1821 (31)
  Ixodes larvae (279, 280)
Peromyscus leucopus
  Dermacentor andersoni Stiles, 1908 (31)
  D. variabilis (Say, 1821) (11, 52, 74, 277)
  Ixodes dentatus Marx, 1899 (52)
  I. muris Bishopp and Smith, 1937 (11, 30, 31, 52, 229)
  I. scapularis Say, 1821 (29, 31)
  Ixodes sp. (275)
Peromyscus maniculatus
  Dermacentor andersoni Stiles, 1908 (12, 122, 139)
  D. occidentalis Marx, 1892 (31, 129)
  D. parumapertus Neumann, 1901 (23, 107)
  D. variabilis (Say, 1821) (170)
  Dermacentor sp. (15)
  Ixodes angustus Neumann, 1899 (9, 23, 31, 59, 170)
  I. cookei Packard, 1869 (59)
  I. kingi Bishopp, 1911 (9, 23)
  I. marmoratae Cooley and Kohls, 1938 (9)
  I. muris Bishopp and Smith, 1937 (9)
  I. ochotonae Gregson, 1941 (9)
  I. pacificus Cooley and Kohls, 1943 (9, 129)
  I. peromysci Augustson, 1939 (59)
  I. sculptus Neumann, 1904 (31)
  I. spinipalpis Hadwen and Nuttall, 1916 (9, 21)
  Ixodes sp. (9, 122, 129, 219)
  Ornithodorus eremicus Cooley and Kohls, 1941 (58)
Peromyscus polionotus
  Dermacentor variabilis (Say, 1821) (192)
  Ixodes scapularis Say, 1821 (31, 52)
Peromyscus truei
  Dermacentor occidentalis Marx, 1892 (129)

D. parumapertus Neumann, 1901 (107)
Ixodes angustus Neumann, 1899 (23, 59, 129)
I. kingi Bishopp, 1911 (9)
I. ochotonae Gregson, 1941 (9)
I. pacificus Cooley and Kohls, 1943 (9)
I. spinipalpis Hadwen and Nuttall, 1916 (9)
Ixodes sp. (129)
Peromyscus sp.
Amblyomma maculatum Koch, 1844 (82)
Dermacentor occidentalis Marx, 1892 (164)
D. parumapertus Neumann, 1901 (80)
D. variabilis (Say, 1821) (29)
Ixodes angustus Neumann, 1899 (154, 155, 156)
I. neotomae Cooley, 1944 (57)
I. ochotonae Gregson, 1941 (114)
I. pacificus Cooley and Kohls, 1943 (59)
I. scapularis Say, 1821 (59)

FLEAS.—There are many reports of fleas on *Peromyscus;* probably more than appear on the accompanying list, although an attempt has been made to include all.

There are several good general references on fleas. Holland (1949) has summarized information on the fleas of Canada, Hubbard (1947) on the fleas of western North America, Fox (1940) on the fleas of the eastern United States, and Ewing and Fox (1943) on North American fleas. Jellison and Good (1942) brought together the early literature on fleas.

Only adult fleas are found as parasites of mammals; the eggs are usually laid in the nest or litter of the host or sometimes in the hair. (In the latter case they normally drop off before hatching.) The larva is legless and is similar to that of a Dipteran. The larvae feed on detritus in the nest or litter and often eat the fecal pellets of the adults and particles of dried blood. Pupation occurs in the nest, with emergence taking place from four weeks to several months later. Not all the members of a single group of eggs emerge from the cocoon at the same time; rather they are staggered. The adults feed on the blood of the host.

Fleas are very common on *Peromyscus,* often several being found on a single individual. Various species of *Peromyscus* are the type hosts for at least 19 species of fleas.

No effort has been made here to summarize information for the fleas found in nests of *Peromyscus.* Hubbard (1947) states that

"There is little doubt that the nests of deer mice are a far better criterion of the fleas of this mouse than is the mouse itself. While an average of 4 fleas may be removed from these mice, a nest will often produce 100 or more." Hubbard (1947:470–475) presents a good deal of information concerning the fleas taken from nests of *P. maniculatus* and *P. truei* from Oregon. Miles and Kinney (1957) described a method for collecting fleas from rodent nests.

Records of fleas from *Peromyscus* are summarized below.

Peromyscus boylei
  Atyphloceras echis Jordan and Rothschild, 1915 (194)
  A. felix Jordan, 1933 (137)
  A. multidentatus (Fox, 1909) (143, 234)
  Catallagia decipiens Rothschild, 1915 (234)
  C. mathesoni Jameson, 1950 (143)
  Catallagia sp. (24, 234)
  Corypsylla kohlsi Hubbard, 1940 (143)
  Epitedia stanfordi Traub, 1944 (234)
  Hystrichopsylla dippiei Rothschild, 1902 (24)
  H. gigas (Baker, 1895) (2)
  Hystrichopsylla sp. (143)
  Hoplopsyllus anomalus (Baker, 1904) (194)
  Malaraeus sinomus (Jordan, 1925) (24, 131, 234)
  M. telchinum (Rothschild, 1905) (137, 143, 234)
  Megarthroglossus procus Jordan and Rothschild, 1915 (143)
  Monopsyllus eumolpi (Rothschild, 1905) (234)
  M. wagneri (Baker, 1904) (131, 143, 194)
  Opisodasys keeni (Baker, 1896) (143, 234)
  Orchopeas howardi (Baker, 1895) (137)
  O. leucopus (Baker, 1904) (194)
  O. sexdentatus (Baker, 1904) (194)
  Peromyscopsylla ebrighti (Fox, 1926) (137)
  P. hemisphaerium Stewart, 1940 (137, 238)
  P. hesperomys (Baker, 1904) (143)
  P. adelpha (Rothschild, 1915) (24)
  Pleochaetis sibynus (Jordan, 1925) (24)
  Rhadinopsylla sectilis Jordan and Rothschild, 1923 (234)
  Stenoponia ponera Traub and Johnson, 1952 (24)
Peromyscus californicus
  Anomiopsyllus congruens Stewart, 1940 (238)
  A. falsicalifornicus Fox, 1929 (137)
  Atyphloceras felix Jordan, 1933 (137, 238)
  A. longipalpus Stewart, 1940 (137)
  Carteretta carteri Fox, 1927 (137)
  Hoplopsyllus anomalus (Baker, 1904) (129)
  Malaraeus sinomus (Jordan, 1925) (137)

M. telchinum (Rothschild, 1905)   (129, 137)

Monopsyllus wagneri (Baker, 1904)   (137)

Nosopsyllus fasciatus (Bosc, 1801)   (137)

Opisodasys nesiotus Augustson, 1941*b*   (137)

Orchopeas caedens (Jordan, 1925)   (137)

O. howardi (Baker, 1895)   (137)

O. leucopus (Baker, 1904)   (137)

O. sexdentatus (Baker, 1904)   (137)

Peromyscopsylla ebrighti (Fox, 1926)   (137)

P. hemisphaerium Stewart, 1940   (129, 238)

Thrassis arizonensis (Baker, 1898)   (137)

Peromyscus crinitus

Anomiopsyllus falsicalifornicus Fox, 1929   (137)

Atyphloceras longipalpus Stewart, 1940   (137)

Callistopsyllus deuterus Jordan, 1937   (22)

Carterella carteri Fox, 1927   (22, 137)

Hoplopsyllus anomalus (Baker, 1904)   (234)

Malaraeus sinomus (Jordan, 1925)   (22, 137)

M. telchinum (Rothschild, 1905)   (137)

Megarthroglossus divisus (Baker, 1898)   (234)

Meringis hubbardi Kohls, 1938   (137, 165)

M. parkeri Jordan, 1937   (137)

Miochaeta macrodactyla (Good, 1942)   (234)

Monopsyllus wagneri (Baker, 1904)   (2, 137, 234)

O. sexdentatus (Baker, 1904)   (2, 22, 137, 234)

Peromyscopsylla ebrighti (Fox, 1926)   (137)

P. hesperomys (Baker, 1904)   (22)

Stenistomera alpina (Baker, 1895)   (22)

Thrassis aridus Prince, 1944   (22, 234)

T. bacchi (Rothschild, 1905)   (22)

Peromyscus eremicus

Anomiopsyllus amphibolus (Wagner, 1936)   (22)

A. nudatus (Baker, 1898)   (137)

Atyphloceras felix Jordan, 1933   (137)

A. multidentatus (Fox, 1909)   (137)

Callistopsyllus terinus (Rothschild, 1905)   (234)

Echidnophaga gallinacea (Westwood, 1875)   (234)

Jellisonia bullisi (Augustson, 1944)   (250)

Jordanopsylla allredi Traub and Tipton, 1951   (234)

Malaraeus sinomus (Jordan, 1925)   (137, 234)

M. telchinum (Rothschild, 1905)   (137)

Meringis dipodomys Kohls, 1938   (234)

Orchopeas caedens (Jordan, 1925)   (137)

O. howardi (Baker, 1895)   (137)

O. leucopus (Baker, 1904)   (137, 234)

O. sexdentatus (Baker, 1904)   (137)

Peromyscopsylla ebrighti (Fox, 1926) (137)
Thrassis gladiolis (Jordan, 1925) (137)
Peromyscus floridanus
  Ctenopthalmus pseudagyrtes Baker, 1904 (148, 172)
  Hoplopsyllus affinis (Baker, 1904) (148, 172)
  Polygenis floridanus Johnson and Layne, 1961 (148, 172)
  P. gwyni (Fox, 1914) (148, 172)
Peromyscus gossypinus
  Ctenocephalides felis (Bouche, 1835) (192)
  Ctenopthalmus pseudagyrtes Baker, 1904 (192)
  Leptopsylla segnis (Schonherr, 1811) (101, 192)
  Orchopeas leucopus (Baker, 1904) (101, 192)
  O. howardi (Baker, 1895) (101)
  Polygenis gwyni (Fox, 1914) (172, 192, 279, 280)
  Stenoponia americana (Baker, 1898) (101)
  Xenopsylla cheopis (Rothschild, 1903) (192)
Peromyscus guatemalensis
  Ctenopthalmus sanborni Traub, 1950 (250)
  Kohlsia osgoodi Traub, 1950 (250)
  Pleochaetis vermiformis Traub, 1950 (250)
Peromyscus hylocetes
  Jellisonia hayesi Traub, 1950 (250)
  Pleochaetis mathesoni Traub, 1950 (250)
  P. mundus (Jordan and Rothschild, 1922) (250)
  P. parus Traub, 1950 (250)
  Strepsylla fautini Traub, 1950 (250)
Peromyscus leucopus
  Anomiopsylla nudatus (Baker, 1898) (194)
  Catallagia charlottensis (Baker, 1898) (18, 45)
  C. decipiens Rothschild, 1915 (137, 210)
  Corrodopsylla curvata (Rothschild, 1915) (277)
  Ctenopthalmus pseudagyrtes Baker, 1904 (26, 28, 62, 101, 105, 108, 171, 197, 202, 210, 236, 237, 259, 270, 277)
  Doratopsylla blarinae (Fox, 1914) (101, 149, 245)
  Echidnophaga gallinacea (Westwood, 1875) (194)
  Epitedia stanfordi Traub, 1944 (194)
  E. wenmanni (Rothschild, 1904) (18, 27, 28, 45, 53, 62, 101, 105, 108, 137, 149, 150, 151, 171, 202, 210, 216, 224, 235, 237, 245, 259, 270, 275, 276)
  Hystrichopsylla dippiei Rothschild, 1902 (264)
  H. tahavuana Jordan, 1929 (108, 151)
  Malaraeus sinomus (Jordan, 1925) (194)
  Megabothris guirini (Rothschild, 1905) (28, 108)
  M. vison (Baker, 1904) (105)
  Megarthroglossus adversus (Wagner, 1936) (231)
  M. divisus (Baker, 1898) (193, 194)
  Monopsyllus exilis (Jordan, 1937) (194)
  M. wagneri (Baker, 1904) (84, 130, 131, 178, 194, 210, 259)

Nosopsyllus fasciatus (Bosc, 1801) (149)
Orchopeas howardi (Baker, 1895) (101, 108, 235, 237)
O. leucopus (Baker, 1904) (17, 28, 53, 62, 101, 105, 108, 130, 131, 149, 150, 152, 171, 194, 197, 202, 210, 224, 245, 259, 270, 274, 275, 276)
Peromyscopsylla draco Hopkins, 1951 (194)
P. hesperomys (Baker, 1904) (28, 53, 105, 108, 130, 152, 194, 197, 235, 274, 275, 276)
P. scotti Fox, 1939 (62, 99, 101, 108, 171, 270, 276)
Stenoponia americana (Baker, 1898) (51, 62, 101, 108, 171, 197, 224, 245, 270, 276)
Thrassis aridus Prince, 1949 (194)
T. pansus (Jordan, 1925) (194)
T. setosis Prince, 1944 (203)
Peromyscus maniculatus
Anomiopsyllus amphibolus Wagner, 1936 (22, 234)
A. nudatus (Baker, 1898) (234)
Atyphloceras artius Jordan, 1933 (134)
A. echis Jordan and Rothschild, 1915 (234)
A. felix Jordan, 1933 (137)
A. multidentatus Fox, 1909 (134, 137, 143, 234)
Callistopsyllus deuterus Jordan, 1937 (12, 137)
C. terinus (Rothschild, 1905) (122, 132, 134, 137, 138, 194)
Carteretta carteri Fox, 1927 (129, 137)
Catallagia chamberlini Hubbard, 1940 (134, 137)
C. charlottensis (Baker, 1898) (18, 77, 137, 264)
C. decipiens Rothschild, 1915 (2, 122, 134, 137, 138, 194)
C. mathesoni Jameson, 1950 (143)
C. motei Hubbard, 1940 (137)
C. rutherfordi Augustson, 1941 (13, 137)
C. sculleni Hubbard, 1940 (137, 143)
C. vonbloekeri Augustson, 1941*b* (13)
Catallagia sp. (24)
Catallagia sp. (*borealis* Ewing, 1929, or *decipiens* Rothschild, 1915) (47)
Ceratophyllus pelacani Augustson 1942 (137)
Conorhinopsylla nidicola Jellison, 1945 (202)
Corrodopsylla curvata (Rothschild, 1915) (122)
Corypsylla jordani Hubbard, 1940 (137)
C. ornata Fox, 1908 (137)
Ctenopthalmus pseudagyrtes Baker, 1904 (28, 47, 108, 134, 170, 171, 202, 259, 270)
Dactylopsylla ignota (Baker, 1895) (2, 137, 234)
Delotelis hollandi Smit, 1952 (143)
D. telegoni (Rothschild, 1905) (12, 134, 137)
Diamanus montanus (Baker, 1895) (2, 24, 137, 234)
Doratopsylla jellison Hubbard, 1940 (137)
Echidnophaga gallinacea (Westwood, 1875) (24)
Epitedia jordani Hubbard, 1940 (137)

E. scapani (Wagner, 1936) (132, 134, 143)

E. stanfordi Traub, 1944 (83, 137, 194, 234)

E. wenmanni (Rothschild, 1904) (2, 12, 24, 28, 47, 77, 101, 108, 134, 137, 143, 170, 171, 202, 210, 219, 234, 259, 270, 276)

Hoplopsyllus anomalus (Baker, 1904) (268)

Hystrichopsylla dippiei Rothschild, 1902 (24, 47, 134, 234)

H. gigas (Baker, 1895) (2, 137, 138)

H. occidentalis Holland, 1949 (137)

Hystrichopsylla sp. (143)

Malaraeus bitterrootensis Dunn and Parker, 1923 (12, 137, 234)

M. euphorbi (Rothschild, 1905) (18, 22, 83, 122, 134, 137, 194, 234, 264)

M. penicilliger Grube, 1852 (137, 151)

M. sinomus (Jordan, 1925) (22, 137, 138, 194, 233, 234)

M. telchinum (Rothschild, 1905) (2, 22, 77, 122, 129, 134, 137, 138, 143, 194, 234, 268)

Megabothris abantis (Rothschild, 1905) (2, 12, 122, 134, 137, 234)

M. adversus (Wagner, 1936) (137)

M. asio (Baker, 1904) (134, 170, 277)

M. lucifer (Rothschild, 1905) (134)

M. guirini (Rothschild, 1905) (47, 134, 170)

Megarthroglossus bisetis (Jordan and Rothschild, 1915) (273)

M. divisus (Baker, 1895) (134, 194)

M. procus Jordan and Rothschild, 1915 (12, 137, 143)

Meringis cummingi (Fox, 1926) (137)

M. dipodomys Kohls, 1938 (137, 234)

M. hubbardi Kohls, 1938 (122, 137)

M. nidi Williams and Hoff, 1951 (194)

M. parkeri Jordan, 1937 (22, 122, 137, 194, 234)

M. rectus Morlan, 1955 (194)

M. shannoni (Jordan, 1929) (134, 137)

Miochaeta macrodactyla (Good, 1942) (234)

Monopsyllus ciliatus (Baker, 1904) (132, 137, 143)

M. eumolpi (Rothschild, 1905) (2, 22, 122, 137, 194, 234)

M. exilis (Jordan, 1937) (234)

M. thambus (Jordan, 1929) (133)

M. vison (Baker, 1904) (134)

M. wagneri (Baker, 1904) (2, 12, 22, 24, 45, 47, 77, 83, 122, 129, 130, 131, 134, 137, 138, 143, 170, 194, 210, 233, 234, 259, 264)

Nearctopsylla hamata Holland and Jameson, 1949 (149)

N. princei Holland and Jameson, 1949 (138, 143)

Neopsylla inopina Rothschild, 1915 (138)

Nosopsyllus fasciatus (Bosc, 1801) (2, 134, 137, 234)

Opisodasys keeni (Baker, 1896) (12, 16, 17, 122, 125, 134, 137, 138, 144, 234, 264)

O. nesiotus Augustson, 1941*b* (13, 129)

Orchopeas caedens (Jordan, 1925) (137)

O. howardi (Baker, 1895) (101, 102, 108, 234, 237)

O. leucopus (Baker, 1904) (28, 45, 47, 53, 83, 101, 108, 131, 134, 137, 138, 157, 170, 171, 194, 202, 210, 219, 234, 259, 270, 277)

O. neotomae Augustson, 1943 (234)

O. nepos (Rothschild, 1905) (12, 137)

O. sexdentatus (Baker, 1904) (2, 137, 194, 234)

Oropsylla idahoensis (Baker, 1904) (2, 122, 234)

Peromyscopsylla catatina (Jordan, 1928) (170)

P. draco Hopkins, 1951 (194)

P. ebrighti (Fox, 1926) (137)

P. hesperomys (Baker, 1904) (12, 22, 24, 28, 101, 108, 130, 134, 137, 143, 194, 234)

P. scotti Fox, 1939 (202)

P. selenis (Rothschild, 1906) (134, 137, 216, 224, 234, 237, 264)

Pleochaetis sibynus (Jordan, 1925) (24, 137)

Rhadinopsylla fraterna (Baker, 1895) (134)

R. goodi (Hubbard, 1941) (83, 137)

R. sectilis Jordan and Rothschild, 1923 (77, 134, 137, 194, 234, 264)

Stenistomera alpina (Baker, 1895) (137, 234)

Stenoponia americana (Baker, 1899) (134, 171, 202, 270)

S. ponera Traub and Johnson, 1952 (24)

Thrassis acamantis (Rothschild, 1905) (137, 234)

T. aridus Prince, 1944 (194, 234)

T. campressus Prince, 1944 (203)

T. francisi (Fox, 1927) (234)

T. pandorae Jellison, 1937 (234)

T. pansus (Jordan, 1925) (194)

Peromyscus melanotis

  Ctenopthalmus haagi Traub, 1950 (250)

  Pleochaetis mundus (Jordan and Rothschild, 1922) (250)

  P. sibynus (Jordan, 1925) (250)

  Strepsylla mina Traub, 1950 (250)

Peromyscus nasutus

  Atyphloceras echis Jordan and Rothschild, 1915 (194)

  Diamanus montanus (Baker, 1895) (24)

  Epitedia stanfordi Traub, 1944 (194)

  Malaraeus sinomus (Jordan, 1925) (194)

  Megarthroglossus bisetis Jordan and Rothschild, 1915 (193, 194)

  Monopsyllus vison (Baker, 1904) (194)

  M. wagneri (Baker, 1904) (194)

  Peromyscopsylla hesperomys (Baker, 1904) (194)

  Rhadinopsylla sectilis Jordan and Rothschild, 1923 (194)

Peromyscus pectoralis

  Pleochaetoides bullisi Augustson, 1944 (14)

Peromyscus polionotus

  Ctenopthalmus pseudagyrtes Baker, 1904 (105, 192)

  Echidnophaga gallinacea (Westwood, 1875) (192)

Polygenis gwyni (Fox, 1914) (148, 192)
Xenopsylla cheopis (Rothschild, 1903) (192)
Peromyscus truei
  Anomiopsyllus amphibolus Wagner, 1936 (22)
  A. nudatus (Baker, 1898) (194)
  Atyphloceras echis Jordan and Rothschild, 1915 (194)
  A. felix Jordan, 1933 (137, 238)
  A. multidentatus Fox, 1909 (137)
  Callistopsyllus terinus (Rothschild, 1905) (194)
  Carteretta carteri Fox, 1927 (137, 234)
  Catallagia chamberlini Hubbard, 1940 (137)
  C. decipiens Rothschild, 1915 (22)
  Diamanus montanus (Baker, 1895) (130, 194)
  Epitedia stanfordi Traub, 1944 (194, 234)
  E. wenmanni (Rothschild, 1904) (137, 233, 234)
  Hoplopsyllus affinis (Baker, 1904) (130, 194)
  Malaraeus bitterrootensis Dunn and Parker, 1923 (234)
  M. euphorbi (Rothschild, 1905) (194)
  M. sinomus (Jordan, 1925) (22, 137, 194, 234)
  M. telchinum (Rothschild, 1905) (2, 129, 137, 234)
  Megarthroglossus bisetis Jordan and Rothschild, 1915 (194)
  Meringis jamesoni Hubbard, 1943 (194)
  M. nidi Williams and Hoff, 1951 (194)
  M. rectus Morlan, 1955 (194)
  Monopsyllus wagneri (Baker, 1904) (2, 22, 130, 131, 137, 194, 234)
  Opisodasys keeni (Baker, 1896) (234)
  O. nesiotus Augustson, 1941*b* (129)
  Orchopeas caedens (Jordan, 1925) (137, 233, 234)
  O. leucopus (Baker, 1904) (130, 131, 194, 234)
  O. sexdentatus (Baker, 1904) (194)
  Peromyscopsylla draco Hopkins, 1951 (194)
  P. hamifer (Rothschild, 1906) (234)
  P. hemisphaerium Stewart, 1940 (137, 238)
  P. hesperomys (Baker, 1904) (130, 194, 233, 234)
  Stenistomera alpina Baker, 1895 (137, 234)
  Thrassis aridus Prince, 1944 (203)
  T. bacchi (Rothschild, 1905) (22)
  Trirachipsylla digitiformis Stewart, 1940 (238)
Peromyscus sp.
  Atyphloceros atrius Jordan, 1933 (137)
  Callistopsyllus deuterus Jordan, 1937 (153)
  Catallagia chamberlini Hubbard, 1940 (137)
  C. decipiens Rothschild, 1915 (2, 137, 146)
  C. sculleni Hubbard, 1940 (137)
  Ctenopthalmus sanborni Traub, 1950 (250)
  Dactylopsylla ignota (Baker, 1895) (2, 137, 233)
  Epitedia wenmanni (Rothschild, 1904) (51, 137, 146, 264)

Hoplopsyllus anomalus (Baker, 1904) (2)
Kohlsia cora Traub, 1950 (250)
K. graphis erana Traub, 1950 (250)
K. gammonsi Traub, 1950 (250)
K. uniseta Traub, 1950 (250)
Malaraeus jordani Fox, 1939 (100)
M. penicilliger Jordan, 1938 (264)
M. sinomus (Jordan, 1925) (82)
M. telchinum (Rothschild, 1905) (2, 146)
Megarthroglossus procus Jordan and Rothschild, 1915 (137, 264)
Meringis hubbardi Kohls, 1938 (165)
Monopsyllus eumolpi (Rothschild, 1905) (137)
M. wagneri (Baker, 1904) (2, 17, 51, 137, 146, 230, 235)
Opisodasys keeni (Baker, 1896) (144, 146, 230)
Orchopeas leucopus (Baker, 1904) (82, 137, 146, 154)
Peromyscopsylla hesperomys (Baker, 1904) (17, 137)
Pleochaetis sibynus Jordan, 1925 (100, 250)
P. vermiformis Traub, 1950 (250)
Rhadinopsylla goodi (Hubbard, 1941) (137)
R. sectilis Jordan and Rothschild, 1923 (137, 146, 230)
Stenistomera alpina (Baker, 1895) (233)
Stenoponia americana (Baker, 1899) (146)
Strepsylla devisae Traub and Johnson, 1952 (251)
Thrassis acamantis (Rothschild, 1905) (146)
T. pandorae Jellison, 1937 (2)
T. pansus (Jordan, 1925) (203)

LICE.—There are two groups of lice, the Mallophaga, or biting lice, most common on birds but also found on mammals, and the Siphunculata or Anoplura, the sucking lice, found only on mammals. Only one record was found of biting lice on *Peromyscus;* hence, the discussion will relate only to the sucking lice.

Anoplurans feed solely on blood which they obtain by piercing the skin of the host and sucking. Small numbers of lice on *Peromyscus* probably cause little or no real harm to the host, but when in large numbers they may cause considerable irritation. They may, of course, transmit many of the blood diseases or parasites. Transfer from host to host probably takes place during direct contact between hosts while the young are still in the nest, or possibly during mating. Lice generally undergo their entire life cycle on one host; there being no secondary host and no free living stage involved. This has allowed sucking lice to become rather highly specialized to the living conditions on the host. It is probably for this reason that most lice are very highly host specific.

In the case of fleas and ticks, there is a free living stage; it would be disadvantageous for fleas to have to find one particular species of host at the end of a free-living stage, which probably accounts for the rather small amount of host specificity in fleas.

Some of the major references concerning lice are those of Hopkins (1949), Ferris (1951), and Kim (1965). Records of lice from *Peromyscus* are summarized below.

Anoplura:

Peromyscus boylei
    Hoplopleura ferrisi Cook and Beer, 1959 (24, 55, 95, 97, 162, 195)
    Polyplax auricularis Kellogg and Ferris, 1915 (24)

Peromyscus californicus
    Hoplopleura hesperomydis (Osborn, 1891) (97, 163)
    Polyplax auricularis Kellogg and Ferris, 1915 (135)

Peromyscus crinitus
    Hoplopleura difficilis Kim, 1965 (163)

Peromyscus eremicus
    Hoplopleura ferrisi Cook and Beer, 1959 (24, 55)
    Polyplax auricularis Kellogg and Ferris, 1915 (24)

Peromyscus floridanus
    Hoplopleura hirsuta Ferris, 1916 (172)

Peromyscus gossypinus
    Hoplopleura hesperomydis (Osborn, 1891) (163, 192)
    H. hirsuta Ferris, 1916 (163)
    Polyplax spinulosa (Burmeister, 1839) (163)

Peromyscus guatemalensis
    Hoplopleura ferrisi Cook and Beer, 1959 (163)

Peromyscus leucopus
    Haemodipsus sp. (84)
    Hoplopleura hesperomydis (Osborn, 1891) (25, 55, 162, 163, 195, 196, 197, 217,
        245, 276, 277)
    Polyplax auricularis Kellogg and Ferris, 1915 (217)

Peromyscus maniculatus
    Hoplopleura acanthopus (Burmeister, 1839) (12, 122, 170)
    H. erratica (Osborn, 1891) (170)
    H. hesperomydis (Osborn, 1891) (12, 55, 83, 94, 122, 162, 163, 170, 195, 217,
        219, 277)
    H. sciuricola Ferris, 1921 (170)
    Neohaematopinus laeviusculus (Grube, 1851) (12)
    Polyplax auricularis Kellogg and Ferris, 1915 (12, 24, 83, 94, 97, 122, 162, 195,
        217)

Peromyscus mexicanus
    Hoplopleura ferrisi Cook and Beer, 1959 (163)

Peromyscus nasutus
    Hoplopleura ferrisi Cook and Beer, 1959 (24, 55, 195)

H. hesperomydis (Osborn, 1891) (163)
Polyplax auricularis Kellogg and Ferris, 1915 (24, 195)
Peromyscus nudipes
Hoplopleura ferrisi Cook and Beer, 1959 (163)
Peromyscus polionotus
Hoplopleura hesperomydis (Osborn, 1891) (192)
H. hirsuta Ferris, 1916 (192)
Polyplax spinulosa (Burmeister, 1839) (192)
Peromyscus sitkensis
Polyplax auricularis Kellogg and Ferris, 1915 (162)
Peromyscus truei
Hoplopleura hesperomydis Kellogg and Ferris, 1915 (163, 195)
Polyplax auricularis Kellogg and Ferris, 1915 (195)
Peromyscus sp.
Polyplax auricularis Kellogg and Ferris, 1915 (96)
Mallophaga:
Peromyscus maniculatus
Strigiphilus ceblebrachys (Denny, 1842) (122)

DIPTERA: CUTEREBRIDAE.—A rather large amount of information has accumulated concerning parasitism by botfly larvae, *Cuterebra*, probably because these forms are large and obvious, and because they are easily identified at least to genus. Botfly larvae or "bots" as they are commonly called, are usually found in the inguinal and scrotal region of mice; they reach a length of about 25 mm and a width of about 18 mm in *C. angustifrons*. They are dark reddish brown. When about to emerge from the host they often protrude from the opening of their subdermal cavity. Unfortunately, relationships within the genus are not well understood and the species are difficult to identify by the larvae.

Cuterebra angustifrons Dalmat, 1942, is the common bot of the woodland deermouse, *Peromyscus maniculatus gracilis*. Blair (1942) found two bot larvae in *P. maniculatus gracilis* in northern Michigan, as well as in other woodland forms, *Tamias* and *Napaeozapus*. Farther south, he found that *Peromyscus maniculatus bairdi*, a field form, was not subject to parasitism by bots although *P. leucopus*, again a woodland form, had a heavy infestation. He hypothesized that the species of botfly involved was a woodland form. In Vigo County, west central Indiana, of 276 *Peromyscus leucopus* examined, 3 harbored *Cuterebra*, while none of 454 *Peromyscus maniculatus bairdi* did. Thus, these observations support the conclusion of Blair. It is possible, of course, that the two subspecies of *Pero-*

myscus maniculatus are enough different behaviorly or structurally, so that *P. m. bairdi* is either not in the right place under the right conditions to be parasitized, or else bots simply will not parasitize this form. Experimental work is desirable to solve the second problem, and additional life history work on both the hosts and parasites is desirable to solve the first.

Most workers have found *P. leucopus* much more heavily infested than *P. maniculatus* where both forms have been studied, that the fall was the most important time of infestation, and that most larvae were inguinal or scrotal in position. Brown (1965), working with *P. leucopus* and *P. boylei* in Missouri, found that only 0.3 per cent of 661 *P. leucopus* were infested as opposed to 7.7 per cent of 755 *P. boylei,* that there was a spring infestation period as well as one in the fall, and that in *P. boylei* the principal site of the parasite was on the back. He concluded that the form infesting *P. boylei* was probably a different species of *Cuterebra.*

Sillman (1959) provides information on the life history of *C. angustifrons.* Eggs are laid on vegetation where the mice may come in contact with them. The parasite then actively invades the host at an early stage of larval development. First instar larvae migrate beneath the skin, the majority coming to rest in the inguinal region. A small hole through which excretion and respiration occur is then made in the skin by the larva. The larvae mature in 23 to 30 days, at which time they force themselves through the opening to the outside by alternate expansion and contraction of the body. The larvae then drop to the ground, burrow into the soil and pupate. The pupae overwinter and the adult emerges the following summer. Penner and Pocius (1956) found that invasion could occur through the nostrils. Also, experimentally, they were able to make *Cuterebra (fontinella?)* invade *P. leucopus* but not *P. maniculatus.* Radovsky and Catts (1960) found nasal invasion by *C. latifrons,* a parasite of *Neotoma fuscipes.* Dalmat (1943) has suggested that an added generation might occur during the summer months. Wecker (1962) placed a mature larva in an insect cage on August 1; it pupated and emerged as an adult on September 6. Wecker felt that if this occurrence was rather frequent it could lead to the high numbers of cuterebrid larvae often found during the months of July to October. Some of the higher percentages of animals infested were recorded by Wecker (1962) who found 14.1, 40.9,

65.9, and 38.8 per cent of the mice, *Peromyscus leucopus,* infested in July, August, September, and October, respectively.

There has been a good deal of discussion about whether *Cuterebra* harms the host. It has long been hypothesized that since they are in the inguinal and scrotal region they might affect reproduction. Wecker (1962) found that the testes of 72.5 per cent of 251 mice not parasitized by *Cuterebra* were descended, while in the group of 73 males in which inguinal or scrotal cuterebrids were present, only 20.5 per cent had the testes descended. Dalmat (1942*a*) found 75 per cent emasculization in the males infested, and Sillman (1955) found 3 cases in which parasitized females failed to nurse their young. Wecker, likewise, found that parasitized mice were less likely to emigrate. He did not find increased mortality in those cases when only one larva was involved, but mortality apparently was increased in mice that harbored two or more larvae. It is possible that decreased agility leads to increased mortality.

Sealander (1961) found significant differences in blood values between parasitized and nonparasitized mice. In infested mice, the hemoglobin concentration was lower, the hematocrit per cent was higher, and the mean corpuscular hemoglobin concentration percentage was lower.

Four mice with fresh scar tissue showed blood values within the normal range for unparasitized mice. One female among the parasitized mice showed evidence of resorption of embryos.

Payne and Cosgrove (1966) described the histological changes during the infestation and found that healing was very rapid, being nearly complete in nine days.

A simple key to the larvae of *Cuterebra* is needed so that mammalogists can identify their own material to species.

Dipteran parasite records from *Peromyscus* are summarized below. All are *Cuterebra* except the one of *Atrypoderma* of *P. leucopus.*

Peromyscus boylei
   Cuterebra sp. (46*a*)
Peromyscus floridanus
   Cuterebra sp. (172)
Peromyscus gossypinus
   Cuterebra angustifrons Dalmat, 1942*b* (or very similar to it) (172)
Peromyscus leucopus
   Atrypoderma sp. (192)

Cuterebra angustifrons Dalmat, 1942*b*   (48, 64, 76, 198, 221, 226, 243, 267)
C. frontinella Clark (124, 127, 147, 220)
C. griseus (226)
C. peromysci Dalmat, 1942*a* (64, 65)
Cuterebra sp. (1, 32, 46*a*, 121, 140, 190, 199, 225, 226)
Peromyscus maniculatus
  Cuterebra americana complex (117)
  Cuterebra sp. (32, 170, 219)
Peromyscus melanophrys
  Cuterebra sp. (66)
Peromyscus nasutus
  Cuterebra sp. (190)
Peromyscus truei
  Cuterebra sp. (190)
Peromyscus sp. (probably both *P. leucopus, P. maniculatus* included)
  Cuterebra sp. (109)

Discussion

Host Specificity.—Some parasites are host specific while others are tolerant of the living conditions on or in several or many different hosts. In general, fleas, ticks, and many of the mites are rarely host specific, whereas sucking lice and many of the nematodes are.

In *Peromyscus,* the flea, *Orchopeas leucopus,* is common on *P. leucopus* in the eastern United States as well as on other species of *Peromyscus,* but it has also been taken on many other genera of mammals including, *Blarina, Didelphis, Microtus, Mus, Clethrionomys, Tamiasciurus, Sciurus, Marmota, Zapus,* and *Napaeozapus. Polygenis floridana,* on the other hand, a flea of *Peromyscus floridanus,* appears to be restricted to that species. Some other fleas which appear to be *"Peromyscus* fleas" are *Opisodasys keeni, Monopsyllus wagneri, Epitedia wenmanni,* and *Malaraeus euphorbi,* although none of these are entirely restricted to *Peromyscus.*

Some other parasites of *Peromyscus* that appear to be rather host specific are *Brachylaima chiapensis* and *B. peromysci,* trematodes of *P. guatemalensis* and *P. leucopus,* respectively, and *Syphacia peromysci,* a nematode of *P. leucopus* and *P. maniculatus. Rictularia coloradensis* was originally described from *Eutamias* but it and other species of *Rictularia* are very common in *Peromyscus.*

Relatively little information is available concerning host specificity for the mites, but Allred and Beck (1966) list *Dermanyssus becki* as a species of *Neotoma* and *Peromyscus, Eubrachylaelaps* as a species closely associated with *P. crinitus* (but often occurring on *P. maniculatus*), *E. circularis* as a species of *P. truei* (and *P. maniculatus*); and *E. debilis* as a species of *P. maniculatus. Androlaelaps casalis* is a parasite of *P. boylei* and *Thomomys,* and *A. fahrenholzi* is found on many species but is particularly abundant on *Peromyscus.* The chigger, *Euschoengastia peromysci,* is primarily a parasite of *P. leucopus; Euschoengastia criceticola* is a parasite of *Peromyscus,* particularly *P. maniculatus.*

EFFECTS ON PARASITE OF AGE AND HABITAT OF HOST, AND OF SEASON.—A great deal of the information on parasites of *Peromyscus* is concerned only with the species of host and parasite and the location of the parasite on the host. Relatively few papers have discussed the relationship of the parasite to the age or habitat of the host, or to season.

Grundmann and Frandsen (1960) present information on the helminth parasites of *Peromyscus maniculatus* from six different habitats. Farrell (1956), working with parasitic chiggers in North Carolina, presented information on the habitat and seasonal distribution of *Euschoengastia. E. peromysci* in this area was taken in hardwoods and in the narrow bottomlands along streams. The free-living stages lived in the soil, presumably in the ecological situation in which *P. leucopus,* the principal host of this form, spent much time. *E. peromysci* was most abundant in the colder part of the year when soil moisture was at its greatest. Layne (1963) has accumulated information of ecological interest concerning the parasites of *Peromyscus floridanus.* He found no relation between density of hosts and number of parasites per host, that young *P. floridanus* harbored fewer parasites than older ones (especially with regard to the internal parasites), and that most external parasites showed a correlation with habitat. Most of them were more abundant in moist, woodland situations, while the fleas were more abundant in dry habitats. Some species of internal parasites also showed strong correlations with certain types of habitat. Many species of parasites reached their greatest abundance in the fall and winter.

PARASITES AND DISEASES.—There is little information available concerning parasites of *Peromyscus* as disease organisms in man. In most cases it is assumed that they are not.

Allred (1952) lists mammals of the western United States which have been reported as having plague bacillus-infected tissue or have acted as hosts for plague-infected fleas. Included are *Peromyscus boylei, P. leucopus, P. maniculatus,* and *P. truei.* Allred also lists 30 species of fleas as "potential or capable" vectors of plague. Of the 30, at least 18 have been reported as occurring on one or more species of *Peromyscus.*

Loomis (1956) discusses chiggers as vectors for various Old World diseases, but states that there are no known cases of chiggers acting as vectors for diseases in the New World.

One species of chigger sometimes occurring on *Peromyscus, Eutrombicula alfreddugesi,* is a common pest of man. It is one of the species that causes severe irritation and swelling when larvae attach themselves to man, most often where clothes are pressed tight against the skin. It is the "chigger" inflammation especially common in southern United States.

Rocky Mountain spotted fever (and related diseases) is the most important disease of man transmitted by ticks. This disease has been found in most of the western states, although it is most common in Montana and Idaho. It is a disease of wild rodents and is caused by the rickettsia, *Dermacentroxenus rickettsi,* found in the cells of the ticks and hosts. The rickettsia is passed directly from the tick to her eggs and overwinters in the cells of the larval or nymphal ticks. *Dermacentor andersoni* and *Haemaphysalis leporispalustris* are the ticks most often transmitting the disease. Both of these ticks have been reported from *Peromyscus.* Eastern spotted fever, much milder than Rocky Mountain spotted fever, is widely distributed in eastern United States. It is transmitted by *Dermacentor variabilis* and is most common in the Allegheny Mountains.

In the Rocky Mountain area the bite of *Dermacentor andersoni,* a tick commonly found on *Peromyscus* and on many other hosts, may cause a progressive paralysis, called tick paralysis, which may result in death if neglected, but rapidly disappears upon removal of the tick.

Another disease transmitted by ticks (or fleas) is tularemia, caused by the bacterium, *Pasteurella tularensis,* but I know of no direct evidence linking this disease with mice of the genus *Peromyscus.*

Bubonic plague is a bacterial disease caused by *Pasteurella pestis.* The disease is transmitted by the bite of a flea that has bitten an infected animal. The bacteria multiply rapidly and cause subcutaneous hemorrhaging, forming black patches, hence the name "black death."

The first known case of plague in North America occurred in San Francisco in 1900, and between 1900 and 1904 there were 121 cases in the oriental sections of that city, of which 113 terminated in death. Presumably, the bacillus had entered this country on Oriental rat fleas, *Xenopsylla cheopis,* on rats from ships from the Orient. From there it spread to other species of mammals of the west, especially ground squirrels.

Fleas are also capable of serving as vectors for typhus, a disease caused by *Rickettsia prowazekii.* The murid rodent fleas *Xenopsylla cheopis, Echidnophaga gallinacea,* and *Leptopsylla segnis,* all have been found to harbor this virus, but I know of no cases of this disease in *Peromyscus,* although two of these species of fleas have been found on them.

It should be emphasized that the latter two diseases are primarily Old World diseases of murid rodents, or specifically of *Rattus.* Native deermice, *Peromyscus,* are not to be considered as major sources of human diseases; indeed, they appear to be very clean little animals.

RELATION OF PARASITE TO HOST.—We usually define a parasite as an organism which lives at the expense of another organism without killing it. Parasites usually live on, or in, the host organism, and are usually much smaller than the host. They are usually thought to use the host as their food supply, thus to harm the host.

Under this definition, many of the forms of animals we commonly speak of as parasites are not really parasites at all. Some of the mites do not feed, while others feed on the dead skin and detritus that accumulates in the fur of mice. These cannot be thought of as harming the host, if indeed, they affect the host at all.

Other parasites, for example, sucking lice, ticks, and other mites, suck blood or body juices and thus adequately fit the definition.

Most internal parasites probably do live at the expense of the host at least to a degree. They live upon its blood, its body juices, or its food supply. However, it is often debatable how much damage many of the parasites do to their hosts. In many cases, it is probably very little.

In reality, it is disadvantageous to the parasite to harm its host, since it is essentially harming its habitat. There is probably a distinct evolutionary advantage to the parasite that does the least harm to its host. In fact, we might speak of "perfect parasitism" as that case where no harm is done.

Conclusion

From a brief glance at the amount of information on the preceding pages it might be assumed that not much yet remains to be learned about the parasites of *Peromyscus*. It is hoped that actually the reader has gotten the reverse impression, that we really know little about the parasites of *Peromyscus* and that there is still plenty to do. We know nothing of the parasites of the majority of species of *Peromyscus*. Little or nothing is known of the life history and relation to the host for most of the parasites that have been found. Little information is available concerning the factors involved in differences in density of the parasites between different species of host, and between different habitats, ages, and densities of hosts.

The mammalogist, by diligently examining large series of the various species of *Peromyscus,* and by noting the numbers of parasites found, their location, and the ecological situation and condition of the host, can make major contributions to our knowledge of the parasites of this interesting genus of mice.

Literature Cited

1. ABBOTT, H. G., AND M. A. PARSONS. 1961. *Cuterebra* infestation in *Peromyscus*. Jour. Mamm., 42:383–385.
2. ALLRED, D. M. 1952. Plague important fleas and mammals in Utah and the Western United States. Great Basin Nat., 12:67–75.
3. ——— 1956. Mites found on mice of the genus *Peromyscus* in Utah. I. General Infestation. *Ibid.*, 16:23–31.
4. ——— 1957a. Mites found on mice of the genus *Peromyscus* in Utah. III. Family Dermanyssidae. Amer. Midland Nat., 57:450–460.
5. ——— 1957b. Mites found on mice of the genus *Peromyscus* in Utah. II. Family Haemogamasidae. Proc. Ent. Soc. Wash., 59:31–39.
6. ——— 1957c. Mites found on mice of the genus *Peromyscus* in Utah. V. Trombiculidae and miscellaneous families. Great Basin Nat., 17:95–102.

7. ———— 1958. Mites found on mice of the genus *Peromyscus* in Utah. IV. Families Laelaptidae and Phytoseiidae. Pan-Pacific Ent., 34:17–32.

8. ALLRED, D. M., AND D. E. BECK. 1966. Mites of Utah Mammals. Brigham Young Univ. Sci. Bull., Biol. Ser., 8 (1) :1–123.

9. ALLRED, D. M., D. E. BECK, AND L. D. WHITE. 1960. Ticks of the genus *Ixodes* in Utah. Brigham Young Univ. Sci. Bull., Biol. Ser., 1 (4) :1–42.

10. ALLRED, D. M., AND M. A. GOATES. 1964. Mites from mammals at the Nevada test site. Great Basin Nat., 24:71–73.

11. ANASTOS, G. 1947. Hosts of certain New York ticks. Psyche, 54:178–180.

12. AUGUSTSON, G. F. 1941a. Ectoparasite-host records from the Sierran region of east-central California. Bull. S. Cal. Acad. Sci., 40:147–157.

13. ———— 1941b. Contributions from the Los Angeles Museum Channel Islands Biological Survey, No. 20. Three new fleas (Siphonaptera). *Ibid.*, 40:101–107.

14. ———— 1944. A new mouse flea, *Pleochaetoides bullisi,* n. gen., n. sp., from Texas. Jour. Parasit., 30:366–368.

15. BACON, M., C. H. DRAKE, AND N. G. MILLER. 1959. Ticks (Acarina: Ixodoidea) on rabbits and rodents of eastern and central Washington. Jour. Parasit., 45:281–286.

16. BAKER, C. F. 1896. A new *Pulex* from Queen Charlotte Islands. Canad. Ent., 28:234.

17. ———— 1904. A revision of American Siphonaptera, or fleas; together with a complete list and bibliography of the group. Proc. U. S. Natl. Mus., 27:365–469.

18. ———— 1905. The classification of the American Siphonaptera. *Ibid.*, 29:121–170.

19. BAKER, E. W., AND G. W. WHARTON. 1952. An introduction to Acarology. New York: Macmillan, 465 pp.

20. BANKS, N. 1905. Descriptions of some new mites. Proc. Ent. Soc. Wash., 7:133–142.

21. BECK, D. E. 1955. Some unusual distribution records of ticks in Utah. Jour. Parasit., 41:1–4.

22. BECK, D. E., AND D. M. ALLRED. 1966. Siphonaptera (fleas) of the Nevada test site. Brigham Young Univ. Sci. Bull., Ser. 7 (2) :27 pp.

23. BECK, D. E., D. M. ALLRED, AND E. P. BRINTON. 1963. Ticks of the Nevada test site. Brigham Young Univ. Sci. Bull. Biol., Ser. 4 (1) :1–11.

24. BEER, J. R., E. F. COOK, AND R. G. SCHWAB. 1959. The ectoparasites of some mammals from the Chiricahua Mountains, Arizona. Jour. Parasit., 45:605–613.

25. BELL, J. F., AND W. S. CHALGREN. 1943. Some wildlife diseases in the eastern United States. Jour. Wildlife Mgt., 7:270–278.

26. BENTON, A. H., AND H. J. ALTMANN. 1964. A study of fleas found on *Peromyscus* in New York. Jour. Mamm., 45:31–36.

27. BENTON, A. H., AND R. H. CERWONKA. 1960. Host relationships of some eastern Siphonaptera. Amer. Midland Nat., 63:383–391.

298   *Whitaker*

28. BENTON, A. H., AND R. F. KRUG. 1956. Mammals and siphonapterous parasites of Rensselaer County, N. Y. N. Y. State Mus. and Sci. Service Bull., 353:1–22.
29. BEQUAERT, J. C. 1946. The ticks, or Ixodoidea, of the northeastern United States and eastern Canada. Entomologica Americana, 25:121–232.
30. BISHOPP, F. C., AND C. N. SMITH. 1937. A new species of *Ixodes* from Massachusetts. Proc. Ent. Soc. Wash., 39:133–138.
31. BISHOPP, F. C., AND H. L. TREMBLEY. 1945. Distribution and hosts of certain North American ticks. Jour. Parasit., 31:1–54.
32. BLAIR, W. F. 1942. Size of home range and notes on the life history of the woodland deer-mouse and eastern chipmunk in northern Michigan. Jour. Mamm., 23:27–36.
33. BOSMA, N. J. 1934. The life history of the trematode *Alaria mustelae,* Bosma, 1931. Trans. Amer. Micr. Soc., 53:116–153.
34. BRENNAN, J. M. 1946. A new genus and species of chigger, *Chatia setosa* (Trombiculidae, Acarina) from northwestern United States. Jour. Parasit., 32:132–135.
35. ———— 1948. New North American chiggers (Acarina, Trombiculidae). *Ibid.,* 34:465–478.
36. ———— 1952. The genus *Pseudoschoengastia* Lipovsky, 1951, with the description of two new species and a key to the world species, also *Neoschoengastia paenitens,* new name for *Neoschoengastia kohlsi* Brennan, 1951, preoccupied. Proc. Ent. Soc. Wash., 54:133–137.
37. ———— 1959. Synonymy of *Odontacarus* Ewing, 1929, and *Acomatacarus* Ewing, 1942, with redescriptions of *O. dentatus* (Ewing) and *O. australis* (Ewing), also descriptions of three new species from southern United States (Acarina: Trombiculidae). Ann. Ent. Soc. Amer., 52:1–6.
38. ———— 1960. Eight new species of *Pseudoschoengastia* from Mexico and Panama with a revised key to species (Acarina: Trombiculidae). Acarologia, 2:480–492.
39. ———— 1965. Two new species and other records of chiggers from Texas (Acarina: Trombiculidae). *Ibid.,* 7:79–83.
40. BRENNAN, J. M., AND D. E. BECK. 1955. The chiggers of Utah (Acarina: Trombiculidae). Great Basin Nat., 15:1–26.
41. BRENNAN, J. M., AND E. K. JONES. 1954. A report on the chiggers (Acarina: Trombiculidae) of the Frances Simes Hastings Natural History Reservation, Monterey County, California. Wasmann. Jour. Biol., 12:155–194.
42. ———— 1959. Keys to the chiggers of North America with synonymic notes and descriptions of two new genera (Acarina: Trombiculidae). Ann. Ent. Soc. Amer., 52:7–16.
43. ———— 1961. New genera and species of chiggers from Panama (Acarina: Trombiculidae). Jour. Parasit., 47:105–124.

44. BRENNAN, J. M., AND G. W. WHARTON. 1950. Studies on North American chiggers No. 3. The subgenus *Neotrombicula*. Amer. Midland Nat., 44:153–197.

45. BROWN, J. H. 1944. The fleas (Siphonaptera) of Alberta, with a list of the known vectors of sylvatic plague. Ann. Ent. Soc. Amer., 37: 207–213.

46. BROWN, J. H., AND J. M. BRENNAN. 1952. A note on the chiggers (Trombiculidae) of Alberta. Canad. Jour. Zool., 30:338–343.

46a. BROWN, L. N. 1965. Botfly parasitism in the brush mouse and white-footed mouse in the Ozarks. Jour. Parasit., 51:302–304.

47. BUCKNER, C. H. 1964. Fleas (Siphonaptera) of Manitoba mammals. Canad. Entomologist, 96:850–856.

48. BURT, W. H. 1940. Territorial behavior and populations of some small mammals in Southern Michigan. Misc. Pub. Univ. Mich. Mus. Zool., 45:41–42.

49. CHANDLER, A. C. 1946. *Trichuris peromysci* n. sp. from *Peromyscus californicus,* and further notes on *T. perognathi* Chandler, 1945. Jour. Parasit., 32:208.

50. CHANDLER, A. C., AND D. M. MELVIN. 1951. A new cestode, *Oochoristica pennsylvanica,* and some new or rare helminth host records from Pennsylvania mammals. Jour. Parasit., 37:106–109.

51. CHAPIN, E. A. 1919. New species of North American Siphonaptera. Bull. Brooklyn Ent. Soc., N. S., 14:49–62.

52. CLIFFORD, C. M., G. ANASTOS, AND A. ELBL. 1961. The larval ixodid ticks of the eastern United States (Acarina–Ixodidae). Misc. Publ. Ent. Soc. Amer., 2:215–237.

53. CONNOR, P. F. 1960. The small mammals of Otsego and Schoharie counties, New York. N. Y. State Museum and Sci. Service Bull., 382, 84 pp.

54. COOK, E. F., AND J. R. BEER. 1958. A study of louse populations on the meadow vole and deer mouse. Ecology, 39:645–659.

55. ——— 1959. The immature stages of the genus *Hoplopleura* (Anoplura: Hoplopleuridae) in North America, with descriptions of two new species. Jour. Parasit., 45:405–416.

56. COOLEY, R. A. 1938. The genera *Dermacentor* and *Octocentor* (Ixodidae) in the United States, with studies in variation. Nat. Inst. Health Bull., 171.

57. ——— 1944. *Ixodes neotomae,* a new species from California. (Acarina: Ixodidae). Pan-Pacific Ent., 20:7–12.

58. COOLEY, R. A., AND G. M. KOHLS. 1941. Three new species of *Ornithodoros* (Acarina: Ixodoidea). Publ. Health Reports, 56:587–594.

59. ——— 1945. The genus *Ixodes* in North America. NIH Bull., 184: 1–246.

60. CORRINGTON, J. D. 1941. Working with the microscope. New York: McGraw-Hill, 418 pp.

61. CROSSLEY, D. A., AND L. J. LIPOVSKY. 1954. Two new chiggers from the central states (Acarina, Trombiculidae). Proc. Ent. Soc. Wash., 56:240–246.

62. CUMMINGS, E. 1954. Notes on some Siphonaptera from Albany County, N. Y. Jour. N. Y. Ent. Soc., 62:161–165.

63. DALMAT, H. T. 1942a. A new parasitic fly (Cuterebridae) from the northern white footed mouse. Jour. New York Ent. Soc., 50: 45–59.

64. ———— 1942b. A new *Cuterebra* from Iowa with notes on certain facial structures. Amer. Midland Nat., 27:418–421.

65. ———— 1943. A contribution to the knowledge of the rodent warble flies (Cuterebridae). Jour. Parasit., 29:311–318.

66. DAVIS, W. B. 1944. Notes on Mexican mammals. Jour. Mamm., 25:370–403.

67. DAWES, B. 1946. The Trematoda. Cambridge Univ. Press, 644 pp.

68. DICKMANS, G. 1935. The nematodes of the genus *Longistriata* in rodents. Jour. Wash. Acad. Sci., 25:72–81.

69. DORAN, D. J. 1954a. A catalogue of the protozoa and helminths of North American rodents. I. Protozoa and Acanthocephala. Amer. Midland Nat., 52:118–128.

70. ———— 1954b. A catalog of the protozoa and helminths of North American rodents. II. Cestoda. *Ibid.*, 52:469–480.

71. ———— 1955a. A catalogue of the protozoa and helminths of North American rodents. III. Nematoda. *Ibid.*, 53:162–175.

72. ———— 1955b. A catalogue of the protozoa and helminths of North American rodents. Part IV. Trematoda. *Ibid.*, 53:446–454.

73. DOUTHITT, H. 1915. Studies on the cestode family Anoplocephalidae. Ill. Biol. Monogr., 1:1–96.

74. DRUMMOND, R. O. 1957a. Ectoparasitic Acarina from small mammals of the Patuxent refuge, Bowie, Maryland. Jour. Parasit., 43:50.

75. ———— 1957b. Observations on fluctuations of Acarine populations from nests of *Peromyscus leucopus*. Ecol. Monogr., 27:137–152.

76. DUNAWAY, P. B., J. A. PAYNE, L. L. LEWIS, AND J. D. STORY. 1967. Incidence and effects of *Cuterebra* in *Peromyscus*. Jour. Mamm., 48: 38–51.

77. DUNN, L. H., AND R. R. PARKER. 1923. Fleas found on wild animals in the Bitter Root Valley, Montana. Publ. Health Reports, 38 (47): 2763–2775.

78. EADS, R. B. 1951. New mites of the genus *Androlaelaps* Berlese. Jour. Parasit., 37:212–216.

79. EADS, R. B., AND B. G. HIGHTOWER. 1952. Blood parasites of southwest Texas rodents. Jour. Parasit., 38:89–90.

80. EADS, R. B., G. C. MENZIES, AND V. I. MILES. 1952. Acarina taken during west Texas plague studies. Proc. Ent. Soc. Wash., 54:250–253.

81. EDWARDS, R. L., AND W. H. PITTS. 1952. Dog locates winter nests of mammals. Jour. Mamm., 33:243–244.

82. ELLIS, L. L., JR. 1955. A survey of the ectoparasites of certain mammals in Oklahoma. Ecology, 36:12–18.

83. ELZINGA, R. J., AND D. M. REES. 1964. Comparative rates of ectoparasite infestation on deer and harvest mice. Proc. Utah Acad. Sci., 41: 217–220.

84. ERICKSON, ARNOLD B. 1938. Parasites of some Minnesota rodents. Jour. Mamm., 19:252–253.

85. ———— 1938. Parasites of some Minnesota Cricetidae and Zapodidae, and a host catalogue of helminth parasites of native American mice. Amer. Midland Nat., 20:575–589.

86. EVANS, G. O., AND W. M. TILL. 1962. Studies on the British Dermanyssidae (Acari: Mesostigmata). Part II. Classification. Bull. Brit. Mus. (Nat. Hist.), 14:109–370.

87. EWING, H. E. 1923. The dermanyssid mites of North America. Proc. U. S. Natl. Mus., 32 (13) :1–26.

88. ———— 1931. A catalogue of the Trombiculinae, or chigger mites, of the New World with new genera and species and a key to the genera. *Ibid.,* 80 (8) :1–19.

89. ———— 1933. New genera and species of parasitic mites of the super family Parasitoidea. *Ibid.,* 82 (30) :1–14.

90. ———— 1937. New species of mites of the subfamily Trombiculinae, with a key to the new world larvae of the Akamushi group of the genus *Trombicula.* Proc. Biol. Soc. Wash., 50:167–174.

91. ———— 1942. Remarks on the taxonomy of some American chiggers (Trombiculinae), including the descriptions of new genera and species. Jour. Parasit., 28:485–493.

92. EWING, H. E., AND I. FOX. 1943. The fleas of North America. U. S. D. A. Misc. Publ. No. 500:142 pp.

93. FARRELL, C. E. 1956. Chiggers of the genus *Euschoengastia* (Acarina: Trombiculidae) in North America. Proc. U. S. Natl. Mus., 106: 85–235.

94. FERRIS, G. F. 1916. Notes on Anoplura and Mallophaga, from mammals, with description of four new species and a new variety of Anoplura. Psyche, 23:97–120.

95. ———— 1921. Contributions toward a monograph of the sucking lice. Part II. Stanford Univ. Press, 133 pp.

96. ———— 1923. Contributions toward a monograph of the sucking lice. Part IV. *Ibid.,* 270 pp.

97. ———— 1951. The sucking lice. Mem. Pacific Coast Ent. Soc., 1:1–320.

98. FLORSCHUTZ, O., JR., AND R. F. DARSIE, JR. 1960. Additional records of ectoparasites on Delaware mammals. Ent. News, 21:45–52.

99. FOX, I. 1939. New species and a new genus of Nearctic siphonaptera. Proc. Ent. Soc. Wash., 41:45–49.

100. ———— 1939. New species and records of Siphonaptera from Mexico. Jour. Sci., Iowa State College, 13:335–339.

101. ———— 1940. Fleas of eastern United States. Ames: Iowa State College Press, 191 pp.

102. FRANDSEN, J. C., AND A. W. GRUNDMANN. 1960. *Brachylaime microti* from the deer mouse in Utah. Jour. Parasit., 46:314.

103. FREEMAN, R. S. 1956. Life history studies on *Taenia mustelae* Gmelin, 1790, and the taxonomy of certain taenioid cestodes from Mustelidae. Canad. Jour. Zool., 34:219–242.

104. FREEMAN, R. S., AND K. A. WRIGHT. 1960. Factors concerned with the epizootiology of *Capillaria hepatica* (Bancroft, 1893) (Nematoda) in a population of *Peromyscus maniculatus* in Algonquin Park, Canada. Jour. Parasit., 46:373–382.

105. FULLER, H. S. 1943. Studies on Siphonaptera of eastern United States. Bull. Brooklyn Ent. Soc., 38:18–23.

106. FURMAN, D. P. 1955. Revision of the genus *Eubrachylaelaps* (Acarina: Laelaptidae) with the descriptions of two new species. Ann. Ent. Soc. Amer., 48:51–59.

107. GASTFRIEND, A. 1955. New host records for the immature stages of the tick *Dermacentor parumapertus*. Jour. Parasit., 41:63–65.

108. GEARY, J. M. 1959. The fleas of New York. Cornell Univ. Agric. Exp. Stat. Mem., 355:104 pp.

109. GOERTZ, J. W. 1966. Incidence of warbles in some Oklahoma rodents. Amer. Midland Nat., 75:242–245.

110. GOULD, O. J. 1956. The larval Trombiculid mites of California (Acarina: Trombiculidae). Univ. Cal. Publ. Ent., 11:1–116.

111. GRANDJEAN, P. H., AND J. R. AUDY. 1965. Revision of the genus *Eutrombicula* Ewing 1938 (Acarina: Trombiculidae). Acarologia, 7 (Suppl.):280–294.

112. GREENBERG, B. 1951. A new subgenus of *Acomatacarus* from Kansas (Acarina: Trombiculidae). Jour. Parasit., 37:525–527.

113. ———— 1952. A review of the New World *Acomatocarus* (Acarina: Trombiculidae). Ann. Ent. Soc. Amer., 45:473–491.

114. GREGSON, J. D. 1941. Two new species of ticks from British Columbia (Ixodidae). Canad. Ent., 73:220–228.

115. GRUNDMANN, A. W. 1957. Nematode parasites of mammals of the Great Salt Lake Desert of Utah. Jour. Parasit., 43:105–112.

116. ———— 1958. Cestodes of mammals from the Great Salt Lake Desert region of Utah. *Ibid.*, 44:425–429.

117. GRUNDMANN, A. W., AND J. C. FRANDSEN. 1960. Definitive host relationships of the helminth parasites of the deer mouse, *Peromyscus maniculatus*, in the Bonneville Basin of Utah. Jour. Parasit., 46:673–677.

118. HALL, M. C. 1912. The parasite fauna of Colorado. Colo. Coll. Publ. Sci., 20:329–383.

119. ———— 1916. Nematode parasites of mammals of the orders Rodentia, Lagomorpha, and Hyracoidea. Proc. U. S. Natl. Mus., 50:258 pp.

120. HALL, J. E., B. SONNENBERG, AND J. R. HODES. 1955. Some helminth parasites of rodents from localities in Maryland and Kentucky. Jour. Parasit., 41:640–641.

121. HAMILTON, W. J., JR. 1930. Notes on the mammals of Breathitt County, Kentucky. Jour. Mamm., 11:306–311.
122. HANSEN, C. G. 1964. Ectoparasites of mammals from Oregon. Great Basin Nat., 24:75–81.
123. HANSEN, M. F. 1950. A new dilepidid tapeworm and notes on other tapeworms of rodents. Amer. Midland Nat., 43:471–479.
124. HARKEMA, R. 1936. The parasites of some North Carolina rodents. Ecol. Monogr., 6:153–232.
125. HARVEY, R. V. 1907. British Columbia fleas. Bull. Ent. Soc. Brit. Columbia, 7:1.
126. HAYS, K. L., AND F. E. GUYTON. 1958. Parasitic mites (Acarina: Mesostigmata) from Alabama mammals. Jour. Econ. Ent., 51:259–260.
127. HIRTH, H. F. 1959. Small mammals in old field succession. Ecology, 40: 417–425.
128. HOFFMAN, A. 1951. Contribuciones al conocimiento de los trombiculidas Mexicanos. 3a parte. Ciencia, 11:29–36.
129. HOLDENRIED, R., F. C. EVANS, AND D. S. LONGANECKER. 1951. Host-parasite-disease relationships in a mammalian community in the central coast range of California. Ecol. Monogr., 21:1–18.
130. HOLDENRIED, R., AND H. B. MORLAN. 1955. Plague infected fleas from northern New Mexico wild rodents. Jour. Infect. Dis., 96:133–137.
131. ——— 1956. A field study of wild mammals and fleas of Santa Fe County, New Mexico. Amer. Midland Nat., 55:369–381.
132. HOLLAND, G. P. 1941. Further records of Siphonaptera for British Columbia. Proc. Ent. Soc. Brit. Col., 37:10–14.
133. ——— 1944. Notes on some northern Canadian Siphonaptera, with the description of a new species. Canad. Ent., 76:242–246.
134. ——— 1949. The Siphonaptera of Canada. Dom. of Canada. Dept. Agric. Publ., 817, Tech. Bull., 70:1–306.
135. HOPKINS, G. H. E. 1949. The host associations of the lice of mammals. Proc. Zool. Soc. London, 119:387–604.
136. HOWELL, J. F., AND R. J. ELZINGA. 1962. A new *Radfordia* (Acarina: Myobidae) from the kangaroo rat and a key to the known species. Ann. Ent. Soc. Amer., 55:547–555.
137. HUBBARD, C. A. 1947. The fleas of Western North America. Iowa State College Press, 533 pp.
138. ——— 1949. Fleas of the state of Nevada. Bull. S. Cal. Acad. Sci., 48: 115–128.
139. HUNTER, W. D., AND F. C. BISHOPP. 1911. The Rocky Mountain spotted fever tick, with special reference to the problem of its control in the Bitterroot Valley in Montana. Bull. Bur. Entom., USDA, 105:47 pp.
140. JAMESON, E. W., JR. 1943. Notes on the habits and siphonapterous parasites of the mammals of Welland County, Ontario. Jour. Mamm., 24:194–197.

304   *Whitaker*

141. ———— 1950. *Hirstionyssus obsoletus,* a new mesostigmatic mite from small mammals of the Western United States. Proc. Biol. Soc. Wash., 63:31–34.
142. ———— 1952. *Euhaemogamasus keegani,* new species, a parasitic mite from Western North America. Ann. Ent. Soc. Amer., 45:600–604.
143. JAMESON, E. W., JR., AND J. M. BRENNAN. 1957. An environmental analysis of some ectoparasites of small forest mammals in the Sierra Nevada, California. Ecol. Monogr., 27:45–54.
144. JELLISON, W. L. 1939. *Opisodasys* Jordan, 1933, a genus of Siphonaptera. Jour. Parasit., 25:413–420.
145. JELLISON, W. L., AND N. E. GOOD. 1942. Index to the literature of Siphonaptera of North America. NIH Bull., 178:1–193.
146. JELLISON, W. L., G. M. KOHLS, AND H. B. MILLS. 1943. Siphonaptera–species and host list of Montana fleas. Mont. State Bd. Entom., Misc. Publ., 2:22 pp.
147. JOHNSON, C. W. 1930. A bot fly from the white footed mouse. Psyche, 37:283–284.
148. JOHNSON, P. T., AND J. N. LAYNE. 1961. A new species of *Polygenis* Jordan from Florida, with remarks on its host relationships and zoogeographic significance. Proc. Ent. Soc. Wash., 63:115–123.
149. JORDAN, K. 1928. Siphonaptera collected during a visit to the eastern United States of North America in 1927. Novitates Zool., 34:178–188.
150. ———— 1929. On a small collection of Siphonaptera from the Adirondacks, with a list of the species known from the State of New York. *Ibid.,* 35:168–177.
151. ———— 1932. Siphonaptera collected by Mr. Harry Swarth at Atlin in British Columbia. *Ibid.,* 38:253–255.
152. ———— 1933. Records of Siphonaptera from the State of New York. *Ibid.,* 39:62–65.
153. ———— 1937. On some North American Siphonaptera. *Ibid.,* 40:262–271.
154. JUDD, W. W. 1950. Mammal host records of Acarina and insects from the vicinity of Hamilton, Ontario. Jour. Mamm., 31:357–358.
155. ———— 1953. Mammal host records of Acarina and insects from the vicinity of London, Ontario. Jour. Mamm., 34:137–139.
156. ———— 1954. Some records of ectoparasitic Acarina and insecta from mammals in Ontario. Jour. Parasit., 40:483–484.
157. ———— 1955. A collection of fleas from the vicinity of Fort Simpson, Northwest Territories, Canada. *Ibid.,* 41:441–442.
158. KARDOS, E. H. 1954. Biological and systematic studies on the subgenus *Neotrombicula* (genus Trombicula) in the central United States (Acarina, Trombiculidae). Univ. Kansas Sci. Bull., 36 (Pt. I) No. 4:69–123.
159. KEEGAN, H. L. 1951. The mites of the subfamily *Haemogamasinae* (Acari: Laelaptidae). Proc. U. S. Natl. Mus., 101:203–268.

160. ———— 1953. Collections of parasitic mites from Utah. Great Basin Nat., 13:35–42.
161. ———— 1956. *Ischyropoda furmani* n. sp., a new ectoparasitic mite from Utah. Jour. Parasit., 42:311–315.
162. KELLOGG, V. L., AND G. F. FERRIS. 1915. The Anoplura and Mallophaga of North American mammals. Stanford Univ. Publ., 74 pp.
163. KIM, K. C. 1965. A review of the *Hoplopleura hesperomydis* complex (Anoplura, Hoplopleuridae). Jour. Parasit., 51:871–887.
164. KOHLS, G. M. 1937. Hosts of the immature stages of the Pacific Coast tick, *Dermacentor occidentalis* Neum. (Ixodidae). Pub. Health Repts., 52 (16) :490–496.
165. ———— 1938. Two new species of *Meringis* Jordan. *Ibid.*, 53:1216–1220.
166. KOMAREK, E. V. 1939. A progress report on southeastern mammal studies. Jour. Mamm., 20:292–299.
167. KRUIDENIER, F. J., AND V. GALLICCHIO. 1959. The orthography of the Brachylaimidae (Joyeux and Foley, 1930) ; *Brachylaime microti* sp. nov.; *B. rauschi* McIntosh, 1950; and an addendum to Dollfus' (1935) list of Brachylaime (Trematoda: Digenea). Trans. Amer. Micro. Soc., 78:428–441.
168. KRUIDENIER, F. J., AND C. R. PEEBLES. 1958. *Gongylonema* of rodents: *G. neoplasticum* (redefinition) , *G. dipodomysis* n. sp. and *G. peromysci* n. sp. Trans. Amer. Micro. Soc., 77:307–315.
169. KRULL, W. H. 1934. The white footed mouse, *Peromyscus leucopus noveboracensis,* a new host for *Entosiphonus thompsoni* Sinitsin 1931; Brachylaemidae. Jour. Parasit., 20:98.
170. LAWRENCE, W. H., K. L. HAYS, AND S. A. GRAHAM. 1965. Arthropodous ectoparasites from some northern Michigan mammals. Occ. Pap. Mus. Zool. Univ. Mich., 639:1–7.
171. LAYNE, J. N. 1958. Records of fleas (Siphonaptera) from Illinois mammals. Nat. Hist. Misc., Chi. Acad. Sci., 162:1–72.
172. ———— 1963. A study of the parasites of the Florida mouse, *Peromyscus floridanus,* in relation to host and environmental factors. Tulane Stud. Zool., 11:1–27.
173. LAYNE, J. N., AND J. V. GRIFFO. 1961. Incidence of *Capillaria hepatica* in populations of the Florida deer mouse, *Peromyscus floridanus.* Jour. Parasit., 47:31–37.
174. LEIBY, P. D. 1962. Helminth parasites recovered from some rodents in Southeastern Idaho. Amer. Midland Nat., 67:250.
175. LIPOVSKY, L. J. 1951a. A new genus of Walchiinae (Acarina: Trombiculidae). Jour. Kans. Ent. Soc., 24:95–102.
176. ———— 1951b. A washing method of ectoparasite recovery with particular reference to chiggers (Acarina: Trombiculidae). *Ibid.,* 24: 151–156.
177. LIPOVSKY, L. J., AND R. B. LOOMIS. 1954. A new chigger mite (genus *Euschoengastia*) from the central United States. Jour. Parasit., 40:407–409.

306   *Whitaker*

178. Loomis, R. B. 1954. A new subgenus and six new species of chigger mites (genus *Trombicula*) from the central United States. Univ. Kans. Sci. Bull., 36 (II) No. 13:919–941.

179. ———— 1955. *Trombicula gurneyi* Ewing and two new related chigger mites (Acarina, Trombiculidae). *Ibid.*, 37 (I) No. 9:251–267.

180. ———— 1956. The chigger mites of Kansas. *Ibid.*, 37 (II) No. 19:1195–1443.

181. ———— 1963. The discovery of chiggers (Acarina, Trombiculidae) in the nasal passages of cricetid rodents from California, with descriptions of two new species. Jour. Parasit., 49:330–333.

182. Loomis, R. B., and M. Bunnell. 1962. A new species of chigger, genus *Euschoengastia* (Acarina, Trombiculidae), with notes on other species of chiggers from the Santa Ana Mountains, California. Bull. S. Calif. Acad. Sci., 61 (Pt. 3) :177–184.

183. Loomis, R. B., and D. A. Crossley, Jr. 1963. New species and new records of chiggers (Acarina: Trombiculidae) from Texas. Acarologia, 5:371–383.

184. Loomis, R. B., and L. J. Lipovsky. 1954. Two new chigger mites (genus *Trombicula*) from the central United States. Jour. Parasit., 27: 47–53.

185. Lubinsky, G. 1957. List of helminths from Alberta rodents. Can. Jour. Zool., 35:623–627.

186. McDaniel, B. 1965. The subfamily Listrophorinae Gunther with a description of a new species of the genus *Listrophorus* Pagenstecher from Texas (Acarina, Listrophoridae). Acarologia, 7:704–712.

187. McIntosh, A. 1935. New host records of parasites. Proc. Helminth. Soc. Wash., 2:80.

188. ———— 1937. New host records for *Diphyllobothrium mansonoides* Mueller, 1935. Jour. Parasit., 23:313–315.

189. ———— 1939. A new dicrocoelid trematode, *Eurytrema komareki* n. sp. from a white footed mouse. Proc. Helminth. Soc. Wash., 6: 18–19.

190. Manville, R. H. 1961. Cutaneous miasis in small mammals. Jour. Parasit., 47:646.

191. Miles, Virgil I., and A. R. Kinney. 1957. An apparatus for rapid collection of fleas from rodent nests. Jour. Parasit., 43:656–658.

192. Morlan, H. B. 1952. Host relationships and seasonal abundance of some southwest Georgia ectoparasites. Amer. Midland Nat., 48:74–93.

193. ———— 1954. Notes on the genus *Megarthroglossus* (Siphonaptera, Hystrichopsyllidae) in Santa Fe County, New Mexico. Jour. Parasit., 40:446–448.

194. ———— 1955. Mammal fleas of Santa Fe County, New Mexico. Tex. Repts. Biol. Med., 13:93–125.

195. Morlan, H. B., and C. C. Hoff. 1957. Notes on some Anoplura from New Mexico and Mexico. Jour. Parasit., 43:347–351.

196. OSBORN, H. 1891. The Pediculi and Mallophaga affecting man and the lower animals. Bull. U. S. Bur. Ent., 7:1–56.

197. PARSONS, M. A. 1962. A survey of the ectoparasites of the wild mammals of New England and New York State. Unpubl. Masters Thesis, Univ. of Amherst, Amherst, Mass.

198. PAYNE, J. A., AND G. E. COSGROVE. 1966. Tissue changes following *Cuterebra* infestation in rodents. Amer. Midland Nat., 75:205–213.

199. PEARSON, O. P., AND A. K. PEARSON. 1947. Owl predation in Pennsylvania, with notes on the small mammals of Delaware County. Jour. Mamm., 28:137–147.

200. PENNER, L. R. 1938. A hawk tapeworm which produces a proliferating cysticercus in mice. Jour. Parasit., 24 (Suppl.) :25.

201. PENNER, L. R., AND F. P. POCIUS. 1956. Nostril entry as the mode of infection by the first stage larvae of a rodent *Cuterebra*. Jour. Parasit., 42 (Suppl.) :42.

202. POORBAUGH, J. H., AND H. T. GIER. 1961. Fleas (Siphonaptera) of small mammals in Kansas. Jour. Kans. Ent. Soc., 34:198–204.

203. PRINCE, F. M. 1944. Descriptions of three new species of *Thrassis* Jordan and the females of *T. bacchi* (Roths.) and *T. pansus* (Jordan). Pan-Pacific Ent., 20:13–19.

204. RADFORD, C. D. 1942. The larval Trombiculinae (Acarina, Trombididae) with descriptions of twelve new species. Parasit., 34:55–81.

205. ———— 1950. The mites (Acarina) parasitic on mammals, birds and reptiles. *Ibid.*, 40:366–394.

206. ———— 1954. The larval genera and species of "Harvest Mites" (Acarina: Trombiculidae). Parasitology, 44:247–276.

207. RADOVSKY, F. J., AND E. P. CATTS. 1960. Observations on the biology of *Cuterebra latifrons* Coquillet (Diptera: Cuterebridae). Jour. Kans. Ent. Soc., 33:31–36.

208. RANKIN, J. S. 1945. Ecology of the helminth parasites of small mammals collected from Northrup Canyon, Upper Grand Coulee, Washington. Murrelet, 26:11–14.

209. RAPP, W. F., JR. 1962. Distributional notes on parasitic mites. Acarologia, 4:31–33.

210. RAPP, W. F., AND D. B. GATES. 1957. A distributional checklist of the fleas of Nebraska. Jour. Kans. Ent. Soc., 30:50–53.

211. RAUSCH, R., AND J. D. TINER. 1949. Studies on the parasitic helminths of the North Central States. II. Helminths of voles (Microtus spp.) preliminary report. Amer. Midland Nat., 41:665–694.

212. READ, C. P. 1949. Studies on North American helminths of the genus *Capillaria* Zeder, 1800 (Nematoda). II. Additional capillarids from mammals with keys to the North American mammalian species. Jour. Parasit., 35:230–233.

213. REIBER, R. J., AND E. E. BYRD. 1942. Some nematodes from mammals of Reelfoot Lake in Tennessee. Jour. Tenn. Acad. Sci., 17:78–89.

214. REYNOLDS, B. D. 1938. *Brachylaemus peromysci* n. sp. (Trematoda) from the deer mouse. Jour. Parasit., 24:245–248.

308   *Whitaker*

215. Robinson, E. J. 1949. The life history of *Postharmostomum helicis* (Leidy, 1847) n. comb. (Trematoda: Brachylaemidae). Jour. Parasit., 35:513–533.
216. Rothschild, N. C. 1906. Three new Canadian fleas. Canad. Ent., 38: 321–325.
217. Scanlon, J. E. 1960. The Anoplura and Mallophaga of the mammals of New York. Wildlife Diseases No. 5. 3 microcards.
218. Schad, G. A. 1954. Helminth parasites of mice in northern Quebec and the coast of Labrador. Canad. Jour. Zool., 32:215–224.
219. Scholten, T. H., K. Ronald, and D. M. McLean. 1962. Parasite fauna of the Manitoulin Island Region. 1. Arthropoda Parasitica. Canad. Jour. Zool., 40:605–606.
220. Scott, T. G., and E. Snead. 1942. Warbles in *Peromyscus leucopus noveboracensis*. Jour. Mamm., 23:94–95.
221. Sealander, J. A. 1961. Hematological values in deermice in relation to bot fly infection. Jour. Mamm., 42:57–60.
222. Self, J. T., and F. M. McMurray. 1948. *Porocephalus crotali* Humboldt (Pentastomida) in Oklahoma. Jour. Parasit., 34:21–23.
223. Senger, C. M., and R. W. Macy. 1952. Helminths of northwest mammals. Part III. The description of *Euryhelmis pacificus* n. sp. and notes on its life cycle. Jour. Parasit., 38:481–486.
224. Shaftesbury, A. D. 1934. The Siphonaptera (fleas) of North Carolina with special reference to sex ratios. Jour. Elisha Mitchell Sci. Soc., 49:247–263.
225. Siegmund, O. H. 1964. Further notes on cuterebrid infestation. Jour. Mamm., 45:149–150.
226. Sillman, E. J., and M. V. Smith. 1959. Experimental infestation of *Peromyscus leucopus* with larvae of *Cuterebra angustifrons*. Science, 130:165–166.
227. Skinker, M. S. 1935. Two new species of tapeworms from carnivores and a redescription of *Taenia laticollis* Rudolphi, 1819. Proc. U. S. Natl. Mus., (2980) 83:211–220.
228. Smith, C. F. 1954. Four new species of cestodes of rodents from the high plains, central and southern Rockies and notes on *Catenotaenia dendritica*. Jour. Parasit., 40:245–254.
229. Smith, C. N. 1944. Biology of *Ixodes muris* Bishopp and Smith (Ixodidae). Ann. Ent. Soc. America, 37:221–234.
230. Spencer, G. J. 1936. A checklist of the fleas of British Columbia, with a note on fleas in relation to sawdust in homes. Proc. Ent. Soc. Brit. Col., 32:11–17.
231. ———— 1938. Further notes on the fleas of British Columbia. *Ibid.*, 34:36–38.
232. ———— 1940. Ectoparasites of birds and mammals in British Columbia. *Ibid.*, 1940:14–18.
233. Stanford, J. S. 1931. A preliminary list of Utah Siphonaptera. Proc. Utah Acad. Sci., 8:153–154.

234. Stark, H. E. 1958. The Siphonaptera of Utah. U. S. Dept. Health, Education & Welfare Communicable Disease Center, Atlanta, Georgia, 239 pp.

235. Stewart, M. A. 1928a. Two new Siphonaptera from Colorado. Canad. Ent., 60:148–151.

236. ———— 1928b. *In*: Leonard, M. D., 1928, (Pp. 868–869). A list of the insects of New York. Cornell Univ. Agric. Exp. Sta. Mem., 101: 1121 pp.

237. ———— 1933. Revision of the list of Siphonaptera from New York State. Jour. N. Y. Ent. Soc., 41:253–262.

238. ———— 1940. New Siphonaptera from California. Pan-Pacific Ent., 16: 17–28.

239. Strandtmann, R. W. 1949. The blood sucking mites of the genus *Haemolaelaps* (Acarina: Laelaptidae) in the United States. Jour. Parasit., 35:325–352.

240. Strandtmann, R. W., and D. M. Allred. 1956. Mites of the genus *Brevisterna* Keegan, 1949 (Acarina-Haemogamasidae). Jour. Kans. Ent. Soc., 29:113–132.

241. Strandtmann, R. W., and H. B. Morlan. 1953. A new species of *Hirstionyssus* and a key to the known species of the world. Tex. Repts. Biol. Med., 11:627–637.

242. Strandtmann, R. W., and G. W. Wharton. 1958. A manual of mesostigmatid mites parasitic on vertebrates. Inst. Acar., Contib. 4:330 pp.

243. Test, F. H., and A. R. Test. 1943. Incidence of dipteran parasitosis in populations of small mammals. Jour. Mamm., 24:506–508.

244. Till, W. M. 1963. Ethiopian mites of the genus *Androlaelaps* Berlese s. lat. (Acari: Mesostigmata). Bull. Brit. Mus. (Nat. Hist.) Zool., 10:1–104.

245. Tindall, E. E., and R. F. Darsie, Jr. 1961. New Delaware records for mammalian ectoparasites including Siphonaptera host list. Bull. Brooklyn Ent. Soc., 56:89–99.

246. Tiner, J. D. 1948a. *Syphacia eutamii* n. sp. from the least chipmunk, *Eutamias minimus*, with a key to the genus (Nematoda: Oxyuridae). Jour. Parasit., 34:87–92.

247. ———— 1948b. Observations on the *Rictularia* (Nematoda: Thelaziidae) of North America. Trans. Amer. Micr. Soc., 67:192–200.

248. ———— 1953. Fatalities in rodents caused by larval *Ascaris* in the central nervous system. Jour. Mamm., 34:153–167.

249. ———— 1954. The fraction of *Peromyscus leucopus* fatalities caused by raccoon ascarid larvae. *Ibid.*, 35:589–592.

250. Traub, R. 1950. Siphonaptera from Central America and Mexico. Fieldiana, Zoology Memoirs, 1:127 pp.

251. Traub, R., and P. T. Johnson. 1952. Four new species of fleas from Mexico (Siphonaptera). Amer. Mus. Novitates, 1598:1–28.

252. TRAUB, R., AND M. NADCHATRAM. 1966. A revision of the genus *Chatia* Brennan, with synonymic notes and descriptions of two new species from Pakistan (Acarina: Trombiculidae). Jour. Med. Ent., 2:373–383.

253. TRAVASSOS, L. 1937. Revisao da familia Trichostrongylidae Leiper, 1912. Monogr. Inst. Oswaldo Cruz, Rio de Janeiro, 512 pp.

254. ———— 1944. Revisao da familia Dicrocoeliidae Odhner, 1910. *Ibid.*, 357 pp.

255. TRAVIS, B. V. 1941. Examinations of wild animals for the cattle tick, *Boophilus annulatus microlophus* (Canestrini) in Florida. Jour. Parasit., 27:465–467.

256. UBELAKER, J. E., AND M. D. DAILEY. 1966. Taxonomy of the genus *Brachylaima* Dujardin (Trematoda: Digenea) with description of *B. chiapensis* sp. n. from *Peromyscus guatemalensis*. Jour. Parasit., 52:1062–1065.

257. ULMER, M. J. 1951. *Postharmostomatum helicis* (Leidy, 1847) Robinson, 1949 (Trematoda), its life history and a revision of the family Brachylaemidae. I. Trans. Amer. Micr. Soc., 70:189–238.

258. VAN CLEAVE, H. J. 1953. Acanthocephala of North American mammals. Ill. Biol. Monogr., 23 (1–2) :1–179.

259. VERTS, B. J. 1961. Observations on the fleas (Siphonaptera) of some small mammals in Northwestern Illinois. Amer. Midland Nat., 66:471–476.

260. VILLELA, J. B. 1953. The life history of *Entosiphonus thompsoni* Sinitsin, 1931 (Trematoda: Brachylaemitidae). Jour. Parasit., 39 (Suppl.) : 20.

261. VOGE, M. 1952. Variation in some unarmed Hymenolepididae (Cestoda) from rodents. Univ. Calif. Publ. Zool., 57 (1) :1–51.

262. ———— 1955. A list of cestode parasites from California mammals. Amer. Midland Nat., 54:413–417.

263. ———— 1956. A list of Nematode parasites from California mammals. *Ibid.*, 56:423–429.

264. WAGNER, J. 1936. The fleas of British Columbia. Canad. Ent., 68:193–207.

265. WARDLE, R. A., AND I. A. MCLEOD. 1952. The zoology of tapeworms. Univ. Minn., 780 pp.

266. WEBSTER, J. D. 1949. Fragmentary studies on the life history of the cestode *Mesocestoides latus*. Jour. Parasit., 35 (1) :83–90.

267. WECKER, S. C. 1962. The effects of botfly parasitism on a local population of the white-footed mouse. Ecology, 43:561–565.

268. WHELDON, R. M. 1941. Some mammalian ectoparasites. Jour. Mamm., 22:202–203.

269. WHITAKER, J. O., JR., AND N. WILSON. In Press. Mites of small mammals of Vigo County, Indiana. Amer. Midland Nat.

270. WHITAKER, J. O., JR., AND K. W. CORTHUM, JR. 1967. Fleas of Vigo County, Indiana. Proc. Ind. Acad. Sci. for 1966, 76:431–440.

271. WHITAKER, J. O., JR., AND V. RUPES. Unpublished MS.

272. WHITEHEAD, W. E. 1934. Records of some Quebec Mallophaga and Ano-
     plura. Rep. Quebec Soc. Prot. Pl., 26:25–26, 84–87.
273. WILLIAMS, L. A., AND C. C. HOFF. 1951. Fleas from the Upper Sonoran
     Zone near Albuquerque, New Mexico. Proc. U. S. Natl. Mus.,
     101:305–313.
274. WILSON, L. W. 1943. Some mammalian ectoparasites from West Virginia.
     Jour. Mamm., 24:102.
275. ――― 1945. Parasites collected from wood mouse in West Virginia.
     Ibid., 26:200.
276. WILSON, N. 1957. Some ectoparasites from Indiana mammals. Jour
     Mamm., 38:281–282.
277. ――― 1961. The ectoparasites (Ixodides, Anoplura and Siphonaptera)
     of Indiana mammals. Unpublished Thesis, Purdue University.
278. WOLFENBARGER, K. A. 1952. Systematic and biological studies on North
     American chiggers of the genus *Trombicula,* subgenus *Eutrom-
     bicula* (Acarina, Trombiculidae). Ann. Ent. Soc. Amer., 45:
     645–677.
279. WORTH, C. B. 1950a. A preliminary host-ectoparasite register for some
     small mammals of Florida. Jour. Parasit., 36:497–498.
280. ――― 1950b. Observations on ectoparasites of small mammals in
     Everglades National Park and Hillsborough County, Florida.
     Ibid., 36:326–335.
281. YAMAGUTI, S. 1958–1963. Systema helminthum. New York: Interscience,
     5 volumes.
282. YORKE, W., AND P. A. MAPLESTONE. 1962. The nematode parasites of
     vertebrates. New York: Hafner, 536 pp.
283. YUNKER, C. E. 1961. A sampling technique for intranasal chiggers (Trom-
     biculidae). Jour. Parasit., 47:720.

8
ENDOCRINOLOGY

BASIL E. ELEFTHERIOU

Introduction

ADDITIONAL COMPARATIVE information is becoming increasingly important in endocrinology because it will lead to the elucidation of general hormonal regulatory systems. The genus *Peromyscus,* with over fifty species, provides the comparative endocrinologist a wide range of phylogenetic divergence within a single genus. Comparisons can be made with the thoroughly investigated laboratory rat and mouse of a different group of rodents.

Although, endocrine investigations in *Peromyscus* are limited, the following summary, indicates the value of the comparative approach and some promising leads.

Thyroid Gland

Thyroid involvement during growth and maturation has been amply demonstrated during postnatal life in the rat (Scow and Simpson, 1945; Salmon, 1936, 1938), mouse (Smith and Starkey, 1940; Hurst and Turner, 1947), guinea pig (Koger and Turner, 1943), rabbit (Koger and Turner, 1943), sheep (Simpson, 1924), goat (Simpson, 1924), dog (Dye and Manghan, 1929), cow (Brody and Kibler, 1941), pig (Moussu, 1892; Caylor and Schlotthauer, 1929), rhesus monkey (Fleischmann and Schumacher, 1943), and man (Wilkins and Fleischmann, 1941; Hertz and Galli-Mainini, 1941).

Similar studies in *Peromyscus maniculatus bairdi* and *P. m. gracilis* have shown that the structure of the thyroid gland appears alike for both subspecies at birth (Fig. 1), but that at 70 days of age, *bairdi* exhibits a significantly higher thyroid activity than

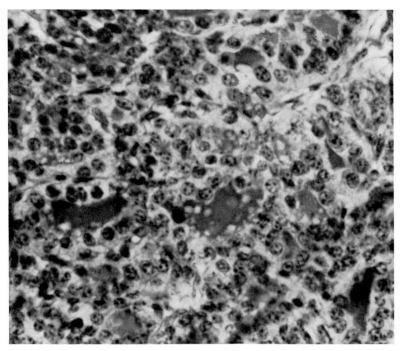

FIG. 1. Thyroid gland of *P. m. bairdi* at birth (60×). Thyroid glands from *P. m. gracilis* are similar in appearance. (From Eleftheriou, 1961.)

gracilis (Fig. 2; Eleftheriou, 1961). The higher thyroid activity in *bairdi* as indicated by histology was later confirmed by the thyroxine-secretion rate (TSR) in adults. During the month of July, the TSR of *bairdi* was 1.45 as compared with 0.75 for *gracilis* (Eleftheriou and Zarrow, 1961a; Eleftheriou and Zarrow, 1962a). *P. m. rubidus* also possesses a thyroid gland with large, distended follicles indicating lesser activity than the highly active thyroid glands of *P. m. sonoriensis* which exhibit cuboidal epithelium and very little colloid (Yocum and Huestis, 1928).

Whereas the structure of the thyroid gland in *bairdi* and in *gracilis* is similar at birth, higher radioiodine uptake (I^{131}, Fig. 4) indicates that the thyroid of *bairdi* is actually more active throughout the entire developmental period from birth to 70 days of age (Eleftheriou and Zarrow, 1961b), independent of the significantly greater body weight of *gracilis* (Fig. 3). Thyroids from both sub-

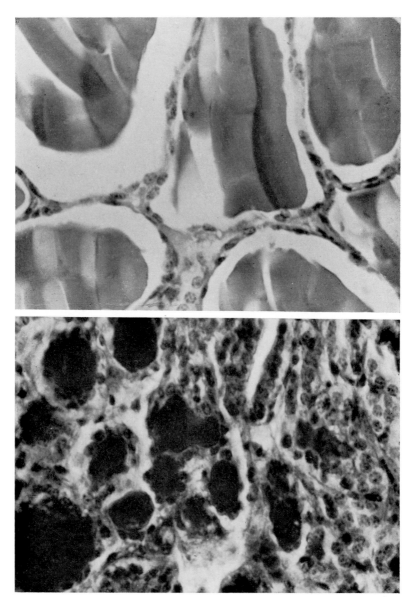

FIG. 2. Thyroid glands of adult *bairdi* (bottom) and adult *gracilis* (top) (60×). Thyroid gland of *gracilis* is composed of much larger follicles with greater colloid as compared with *bairdi*. (From Eleftheriou and Zarrow, 1961*b*.)

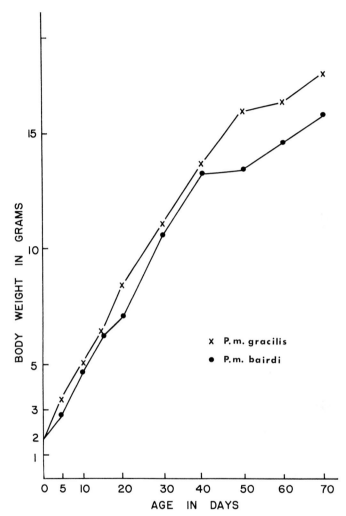

FIG. 3. Mean body weight during growth and development in two subspecies of deermice (*P. m. bairdi* and *P. m. gracilis*). (From Eleftheriou and Zarrow, 1961*b*.)

species weigh approximately the same during the same period (Fig. 5). Furthermore, the higher activity of the thyroid gland contributes to the earlier onset of external morphological characteristics such as pigmentation, opening of the eyes, general loco-

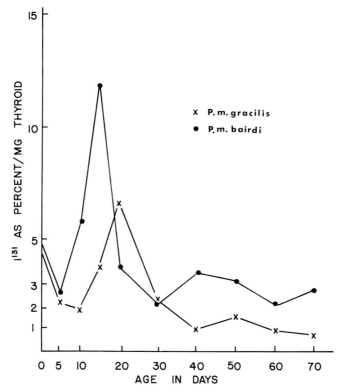

FIG. 4.   Mean radioiodine uptake by the thyroid gland during growth and development in two subspecies of deermice. I^{131} uptake expressed as per cent of total dose. (From Eleftheriou and Zarrow, 1961*b*.)

motor activity, and opening of external auditory meatus in *bairdi* than in *gracilis* (Eleftheriou, 1961). The general conclusion from these studies is that the significantly higher thyroid activity of *bairdi* enables it to mature more rapidly than *gracilis,* despite the latter's greater body weight.

BRAIN DEVELOPMENT.—In all probability, no other tissue suffers more severely than the brain as a consequence of lack of thyroxine during prenatal and postnatal development. The changes that take place in the central nervous system vary according to the time of onset of thyroid deficiency. In the cretin, thyroid deficiency results in mental retardation, whereas in myxedema the retardation is less

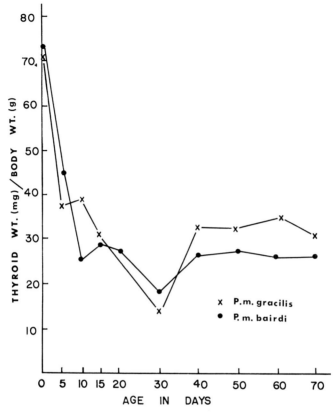

Fɪɢ. 5. Mean thyroid gland weight during growth and development in two subspecies of deermice. (From Eleftheriou and Zarrow, 1961*b*.)

severe and myxedematous animals show a reduced alertness. Thy-roxine replacement therapy usually can restore cerebral function in the adult life of myxedematous animals, but this treatment is not effective in the congenital or early thyroidectomized animals.

Usual symptoms of early thyroidectomy are: reduced water con-tent of the brain (Hammett, 1926), vascular stasis, edema, impair-ment of myelination, and loss of cortical cells (Barrnett, 1949). Such effects resemble effects of chronic anoxia and have been attributed to lowered tissue metabolism. In addition, thyroidec-tomized animals are retarded in the age of acquisition of air-right-

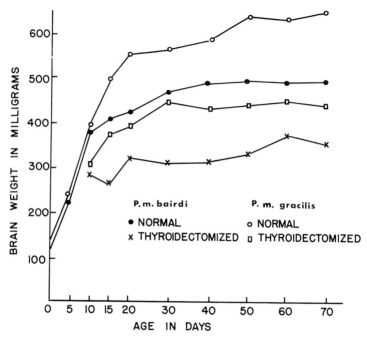

Fɪɢ. 6. Brain weight during growth and maturation and with thyroidectomy at birth for two subspecies of deermice. (From Eleftheriou, 1961.)

ing response and of the placing reflex. Their brains weigh significantly less and the pyramidal cells of the sensorimotor cortex are more reduced than in normal intact animals (Kendall and Simpson, 1928). Primary lesions in the capillary endothelium of the vasa vasorum have been attributed to thyroid deficiency (Hunt, 1923), and are believed to be caused by hypercholesterolemia and disturbed fat metabolism. Early thyroidectomy results in general reduction in the axons of the infragranular layer of the cerebral cortex (Riddle, 1927). Since growth of axonal network has been assumed to be a function of the capacity of the individual to synthesize protein, it has been suggested that thyroidectomy leads to impaired protein synthesis in the brain. Such impairment may be owing to reduction of oxidative processes which lead either to inactivation or to faulty development of necessary enzyme systems. In the absence of thyroid hormones, the neurons are unable to

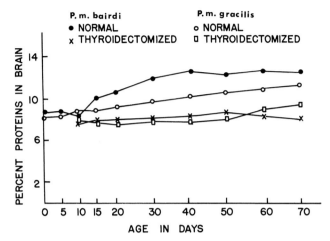

FIG. 7. Per cent proteins in the brain of two subspecies of deermice during growth and maturation under normal conditions and after thyroidectomy. (From Eleftheriou, 1961.)

absorb the nutrients transported in the blood stream which are essential for protein synthesis; this could explain the irreversibility of retarded mental and behavioral development in cretins and hypothyroid animals. The influence of hypothyroidism upon enzymes of the cerebral cortex has been found to result in reduction of several oxidative enzymes of the tricarboxylic acid cycle (Hoffman and Zarrow, 1958).

Such extensive research has not been carried out with members of the genus *Peromyscus.* However, data on normal development and development after radiothyroidectomy at birth of certain brain constituents have been gathered on two subspecies, *gracilis* and *bairdi* (Eleftheriou, 1961; Rawson and Money, 1949). It has been found that *gracilis,* the larger of the two, has a significantly greater brain weight than *bairdi* as early as 15 days of age. The adult brain weight of *gracilis* reaches an average maximum of 584 milligrams whereas *bairdi* attains a brain weight of 470 milligrams (Fig. 6). In spite of this difference in weight between the two subspecies, it has been found that on a per cent basis, the deposition of total solids, proteins (Fig. 7), lipids (Fig. 8), and cholesterol is significantly earlier in *bairdi* than in *gracilis,* and neurokeratin content

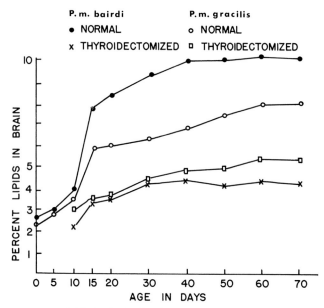

FIG. 8.   Per cent lipids in the brain of two subspecies of deermice with normal growth and after thyroidectomy. (From Eleftheriou, 1961.)

of the brain is about equal for both subspecies. In addition, these brain constituents were found to be greater in the brain of *bairdi* than in *gracilis* up to 70 days of age. Although this study was not carried beyond 70 days of age, these determinations indicate that the brain of *bairdi* matures at a faster rate than the brain of *gracilis*. The earlier maturation of the brain in *bairdi* was attributed to the higher activity of the thyroid gland in this subspecies than in *gracilis*.

In the absence of the thyroid gland, the aforementioned constituents of the brain were significantly reduced. It is interesting to point out that with radiothyroidectomy the lipid content in the brain of *bairdi* was reduced by 42 per cent and the protein content was reduced by 33 per cent while the same constituents were reduced by 31 and 15 per cent, respectively, in *gracilis*. That *bairdi* depends on the thyroid gland to a greater extent than *gracilis*, and that the former subspecies is more sensitive to thyroxine administration, is clearly indicated (Eleftheriou, 1961).

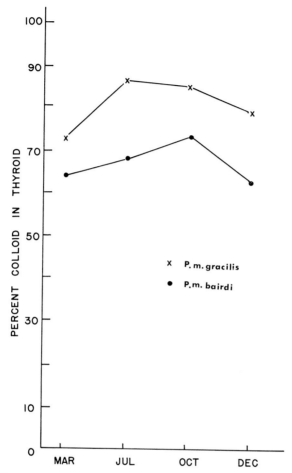

F<small>IG</small>. 9. Per cent colloid in thyroid gland in two subspecies of *Peromyscus maniculatus* during four seasons of the year. (From Eleftheriou and Zarrow, 1962*a*.)

S<small>EASONAL</small> V<small>ARIATION</small>.—As early as 1928 it was demonstrated in *Rattus* that thyroxine and per cent total iodine in the thyroid gland varied according to season, with the greatest content present during the months of July and August, and the least during the months of December and April (Kendall and Simpson, 1928). In general, the seasonal variations exhibited by the thyroid gland

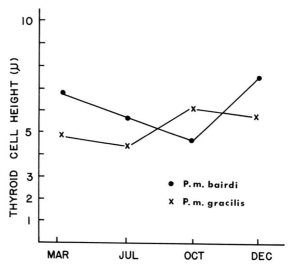

Fig. 10.   Cell height of thyroid gland in two subspecies of *Peromyscus mani-culatus* during four seasons of the year. (From Eleftheriou and Zarrow, 1962*a*.)

result from hypertrophy of the gland during autumn and winter months and a progressive decrease in size during the months of spring and summer (Hunt, 1923; Riddle, 1927). Of great interest is a recent comparative study on the rat (*Rattus*) and ground squirrel (*Citellus*), in which the thyroid gland underwent seasonal fluctuations in activity under constant temperature and lighting conditions (Hoffman and Zarrow, 1958). This study established the fact that seasonal changes in thyroid gland can occur without the benefit of changes in ambient temperature. The response of the thyroid gland to cold also varies with season; the maximum in animals exposed to cold in December and the minimum in animals exposed to cold in July or August (Hoffman and Zarrow, 1958). In general, thyroid gland activity, when at a minimum, is reflected by a high colloid content of the gland, low height of the follicular cells, and low thyroxine-secretion rate. Maximum activity is reflected in low colloid content, high follicular cell height, and increased thyroxine-secretion rates. These changes occur with exposure to cold as well as with seasonal changes, thus confirming the hypothesis that an iodine loss or depletion of colloid takes

TABLE 1

THYROXINE SECRETION RATES DURING 4 SEASONS IN
2 SUBSPECIES OF *Peromyscus maniculatus*
(From B. E. Eleftheriou and M. X. Zarrow, 1962)

Month	bairdi			gracilis		
	No. of animals	TSR range*	Mean*	No. of animals	TSR range*	Mean*
March	12	1.0–2.5	1.62	12	0.5–1.5	0.92
July	12	1.0–2.0	1.45	12	0.5–1.0	0.75
October	16	0.5–1.5	0.81	16	1.0–2.0	1.50
December	8	0.5–2.5	1.93	8	0.5–1.5	1.12

* In μg L-thyroxime/100 g body weight.

place prior to any increase in cellular activity or enlargement of the gland (Rawson and Money, 1949).

Seasonal variation in thyroid gland activity has been recently reported in two subspecies of *Peromyscus maniculatus* (Eleftheriou, 1961; Eleftheriou and Zarrow, 1961a; Eleftheriou and Zarrow, 1962a, 1962b). Thyroid gland activity as measured by per cent colloid, cell height, I^{131} uptake, and thyroxine secretion rates (TSR) was studied in male *gracilis* and *bairdi* during spring, summer, fall, and winter (Figs. 9–11, Table 1). A seasonal difference in activity of the thyroid gland was noted in both subspecies. However, time of maximum and minimum activity also varied in these two subspecies. Thyroid cell height and TSR were found to be at a maximum during December-March, and a minimum in October for *bairdi*—at a maximum during October-December and a minimum during March-July for *gracilis*. Radioiodine uptake by the thyroid glands of both subspecies were high during July and December and low during March and October. The thyroid gland of *bairdi* was found to be more active than that of *gracilis* throughout the year, except during the October period.

GENERAL EVALUATION.—In desert-inhabiting *P. m. sonoriensis,* activity of the thyroid gland was found to be greater than in *P. m. rubidus* of a coastal area (Yocum and Huestis, 1928). In addition, *P. m. bairdi* from midwestern fields were found to exhibit higher thyroid activity than *P. m. gracilis*, a forest inhabitant (Eleftheriou, 1961). The greater activity and efficiency of the thyroid gland

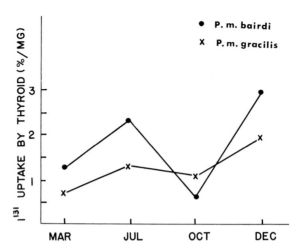

Fɪɢ. 11. Radioactive iodine uptake of thyroid gland in two subspecies of *Peromyscus maniculatus* during four seasons of the year. (From Eleftheriou and Zarrow, 1962a.)

exhibited by *bairdi* when compared with *gracilis* also corresponds to its greater sensitivity to exogenous and endogenous thyrotropin than either *gracilis* or the albino-Swiss mouse. The thyroid glands of both subspecies of deermice, however, are more sensitive to exogenous thyrotropin than are those of the albino-Swiss mouse.

Adrenal Gland

Isolation, characterization, and measurement of corticosteroids in the adrenals of the genus *Peromyscus* have been rather limited. In two subspecies, *bairdi* and *gracilis,* analyses of adrenal and plasma have revealed that corticosterone is the only glucocorticosteroid (Eleftheriou, 1964). In this respect these two subspecies of deermice are like the laboratory mouse and rat. There appears to be a significant difference in the level of circulating corticosterone in the two subspecies. Under normal controlled laboratory conditions, total corticosterone in *bairdi* is 25 μg/100 ml of plasma while in *gracilis* it is 57.5 μg/100 ml of plasma. The corticosterone found in *bairdi* is in its bound form whereas in *gracilis* 17.5 μg/100 ml is free and 40 μg/100 ml is plasma-bound. The high release of corticosterone by *gracilis* may indicate that it is unduly stressed under laboratory conditions. Although these animals were bred and raised in the laboratory for many generations, they may not have adapted completely to captivity.

Exposure to cold stress at 2° C for 15 days has resulted in temporary reduction of adrenal cortical activity, which is eventually overcome. This decline in adrenocortical function in *gracilis* may have resulted from a reduction of ACTH secretion in favor of that of thyrotropin as has been previously proposed for other species (Zarrow *et al.*, 1957; McCarthy *et al.*, 1959). However, the possibility that cold induced an increase in thyroid activity which accelerated the reduction of steroids by the liver (Yates *et al.*, 1958) to a greater extent in *gracilis* than in *bairdi* can be excluded because it was previously found (Eleftheriou, 1961) that cold did not appreciably alter thyroid activity in *gracilis,* but altered it significantly in *bairdi.* Conceivably, then, the decrease in the adrenal output of corticosterone in *gracilis* may result from an inability of the adrenal gland to respond to stress. The basic difference between the two subspecies appears to be that *bairdi* has a more sensitive pituitary-adrenal system with greater ability to adaptation and acclimatization than *gracilis.* In connection with the higher adrenal output of *gracilis,* it should be pointed out that *bairdi* is wilder, and more timid and nervous than *gracilis,* which is placid and can be handled with ease (King, 1958).

P. leucopus is particularly interesting because of its exceptionally large adrenals (Southwick, 1964). The adrenals average 10.5 to 16.5 mg/50.6 to 93.5 mg/100 g body weight. These values are about three times as high as those that were reported for *bairdi* and *gracilis* (Bronson and Eleftheriou, 1963; Wilson and Dewees, 1962). In addition, males of *P. leucopus* possess larger adrenals than females which is the reverse from all other rodent species studied with the exception of the nutria (Wilson and Dewees, 1962). Although further research is needed to evaluate and assess these differences, it is possible that males of this species are more responsive than females to environmental stress.

Adrenal weights in *Peromyscus* change with environmental and social stress. In this respect *bairdi* exhibit increased adrenal weight and ascorbic acid depletion following exposure to trained fighters, thus indicating that they respond to aggression and agonistic behavior like C57BL/10J mice (Bronson and Eleftheriou, 1963*a*). Similar results have been obtained with *leucopus* which exhibits tolerance to high cage densities among social congeners and significant intolerance among social strangers (Southwick, 1964).

Other Hormones

RELAXIN.—Two subspecies of *P. maniculatus, bairdi* and *gracilis,* exhibit separation of the pubic symphysis during pregnancy and after treatment with the hormone relaxin (Zarrow *et al.,* 1961). Change in the pubic symphysis involves a separation of the bones in the guinea pig, laboratory mouse, deermouse, and human. It is apparent that relaxin is involved in this phenomenon in the guinea pig, laboratory mouse, and the deermouse.

As in the laboratory mouse and guinea pig, the pubic ligament of the deermouse elongates during middle to late pregnancy to a maximum of 5 mm on the day of parturition and thereafter decreases rapidly to the pre-pregnancy size of 0.15 mm within ten days. Treatment with relaxin also causes a marked separation of the pubic bones, but in immature *bairdi* and *gracilis* the separation returns to normal within 60 hours, whereas the separation in the Swiss mouse is not only maintained but increased slightly for a period of as long as 168 hours (Zarrow *et al.,* 1961). Thus, in comparison with the laboratory mouse, deermice are less sensitive to hormonal regimen with relaxin. Nevertheless, *bairdi* and *gracilis* must be added to the small list of animals exhibiting the phenomenon of separation of the pubic symphysis that involves relaxin.

ENVIRONMENTAL STIMULI AND IMPLANTATION.—When recently inseminated female laboratory mice were exposed to strange males, the incidence of pregnancy decreased (Bruce, 1959). The olfactory stimuli produced by the strange males apparently caused a reduction in implantation. A similar response was found in deermice. Unlike laboratory mice, however, changes in either physical or social environments produce changes in the incidence of implantation (Eleftheriou *et al.,* 1962; Bronson and Eleftheriou, 1963*b*; Bronson *et al.,* 1964).

Inseminated *bairdi* were exposed for 24 hours to a strange male or female: *bairdi, gracilis,* or C57BL/10J mice (Eleftheriou *et al.,* 1962). In this experiment, implantation was significantly affected except when exposed to females of *bairdi* or C57BL/10J. Exposure

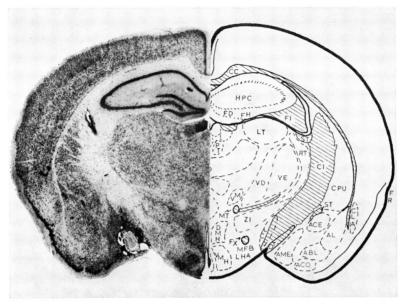

FIG. 12. Photographic and diagrammatic representation of the diencephalon of *P. m. bairdi*. Left hemisphere represents location of lesions. Right hemisphere identifies location of various nuclei: ABL—basolateral amygdaloid nucleus; AME —medial amygdaloid nucleus; ACE—central amygdaloid nucleus; ACO—cortical amygdaloid nucleus; AL—lateral amygdaloid nucleus. (Eleftheriou, Zolovick, and Pearse, 1966).

to male *bairdi* or male *gracilis* caused the highest incidence of blocked pregnancies. Differences between the sexes in producing implantation failures in inseminated *bairdi* were marked in *bairdi* and negligible in *gracilis* and C57BL/10J mice. Furthermore, it was shown that no relationship exists between successful implantation and the number of strange males to which a female is simultaneously exposed. This seems to suggest that social stress as defined by pituitary-adrenal-gonadal responses to crowding (Christian, 1960) is probably not a factor in the intraspecific strange male effect in *bairdi*. In interspecific exposure there was only a slight tendency for males to be more effective blocking agents than females. This suggests that a somewhat different mechanism was involved in the latter case. Enforced habitation with an animal of a different species might prove to be socially stressful (Christian,

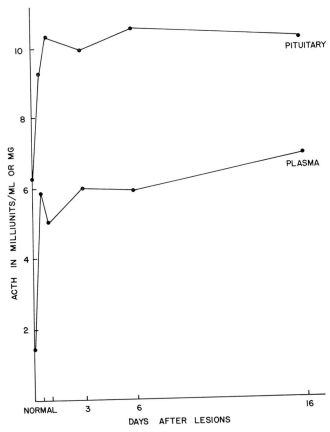

FIG. 13.   Pituitary and plasma ACTH levels (in mU/mg or mU/ml) in normal male deermice and in those with lesions in the medial amygdaloid complex for 12 hours and 1, 3, 6, and 16 days. (From Eleftheriou, Zolovick, and Pearse, 1966.)

1960). If so, it would be expected that such exposure would be accompanied by some degree of failure in implantation. Further research into these relationships between social and physical environments and implantation is being pursued.

AMYGDALOID REGULATION OF HORMONAL SECRETION.—The role of the hypothalamus in the regulation of pituitary hormonal secretion is well documented. However, our knowledge of regulation of

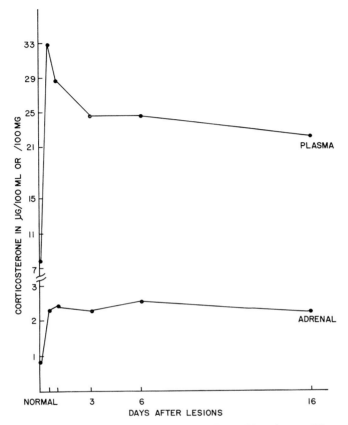

FIG. 14. Adrenal and plasma corticosterone (in $\mu g/100$ ml or $\mu g/100$ mg) in normal, without lesions, male deermice and at 12 hours and 1, 3, 6, and 16 days after lesions were placed in the medial amygdaloid complex. (From Eleftheriou, Zolovick, and Pearse, 1966.)

hormonal secretion by the amygdaloid complex is indeed very limited (Kovacs, Sandor, Vertes, and Vertes, 1965; Mason, 1959; Bovard and Gloor, 1961; Setekleiev, Skoug, and Kaada, 1961; Mason, Nauta, Brady, and Robinson, 1959) although extensive research has been conducted on the behavioral effects of lesions and electrical stimulation of the amygdala (see Altman, 1966, for review).

A number of recent investigations into the role of the amygdaloid complex in the regulation of certain hormones have been conducted with *P. m. bairdi*. One of these, by Eleftheriou, Zolovick, and Pearse (1966), explored the effect of bilateral amygdaloid lesions on the pituitary-adrenal axis in male *P. m. bairdi*. The results indicate that bilateral lesions in the medial amygdaloid complex (Fig. 12) are followed by significant increases in plasma ACTH levels. Simultaneous with the plasma increases, pituitary ACTH content also increases, reflecting a pituitary ACTH synthesis increase (Fig. 13).

The increased synthesis and circulatory levels of ACTH are best reflected in increased production of adrenal corticosterone, which is accompanied by significant increases in free plasma corticosterone (Fig. 14). Thus, the increases in adrenal and plasma corticosterone reflect adrenal activation as a result of increased circulatory levels of ACTH.

The results, together with previous work, indicate the possible existence of a separate inhibitory center regulating the pituitary-adrenal axis. The possibility also exists that injury from lesions produces a focal point of irritation which may be the starting point of potentials that spread to other areas of the brain, and mainly to the hypothalamus as reported by Taleisnik, Coligaris, and DeOlmos (1962). However, in the opinion of the authors it seemed unlikely that such an irritation would produce sustained stimulation over 16 days.

The effects of bilateral lesions in the basolateral group on the plasma and pituitary levels of luteinizing hormone and oestrous behavior were investigated in female *P. m. bairdi* by Eleftheriou and Zolovick (1966, 1967). Results of daily vaginal smears indicated that all lesioned animals went into diestrus immediately after lesions were placed in the basolateral amygdaloid nucleus (Fig. 14) and remained in this phase throughout the 21-day period of this experiment.

During the estrous cycle, plasma levels of LH increased significantly (p < 0.01) from a diestrus value of 0.82 to an estrus level of 2.0 milliunits/ml (Fig. 15). In this period, pituitary levels of LH decreased from a diestrus level of 0.61 to an estrus level of 0.06 milliunits/mg of pituitary tissue. These values for plasma and pituitary levels, when corrected to milliunits/pituitary/animal, and the reciprocal relationship of pituitary to plasma levels, are con-

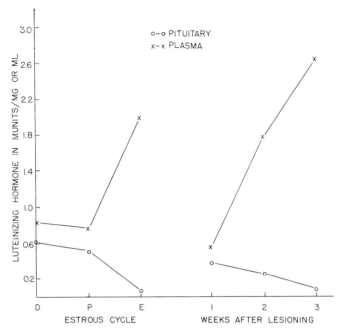

F**ig. 15.** Plasma and pituitary levels of luteinizing hormone (in milliunits/ mg or ml) during three phases of the estrous cycle and during three weeks after lesions of the basolateral amygdaloid group. (From Eleftheriou and Zolovick, 1967.)

sistent with previous findings in the rat by Schwartz and Caldarelli (1965) and Maric, Matsuyama, and Lloyd (1965).

Although after lesions were placed in the basolateral amygdaloid complex the vaginal smears indicated animals to be in diestrus, the plasma and pituitary levels of LH exhibited the same reciprocal relationship that was apparent during the estrous cycle (Fig. 15). Thus, from a value of 0.51 milliunits/ml, at one week, the plasma level of LH increased significantly to a level of 2.63 milliunits/ml at three weeks after treatment. At three weeks after lesioning of the basolateral amygdaloid group, the plasma level of LH was a significant ($p < 0.01$) 31 per cent greater than that during estrus. The pituitary exhibited a gradual decline in LH content during the three-week period of this experiment. The LH level of sham-operated animals indicated them to be in the diestrus phase.

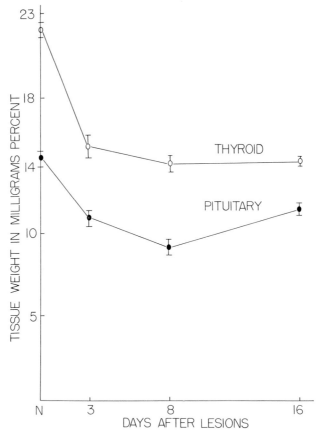

Fig. 16. Weight of pituitary and thyroid glands (in mgs per gram body weight) in normal, without lesions, male *P. m. bairdi* and in those lesioned in the medial amygdaloid complex at 3, 8, and 16 days. (From Eleftheriou and Zolovick, in press.)

Although additional data are necessary for complete interpretation of these results, these data, together with previous work by Yamada and Greer (1960); Shealy and Peele (1957); Koikegami, Yamada, and Usui (1953); and Koikegami, Fuse, Yokoyama, Watanobe, and Watanobe (1955) indicate the possible existence of a separate inhibitory center in *P. m. bairdi* for the regulation of luteinizing hormone. Female *bairdi* that were sham-operated by placing electrodes in the amygdala, but without administration of current, did

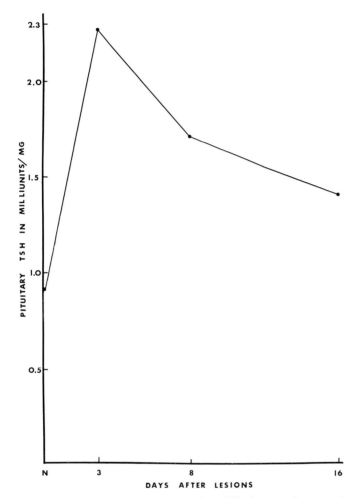

F_IG._ 17. Pituitary thyrotropin content (in milliunits/mg) in normal male
P. m. bairdi and in those with lesions in the medial amygdaloid complex at
3, 8, and 16 days. (From Eleftheriou and Zolovick, in press.)

not exhibit any significant changes in LH. This latter finding is
in agreement with previous findings of Bunn and Everett (1957)
in that mere insertion of an electrode does not serve as an adequate
stimulus for the release of LH.

These studies were conducted with the aid of a stereotaxic atlas of the brain of *P. m. bairdi* constructed by Eleftheriou and Zolovick (1965).

Eleftheriou and Zolovick (in press) also explored the effects of lesions in the medial amygdaloid nuclei (Fig. 12) on the pituitary-thyroid axis in male *bairdi*. After lesions were placed in the medial amygdaloid complex, the weight of the thyroid gland declined significantly (Fig. 16), and remained significantly lower when compared with either normal, unlesioned control or sham-operated *bairdi*. Similar findings were reported for the pituitary gland. Simultaneous with this decline in the weight of the thyroid and pituitary glands, thyrotropin content in the pituitary increased significantly (Fig. 17). This significant increase in pituitary TSH content possibly may reflect a decline in circulating levels of TSH especially when the decline in thyroid weight is taken into consideration.

The finding that, after lesions, the weight of the pituitary gland declined is surprising inasmuch as one would expect an increase. However, the decline in the pituitary weight may be attributed to the great loss of pituitary LH, ACTH, and LTH. Thus, the increase in pituitary TSH is counterbalanced by a decrease in the other tropic hormones.

Further work is needed to clarify the role of the amygdaloid complex in the regulation of these tropic hormones. However, in *bairdi* the medial amygdaloid nuclei possibly may exert an inhibitory effect on the pituitary-adrenal axis while exerting a possible stimulatory effect on the pituitary-thyroid axis. In addition, the basolateral and possibly the medial amygdaloid complex may, under normal conditions, exert an inhibitory effect on the pituitary for the secretion of LH and possibly LTH.

This work in *bairdi* represents the first systematic approach in an attempt to clarify the role of the amygdala in the regulation of hormonal secretory activity. If further investigations support this early work in *bairdi,* then the amygdala, in addition to the hypothalamus and other brain areas, must be considered as a regulatory complex in the secretion of certain hormones.

Summary

Knowledge of the endocrinology in the genus *Peromyscus* is indeed limited. Of the fifty-odd species, only two, *Peromyscus maniculatus* and *Peromyscus leucopus,* have been investigated by endocrinologists. Thyroid gland activity has been found to be higher in *P. m. sonoriensis* than in *P. m. rubidus,* and higher in *P. m. bairdi* than in *P. m. gracilis.* Unfortunately, no comparison can be made among these four subspecies. It has been shown, however, that, during maturation, the higher activity of the thyroid gland in *bairdi* is conducive to significantly earlier body growth and maturation of the central nervous system in this subspecies when compared with *gracilis.* Maximum thyroid activity was noted for *bairdi* during December-March and minimum in October, while maximum for *gracilis* was in October-December and minimum during March-July.

Corticosterone has been found to be the only glucocorticosteroid in *bairdi* and *gracilis,* with significantly higher circulating levels in *gracilis* than in *bairdi.* A basic difference between these two subspecies has been found in that *bairdi* has a more sensitive pituitary-adrenal system with greater adaptation and acclimatization to cold than *gracilis.* The weight of adrenal glands of *P. leucopus* are about three times those of *bairdi* and *gracilis.*

The two subspecies, *bairdi* and *gracilis,* exhibit the phenomenon of separation of the pubic symphysis during pregnancy that involves the hormone relaxin.

Recently inseminated female *bairdi* when exposed to changes in either physical or social environments exhibit a failure in the incidence of implantation. Indications are that the integrity of the basolateral amygdaloid nucleus (part of the rhinencephalic-limbic system) is a significant factor in governing normal mating behavior of the female *bairdi* and may be involved in this phenomenon of failure of implantation.

Literature Cited

ALTMAN, JOSEPH. 1966. Organic foundations of animal behavior. New York: Holt, Rinehart, and Winston.

BARRNETT, R. J. 1949. Quoted in Pincus, G., and K. V. Thimann. The Hormones. New York: Academic Press.

BOVARD, E. W., AND P. GLOOR. 1961. Effect of amygdaloid lesions on plasma corticosteroid response of the albino rat to emotional stress. Experientia, 17:521.

BRODY, S., AND H. H. KIBLER. 1941. Growth and development. Relation between organ weight and body weight in growing and mature animals. Mo. Agric. Exp. Sta. Res. Bull. No. 328.

BRONSON, F. H., AND B. E. ELEFTHERIOU. 1963a. Adrenal responses to crowding in *Peromyscus* and C57BL/10J mice. Physiol. Zoöl., 36:161–166.

———— 1963b. Influence of strange males on implantation in the deermouse. Gen. Comp. Endocrinol., 3:515–518.

BRONSON, F. H., B. E. ELEFTHERIOU, AND E. C. GARICK. 1964. Effects of intra- and interspecific social stimulation on implantation in deermice. Jour. Reprod. Fertil., 8:23–27.

BRUCE, H. M. 1959. Olfactory stimuli and pregnancy block. Nature, 180:105.

CAYLOR, H. D., AND C. F. SCHLOTTHAUER. 1929. The influence of diet on the development of myxedema in thyroidectomized pigs. Amer. Jour. Physiol., 89:596–600.

CHRISTIAN, J. J. 1960. Endocrine adaptive mechanism and the physiologic regulation of population growth. Nav. Med. Res. Inst. Lect., Res. Series No. 49:60–62.

DYE, J. A., AND G. H. MANGHAN. 1929. Further studies of the thyroid gland. V. The thyroid gland as a growth-promoting and form-determining factor in the development of the animal body. Amer. Jour. Anat., 44:331–380.

ELEFTHERIOU, B. E. 1961. The role of the thyroid gland during growth and maturation in two subspecies of *Peromyscus maniculatus*. Ph.D. Thesis, Purdue University, Lafayette, Indiana.

———— 1964. Bound and free corticosteroids in the plasma of two subspecies of deermice (*Peromyscus maniculatus*) after exposure to a low ambient temperature. Jour. Endocrinol., 31:75–80.

ELEFTHERIOU, B. E., F. H. BRONSON, AND M. X. ZARROW. 1962. Interaction of olfactory and other environmental stimuli on implantation in the deer mouse. Science, 137:764.

ELEFTHERIOU, B. E., AND M. X. ZARROW. 1961a. Influence of season and exposure to cold on TSR in two subspecies of *Peromyscus maniculatus* and the Swiss-albino mouse. The Physiologist, 4:33 (Abstract).

———— 1961b. A comparison of body weight and thyroid gland activity in two subspecies of *Peromyscus maniculatus* from birth to 70 days of age. Gen. Comp. Endocrinol., 1:534–540.

———— 1962a. Seasonal variation in thyroid gland activity in deermice. Proc. Soc. Exptl. Biol. Med., 110:128–131.

———— 1962b. Sensitivity to thyrotropin in two subspecies of deermice (*Peromyscus maniculatus*) and Swiss-albino mouse as measured by I^{131} depletion. Gen. Comp. Endocrinol., 2:441–445.

ELEFTHERIOU, B. E., AND A. J. ZOLOVICK. 1965. The forebrain of the deermouse in stereotaxic coordinates. Kans. Agric. Exp. Sta., Tech. Bull. No. 146.

———— 1966. Effect of amygdaloid lesions on oestrous behavior in the deermouse. Jour. Reprod. Fertil., 11:451–453.

———— 1967. Effect of amygdaloid lesions on plasma and pituitary levels of luteinizing hormone. Jour. Reprod. Fertil., 14:33–37.

———— In press. Effect of amygdaloid lesions on pituitary thyrotropin content in the deermouse. Proc. Soc. Exptl. Biol. and Med.

ELEFTHERIOU, B. E., A. J. ZOLOVICK, AND R. PEARSE. 1966. Effect of amygdaloid lesions on the pituitary-adrenal axis in the deermouse. Proc. Soc. Exptl. Biol. and Med., 122:1259–1262.

FLEISCHMANN, W., AND H. B. SCHUMACHER, JR. 1943. Influence of age on the effects of thyroidectomy in the rhesus monkey. Endocrinol., 32: 238–246.

HAMMETT, F. S. 1926. Studies of the thyroid apparatus. XXXII. The role of the thyroid apparatus in the solids-water content of the central nervous system during growth. Jour. Comp. Neurol., 41:205–222.

HERTZ, S., AND GALLI-MAININI. 1941. Effects of thyroid hormone on growth in thyrotoxic and myxedematous children and adolescents. Jour. Clin. Endocrinol., 1:518–522.

HOFFMAN, R. A., AND M. X. ZARROW. 1958. A comparison of seasonal changes and the effect of cold on the thyroid gland of the male rat and ground squirrel (*Citellus tridecimleatus*). Acta Endocrinologica, 27: 77–84.

HUNT, R. 1923. The acetonitril test for thyroid and of some alterations of metabolism. Amer. Jour. Physiol., 63:257–299.

HURST, V., AND C. W. TURNER. 1947. The thyroid secretion rate of growing and mature mice. Amer. Jour. Physiol., 150:686–692.

KENDALL, E. C., AND D. G. SIMPSON. 1928. Seasonal variations in thyroxine content of the thyroid gland. Jour. Biol. Chem., 80:357–377.

KING, J. A. 1958. Maternal behavior and behavioral development in two subspecies of *Peromyscus maniculatus*. Jour. Mamm., 39:177–190.

KOGER, M., AND C. W. TURNER. 1943. The effects of mild hyperthyroidism on growing animals of four species. Mo. Agric. Exp. Sta. Res. Bull. No. 377.

KOIKEGAMI, H., S. FUSE, T. YOKOYAMA, T. WATANOBE, AND H. WATANOBE. 1955. Contribution to the comparative anatomy of the amygdaloid nuclei of mammals with some experiments of their destruction or stimulation. Folia Psychiat. Neurol. Jap., 8:336.

KOIKEGAMI, H., T. YAMADA, AND K. USUI. 1953. Stimulation of amygdaloid nuclei and periamygdaloid cortex with special reference to its effects on uterine movements and ovulation. Folio Psychiat. Neurol. Jap., 8:7.

KOVACS, S., A. SANDOR, Z. VERTES, AND M. VERTES. 1965. The effect of lesions and stimulation of the amygdala on pituitary-thyroid function. Acta Physiol. Hug, 27:221.

McCARTHY, J. L., R. C. CORLEY, AND M. X. ZARROW. 1959. Effect of goitrogens on adrenal gland of the rat. Amer. Jour. Physiol., 197:693–698.

MARIC, D. K., E. MATSUYAMA, AND C. W. LLOYD. 1965. Gonadotropin content of pituitaries of rats in constant estrus induced by continuous illumination. Endocrinology, 77:529.

MASON, J. W. 1959. Plasma 17-hydroxycorticosteroid levels during electrical stimulation in amygdaloid complex in conscious monkeys. Amer. Jour. Physiol., 196:44.

MASON, J. W., W. J. NAUTA, J. V. BRADY, AND J. A. ROBINSON. 1959. Limbic system influences on the pituitary-adrenal cortical system. Forty-first Meeting, Endocrine Soc., p. 29 (Abstract).

MOUSSU, M. 1892. Sur la function thyroidienne. Cretinisme experimental sous les deux formes typiques. C. R. Soc. Biol., Paris, 44:972–976.

RAWSON, R., AND W. MONEY. 1949. Physiologic reaction of the thyroid stimulating hormone. Rec. Prog. Horm. Res., 4:397–418.

RIDDLE, O. 1927. Studies on thyroids. Endocrinology, 11:161–172.

SALMON, T. N. 1936. Effect of thyro-parathyroidectomy in the newborn rat. Proc. Soc. Exptl. Biol. and Med., 35:489–491.

————— 1938. The effect on the growth rate of thyro-parathyroidectomy in newborn rats and of subsequent administration of thyroid, parathyroid and anterior hypophysis. Endocrinology, 23:446–457.

SCHWARTZ, N. B., AND D. CALDARELLI. 1965. Plasma LH in cyclic female rats. Proc. Soc. Exptl. Biol. and Med., 119:16.

SCOW, R. O., AND M. E. SIMPSON. 1945. Thyroidectomy in the newborn rat. Anat. Rec., 91:209–226.

SETEKLEIEV, J., O. E. SKOUG, AND B. R. KAADA. 1961. Plasma hydroxycorticosteroids and amygdaloid lesions. Jour. Endocrinology, 22:119.

SHEALY, N. C., AND T. L. PEELE. 1957. Studies on amygdaloid nucleus of the cat. Jour. Neurophysiol., 20:125.

SIMPSON, S. 1924. The effect of thyroidectomy on growth in the sheep and goat as indicated by body weight. Quart. Jour. Exptl. Physiol., 14:161–165.

SMITH, R. D., AND W. F. STARKEY. 1940. Histological and quantitative study of age changes in the thyroid of the mouse. Endocrinology, 27:621–627.

SOUTHWICK, C. 1964. *Peromyscus leucopus*: An interesting subject for studies of socially induced stress responses. Science, 143:55.

TALEISNIK, S., L. COLIGARIS, AND J. DEOLMOS. 1962. Luteinizing hormone release by cerebral cortex stimulation in rats. Amer. Jour. Physiol., 203:1109.

WILKINS, L., AND W. FLEISCHMANN. 1941. Hypothyroidism in childhood. III. The effect of withdrawal of thyroid therapy upon the serum cholesterol. Relationships of cholesterol, BMR, weight and clinical symptoms. Jour. Clin. Endocrinology, 1:91–95.

WILSON, E. D., AND A. A. DEWEES. 1962. Body weights, adrenal weights and oestrous cycles of Nutria. Jour. Mamm., 43:362–368.

YAMADA, T., AND M. A. GREER. 1960. The effect of bilateral ablation of the amygdala on endocrine function in the rat. Endocrinology, 66:565.

YATES, F. E., J. URQUHART, AND A. L. HERBST. 1958. Effects of thyroid hormones on ring A reduction of cortisone by liver. Amer. Jour. Physiol., 195:373–380.

YOCOM, H. B., AND R. R. HUESTIS. 1928. Histological difference in the thyroid gland from two subspecies of *Peromyscus maniculatus*. Anat. Rec., 39:57–68.

ZARROW, M. X., B. E. ELEFTHERIOU, G. L. WHITECOTTEN, AND J. A. KING. 1961. Separation of the pubic symphysis during pregnancy and after treatment with relaxin in two subspecies of *Peromyscus maniculatus*. Gen. and Comp. Endocrinology, 1:386–391.

ZARROW, M. X., L. M. HORGER, AND J. L. MCCARTHY. 1957. Atrophy of adrenal gland following thiouracil and vitamin B_{12}. Proc. Soc. Exptl. Biol. and Med., 94:348–351.

ZOLOVICK, A. J., B. E. ELEFTHERIOU, AND R. PEARSE. 1966. Effect of amygdaloid lesions on the pituitary-adrenal axis in the deermouse. Proc. Soc. Exp. Biol. Med., 122:1259–1262.

ADDENDUM

BRONSON, F. H., AND S. H. CLARKE. 1966. Effect of adrenalectomy on coat color in deermice. Science, 154:1349–1350.

BRONSON, F. H., B. E. ELEFTHERIOU, AND H. E. DEZELL. In press. Melanocyte-stimulating activity following adrenalectomy in deermice. Proc. Soc. Exptl. Biol. and Med.

DAWSON, W. D. 1966. Comparative thyroid activity in two species of *Peromyscus*. Gen. Comp. Endocrinology, 8:267–272.

ELEFTHERIOU, B. E. 1967. Effect of amygdaloid lesions on hypothalamic LH-RF in the male deermouse. Jour. Endocrinol., 38:479–480.

ELEFTHERIOU, B. E., AND M. L. PATTISON. 1967. Effect of amygdaloid lesions on hypothalamic FSH-RF in the male deermouse. Jour. Endocrinol., 39: 613–614.

HAY, D. A. 1967. A histochemical study of the distribution of oxidation and hydrolytic enzymes in the brain of the adult *Peromyscus polionotus*. M. S. dissertation, Mich. State Univ.

WALTON, R. E., AND W. D. DAWSON. 1967. Thyroid effect on birthweight in C57Bl mice and *Peromyscus*. Proc. Soc. Exp. Biol. Med., 124:1067–1069.

9

GENETICS

David I. Rasmussen

Introduction

MAMMALIAN GENETICS (not including that of man) has been and will probably continue to be an area of study dominated by a single genus, *Mus*. Similarly, entomological genetics is dominated by the genus *Drosophila*. However, other species offer unique genetic problems for the investigator and expand our general knowl- edge of genetic phenomena. The English work with the lepidopterans, as well as the recent work with the house fly and flour beetles, has contributed to the basic understanding of certain genetic phenomena not easily assessable in the study of the genus *Drosophila*. It can be hoped that geneticists will expand their view of mammalian genetics with increased research activity within groups such as *Peromyscus* which offer unique rewards for certain genetic problems.

The genus *Peromyscus* is represented by a diversity of forms of such ubiquity and variation that it cannot help but promise excitement to the investigator interested in the phenomenon of variation, the indispensable ingredient in the study of genetics and evolution. Coupled with this spectrum of variation is the mass of information on other aspects of the biology of this genus to which this volume attests.

A few problems limit the usefulness of the genus for genetic investigations. The knowledge of genetics of the house mouse greatly exceeds that of native mice and, therefore, the house mouse is naturally preferred for many areas of investigation. The lack of inbred lines in *Peromyscus* precludes certain experimental replications that the many inbred stocks of the house mouse offer. Inbred lines in *Peromyscus* would be a valuable aid to further genetic investigations, but to date the development of such stocks, although attempted, has not been successful. Therefore, the geneticist must deal with individual mice as unique organisms rather than being able to assume genetic homogeneity of a given stock. However, a wide understanding of biological phenomena cannot be gained with inbred stocks alone and organisms such as *Peromyscus* offer an approach to various problems not available with inbred stocks. Kavanau (1964) pointed out such limitations of domestic animals in the study of behavior. The reproductive performance of some stocks of mice of the genus *Peromyscus* within a laboratory appears more sporadic and irregular than that of certain laboratory stocks of the house mouse.

The interaction of genetic analyses of laboratory animals and that of natural populations may well be the true treasure of the genus *Peromyscus*. The interesting ramifications of the analysis of the T-locus polymorphism in the house mouse (Dunn, 1964) indicates that the study of natural populations of mice offers numerous rewards.

Finally it should be noted that although genetic data have accumulated for the genus, these data are largely limited to the species *P. maniculatus,* and to a lesser extent *P. leucopus* and *P. polionotus.* Therefore, most of the following discussion will be confined to *P. maniculatus.*

Breeding Procedure

The general protocol, marking, record keeping, and care of animals during breeding, is discussed in the symposium text edited by Burdette (1963). Each litter from single-pair matings of *P. maniculatus, P. boylei, P. leucopus,* and *P. polionotus* can be successfully reared in a commercial mouse cage with a floor space of approximately 80 square inches, furnished with a sawdust, ground corn cobs, beet pulp, San-i-cel, or cedar shavings covering and a piece of cotton for nest material. Commercial mouse chow can serve as the basic diet, although some trouble with obesity may result from certain chows with high fat content. Some investigators suggest a grain supplement such as chicken "scratch feed." Several laboratories have found a "hanging drop" water supply using inverted 4 to 8 oz shoulderless, medicine bottles with a small hole bored in the plastic screw caps to be an inexpensive, readily cleanable watering unit.

The female ovulates immediately after parturition, and if mating takes place at that time, the next litter will be born 25 to 26 days later, so that the age at weaning should be about three weeks. The age of weaning depends somewhat on litter size; larger litters mature more slowly. Litter survival is erratic with wild-trapped parents, but successful weanings usually follow in those matings which produce and raise an initial laboratory litter. Controlled lighting (i.e., 18 hours light, 6 hours dark) appears to enhance the fertility of stocks as well as speed the arrival of litters in newly mated pairs (Price, 1966). Reproductive performance tends to decline after three to five litters or in mice more than 18 months old.

Cytogenetics

In numerous species and subspecies of the genus *Peromyscus,* the observed diploid number of chromosomes appears to be 48 (Cross, 1931, 1938; Harpst and Martin, 1950; Makino, 1953*a*, *b*; and Tamsitt, 1960). An exception to 48 chromosomes has been reported in *P. maniculatus hollisteri,* in which Cross (1938) counted 52 chromosomes in spermatogonia cells of one individual of this subspecies although he considered this material as possibly unusual. Tamsitt (1960) considers chromosomal fragmentation to be the likely explanation for this observation. Cytogenetic examination of spleen and testicular tissues of three males of *P. m. hollisteri* by Sparkes and Arakaki (personal communication) has revealed a consistent chromosome number of 48. It is therefore believed that the diploid number of 52 as reported by Cross is not representative of this subspecies.

Cross (1938), Moree (1946, 1948), and Tamsitt (1960) reported 48 chromosomes in *P. truei, P. nasutus,* and *P. comanche.* Makino (1953*a*, *b*) stated that *P. truei* has 48 chromosomes and *P. eremicus* has 52. Cross (1938) reported 58 chromosomes for *P. eremicus,* but Tamsitt (1960) questions these assignments.

Warranted re-investigations of the comparative cytogenetics of the genus *Peromyscus* are presently being undertaken, based on advances in cytogenetic techniques. (A discussion of recent cyto-genetic techniques is given by G. Yerganian, p. 469–510 in Burdette, 1963.) Investigations of karyotypic patterns and morphology of chromosomes within populations promise interest in view of the preliminary studies by Sparkes and Arakaki (personal communica-tion). Their examination of individual mice representing three subspecies of *P. maniculatus* indicates a constant diploid number of 48, but a marked karyotype variation between individuals. Although each animal appeared to have a constant karyotype in the different tissues examined, the karyotypes varied from animal to animal. The major karyotypic differences were found in the variable number of acrocentric chromosomes. The extent of karyotypic variation precluded a single classification of individual chromosomes for the species. These chromosomal polymorphisms included almost continuous variation of chromosomal size as well as differences related to centromere positions. This raises the possibility of genetically homologous chromosomes exhibiting variable centromere positions and translocations of genetically homologous segments to various chromosomes. Pericentric inversion seems the best explanation for most of the polymorphism, which was observed in animals of single subspecific origin as well as those with mixed subspecific background.

Makino (1953*b*) reported the male as the heterogametic sex in the genus *Peromyscus* and invariably observed a heteromorphic bivalent considered as the XY pair. The X-chromosome is one of the larger chromosomes, somewhat J-shaped, and faintly stained at one end. The Y-chromosome is a short rod-shaped, uniformly staining body, occasionally exhibiting a small condensed portion at one end. At metaphase the densely stained portion of the X and Y lie together and disjoin at first anaphase; these chromosomes pass to opposite poles apparently more rapidly than the autosomes.

Genetic Markers and Linkage Relationships

A list of loci, names, symbols, effects and known linkage relationships are given in Tables 1, 2, and 3. This listing of identified loci and linkage groupings in the species *Peromyscus maniculatus* consists of genes which have been identified as mutant alleles in laboratory stocks or as naturally occurring polymorphic or racial characters. Two loci, white cheek and spinner, although originally discerned in *P. polionotus,* have since been transferred to *P. maniculatus* stocks by laboratory hybridization.

The gene effects are coded following the classification of principal or most diagnostic effect used in the MOUSE NEWS LETTER (Laboratory Animals Bureau, Medical Research Council Laboratories, Holly Hill, Hampstead, London, N. W. 3) .

1. Pelage coloration
2. Skin and hair texture
3. Skeleton
4. Tail structure
5. Eye

6. Ear and circling behavior
7. Neuromuscular
8. Metabolic defects
9. Enzymes, antigens, proteins
10. Miscellaneous

In a few cases the choice of symbols has been arbitrary. There exist some ambiguous sets of gene symbols as they appear in literature: Clark (1938) gives the symbol G for Sumner's trait grizzled while Blair (1947b) used g for gray. Feldman (1936) used the symbol p with subscripts to designate minor white spotting loci, while p has also been used to designate pink-eye or pallid (Sumner, 1917). Egoscue (1962) used the symbol n for naked, although Clark (1938) had previously assigned this symbol to nude. Therefore, naked is here symbolized by nk. Barto (1955) designated spinner by sp, while Huestis' group (Huestis *et al.*, 1956) has used sp for spherocytosis. Since this is perhaps the most serious case of ambiguity, and Barto's sp has priority, the symbol sph is suggested for spherocytosis.

A number of loci have been designated by different symbols. Among these are the waltzing traits designated v_1, v_2, etc. by Clark (1938), but other workers have applied different symbols—w by Watson (1939), sp by Barto (1955). Gray has been designated g by Blair (1947b), although Clark (1938) considered it to be an allele at the agouti locus, A^b. Confusion in the literature between "buff," the dominant allele of gray, and wide-band, a dominant gene at a separate locus, has been discussed by McIntosh (1956a), who is of the opinion that the "buff" of Clark (1938) may be identical to the wide-band gene. The gene designated by Blair (1944, 1947b) as buff is designated as gray (Dice, 1933), a locus independent from wide-band or buff. The symbols si and sl have both been applied to silver, the latter more frequently, and is adopted here. In *P. m. artemisiae* (Watson, 1939) it appears as if epilepsy and waltzing are both caused by the same major factor and this syndrome was designated by Barto (1956) with the symbol EP. The EP syndrome consists of whirling, sound-induced convulsions, and progressive loss of hearing. Where confusion might exist, alternate gene symbols are given in parentheses in the listing of the tables.

TABLE 1

Loci in *Peromyscus maniculatus*

Known linkage associations

Name of locus	Symbol	Effect code	Original description	Collateral description	Tests for linkage
				Linkage Group I	
albino	c	1	Sumner (1922)	Sumner and Collins (1922)	Sumner (1922); Clark (1936, 1938); Feldman (1937); Barto (1942a); Huestis and Lindstedt (1946); Robinson (1964); Barto (1956)
pink-eye (pallid) (partial albino)	p	1	Sumner (1917)	Barto (1942b)	Sumner (1922); Clark (1936, 1938); Feldman (1937); Robinson (1964)
silver	sl (si)	1	Huestis and Barto (1934)		Huestis and Barto (1934); Huestis and Piestrak (1942); Barto (1956); Robinson (1964)
flexed-tail	f	4	Huestis and Barto (1936a)		Huestis and Barto (1936a); Huestis and Piestrak (1942); Huestis and Lindstedt (1946); Huestis et al. (1956); Barto (1956); Robinson (1964)
snub-nose	sb	3	Silliman and Huestis (unpub.)		Silliman and Huestis (unpub.)

TABLE 1 (Continued)

Name of locus	Symbol	Effect code	Original description	Collateral description	Tests for linkage
			Linkage Group II		
dilute	*d*	1	Dice (1933)		Clark (1938) ; Barto (1942*a*, 1956) ; McIntosh (1956*a*) ; Robinson (1964)
brown (brown-tip)	*b*	1	Huestis and Barto (1934)	Blair (1947*a*) ; McIntosh (1956*a*)	Huestis and Barto (1934) Blair (1947*a*) ; Barto (1955, 1956) ; McIntosh (1956*a*) ; Robinson (1964)
			Linkage Group III		
wide-band ("buff" in *nebrascensis*)	*Nb*	1	McIntosh (1956*a*)		Clark (1938) , as "buff"; Robinson (1964)
waltzer (waltzing in *bairdi*) (WZ)	*v* (*w*)	6	Dice (1935)	Watson (1939)	Barto (1942*a*, 1956) ; McIntosh (1956*a*) ; Robinson (1964) ; Dice, Barto, and Clark (1963)

TABLE 2

Loci in *Peromyscus maniculatus*

Independent loci for pairings tested

Name of locus	Symbol	Effect code	Original description	Collateral description	Tests for linkage
dominant spot (white face)	S	1	Feldman (1936)		Feldman (1937); Robinson (1964)
gray (gray-band)	g (A^b)	1	Dice (1933)	Clark (1938); Blair (1947a); McIntosh (1956a)	Blair (1944, 1947a); Robinson (1964)
ivory	i	1	Huestis (1938)	Clark (1938)	Barto (1942a, 1956); McIntosh (1956a); Robinson (1964)
white cheek	Ec	1	Blair (1944)		Blair (1944)
yellow	y	1	Sumner (1917)	Sumner and Collins (1922); McIntosh (1956a)	Sumner (1922); Feldman (1937); Clark (1938); McIntosh (1956a); Barto (1956); Robinson (1964)
hairless	hr	2	Sumner (1924)		Sumner (1924, 1932); Feldman (1937); Clark (1938); Barto (1942a, 1955, 1956); McIntosh (1956a); Robinson (1964)
nude (post-juvenile)	n	2	Clark (1938)		Barto (1942a); Robinson (1964)
audiogenic epilepsy (EP)	e	6	Dice (1935)	Watson (1939); Chance (1954); Chance and Yaxley (1949, 1950); Dice, Barto, and Clark (1963); Ross (1962)	Watson (1939); Barto (1956); Robinson (1964)
osmogenic epilepsy (CV)	CNV (Cn)	6	Dice (1935)	Clark (1938); Watson (1939); Dice, Barto, and Clark (1963)	
boggler	bg	7	Barto (1955)		Barto (1955)
spherocytosis	sph	8	Huestis and Anderson (1954)	Huestis et al. (1956); Motulsky et al. (1956)	Huestis et al. (1956); Robinson (1964)

TABLE 3

LOCI IN *Peromyscus maniculatus*

Linkage not tested

Name of locus	Symbol	Effect code	Original description	Collateral description	Comments
			Described loci		
minor white	p_1	1	Feldman (1936)	Sumner (1932); Barto and Huestis (1933)	
spotting (star, blaze)	p_2	1	Feldman (1936)		
naked	nk	2	Egoscue (1962)		
tremor	t	7	Huestis and Barto (1936b)		
spinner (waltzing in P. p. rhoadsi)	sp (v_2)	6	Watson (1939)	Barto (1954)	
red eye	r	5	Huestis and Willoughby (1950)		
whiteside	wh	1	McIntosh (1956b)		
orange-tan	ot	1	Egoscue and Day (1958)		
A-B erythrocytic antigens	Pm^A, Pm^B	9	Rasmussen (1961a)		Natural polymorphism
lactate dehydrogenases	$L-1^a, L-1^b$	9	Shaw and Barto (1963)		Laboratory polymorphism
erythrocytic esterases	$Es-1^a, Es-1^b$ $Es-1^o$	9	Randerson (1964, 1965)		Natural polymorphism
glucose-6-		9	Shaw and Barto (1965)		Laboratory polymorphism

TABLE 3 (Continued)

Name of locus	Symbol	Effect code	Original description	Collateral description	Comments
phosphate dehydrogenases	not defined	9			
serum albumins	*Al*	9	Welser *et al.* (1965)		Laboratory polymorphism
			Known loci, not yet described in publication		
ashy	(*as*)	1	Silliman and Huestis		
platinum	(*pt*)	1	McIntosh and Barto		
pectoral spots	(*ps*)	1	Huestis		
cataract-webbed (syndactyly)	(*cw*)	2	Huestis		
curly vibrissae	(*cv*)	2	Silliman and Huestis		
spiral tail	(*st*)	4	Huestis and Silliman		
rotator	(*ro*)	6	Barto		
serum transferrins	(*Trf*)	9	Rasmussen and Koehn		Natural polymorphism
			Possibly identified loci, not yet described		
grizzled	G	1	Sumner (1932)		"complex dominant"
chemogenic epilepsy		7	Watson (1939)		"two independent recessive loci"
haptoglobins		9	Rasmussen		Natural polymorphism
hemoglobin		9	Ahl; Foreman; Rasmussen, Jensen, and Koehn		Natural polymorphism
serum esterases		9	Rasmussen		Natural polymorphism

Robinson (1964) has reviewed the knowledge of linkage in the species *P. maniculatus*. Data regarding 18 genetic markers were subjected to the system of maximum likelihood scoring introduced by Fisher (1946). Thirty-seven combinations of genes were analyzed, exhibiting three linkage groups containing a total of eight of the markers. These linkage assignments were the same as those noted by previous investigators as cited in Tables 1 and 2. The published assignment of linkage relationship has not yet suggested the linear order of any loci within the known linkage groups. However, Huestis and Silliman (personal correspondence) have data suggesting that the order of loci of linkage group I is: p (pink-eye); c (albino); sb (snub-nose); sl (silver); and f (flexed-tail). They also suggest that recombination occurs between the two closely linked flexed-tail sites.

The results do not exclude the possibility of linkage of genes listed in Table 2 with any of the other genes given in Table 1 or Table 2 since all combinations have not been tested.

Some contradictions of data and ambiguity of gene designation have been discussed by Robinson (1964). Perhaps the most important contradiction is that Feldman (1937) reported independence between *c* and *p* (Table 1), while the work of Clark (1938) showed a significant lack of independence between these two loci. Clark's thesis of this linkage is supported by Sumner's work (1922).

A marked heterogeneity of estimates for the *f* and *sl* intercept of linkage group I (Table 1) has been noted (Huestis and Lindstedt, 1946; Huestis and Piestrak, 1942; Robinson, 1964), although all authors noted a lack of independence for this pair of loci. Similarly, McIntosh (1956*a*) noted a heterogeneity between four sibships with respect to the linkage between *d* and *b* of the linkage group II. The heterogeneity was owing to the offspring of a single female. These data were rejected by McIntosh and Robinson in assigning crossover values to this pair of loci. Such heterogeneity of linkage estimates may result from chromosomal polymorphisms within the species.

Randerson (1965) mentions an apparent close linkage between the Es-1 and Pm locus (Table 3), but detailed linkage data on this new linkage group is not given.

Hereditary Disease

Several diseases with genetic etiology have been observed within the genus, some of which have rather direct medical implications.

The convulsive mutants known in the genus represent genetic entities of neurological interest with similarities to idiopathic epilepsy in man. As Watson (1939) pointed out, the clonic and tonic phases of seizure as well as the recovery and post convulsive behavior is almost identical in *Peromyscus* and in man. Within these mice the experimenter has the advantage of assessing environmental, biochemical, and behavioral concomitants related to the single complex phenotypic entity—the seizure.

The observations from the single species *P. maniculatus* (Watson, 1939) indicate the complexity of various expressions of the convulsive state which occurs in different stocks and the difficulty of considering the seizure a simple disease entity. Chance and Yaxley (1949) have described behavioral concomitants in the audiogenic strain (EP) of *P. maniculatus*. Chance (1954) has shown that training can modify the expression of the hyperexcitement syndrome in the EP strain. Dice, Barto, and Clark (1963) have measured a number of behavioral traits (e.g., maze learning, response to light, amount of spontaneous activity) in mice subject to audiogenic seizures (EP), osmogenic seizures (CV), and waltzer traits. Some of the concomitant behavioral differences have obvious relationships to these major behavioral defects (e.g., waltzing handicaps success in maze learning). Other differences measured, although significant, do not seem to be directly related to the major defect but result from pleiotropic effects or inherant differences between strains.

Chance and Yaxley (1950) have recorded some biochemical changes of the brain tissue related to seizures in the EP stock. An increase in glycogen is specifically related to occurrences of epileptic discharges in the brain. This increase was noted in all portions of the brain measured, though in varying degrees. Epilepsy has been observed in laboratory stocks from numerous geographically distant wild-caught stocks, suggesting that these genes are widely distributed in the species *P. maniculatus* and occur within other species of the genus. The role of mutation and selection in this apparent ubiquity is not obvious.

A hereditary blood disease described in the deer mouse (*P. maniculatus*) has features essentially identical to hereditary spherocytosis in man (Huestis and Anderson, 1954; Huestis, Anderson, and Motulsky, 1956; Anderson, Huestis, and Motulsky, 1960). The inheritance of the disease in man is attributable to a dominant allele; in the mouse it is a recessively inherited trait. These authors suggest that, should the mouse syndrome be shown biochemically identical to human hereditary spherocytosis, the difference in the genotype-phenotype relationships may be a result of the evolution of dominance (Fisher, 1930) that has occurred during the selective history of the deer mouse (Anderson, *et al.,* 1960). The existence of this inherited syndrome in the deer mouse illustrates the potential value of this species as a convenient experimental tool for the study of a human hereditary disease. With this approach of simultaneous animal and clinical investigations in hereditary disease, information of both medical and genetic interest can be obtained (Motulsky, *et al.,* 1956).

The demonstration of isoimmune responses in deer mice to erthrocytic antigens occurring as a genetic polymorphism within the same populations (Rasmussen, 1961*b*, 1964) raises the possibility of fetal-maternal incompatibility in certain naturally occurring matings or experimental matings of these mice. Since the first recognition of the medical importance of fetal-maternal incompatibility caused by the Rh blood groups in man, fetal-maternal incompatibility has been demonstrated in a number of mammalian species. Rasmussen (1965) demonstrated such incompatibility in *P. maniculatus.* Rather massive maternal immunization appears prerequisite for the isoimmune response and for the sensitization of fetal erythrocytes, such that the incompatibility is not an important naturally occurring selective force affecting the blood group polymorphism in the mouse populations. Therefore, the mouse data would relate little to the population aspects of the human Rh blood group polymorphism, but the deer mouse might afford an organism for investigation of erthroblastosis in experimentally immunized females.

A wide array of genetic entities in the mouse *Peromyscus* have potential importance to medical genetics, but their relation to human inherited diseases is not as obvious as in the examples discussed. Neuromuscular and behavioral mutants (e.g., tremor, waltzer, spinner, boggler) and recently observed protein variants (e.g., lactate dehydrogenase, erythrocytic and serum esterases, hemoglobins, haptoglobins, albumins, and transferrins) in the genus may well develop more direct importance to medical genetics as our knowledge of human and mouse variants increases.

Pigmentation

Sumner's numerous studies of differences in coloration within and between populations in the genus *Peromyscus* (e.g., Sumner, 1930) are remarkably lucid attempts to relate Mendelian inheritance with evolutionary processes in vertebrates. These studies of pigmentation, with *Peromyscus,* were essential in the development of our population concept of taxonomic entities and the gene pool concept of mammalian species. The mere complexity and subtlety of the differences observed by Sumner did not allow for simple genetic explanations of naturally occurring geographic differences, but forced him into recognizing the genetic and selective complexities governing the evolution of various taxonomic groups (especially subspecies of *P. maniculatus* and *P. polionotus*).

Rather discreetly definable genetic pigmentation mutants in laboratory stocks have, on the other hand, furnished tools for the other end of the scale of genetic interests—mammalian physiological and developmental genetics. Analysis of pigmentation mutant strains of house mice have led to an increased understanding of the gene-phenotype interaction in vertebrate cell populations that can be manipulated and experimentally approached. (See symposia edited by Gordon, 1959; Burdette, 1963.)

The physiological genetics of six of the coat color mutants of *P. maniculatus* (albino, brown, ivory, dilute, silver, and pink-eye dilute) have been investigated by Foster and Barto (1963). They suggest that the albino (*C*) locus is the structural gene for the tyrosinase molecule. The other five loci appear to control the synthesis of additional components essential for normal melanogenesis. While these loci perhaps do not directly control the biosynthetic pathway leading to melanin formation, they participate, via their proteins, in the construction of the melanin granule; the organelle which serves as a focus for regulating genic interaction and the final coat color phenotype.

The approach of Foster and Barto lends itself well to future genetic and developmental comparisons between the control of phenotypes in the house mouse and the deer mouse, as the knowledge of melanogeneses increases in both of these rodent groups.

Biochemical Variants

Recent advances in methods of separation, identification, and characterization of macromolecules, especially proteins, by physical-electrical systems (i.e., paper electrophoresis, gel electrophoresis, and agar electrophoresis) has placed a new array of phenotypes within reach of genetic analysis. These advances in methodology, coupled with the advances in the concept of gene action since the original "one-gene one-enzyme" hypothesis of Beadle and Tatum in 1941, have increased our knowledge and interest in the relationship of proteins and genetic units.

Such protein variants can be of interest and application to the whole spectrum of genetic interests, from elucidating the relationship between a structural gene and the specific amino acid sequence of a given protein to a population genetic analysis of the selective forces acting upon the protein variants. The human hemoglobins are an exciting example of this spectrum of genetic interest and the interactions between these various interests (see Ingram, 1963).

The literature of mammalian protein variants is developing extremely rapidly, and that on the genus *Peromyscus* appears to be no exception.

The lactate dehydrogenases (LDH) commonly occur in from one to five molecular forms (isozymes) in various tissues of animal species. It has been theorized that the five isozymes are the five different tetramers that could be obtained by associating two structural subunits (polypeptide chains) in all possible combinations of four. Shaw and Barto (1963) have utilized laboratory stocks of *P. maniculatus* to verify this hypothesized relationship between the functional enzymatic tetramer of LDH and the two different polypeptide subunits. Within the deer mice, the LDH molecule appears to be controlled by two segregating allelic structural genes. Shaw and Barto (1965) have reported a glucose-6-phosphate dehydrogenase (G-6-PD) polymorphism in laboratory stocks of *P. maniculatus*. The variation involves one of the two G-6-PD enzyme forms (designated B) found in most tissues of the mouse. The genes producing the B enzyme occur as two autosomal alleles. The two single B component phenotypes (B-a and B-b) are considered the homozygous forms while the heterozygote produces a three B component phenotype (B-ab) exhibiting a "hybrid" molecule B component with intermediate mobility, as well as the two B components exhibited by the respective homozygotes. The occurrence of this "hybrid" molecule indicates that the B form of the enzyme is probably a dimer composed of randomly associated subunits.

These investigations of protein variants in the deer mouse have aided our interpretation of the structural basis of some animal enzymes and point to the significant contributions that *Peromyscus* genetics can provide toward the interpretation of biochemical structures (Shaw, 1964, 1965).

A number of protein variants have been observed within the genus *Peromyscus* by the use of electrophoretic techniques, but many of these observations have not yet reached publication. The variations that have been observed, however, point to potential use of protein variation to attack numerous problems of varying interest.

Randerson (1964, 1965), utilizing vertical starch gel electrophoresis, has demonstrated at least thirteen erythrocytic esterases in *P. m. gracilis*. In a population of wild deer mice trapped from Northern Michigan a polymorphism has been observed. The observed natural polymorphism appears to be owing to a single segregating locus. Rasmussen and Jensen (unpublished data) have noted a great variety of serum esterase patterns in *P. maniculatus* and *P. boylei* from Arizona, utilizing horizontal starch gel electrophoresis. How much of this complexity is attributable to genetic differences is not known.

Foreman (1964*a*) compared hemoglobins of six species of the genus *Peromyscus*. Several electrophoretically identical hemoglobins were noted in the small samples of several species. However, tryptic peptide patterns (fingerprints) from the six species indicates interspecific differences in primary structure of hemoglobins that are electrophoretically indistinguishable. These observations point out that difficulties of homology and analogy can occur when electrophoretic characterization of macromolecules are used for taxonomic evaluation.

Foreman (1964*a*, 1964*b*, and unpublished data) has observed hemoglobin polymorphisms in the cotton mouse, *P. gossypinus*. Three codominant, autosomal alleles determine the electromigration properties of the hemoglobin molecule, leading to six phenotypes distinguishable by paper electrophoresis. Hybridization experiments suggest that the three alleles in *P. gossypinus* act codominantly and as alleles with the hemoglobin locus in *P. leucopus*. A variation of hemoglobins in *P. maniculatus* has also been observed. Ahl (unpublished data) has observed hemoglobin variation in populations of *P. maniculatus* from Wyoming. Rasmussen, Jensen, and Koehn (unpublished data) have observed haptoglobin, transferrin, and hemoglobin polymorphisms in *P. maniculatus* and *P. boylei* from Arizona.

Welser *et al.* (1965) have noted several serum albumin phenotypes in various species of the genus *Peromyscus* using starch gel electrophoresis. Albumin bands of different mobilities appear to result from allelic differences within *P. maniculatus*. This protein polymorphism has been observed in laboratory stocks of mice at the University of Michigan. Such individual variation questions the validity of utilizing major protein fractions such as albumins to characterize taxonomic categories (Johnson and Wicks, 1959, 1964).

The extent to which unpublished observations have been cited is perhaps indicative of the recent role that electrophoretic techniques have assumed in studies of the genus *Peromyscus* and is also indicative of the apparent wealth of "cryptic" polymorphisms that such techniques are revealing.

Several studies have called to attention interspecies differences of proteins within the genus *Peromyscus* (e.g., Johnson and Wicks, 1959; Gough and Kilgore, 1964). The use of molecular phenotypes to assess taxonomic relationships is a promising area of study. However, serious misinformation may arise if interspecific relationships are interpreted before the extent of intrapopulation or intraspecific variation is assessed. The large amount of protein polymorphism that is being observed within single populations of deer mice indicates that we can not approach "molecular taxonomy" with a typological species or subspecies concept. It is disconcerting that many papers citing protein or antigen-antibody differences between populations (see next section) neglect to state how many individuals were sampled or how many separate individuals were pooled for assessing protein or antigenic differences. Hopefully, greater interest and recognition of individual variation will develop before interspecies molecular differences are assessed and interpreted.

Some other interesting problems are becoming apparent with the analysis of hemoglobins in *P. maniculatus*. Anderson, Huestis, and Motulsky (1960), utilizing paper electrophoresis, observed two hemoglobin bands in all mice tested from their laboratory stocks. Foreman (1964*a, b*) has also noted multiple hemoglobins in his samples of *P. maniculatus*. Rasmussen (unpublished data) tested hemoglobins from 107 wild-trapped mice (*P. m. gracilis*) from Northern Michigan (Rasmussen, 1964), as well as eighteen individuals of the laboratory stock of *P. m. gracilis* at the University of Michigan. Two hemoglobin components were apparent and all individuals exhibited identical phenotypes in the 0.8 M borate-buffered starch gel. These observations suggest that the multiple components within these samples are not attributable to a heterozygous condition but to at least two structural loci within the populations sampled. However, two hemoglobin components occur in individual mice of *P. maniculatus* from Arizona (Rasmussen, Jensen, and Koehn, unpublished data) with other single component mice existing in the same natural population. It becomes apparent, therefore, that to consider the species *P. maniculatus* as having one or multiple hemoglobin components for taxonomic comparisons is invalid. The more interesting questions are: (1) what variations exist within various species and (2) why are some populations conservatively monomorphic and others polymorphic?

Immunogenetics

The blood groups of humans have furnished genetic markers which have served as a foundation for empirical studies of population genetics and allelic frequency differences in populations of man. These studies have indicated geographical patterns or racial differences of allelic frequencies. Since Landsteiner's discovery of the human ABO system in 1900, various blood types in numerous species have been described (Cohen, 1962).

Within the genus *Peromyscus*, studies of erythrocytic antigenic variations have been of two different approaches: first, detection and description of interpopulation (species, subspecies, or different local populations) differences, second, detection and description of individual or intrapopulation differences. Studies which have utilized pooled samples of blood either for immunization or absorption of immune sera have by necessity dealt with interpopulation differences.

Species-specific erythrocytic antigenic differences between mice of the species *P. maniculatus* and *P. leucopus* were demonstrated by Moody (1941), using immune rabbit sera. Cotterman (1944) found that normal human group-B sera, absorbed with pooled erythrocytes from individuals of various species within the genus, possess agglutinins showing quantitatively different reactions. Moody (1948) demonstrated different cellular antigenic components between local populations of the species *P. maniculatus* from the Columbia River Valley. These cellular antigens were demonstrated with the use of immune rabbit sera produced by inoculation of pooled erythrocytes from mice of a given locale. The immune sera were absorbed with pooled blood obtained from several mice. Moody's study (1948) suggested that some of his reagents were detecting rather simple genetic differences, but the pooling of blood plus the sacrifice of mice precluded the definite assignment of antigenic types based on single genetic differences. Moody's study did, however, vividly illustrate the antigenic complexity of populations within a single species.

Rasmussen (1961*a*, *b*) noted agglutinating antibodies which appeared as interspecific agglutinins between the typed individuals of the species *P. maniculatus* and *P. leucopus*. These findings are similar to those noted by Moody (1941), although the immunization and the absorption procedures used by Rasmussen avoided pooling of individual bloods. Rasmussen (1961*b*) noted intraspecies population differences between individuals of *P. m. gracilis* and *P. m. bairdi* from Michigan. These population antigenic differences are similar to those reported by Moody (1948), although the differences reported by Rasmussen were observed using erythrocytes from individual mice for absorption. It should be noted that both the species and population differences existed in the small samples of mice tested and such differences may not exist throughout all populations of these taxa.

Dolyak and Waters (1962) investigated the antigenic relationships between *P. maniculatus gracilis, P. m. abietorum,* and *P.*

leucopus noveboracensis. Naturally occurring isoagglutinins and naturally occurring heteroagglutinins were found absent in the individuals tested from these taxa. This absence of naturally occurring antibodies within the genus has now been noted by several investigators (Moody, 1941; Cotterman, 1944; Rasmussen, 1961*b*) and appears to be a widespread characteristic of *Peromyscus* sera. For the most part non-immune sera of man, horse, cow, sheep, pig, rabbit, and chicken, caused non-specific agglutination with the erythroctye samples of the three kinds of mice (Dolyak and Waters, 1962). Absorption of human type A and type B with human erythrocytes revealed that the remaining hetero-agglutination with the *Peromyscus* cells was distinguishable from the normal human isoagglutinins of the ABO system of man. Dolyak and Waters (1962) also noted that the erythrocytic antigens and soluble antigens within the samples tested indicate close antigenic relationships of the three kinds of mice. Unfortunately, the authors do not state how many individual mice were used to produce immune antisera, and one has no way of assessing the part played by individual variation. Waters (1963) has evaluated antigenic relationships of mice of *P. m. gracilis, P. m. abietorum,* and *P. l. noveboracensis.* Waters' use of "at least ten animals of each subspecies . . . in absorbing immune agglutinins from rabbit sera" could perhaps have caused some variations between his reagents owing to simple segregating antigenic differences within the samples of mice of a given taxon. Such pooling of cells of potentially varied antigenic constitutions may severely hinder the replication and interpretation of results as discussed by Moody (1948).

A problem of relating antigenic to phylogenetic relationships is that antigenicity is but a small aspect of a macromolecule and may not be directly related to many of the normal functions of the molecule. Before relationships of specific attributes of macromolecules are interpreted as phylogenetic or evolutionary, the relationship between identity, analogy, and homology of molecules should be assessed. A lucid discussion of such problems with respect to developmental similarities is given by Nace, Suyama, and Smith (1961).

Rasmussen (1961*a, b*) described erythrocytic antigenic differences between individuals of *P. maniculatus.* The polymorphism was identified using erythrocytes from individual mice for the hetero-immunization of rabbits and the subsequent absorption analysis of the immune rabbit sera. Two unitary complete (saline) agglu-tinins, designated anti-A and anti-B, are reactive with erythrocytic antigens in this species. Isoimmunization apparently demonstrated these same antigenic differences. The two antigenic characters (A and B) defined by the agglutinins are inherited as if simply related to allelic factors determining three phenotypes (A, AB, and B). These two antigens are found in representatives of both subspecies, *P. m. gracilis* and *P. m. bairdi,* from Michigan, and *P. m. gambeli* from Contra Costa County, California, as well as *P. maniculatus* from northern Arizona. The A antigen was observed in the University of Michigan Laboratory stock of *P. polionotus.* Tested individuals of the species *P. leucopus* and various laboratory strains of the house mouse possessed neither antigen. Human erythrocytes possessing the human A, B, M, and N antigens were also nonreac-tive to the agglutinin reagents.

Size Inheritance

Dawson (1965) has recently observed an interesting pattern of size inheritance in hybrid matings of *P. maniculatus* and *P. polio-notus.* F_1 hybrids from the female *P. maniculatus* $\times$ male *P. polio-notus* cross were significantly smaller on the average than *P. polionotus,* the smaller parental form. Hybrids from the reciprocal cross were generally larger than *P. maniculatus,* the larger parent. These results, as well as results of the various backcross matings, were consistent with a model assuming a prenatal maternal influ-ence in size. The size relationships observed were apparent in newborn weights and persisted throughout post-weaning growth of the adult. The physiological nature of the maternal size "ele-ment" is unknown.

Falconer (1955) observed a strong asymmetry in response to size selection in the house mouse which was attributed to maternal effects. In the house mouse the asymmetry resided in the weaning weight and not in the post-weaning growth. Thus, in Falconer's study the asymmetry was attributed to mothering ability rather than to prenatally fixed "foreign" genotype components as proposed by Dawson.

Population Genetics

The formal theoretical expressions of population genetics and the accumulation of empirical data of populations have progressed as complementary approaches to our knowledge of evolutionary events and speciation. The evaluation of evolutionary events and the complexity of species variation has been especially rewarding within the genus *Peromyscus* because many of the investigators of variation within the group have been keenly aware of the genetic implications of population variation. This is especially true of the many investigations of Sumner and Dice and their associates. Upon reading such studies, one is impressed by the appreciation of the concept of a variable gene pool existing within natural populations, and also of the roles that such investigations have had in the development of the gene pool concept from empirical observation of mammalian populations. This close association between analysis of population variation and the genetic interpretation of such variation makes it difficult to draw a line between a review of investigations of population variation and speciation and investigations of so-called "beanbag" population genetics (Haldane, 1964). Since this volume includes a discussion of speciation by Dr. Dice, the discussion of population genetics included here will be somewhat arbitrarily confined to a selection of papers dealing with specific genetic loci or population events of specific genetical interests (e.g., inbreeding and genetic loads).

Blair (1944) discussed the distribution of the white-cheek gene (*Wc*) in populations of *P. polionotus polionotus, P. p. leucocephalus,* and *P. maniculatus* from Alabama, mainland Florida, and Santa Rosa Island, Florida. Although no specific gene frequencies were estimated, the study did imply a variation of the frequencies of a specifically segregating genetic marker in these populations. Blair assessed the color variation of mice of the subspecies *P. m. blandus* from New Mexico (Blair, 1947a) and a number of widely separated subspecies (Blair, 1947c). The color variations discussed in both these papers were primarily attributable to a single segregating loci (the buff-gray alleles: *G, g*). Variations from the two expected phenotypes occurred apparently as a result of independent segregating modifiers affecting the final expression of the major segregating alleles. These modifiers were especially evident in the "hybrid" progeny of the widely separated subspecies. No specific estimates of allelic frequencies were available from the data.

A pair of papers of rather wide interest to population genetics are based on reports that owe much of their content to observations of mice of the genus *Peromyscus*. Haldane (1948) presented a formal mathematical model for the geographical cline of frequencies of different phenotypes. The model developed is a general expression of the relationship of selection, migration, dispersal, and density in maintaining a cline of gene frequencies. The data of Sumner (1929*a, b*) and Blair (1944), with respect to empirical observation of coat color differences of *P. polionotus* in Florida, were part of the basis for Haldane's mathematical treatment of the problem. A review of population dynamics in small mammals, including *Peromyscus,* presented by Blair (1953), called attention to the importance of ecological data of population structure in evaluating problems of population genetics and genetical structure of mammalian populations.

Blair (1947*b*) estimated the allelic frequencies of the buff (*G*) and gray (*g*) alleles in a population of *P. m. blandus* from Tularosa Basin, New Mexico. The zygotic proportions in the samples from these natural populations did not differ significantly from the expectations of a Hardy-Weinberg genetic equilibrium. Significant differences, however, did exist between allelic frequencies at various locations within the basin. These differences in gene frequencies were attributed to the joint effects of: (1) partial ecological isolation of relatively small local populations of deer mice; (2) differential selection of the two alleles by predators on differently colored soils; and (3) the combined effects of ecological isolation and isolation by distance. Populations only four miles apart showed no differentiation even though selection was presumably operating in opposite directions in the two. Differences in allelic frequencies were evident in populations fifteen to eighteen miles apart and living on differently colored soils. The close fit of the observed and expected zygotic proportion of samples from five locations within the basin suggests that each of these samples was taken from a panmictic unit, whereas the allelic frequency differences between the separate locations indicates that the entire basin is not a single panmictic gene pool.

Howard (1949) published an extensive study of dispersal and

demography of a population of prairie deer mice, *P. m. bairdi,* in southern Michigan. During the study, a number of consanguineous matings were observed. Howard estimated "that at least 4 to 10 percent of the 186 litters observed were produced by parent-offspring matings or matings between sibs." These observations suggest that in this population of mice the genetic structure of the population, owing to short range dispersal of individuals between birth and mating, leads to a considerable degree of inbreeding. Dice and Howard (1951) published tabular data on dispersal distances between birth and mating sites of 135 mice from the same study area. These data also suggested a restricted size of the genetic neighborhood or panmictic unit in this population. Rasmussen (1964) utilized the data of Howard (1949) and Dice and Howard (1951) as well as some mathematical models of genetic structure developed by Wright (1943, 1946, 1951) to estimate the size of the panmictic unit within this population. Although numerous assumptions were necessary for applying the empirical data to Wright's mathematical models, the estimate of the size of the panmictic unit varied from $N = 12$ to $N = 117$. These estimates are based on separate observations (estimated extent of consanguinity, and dispersal distance and density estimates), both of which suggested a limited genetic neighborhood size in this natural population.

Rasmussen (1964) described a blood group polymorphism in the woodland deer mouse *P. m. gracilis* from northern Michigan. In a sample of 227 trapped wild mice a significant shortage of the heterozygous phenotype was noted. This shortage of heterozygote was interpreted as owing to inbreeding and the resulting gametic correlation within these populations. The gametic correlation (measured by F; see Li, 1955) was estimated to be of the order of $F = 0.25$. Utilizing models of Wright (1943, 1946, 1951), it was estimated that the size of the panmictic unit or genetic neighborhood in these populations was of the order of $N = 10$ to $N = 75$ mice. This rather restricted neighborhood size within the large continuous mouse populations sampled was interpreted as primarily a result of dispersal behavior from birth to mating sites within the population rather than physiographic and ecologic discontinuities imposing a structure of well-defined "island" subgroups on the population. This interpretation is strengthened by the previously discussed observations of Howard (1949) and Dice and Howard (1951).

A growing amount of dispersal data within vertebrates (e.g., Twitty, 1961; Haskins *et al.*, 1961; Kerster, 1964) suggests that non-random dispersal and homing behavior may have significant roles in the genetic structure of natural populations. The scant knowledge from the genus *Peromyscus* is strongly suggestive of such genetic structures in the populations studied.

The progress in further studies of population genetics in mice of the genus *Peromyscus* will depend upon the recognition and description of suitable genetic markers for such studies. The rapid development of knowledge of biochemical variants in the genus should contribute greatly to future knowledge of selection, migration, and population polymorphism in natural populations.

The evaluation of the concept of genetic loads and their mathematical description have recently been rather vigorously debated (Crow, 1963; Li, 1963; Sandvi, 1963). Unfortunately, the word "load" implies a burden with pernicious effects, and although it is based on the concept of relative or Darwinian fitness the evaluation has been extended to absolute probabilities of population survival. Yet, we have little knowledge of the relationship between genetic loads and net reproductive rates of natural populations.

The critical mutation rate (Muller, 1954) affects the reproduction of a population so that the net reproductive rate falls below unity because of mutational death or sterility. But, density-dependent survival must be considered as a mechanism which defrays the effect of mutation loads on a natural population.

Evidence in localized natural populations of *P. leucopus* and *P. maniculatus* (Blair, 1953; Blair and Kennerly, 1959; Blair, 1961) indicates that a marked increase in the mutational load of these populations did not affect population survival. Although irradiated male individuals did exhibit a reduced fertility when placed back into the natural population, the net reproductive rate of the population remained stable. Such stability appears to be owing to density-dependent survival of individuals without the mutational damage. Errington (1963) suggests "that the death of one individual [by a given agency] may mean little more than improving the chances of living of another one." Before geneticists can adequately evaluate the population effects of reduced intrapopulation fitness, a more thorough amount of empirical data on the relationship of genetic loads and net reproductive rates is needed. The data of

Blair and Kennerly on populations of *Peromyscus* contributes to our knowledge of this relationship in natural mammalian populations.

The knowledge of ecological population dynamics and population genetics have remained rather separate; however, the increasing knowledge of ecological dynamics and the beginning knowledge of population genetics in *Peromyscus* will hopefully interact and lead to more comprehensive concepts of animal populations.

Summary

In one respect our knowledge of the genetics of the genus *Peromyscus* is indeed scant. One is immediately impressed by the number of things that are not known and the genetic attributes of the genus that have not been investigated. On the other hand, apart from man, domestic cattle, and the house mouse, the species *P. maniculatus* is as well known genetically as any mammalian species. Most of the research on genetics within the genus has been on this one species. This is not at all distressing when you recognize that mammalian genetics is an expensive area of investigation both in terms of breeding facilities and time. Genetics is primarily an intraspecific area of study and the building of a study of comparative genetics within the genus can proceed when the knowledge of a few "cornerstone" species have been surveyed. The diversity of forms within the single species as well as within the entire genus promises a rewarding array of comparisons for the future.

The accumulation of genetic entities within various species of the genus is a process that is not easily planned. However, certain new techniques such as electrophoresis are allowing investigators to screen individuals for new groups of phenotypes. As such new biochemical phenotypes become described, their use to investigators from the wide spectrum of genetic interests will become apparent.

The relatively new science of behavioral genetics should likewise profit from the variety of forms that the genus represents.

The mathematical models of quantitative genetics and population genetics are developed beyond our empirical knowledge in the genus and both areas of investigation are within reach of the experimental geneticist who is willing to gather empirical data to evaluate certain existing mathematical models.

The science of genetics has room for biochemists, physiologists, embryologists, naturalists, and mathematical theorists, and its strength and development rests in the interaction of these disciplines. The genetics of *Peromyscus* is no exception.

One impression I have derived from undertaking this review is that the genetics of *Peromyscus* has passed its infancy and that rapid growth of knowledge is promised for the near future.

Acknowledgments

I am deeply indebted to several colleagues for their suggestions and assistance in preparing this review. The lists of genetic markers were based largely on unpublished lists compiled by Dr. Elizabeth Barto of the Mammalian Genetics Center, University of Michigan, and by Dr. Wallace D. Dawson of the Department of Biology, University of South Carolina. For their lists, as well as their corrections and suggestions, I owe my special thanks. Dr. R. R. Huestis' suggestions and aid are also graciously appreciated. This review was prepared during my support by Public Health Service Research Grant GM 12190.

Bibliography

ANDERSON, R., R. R. HUESTIS, AND A. G. MOTULSKY. 1960. Hereditary spherocytosis in the deer mouse. Its similarity to the human disease. Blood, 15:491–504.

BARTO, E. 1942*a*. Independent inheritance of certain characters in the deer mouse, *Peromyscus maniculatus*. Papers Mich. Acad. Sci., Arts, and Letters, 27:195–213.

———— 1942*b*. A second occurrence of the character pink-eye in the deer mouse. Amer. Nat., 76:634–636.

———— 1954. Spinner, an inherited type of abnormal behavior in the beach mouse, *Peromyscus polionotus*. Contrib. Lab. Vert. Biol. Univ. Mich., 66:1–16.

———— 1955. Boggler, an inherited abnormality of the deermouse (*Peromyscus maniculatus*), characterized by a tremor and a staggering gait. *Ibid.*, 71:1–26.

———— 1956. Tests for independence of waltzer and EP sonogenic convulsive from certain other genes in the deer mouse (*Peromyscus maniculatus*). *Ibid.*, 74:1–16.

BARTO, E., AND R. R. HUESTIS. 1933. Inheritance of a white star in the deer mouse. Jour. Heredity, 24:245–248.

BLAIR, W. F. 1944. Inheritance of the white-cheek character in mice of the genus *Peromyscus*. Contrib. Lab. Vert. Bio. Univ. Mich., 25:1–7.

———— 1947a. An analysis of certain genetic variations in pelage color of the Chihuahua deer mouse (*Peromyscus maniculatus blandus*). *Ibid.,* 35:1–18.

———— 1947b. Estimated frequencies of the buff and gray genes (*G, g*) in adjacent populations of deer mice (*Peromyscus maniculatus blandus*) living on soils of different colors. *Ibid.,* 36:1–36.

———— 1947c. The occurrence of buff and gray pelage in widely separated geographical races of the deer mouse (*Peromyscus maniculatus*). *Ibid.,* 38:1–13.

———— 1953. Population dynamics of rodents and other small mammals. Advances in Genetics, 5:1–41. New York: Academic Press.

———— 1961. Effects of radiations on natural populations of vertebrates. Recent Advances in Botany, 2:1377–1381. Toronto: Univ. Toronto Press.

BLAIR, W. F., AND T. E. KENNERLY, JR. 1959. Effects of x-irradiation on a natural population of the wood mouse (*Peromyscus leucopus*). Texas Jour. Sci., 11:147–149.

BURDETTE, W. J. (Ed.) 1963. Methodology in mammalian genetics. San Francisco: Holden-Day Inc.

CHANCE, M. R. A. 1954. The suppression of audiogenic hyperexcitement by learning in *Peromyscus maniculatus*. Brit. Jour. Animal Behav., 11:31–35.

CHANCE, M. R. A., AND D. C. YAXLEY. 1949. New aspects of the behaviour of *Peromyscus* under audiogenic hyperexcitement. Behaviour, 2:96–105.

———— 1950. Central nervous function and changes in brain metabolite concentration; glycogen and lactate in convulsing mice. Jour. Exp. Biol., 27:311–323.

CLARK, F. H. 1936. Linkage of pink-eye and albinism in the deer mouse. Jour. Heredity, 27:259–260.

———— 1938. Inheritance and linkage relations of mutant characters in the deer mouse, *Peromyscus maniculatus*. Contrib. Lab. Vert. Genet. Univ. Mich., 7:1–11.

COHEN, C. (Ed.) 1962. Blood groups in infrahuman species. Ann. N.Y. Acad. Sci., V. 97, Art. 1.

COTTERMAN, C. W. 1944. Serological differences in the genus *Peromyscus* demonstrable with normal human sera. Contrib. Lab. Vert. Biol. Univ. Mich., 29:1–12.

CROSS, J. C. 1931. A comparative study of the chromosomes of rodents. Jour. Morph., 52:372–401.

———— 1938. Chromosomes of the genus *Peromyscus* (deer mouse). Cytologia, 8:408–419.

CROW, J. F. 1963. The concept of genetic load: a reply. Amer. Jour. Human Genetics, 15:310–315.

DAWSON, W. D. 1965. Fertility and size inheritance in a *Peromyscus* species cross. Evolution, 19:44–55.

DICE, L. R. 1933. The inheritance of dichromatism in the deer mouse, *Peromyscus maniculatus blandus*. Amer. Nat., 67:571–574.

———— 1935. Inheritance of waltzing and epilepsy in mice of the genus Peromyscus. Jour. Mamm., 16:25–35.

DICE, L. R., AND W. E. HOWARD. 1951. Distance of dispersal by prairie deer mice from birthplaces to breeding sites. Contrib. Lab. Vert. Biol. Univ. Mich., 50:1–15.

DICE, L. R., E. BARTO, AND P. J. CLARK. 1963. Modifications of behaviour associated with inherited convulsions or whirling in three strains of Peromyscus. Animal Behav., 11:40–50.

DOLYAK, F., AND J. H. WATERS. 1962. An examination of serological relationships of three kinds of New England *Peromyscus*. Bull. No. 27, The Serological Museum, Rutgers.

DUNN, L. C. 1964. Abnormalities associated with a chromosomal region in the mouse. Transmission and population genetics of the t-region. Science, 144:260–263.

EGOSCUE, H. J. 1962. A new hairless mutation in deer mice. Jour. Heredity, 53:192–194.

EGOSCUE, H. J., AND B. N. DAY. 1958. Orange-tan—a recessive mutant in deer mice. Jour. Heredity, 59:189–192.

ERRINGTON, P. L. 1963. The phenomenon of predation. Amer. Scientist, 51:180–192.

FALCONER, D. S. 1955. Patterns of response in selection experiments with mice. Cold Spr. Harb. Symp. Quant. Biol., 20:176–196.

FELDMAN, H. W. 1936. Piebald characters of the deer mouse. Jour. Heredity, 27:300–304.

———— 1937. Segregation of mutant characters of deer mice. Amer. Nat., 71:426–429.

FISHER, R. A. 1930. The genetical theory of natural selection. Oxford: Clarendon Press (Dover Publications, New York, 1958).

———— 1946. A system of scoring linkage data with special reference to pied factor in mice. Amer. Nat., 80:568–578.

FOREMAN, C. W. 1964a. Tryptic peptide patterns of some mammalian hemoglobins. Jour. Cellular Comp. Phys., 63:1–6.

———— 1964b. Hereditary hemoglobin polymorphism in cotton mice. Amer. Zool., 4:331.

FOSTER, M., AND E. BARTO. 1963. Physiological genetic studies of coat color pigmentation in the deer mouse, *Peromyscus maniculatus*. Genetics, 48:889–890.

GORDON, M. (Ed.) 1959. Pigment cell biology. New York: Academic Press.

GOUGH, B. J., AND S. S. KILGORE. 1964. A comparative hematological study of *Peromyscus* in Louisiana and Colorado. Jour. Mamm., 45:421–428.

HALDANE, J. B. S. 1948. The theory of a cline. Jour. Genetics, 48:277–284.

———— 1964. A defense of bean bag genetics. Perspectives in Biol. and Med., 7:345–349.

HARPST, H. C., AND A. MARTIN. 1950. Some chromosomes of rodents. Proc. Penna. Acad. Sci., 24:36–39.

HASKINS, C. P., E. F. HASKINS, J. MCLAUGHLIN, AND R. HEWITT. 1961. Polymorphism and population structure in *Lebistes reticulatus,* an ecological study. Pp. 320–395. *In:* W. F. Blair (Ed.) Vertebrate Speciation. Austin: Univ. Texas Press.

HOWARD, W. E. 1949. Dispersal, amount of inbreeding, and longevity in a local population of prairie deer mice on the George Reserve, southern Michigan. Contrib. Lab. Vert. Biol. Univ. Mich., 43:1–50.

HUESTIS, R. R. 1938. Ivory, a feral mutation in *Peromyscus.* Jour. Heredity, 29:235–237.

HUESTIS, R. R., AND R. ANDERSON. 1954. Inherited jaundice in *Peromyscus.* Science, 120:852–853.

HUESTIS, R. R., R. ANDERSON, AND A. G. MOTULSKY. 1956. Hereditary spherocytosis in *Peromyscus.* I. Genetic Studies. Jour. Heredity, 47:225–228.

HUESTIS, R. R., AND E. BARTO. 1934. Brown and silver deer mice. Jour. Heredity, 25:219–223.

———— 1936a. Flexed-tailed *Peromyscus. Ibid.,* 27:73–75.

———— 1936b. An inherited tremor in *Peromyscus. Ibid.,* 27:436–438.

HUESTIS, R. R., AND G. LINDSTEDT. 1946. Linkage relations of flexed-tail in *Peromyscus.* Amer. Nat., 80:85–90.

HUESTIS, R. R., AND V. PIESTRAK. 1942. An aberrant ratio in *Peromyscus.* Jour. Heredity, 33:289–291.

HUESTIS, R. R., AND R. WILLOUGHBY. 1950. Heterchromia in *Peromyscus.* Jour. Heredity, 41:287–290.

INGRAM, V. M. 1963. The hemoglobins in genetics and evolution. New York: Columbia University Press.

JOHNSON, M. L., AND M. J. WICKS. 1959. Serum protein electrophoresis in mammals—taxonomic implications. Systematic Zoology, 8:95–99.

———— 1964. Serum protein electrophoresis in mammals: Significance in higher taxonomic categories. Chapter 47. Pp. 681–694. *In:* Charles A. Leone (Ed.) Taxonomic biochemistry and serology. New York: Ronald Press.

KAVANAU, J. L. 1964. Behavior: confinement, adaptation and compulsory regimes in laboratory studies. Science, 143:490.

KERSTER, H. W. 1964. Neighborhood size in the rusty lizard, *Sceloporus olivaceus.* Evolution, 18:445–457.

LI, C. C. 1955. Population Genetics. Chicago: University of Chicago Press.

———— 1963. The way the load ratio works. Amer. Jour. Human Genetics, 15:316–321.

MAKINO, S. 1953a. Chromosomal numbers of some American rodents. Science, 118:630.

———— 1953b. Notes on the chromosomes of the *Peromysci* (Rodentia-Cricetidae). Experientia, 9:214–216.

McINTOSH, W. B. 1956a. Linkage in *Peromyscus* and sequential tests for independent assortment. Contrib. Lab. Vert. Biol. Univ. Mich., 73:1–27.

———— 1956b. Whiteside, a new mutation in *Peromyscus.* Jour. Heredity, 47:28–32.

MOODY, P. A. 1941. Identification of mice in the genus *Peromyscus* by a red cell agglutination test. Jour. Mamm., 22:40–47.

———— 1948. Cellular antigens in three stocks of *Peromyscus maniculatus* from the Columbia River Valley. Contrib. Lab. Vert. Biol. Univ. Mich., 39:1–16.

MOREE, R. 1946. Genic sterility in interspecific hybrids of *Peromyscus*. Anat. Record, 96:562.

———— 1948. The bearing of hybrid sterility on the genetic relationship of *Peromyscus t. truei* and *P. n. nasutus*. Genetics, 33:621–622.

MOTULSKY, A. G., R. R. HUESTIS, AND R. ANDERSON. 1956. Hereditary spherocytosis in mouse and man. Acta. Genet. et Stat. Med., 6:240–245.

MULLER, H. J. 1954. The nature of the genetic effects produced by radiation. Pp. 351–473. *In*: A. Hollaender (Ed.) Radiation Biology, Vol. 1. New York: McGraw-Hill.

NACE, G. W., T. SUYAMA, AND N. SMITH. 1961. Early development of special proteins. Symposium germ cells and development. Pp. 564–603, Institut Intern. d'Embryologie and Fondazione A. Baselli (1960).

PRICE, E. O. 1966. Influence of light on reproduction in *Peromyscus maniculatus gracilis*. Jour. Mamm., 47:343–345.

RANDERSON, S. 1964. The inheritance of an erythrocytic esterase component in the deer mouse, *Peromyscus maniculatus gracilis*. Genetics, 50:277.

———— 1965. Erythrocyte esterase forms controlled by multiple alleles in the deer mouse. *Ibid.*, 52:999–1005.

RASMUSSEN, D. I. 1961*a*. Cellular antigenic differences between individuals of the species *Peromyscus maniculatus*. Amer. Nat., 95:59–60.

———— 1961*b*. Erythrocytic antigenic differences between individuals of the deer mouse, *Peromyscus maniculatus*. Genetic Res. (Camb.), 2:449–455.

———— 1964. Blood group polymorphism and inbreeding in natural populations of the deer mouse *Peromyscus maniculatus*. Evolution, 18:219–229.

———— 1965. Fetal-maternal incompatibility in the deer mouse. Amer. Nat., 99:127–128.

ROBINSON, R. 1964. Linkage in *Peromyscus*. Heredity, 19:701–709.

ROSS, M. D. 1962. Auditory pathway of the epileptic waltzing mouse. I. A comparison of the acoustic pathways of the normal mouse and those of the totally deaf epileptic waltzer. Jour. Comp. Neurol., 119:317–339.

SANDVI, L. D. 1963. The concept of genetic load: a critique. Amer. Jour. Human Genetics, 15:298–309.

SHAW, C. R. 1964. The use of genetic variation in the analysis of isozyme structure. Brookhaven Symp. Biol., 17:117–130.

———— 1965. Electrophoretic variation in enzymes. Science, 149:936–943.

SHAW, C. R., AND E. BARTO. 1963. Genetic evidence for the subunit structure of lactate dehydrogenase isozymes. Proc. Nat. Acad. Sci., 50:211–214.

———— 1965. Autosomally determined polymorphism of glucose-6-phosphate dehydrogenase in *Peromyscus*. Science, 148:1099–1100.

SUMNER, F. B. 1917. Several color "mutations" in mice of the genus *Peromyscus.* Genetics, 2:291–300.

———— 1922. Linkage in *Peromyscus.* Amer. Nat., 56:412–417.

———— 1924. Hairless mice. Jour. Heredity, 15:475–481.

———— 1929*a*. The analysis of a concrete case on intergradation between two sub-species. (I). Proc. Nat. Acad. Sci., 50:211–214.

———— 1929*b*. The analysis of a concrete case of intergradation between two sub-species. II. Additional data and interpretations. *Ibid.,* 15:481–493.

———— 1930. Genetic and distributional studies of three subspecies of *Peromyscus.* Jour. Genetics, 23:275–376.

———— 1932. Genetic, distributional, and evolutionary studies of the subspecies of deer mice (*Peromyscus*). Bibliog. Genet., 9:1–106.

SUMNER, F. B., AND H. H. COLLINS. 1922. Further studies of color mutations in mice of the genus *Peromyscus.* Jour. Exp. Zool., 36:289–316.

TAMSITT, J. R. 1960. The chromosomes of the *Peromyscus truei* group of white-footed mice. Texas Jour. Sci., 12:152–157.

TWITTY, V. C. 1961. Experiments on homing behavior and speciation in *Taricha.* Pp. 415–459. *In:* W. F. Blair (Ed.) Vertebrate speciation. Austin: Univ. Texas Press.

WATERS, J. H. 1963. Biochemical relationships of the mouse *Peromyscus* in New England. Systematic Zool., 12:122–133.

WATSON, M. L. 1939. The inheritance of epilepsy and waltzing in *Peromyscus.* Contrib. Lab. Vert. Genet. Univ. Mich., 11:1–24.

WELSER, C. F., H. J. WINKELMAN, E. B. CUTLER, AND E. BARTO. 1965. Albumin variations in the white-footed mouse *Peromyscus.* Genetics, 52:483.

WHITE, M. J. D. 1954. Animal cytology and evolution. Cambridge Univ. Press.

WRIGHT, S. 1943. Isolation by distance. Genetics, 28:114–138.

———— 1946. Isolation by distance under diverse systems of mating. *Ibid.,* 31:39–59.

———— 1951. The genetical structure of populations. Ann. Eugenics, 15:323–354.

ADDENDUM

BRONSON, F. H., AND S. H. CLARKE. 1966. Adrenalectomy and coat color in deer mice. Science, 154:1349–1350.

DAWSON, W. D. 1966. Postnatal development in *Peromyscus maniculatus-polionotus* hybrids. I. Developmental landmarks and litter mortality. Ohio Jour. Sci., 66:518–522.

FOREMAN, C. W. 1966. Inheritance of multiple hemoglobins in *Peromyscus.* Genetics, 54:1007–1012.

FOSTER, M. 1965. Mammalian pigment genetics. Advances in Genetics 13:331–339. New York: Academic Press.

HSU, T. C., AND F. E. ARRIGHI. 1966. Chromosomal evolution in the genus *Peromyscus* (Cricetidae, Rodentia). Cytogenetics, 5:355–359.

MOTULSKY, A. G., R. ANDERSON, R. S. SPARKES, AND R. R. HUESTIS. 1962. Marrow transplantation in newborn mice with hereditary spherocytosis: a model system. Trans. Assoc. Amer. Phys., 75:64–71.

OHNO, S., C. WEILER, J. POOLE, L. CHRISTIAN, AND C. STENIUS. 1966. Autosomal polymorphism due to pericentric inversions in the deer mouse (*Peromyscus maniculatus*), and some evidence of somatic segregation. Chromosoma, 18:177–187.

RASMUSSEN, D. I., AND R. K. KOEHN. 1966. Serum transferrin polymorphism in the deer mouse. Genetics, 54:1353–1357.

SINGH, R. P., AND D. B. McMILLAN. 1966. Karyotypes of three subspecies of *Peromyscus*. Jour. Mamm., 47:261–266.

SPARKES, R. S., AND D. T. ARAKAKI. 1966. Intrasubspecific and intersubspecific chromosomal polymorphism in *Peromyscus maniculatus* (deer mouse). Cytogenetics, 5:411–418.

STEINMULLER, D., AND A. G. MOTULSKY. 1967. Treatment of hereditary spherocytosis in *Peromyscus* by radiation and allogenic bone marrow transplantation. Blood, 29:320–330.

10
HOME RANGE AND TRAVELS

LUCILLE F. STICKEL

Introduction

THE CONCEPT OF HOME and the implications that surround the term in human experience have been applied to the behavior of wild mammals by use of the term "home range."

Size of home range is a significant variable in an animal population, reflecting the effects of a combination of physiological and environmental factors. Habitat, food supply, season, weather, age, sex, population density, and basic activity all form part of an interacting complex. It sometimes has been possible to identify the factor that is critical in limiting size of home range, but more often several factors vary together and others are unknown.

Within the home range, mice of the genus *Peromyscus* go about their daily activities of food gathering and exploration, rear their families, encounter their neighbors, escape their enemies, surmount the exigencies of cold, storm, and flood, and so live out their lives. Many, perhaps most, established adults stay in the same general area throughout their lives, although the extent and directions of their travels are only roughly defined and are continuously responsive to environmental and social changes. Attachment to a home area is shown most convincingly by the efforts of displaced individuals to return to their homes.

Individuals occasionally explore outside their customary ranges, and so become acquainted with rather large areas. Some may never return, taking up new residence elsewhere. Young thus disperse from their birth sites, but adults also may disperse. The proportion of animals in transit varies, and it rarely is possible to know whether such transients are on exploratory trips or whether they are individuals who never have succeeded in establishing ranges.

These topics will be considered under three principal headings: (1) variation in size of home range, (2) behavior within the home range, and (3) travels outside the home range. Finally a brief synopsis will be given of methods of measuring size of home range in relation to the home range concept.

Factors Affecting Variation In Size of Home Range

Home ranges of *Peromyscus* can be expected to vary in size within two magnitudes: approximately from 0.1 acre to ten acres. The results of 34 investigations of size of home range for *Peromyscus* are shown in Table 1. These studies were made in many different habitats, from the seaside dunes of Florida to the Alaskan forests. The table defines the methodology and presents the numerical data as given by the investigator.

Variation in size of home range as it relates to environmental, physiological, and social factors will be considered more fully below.

HABITAT.—Size of home range may be influenced by habitat, although in field studies it usually has not been possible to separate habitat from effects of other factors such as food and population density.

P. leucopus noveboracensis inhabiting young upland pine-oak woods in Maryland where trees were small and ground cover was dense had larger home ranges than did those in the bottomlands, where the ground was open and there was an abundance of large trees (Stickel, 1948b).

In Michigan, *P. l. noveboracensis* were also more active and abundant at local sites with numerous large-diameter trees (Brand, 1955). Mice living in both woods and adjoining grassland had larger home ranges than did those confined to woodland (Blair, 1940; Burt, 1940). The mice that ranged both in woods and grassland had 25 per cent of their area in the forest and 75 per cent in the grassland. In contrast, *P. l. noveboracensis* had home ranges of approximately the same size in three seral stages of an old field succession in Connecticut (Hirth, 1959).

During a Michigan winter, *P. maniculatus bairdi* living in the corn shocks traveled much shorter distances than did those living in the wheat stubble of the adjoining fields (Linduska, 1942). Often mouse tracks were almost completely absent around the shocks, indicating that individuals remained in the same shock for

TABLE 1

SIZE OF HOME RANGE FOR *Peromyscus*

Species	Sex	Age	Locality	Habitat	Season	Area (acres)	Distance (feet)	Method	Captures (no.)	Mice or records (no.)	Source
P. boylei	♂	Ad.	Calif. (Sierra Nev.)	Cutover transition zone forest	Comb.	0.27 (0.04–0.95)	–	Min. area	5+	13	Storer et al., 1944
	♀	Ad.				0.41 (0.06–1.6)	–		5+	15	
P. b. attwateri	♂	–	Mo. (Ozarks)	Cedar glades	Comb.	0.64 (0.34–0.87)	–	Min. area	5+	8	Brown, 1964a
	♀	–				0.38 (0.22–0.56)	–		5+	7	
P. b. rowleyi	♂	–	Durango, Mex.	Mixed veg. in rocky canyon	June–July	0.14 (0.08–0.23)	129	Obs. length and excl. bd. strip	3+	8	Drake, 1958
	♀	–				0.11 (0.06–0.16)	95		3+	9	
P. gossypinus	♂	Ad.	Fla.	Mixed hardwoods	Mar.–Apr.	1.82±0.36 (0.45–4.36)	–	Incl. bd. strip	4+	12	Griffo, 1961
	♀	Ad.				1.44±0.19 (0.22–2.75)	–		4+	13	
P. gossypinus	Comb.	–	La.	Loblolly (unburned)	Jan.–June	1.12	249 (75–416)	Obs. length used as diameter	3+	9	Shadowen, 1963
				Loblolly (burned)		0.74	202 (100–420)		3+	21	
P. gossypinus	♂	Ad.	Tex.	Floodplain forest	Comb.	2.26±0.29	–	Incl. bd. strip	8+	8	McCarley, 1959a
	♀	Ad.				1.48±0.30	–		8+	8	
P. gossypinus	Comb.	–	Tex.	Pine	May–July	1.22	–	Incl. bd. strip	–	11	Stephenson et al., 1963
P. g. gossypinus	♂	–	Fla.	Hardwood forest	Comb.	–	129	Dist. bet. capt.	–	161	Pearson, 1953
	♀	–				–	112		–	88	

TABLE 1 (Continued)

Species	Sex	Age	Locality	Habitat	Season	Area (acres)	Distance (feet)	Method	Captures (no.)	Mice or records (no.)	Source
P. g. gossypinus	Comb.	Comb.	Fla.	Swamp, thicket, and wood land	June-July	–	555 (0; 30–2800)	Dist. from initial capture	–	26	Pournelle, 1950
P. leucopus	♂	–	Ark.	Young oak-pine	Comb.	0.15 (0.07–0.46)	–	Incl. bd. strip	3+	10	Redman and Sea-lander, 1958
	♀	–				0.23 (0.09–0.40)	–		3+	3	
P. leucopus	Comb.	–	Kans.	Woodland	Comb.	0.16 (0.03–0.36)	–	Min. area	6+ stations	20	Fitch, 1958
	Comb.	–				0.4	74.4	Dist. bet. capt. used as radius	–	637	
P. leucopus	Comb.	–	Minn.	Deciduous woods	Winter, with snow	0.72	199 (29–500)	Obs. length used as diam.	2+ stations	27	Beer, 1961
	Comb.	–				0.21 (0.03–0.93)	–	Min. area	3+ stations	20	
P. leucopus	Comb.	Comb.	N.Y.	Mixed hardwoods		0.29–1.17	127.5	Obs. length used as diam.	2+ stations	335	New, 1958
	Comb.	Comb.				0.57–1.7	177.5	Adj. length used as diam.	2+ stations	335	
P. leucopus	♀	Imm.	Tenn.	Overgrown field	July-Aug.	0.29	–	Incl. bd. strip	4	1	Howell, 1954
P. l. fusus	♂	–	Mass (Martha's Vineyard)	Pine and decid. woodland	Comb.	0.28	–	Min. area	5+	9	Bowditch, 1965
	♀	–				0.25	–		5+	7	
P. l. leu-copus	Comb.	Comb.	Tenn.	Cutover pine	Comb.	–	137 (11–292)	Obs. length	3+	16	Yeatman, 1960

TABLE 1 (Continued)

Species	Sex	Age	Locality	Habitat	Season	Area (acres)	Distance (feet)	Method	Captures (no.)	Mice or records (no.)	Source
P. l. noveboracensis	♂	Ad.	Conn.	Grass to woodland (3 stages)	Comb.	–	92, 74, 105	Dist. bet capt.	–	–	Hirth, 1959
	♀	Ad.				–	71, 70, 83		–	–	
P. l. noveboracensis	♂	Comb.	Ill.	Upland decid. woods	July-Aug.	–	229 (100–570)	Obs. length	2+	6	Will, 1962
	♀	Comb.				–	243 (150–320)		2+	3	
P. l. noveboracensis	♂	Comb.	Md.	Bottomland forest	June-Aug.	–	150	Obs. length extrapolated	2–6	79	Stickel, 1948b
	♀	Comb.		Bottomland forest		–	100		2–6	73	
	♂	Comb.		Upland oakwoods (Young)		–	275		2–5	8	
	♀	Comb.				–	250		2–5	7	
P. l. noveboracensis	♂	Ad. and subad.	Md.	Bottomland forest	June 1954	–	258±40 (108–576)	Obs. length	4+	13	Stickel, 1960
	♂	Ad. and subad.			June 1957	–	119±3 (46–207)		4+	28	
	♀	Ad. and subad.			June 1957	–	104±13 (0; 51–162)		4+	10	
P. l. noveboracensis	♂	Ad. and subad.	Md.	Upland pine and decid. woodlot	Comb.	–	Ave. betw. 200 and 400	Obs. length	4+	186	Stickel and Warbach, 1960
	♀	Ad. and subad.			Comb.	–	Ave. betw. 150 and 300		4+	125	
P. l. noveboracensis	Comb.	Comb.	Mich.	Oak-hickory + bluegrass	June	0.81±0.15 (0.3–1.50)	–	Incl. bd. strip	–	10	Blair, 1940

TABLE 1 (Continued)

Species	Sex	Age	Locality	Habitat	Season	Area (acres)	Distance (feet)	Method	Captures (no.)	Mice or records (no.)	Source
P. l. noveboracensis	♂	Ad.	Mich.	Oak-hickory	Comb.	0.27 (0.16–0.54)	–	Similar to excl. bd. strip	–	about 40	Burt, 1940
	♀	Ad.				0.208 (0.06–0.37)	–		–	about 34	
P. l. noveboracensis	♂	–	Ohio	Oak-maple woods	Comb.	0.20 (0.07–0.40)	–	Bd. strip	4+	7	Ruffer, 1961
	♀	–				0.20 (0.05–0.38)	–		4+	12	
P. l. tornillo	♂	Ad.	N. Mex.	Mesquite	Apr.-May	3.1 (1.7–4.2)	–	Incl. bd. strip	7+	4	Blair, 1943
	♀	Ad.				4.5 (3.5–5.5)	–		7+	2	
P. maniculatus	Comb.	–	Alaska (S.E.)	Logged clearing	July-Oct.	–	84 (0; 66–284)	Dist. from init. capt.	–	44	McGregor, 1958
P. maniculatus	♂	–	Ark.	Younk oak-pine	Comb.	0.08 (0.04–0.20)	–	Incl. bd. strip	3+	3	Redman and Sealander, 1958
	♀	–				0.13	–		3+	1	
P. maniculatus	♂	Ad.	Calif.	Cutover transition zone forest	Comb.	0.24 (0.03–1.1)	–	Min. area	5+	29	Storer et al., 1944
	♀	Ad.	Calif.		Comb.	0.21 (0.01–0.66)	–	Min. area	5+	27	
	♂	Yg.	Calif.		Comb.	0.13 (0.01–0.24)	–	Min. area	5+	10	
	♀	Yg.	Calif.		Comb.	0.12 (0.02–0.31)	–	Min. area	5+	12	

TABLE 1 (Continued)

Species	Sex	Age	Locality	Habitat	Season	Area (acres)	Distance (feet)	Method	Captures (no.)	Mice or records (no.)	Source
P. manicu-latus	Comb.	–	Kans.	Grassland	Comb.	0.74 (0.27–2.24)	–	Min. area	4+ stations	18	Fitch, 1958
	♂	–				1	120	Dist. betw. capt. used as radius	–	43	
	♀	–				0.6	92	Dist. betw. capt. used as radius	–	41	
P. manicu-latus	Comb.	Ad.	Ore.	Desert sagebrush	Mar.	–	583 (342–761)	Obs. length	23 ave.	8	Broadbooks, 1961
P. m. abie-torum	Comb.	–	Maine (Mt. Desert Island)	Woods, pine and decid.	June-Aug.	–	225	–	3.2 ave.	47	Johnson, 1927
P. m. abie-torum	Comb.	–	New Brunswick	Coniferous forest	May-Nov.	1.27±0.08 (0.74–1.67)	131	Incl. bd. strip and obs. length	7+	13	Morris, 1955
P. m. bairdi	♂	Ad.	Mich.	Bluegrass	Comb.	0.63±0.04 (0.25–1.67)	–	Incl. bd. strip	7 ave.	70	Blair, 1940
	♀	Ad.				0.51±0.04 (0.12–2.29)	–		7 ave.	64	
	♂	Imm.				0.61±0.11 (0.08–1.37)	–		5.4 ave.	10	
	♀	Imm.				0.64±0.09 (0.17–0.99)	–		5.4 ave.	9	
P. m. bairdi	–	–	Mich.	Old field	Oct.	–	75% < 118 95% < 260	Dist. from center of activity	2+	37 mice 185 records	Dice and Clark, 1953
P. m. bairdi	Comb.	Ad.	Mich.	Bluegrass	Comb.	as much as 5–6	221 (50–350)	Obs. length from nest boxes to traps	2+	28	Howard, 1949

TABLE 1 (Continued)

Species	Sex	Age	Locality	Habitat	Season	Area (acres)	Distance (feet)	Method	Captures (no.)	Mice or records (no.)	Source
P. m. blandus	♂	Ad.	N. Mex.	Mesquite	Mar.-May	4.66±0.33 (2.03-9.92)	–	Incl. bd. strip	18.8 ave.	31	Blair, 1943
	♀	Ad.				4.10±0.39 (1.66-8.70)	–		17.5 ave.	21	
P. m. gambeli	♂	–	Calif.	Grass-shrub	Comb.	–	500	Obs. length	2+	53	Brant, 1962
	♀	–					375		2+	42	
P. m. gracilis	♂	Ad.	Mich.	Beech-maple	Aug.-Sept.	2.31±0.27 (0.88-5.64)	–	Incl. bd. strip	22 ave.	27	Blair, 1942
	♀	Ad.				1.39±0.16 (0.75-3.28)	–		22 ave.	14	
	♂	Imm.				0.89±0.10 (0.45-1.60)	–		14 ave.	12	
	♀	Imm.				1.07±0.17 (0.35-3.39)	–		14 ave.	17	
P. m. gracilis	♂	Ad.	Mich.	Conifer and decid. forest	June-Sept.	0.19 (0.10-0.31)	–	Excl. bd. strip (similar to)	3+	12	Manville, 1949
	♀	Ad.				0.18 (0.12-0.25)	–		3+	5	
	Comb.	Juv.				0.18 (0.10-0.27)	–		3+	17	
P. m. rubidus	♂	Ad.	Ore. (N.W.)	Low veg. after conif. forest burn	Comb.	4.40 (2.01-8.53)	–	Av. dist. fm center used as radius	10+	14	Hooven, 1958
	♀	Ad.				2.65 (0.56-7.42)	–		10+	11	
	♂	Subad.				3.82 (0.57-9.57)	–		10+	32	
	♂	Juv.				6.57 (1.60-13.07)	–		10+	5	

TABLE 1 (Continued)

Species	Sex	Age	Locality	Habitat	Season	Area (acres)	Distance (feet)	Method	Captures (no.)	Mice or records (no.)	Source
P. m. rufinus	♀	Subad.				2.34 (0.16–10.24)	–		10+	28	Williams, 1955
	♀	Juv.				5.1 (3.74–8.59)	–		10+	4	
	♂	Ad.	Colo.	Scanty pines, firs, shrubs; rock	July–Aug.	0.52	–	Min. area	3+	21 (total)	
	♀	Ad.				0.36	–		3+		
	♂	Imm.				0.62	–		3+		
	♀	Imm.				0.31	–		3+		
P. m. sono-riensis	♂	–	Nev.	Various	Comb.	–	524	Obs. length	2+ stations	7	Allred and Beck, 1963
	♀	–				–	332			3	
P. polionotus	Comb.	Ad.	S. Car.	Uncultivated fields	Comb.	0.34±0.02	60–80	Incl. bd. strip and 80% prob. of dist. from center	5+	106	Davenport, 1964
P. p. leucocephalus	Comb.	–	Fla.	Grassy beach dunes	Fall	1.97±0.26	344±41	Incl. bd. strip and obs. length	16 ave.	25	Blair, 1951
	Comb.	–			Spring	5.76±0.74	666±69		17 ave.	28	
	Comb.	–		Interior dunes and marshes	Fall	6.38±0.81	692±55		16 ave.	30	
	Comb.	–			Spring	10.66±1.46	951±62		16 ave.	37	

several days at a time; whereas in the adjoining fields many mice moved several hundred feet from their nests in a single night.

On Santa Rosa Island in Florida *P. polionotus leucocephalus* living in the dense cover of the dunes fronting the gulf beach traveled conspicuously shorter distances than did those living in the more open cover back from the beach (Blair, 1951). Differences in density of cover in uncultivated fields in South Carolina, however, appeared not to affect the size of home range of *P. polionotus* living in the fields (Davenport, 1964).

P. gossypinus had home ranges of very similar size in burned and unburned plots of typical loblolly short-leaf pine in the rolling upland area of northern Louisiana (Shadowen, 1963).

P. boylei attwateri in Missouri had home ranges that directly reflected certain conspicuous habitat features (Brown, 1964*b*) ; they were distinctly elongate, showing that the mice had a tendency to move along the parallel outcrops of dolomite that crossed the glades, but were more reluctant to cross the grassy barren areas between successive ledges. Movements along the horizontal axis of the home range were usually five to six times as great as those up or down the slope of the glades.

Food Supply.—Food supply would be expected to affect size of home range, but a clearcut relationship has rarely been shown, partly because of difficulties in appraising the quantity and quality of food.

On Santa Rosa Island, food for *P. p. leucocephalus* seemed adequate and the mice appeared to be well fed at all seasons (Blair, 1951). They fed principally on fruits of the bluestem (*Andropogon maritimus*) and sea oats (*Uniola laxa*). These foods were much more abundant in the fall, however, and the mice then had significantly smaller ranges than they did in the spring. *P. g. gossypinus* in a Florida hardwood forest had smaller ranges during years of high acorn production and high population density (Pearson, 1953).

Since the amount of food available in different habitats can often be surmized, size of home range associated with habitat may be a function of food supply. For example, mice living in corn shocks, where food was abundant, traveled much shorter distances than did mice living in open fields of wheat stubble (Linduska, 1942). *P. l.*

noveboracensis living in relatively unproductive upland pine oak woods had larger home ranges than mice living in the bottomlands, where more food was produced (Stickel, 1948*b*).

SEASON AND WEATHER.—Seasonal changes in size of home range have been apparent in a number of studies. Greatest activity and largest home ranges appear to occur in the spring, least in the winter. These differences have been attributed to onset of spring breeding activity or to reduced food supply at that time. Low temperatures may explain reduction of activity in winter. These seasonal differences have been reported for *P. l. noveboracensis* (Brand, 1955; Hirth, 1959; Thomsen, 1945), for *P. l. leucopus* (Yeatman, 1960), and for *P. p. leucocephalus* (Blair, 1951).

Storer *et al.* (1944) reported unusual seasonal activity for two species of *Peromyscus* in California. *P. boylei* undertook a noticeable seasonal migration to and from the trapping area in the spring; several mice taken in one area in March and April later established residence in another area after the snow had melted there. These movements were most noticeable in the spring and again in the early autumn when the population showed a definite decrease. In the same area, males of *P. maniculatus* took short trips away from established home ranges and returned shortly thereafter, primarily during the breeding season. Other studies have shown no obvious seasonal changes in size of home range; for example, *P. m. blandus* (Blair, 1943) and *P. polionotus* (Davenport, 1964).

Unexplained changes in size of home range between one spring and the next without correlated changes in population level were reported for *P. m. gracilis* (Brant, 1962) and for *P. p. leucocephalus* (Blair, 1951).

Effects of weather on activity of mice are conspicuous to any trapper, but difficult to establish objectively (See Chapter 14).

The amount of illumination in the environment appeared to be the most potent factor affecting amount of activity in Florida populations of *P. p. leucocephalus* (Blair, 1951). The mice were very active on dark nights but moved about very little on clear nights during the full moon. It was difficult to differentiate between the effects of illumination and the effects of such physical factors as temperature, humidity, wind velocity, and rainfall. In general, the mice were most active on stormy, rainy nights (Blair, 1951).

P. l. noveboracensis apparently responds similarly. In ten different habitats in Michigan, the number of captures increased or decreased with the corresponding increase or decrease in minimum temperature and was greatly reduced after heavy snowfalls (Brand, 1955). Ruffer (1961), in Ohio, related temperature to trap success. In Connecticut, mice were more active as the degree of cloudiness increased (Hirth, 1959). There also appeared to be a correlation between rainfall and activity. Rain early in the evening appeared to decrease the catch whereas rain after midnight did not appear to have this effect. Burt (1940) also reported a general impression that one is likely to have a better catch on dark rainy nights than on clear moonlight nights.

AGE AND SEX.—Immature *Peromyscus* remain in the home ranges of their parents until the dispersal period that precedes or coincides with sexual maturity. Home ranges of young mice, therefore, usually are smaller or approximately the same size as those of adults in *P. maniculatus* (Storer *et al.,* 1944), *P. m. bairdi* (Blair, 1940; Howard, 1949), *P. m. gracilis* (Blair, 1942), *P. m. rubidus* (Hooven, 1958), and *P. m. rufinus* (Williams, 1955).

Adult male *Peromyscus* commonly have larger home ranges than do adult females, although there are exceptions. The tabulation below shows the prevalence of this difference.

Male Ranges Larger	
P. boylei	Drake, 1958
P. b. attwateri	Brown, 1964a
P. gossypinus	Griffo, 1961; Pearson, 1953; Shadowen, 1963
P. leucopus	White, 1964
P. l. noveboracensis	Bendell, 1959 (high pop. density) ; Burt, 1940; Hirth, 1959; Stickel, 1948b, 1960; Stickel and Warbach, 1960
P. maniculatus	Fitch, 1958; Storer *et al.,* 1944
P. m. bairdi	Dice and Clark, 1953
P. m. gambeli	Brant, 1962
P. m. gracilis	Blair, 1941, 1942; Manville, 1949
P. m. rubidus	Hooven, 1958
P. m. rufinus	Williams, 1955
P. m. sonoriensis	Allred and Beck, 1963
Female Ranges Larger	
P. boylei	Storer *et al.,* 1944

Male and Female Ranges Approximately Same Size

P. *gossypinus*	McCarley, 1959*a*
P. *leucopus noveboracensis*	Bendell, 1959 (low pop. density)
P. *maniculatus abietorum*	Morris, 1955
P. *m. blandus*	Blair, 1943
P. *polionotus*	Davenport, 1964
P. *p. leucocephalus*	Blair, 1951

POPULATION DENSITY.—An inverse relationship between population density and size of home range has been shown in several studies and probably is usual. However, this should not be assumed to be invariably true, for in certain studies *Peromyscus* had home ranges of essentially the same size despite density changes. It also is theoretically possible that the members of a high density population would increase their ranges following a rapid depletion of food supply and, thus, reverse the expected relationship.

P. l. noveboracensis in Maryland had larger home ranges in low-density populations of the upland woods than in the high-density populations of the bottomlands (Stickel, 1948*b*; Stickel and Warbach, 1960). In a year when the mice were exceptionally scarce in the bottomlands, their home ranges were correspondingly large, about the same as those in the uplands (Stickel, 1960). In the New Jersey pine barrens home ranges of *P. leucopus* were small when populations were high and large when they were low (White, 1964). In California, the average diameter of home range for *P. m. gambeli* was inversely density dependent (Brant, 1962). *P. g. gossypinus* in Florida showed considerably shorter distances between captures under crowded conditions of summer and fall than under the less crowded conditions that prevailed after mid-December (Pearson, 1953).

In a study of *P. maniculatus* and *P. truei* on a 15-acre cutover area in California, only one mouse was present at one period of the study (Tevis, 1956). This mouse covered the entire area in the course of each four to five days, ranging over a different part each night. Later, however, when there were as many as 14 mice in the same area, all the cutover was scrutinized each night but no mouse included the entire area in its home range.

Home ranges appeared to be relatively unaffected by seasonal changes in population density in a study of *P. polionotus* in South Carolina (Davenport, 1964). *P. m. gambeli* in California moved shorter distances in one spring than in another despite similar population levels (Brant, 1962), as did *P. p. leucocephalus* in Florida (Blair, 1951).

ACTIVITY LEVELS.—Although the ability of a mouse to travel very likely is not the limiting factor on size of home range in the field, basic activity patterns might be influential. An ingenious experimental verification of this view is provided in the studies of *P. crinitus stephensi* (Brant and Kavanau, 1965). Mice introduced to mazes where food and water were supplied in abundance still undertook extensive exploratory travels. When there was opportunity for alternative activity in an exercise wheel, movements and exploration of the maze were greatly reduced. This behavior was interpreted to be an expression of inherent tendencies to explore and to develop wide-ranging adapted locomotor activity, including the establishment of home ranges. It was also interpreted as an indication of a dynamic balance between exploration and general activity that had parallels in field behavior.

Behavior Within the Home Range

Behavior within the home range is here taken to include travel patterns, movements in relation to home sites and refuges, territory, and stability of home range.

Evidence is derived from tracking in snow and sand, from observations of released individuals, from records of captures in nest boxes and traps, and from observations of captive mice.

TRAVEL PATTERNS.—When the ground was covered with snow, *P. leucopus* living near the shore of a Minnesota lake restricted their movements largely to well-defined trails and travel areas (Beer, 1961). The routes were followed again and again, with slight deviations, until some travel lanes were as much as two feet wide. The paths followed along or under logs, brush piles, or other protective cover, centering about three principal areas—a group of brush piles, an oak-hazelbush thicket, and an assemblage of hollow trees and logs. The mice followed lanes bordered by brush instead of going straight across the open field or meadow, and took the shortest route between points of refuge when it was necessary to venture into the open.

Snow trails of the same species in Kansas followed a similar pattern, producing well-worn trails between refuge sites (Fitch, 1958). One mouse made many trips 100 feet up and down a slope, tending to follow the same route but deviating from it for part of the distance so that there was one main trail with several parallel to it frequently merging and separating. There also appeared to be a considerable amount of aimless wandering and meandering. One set of tracks was followed a distance of more than 100 feet from the starting place, but the actual travel distance was several times this great. Snow tracking of *P. l. noveboracensis* in Ohio (Ruffer, 1961) and Wisconsin (Thomsen, 1945) provided similar observations.

P. m. bairdi traveled at night out from their winter nest boxes in a Michigan field, usually traversing different parts of their ranges on successive nights. Return trips followed almost exactly the same routes as those used going out (Howard, 1949).

Tracks made in sand by *P. p. leucocephalus* on Santa Rosa Island, Florida, were less restricted to season, yet the pattern was similar (Blair, 1951). Beaten trails were prevalent, with an admixture of random and exploratory travels. Tracks leading from a particular hole usually included one or more beaten trails to places where food was abundant; places where fruits of sea oats (*Uniola laxa*) were blown down and drifted into windrows were frequently the termini of much-traveled mouse trails. In some areas, tracks indicated that one or more mice had run in circles. In other areas, the trail of a single mouse indicated movement of the animal in circles or in the form of a figure eight. Since all trails of this type were in areas barren of food or cover, it is unlikely that they had anything to do with feeding or food gathering. The movements of the mice also were influenced to some extent by the location of the traps, which sometimes were left open and unset in portions of the plot. At such times, beaten trails indicated that the mice were visiting certain of the traps and storing the sunflower seeds used as bait. Some of the mice learned to follow the trap-line. In one case a trail, presumably made by a resident female, indicated that she moved 110 feet to a trap in the nearest line and then followed this line for a distance of 560 feet, visiting each unset trap.

Tevis (1956) reported a similar response to traps by *P. maniculatus* and *P. truei* in California in a study involving poisoned baits. Even when the mice had learned not to touch the bait, they still visited the traps in the same way as they would visit nooks or crannies in other parts of the home range in their search for natural food.

HOME SITES AND REFUGES.—*Peromyscus* generally use and maintain several different home sites and refuges in various parts of their home ranges, frequently shifting about so that principal use may occur around different sites at different times. A female may move her entire litter to a new location.

Home ranges of *P. p. leucocephalus* on beach dunes each contained an average of 20 mouse holes, of which an average of five were entered one or more times by a mouse upon successive releases from traps (Blair, 1951). Travels about the area resulted in maintenance of these sites, for holes were soon cleaned out and re-opened after a sand blow had drifted them shut. These holes were more than mere refuge places; the mouse would use one hole for food storage and nesting for a time then utilize another within its home range. The patterns of travel were well shown in relation to these refuges. The mouse did not leave the nest and wander about throughout the night, but instead made numerous trips away from and back to the hole. Trails usually radiated out from the hole like spokes of a wheel, of unequal length. The number of trails leading from a hole varied from one to nine and averaged 4.3. The number of trips exceeded the number of trails: for example, one trail to a single food plant might show 10 or more round trips in the course of the night.

P. m. bairdi in Michigan had numerous refuges fairly evenly distributed over the home range (Blair, 1940). Mice of this same race held captive in field enclosures of 0.5 acre behaved similarly, occupying an area with several nest boxes, among which they continued to move (Terman, 1961).

Homes of *P. m. gracilis* were assumed to be distributed, within the home range, from periphery to center (Blair, 1942).

P. l. noveboracensis released from traps also were followed to different refuges and appeared to use several (Blair, 1940). Studies of these mice in nest boxes showed frequent changes of site, most of them within the normal home range of the individual (Nicholson, 1941). This appeared to be a normal procedure, not the result of disturbance, for one female brought three litters to nest boxes during one breeding season, the young being from 18 to 25 days old.

Individuals of *P. leucopus* maintained home bases or shelters where they were insulated from extremes of weather and relatively safe from predators, and alternately used several such shelters (Fitch, 1958; Jackson, 1953). *P. crinitus stephensi* in artificial mazes showed several features similar to those of wild mice, including nesting at more than one site (Brant and Kavanau, 1965).

STABILITY OF HOME RANGES.—Once established, many individuals remain in the same areas for a long time, some for the duration of their lives. The extent of their travels in different directions and the intensity of use of different portions of the home range will vary in response to changes in habitat or loss of neighbors, and sometimes for reasons not readily apparent. Departures from the established home area will be discussed later.

In a Maryland woodlot, 79 per cent of the *P. l. noveboracensis* had relatively stable home ranges; they remained in the same general areas from month to month (Stickel and Warbach, 1960). Stability of home range for a large proportion of the members of a population also was reported for this species by Burt (1940).

P. maniculatus in California (Storer *et al.,* 1944) and Texas (Blair, 1958) generally retained fairly stable home ranges from month to month. *P. m. bairdi* in Michigan (Blair, 1940; Dice and Howard, 1951; Howard, 1949) followed the same pattern. In New Mexico, *P. m. blandus* also remained within their home ranges during three months of trapping, but *P. l. tornillo* stayed shorter lengths of time and appeared to have less permanent ranges (Blair, 1943).

P. b. attwateri, in cedar glades in the Missouri Ozarks, showed no tendency to shift areas of use in succeeding trapping periods (Brown, 1964*b*). Of 58 *P. polionotous* caught in more than one season, 55 showed no evidence of shifting home ranges between trapping periods (Davenport, 1964). Blair (1951) found that *P. p. leucocephalus* tended to establish ranges which were retained for the duration of their lives. Sixteen of 18 established residents showed no shifts in location from January to May, and the other two shifted only slightly.

During periods of flooding, *P. l. noveboracensis* and *P. gossypinus* retained their home ranges (Stickel, 1948*a*; McCarley, 1959*b*; Ruffer, 1961). In a study of *P. g. gossypinus,* Pearson (1953) showed that certain mice shifted their home ranges as water levels fluctuated and as other mice left the area. When a mouse that occupied a northern portion of the plot died the ranges of three others were altered. One moved 200 feet northward to occupy a southern portion of the vacated area and still another moved in to fill the northern, vacated portion. A third mouse, that occupied a range that was partially flooded about the same time, shifted its range to that formerly occupied by the mouse that had moved northward. A pair of *P. p. leucocephalus* also shifted their range in response to flood (Blair, 1951). The principal home of this pair was an area of vegetated flood basin and under a large raft that had drifted onto the island during some past storm. A storm filled the flood basins of the island with rainwater and with wind-driven seawater. The principal home of the mice and a considerable portion of their home range were flooded. The mice shifted to a nearby area of scrub dune which they had previously avoided and continued to live in the new area even after the water had subsided.

Some *P. l. noveboracensis* gradually shifted their home ranges across a Maryland woodlot, and several changed their intensity of use from one portion of the area to another (Stickel and Warbach, 1960).

TERRITORY.—Territorial behavior is difficult to demonstrate for a secretive small mammal. However, various authors have obtained both direct and indirect evidence of the existence of territoriality in some degree among certain species of *Peromyscus.* Evidence for territoriality was presented by Burt (1940) for *P. l. noveboracensis* in southern Michigan. He stated: "I have reason to believe that during the breeding season old females are antagonistic toward one another. During this time they maintain definite territories which apparently they protect from others of their sex. Adult females were not taken together in the same trap except once in December when they were not breeding. One observation of an old female chasing a young female was made. It was evident that the old female was attempting to drive the young one out of the home territory. Old males apparently are more tolerant. Mice are more likely to be taken in groups during autumn and winter and singly during spring and summer" (p. 24).

P. l. noveboracensis, studied later by use of nest boxes in the same area, gave further evidence of mutual avoidance (Nicholson, 1941). In two instances, two nest boxes were located in the same tree within a few feet of each other, yet only one box was occupied at a time. Similarly, only one box was used at a time in two instances when a tree box was supplemented by a ground box at the base of the same tree. Mice usually were found living singly during the breeding season. During the winter, however, males and females were taken together more frequently, and occasionally there were assemblages in which the sexes were unequally represented.

Ranges of female *P. leucopus* did not overlap in the population studied by Redman and Sealander (1958) in Arkansas. In contrast, Fitch (1958) reported that territoriality seemed to be only weakly developed among the *P. leucopus* in his study area in Kansas. Several or many mice inhabited the same area and on one occasion a trap set beside a deserted wood rat house caught five adult mice simultaneously. However, he did not give ages, sex, or date of capture.

A demonstration of avoidance without conflict appeared in a study of *P. l. noveboracensis* confined outdoors in an area 30 feet square in a woodland in Pennsylvania (Orr, 1959). Within a few hours, introduced animals defended the areas near the nest boxes and attacked intruders. Reactions toward caged mice placed in parts of the area were similar to those toward free mice, in the sense that activity increased and detours were made around the caged mice.

P. l. noveboracensis introduced on an island in eastern Ontario had overlapping ranges when the population was high (Bendell, 1959). Homes and home ranges were distributed fairly uniformly over the island despite an uneven distribution of habitat types, suggesting intraspecific strife for space.

A hierarchial or dominance relationship may sometimes be superimposed on a system of home ranges and territories, as suggested from trap observations of a male *P. l. noveboracensis* residing in a woodlot for more than three years (Stickel and Warbach, 1960). In this circumstance, a dominant individual freely traversed the home ranges or territories of others, which occupied partly exclusive areas. This relationship has been more conclusively demonstrated in other groups of animals. Davis (1958) suggested that territorial behavior is a part of a behavioral continuum in which interrelationships may be expressed as hierarchies, territories, and home ranges, with intergradations.

P. maniculatus sometimes evidence considerable tolerance of each other as shown by overlapping ranges and association in traps and nest boxes. In other studies, their mutual avoidance resembles that reported for *P. l. noveboracensis.* In a California population of *P. maniculatus,* about 90 per cent of the resident adults of the same sex had home ranges that met but overlapped only slightly (Storer *et al.,* 1944). Boundaries were apparently rather elastic, varying from month to month. Exclusive possession of a home range evidently could be maintained only by constant activity on the part of the inhabitant. Invasion seemed to occur chiefly when the owner happened to be in another part of his range.

Terman (1961) found that *P. m. bairdi* were spatially distributed in a pattern suggesting mutual exclusion by the individuals of the population. Furthermore, during the breeding season the mice were usually found alone or in bisexual pairs and were rarely found in other combinations. In homing studies of confined *P. m. bairdi,* the mice selected empty nest boxes in their own ranges in preference to those into which other mice had been introduced during their absence (Terman, 1962).

Howard (1949) stated that *P. m. bairdi* apparently exhibit a negative response to crowding beyond a certain density within a limited area—a characteristic common to many kinds of animals. Winter tracking in snow showed that usually no two members of the same aggregation of mice traveled over the same grounds during the same night. Additional mice did not occupy the extra nest boxes which were placed close to one another. This avoidance of crowding did not appear to be the result of antagonistic behavior between individuals, for different numbers of mice of both sexes occasionally occurred together in the same nest boxes during the breeding season.

Blair (1940) reported extensive overlapping of ranges of breeding *P. m. bairdi,* both of females and of males. He also noted aggregations of more than one breeding animal of each sex among *P. maniculatus* in Texas (Blair, 1958). In contrast, northern

Michigan *P. m. gracilis* showed little or no overlap of ranges of adults of the same sex, and there were no multiple catches of adult females (Manville, 1949). "How these territories are defended, if indeed they are, for mice may instinctively recognize the property rights of others, I cannot state" (p. 41). Blair (1942) reported a great overlap of ranges among *P. m. gracilis* and no exclusive home range occupancy; however, adult females were never trapped together. Manville (1949) believed the difference in the two studies may have been at least partially the result of different methods of mapping the ranges.

Extensively overlapping ranges have been reported for *P. m. abietorum* (Morris, 1955), and *P. m. blandus* (Blair, 1943). Among *P. gossypinus,* the ranges of males overlapped but those of females did not (Pearson, 1953). Only slight overlap of ranges showed in maps of ranges of *P. boylei* (Drake, 1958). In another study of *boylei,* there appeared to be a spacing out of the individuals although the separation of ranges was not as distinct as among *P. maniculatus* in the same area (Storer *et al.,* 1944).

Among *P. polionotus,* mice of the same sex were distributed at random and distances between them tended to vary inversely with population density, however, ranges frequently overlapped (Davenport, 1964). Overt aggression in *P. p. leucocephalus* in the field was observed by Blair (1951). On three occasions, immature or young sexually mature females were chased from the holes they entered after release from traps; they were evicted by resident females which were subsequently released. On one occasion a resident male went into a hole and nothing happened to indicate that the hole was already occupied. A resident female that associated regularly with this male went there a few minutes later and evicted a transient immature female. On another occasion a transient sexually mature male was released and went to a hole into which a resident male had gone a few minutes earlier. The resident male drove the transient male out of the hole. Nevertheless, both males and females had overlapping ranges.

Caged females (*P. m. gambeli* and *P. truei*) suckling litters showed a defensive behavior at the nest; but not *P. californicus,* which were tolerant. Similar conditions prevailed in the field (McCabe and Blanchard, 1950).

In field studies of *P. maniculatus* and *P. truei,* Evans and Holdenreid (1943) found evidence of fighting in only one of ten instances when both species of any sex were caught together. In laboratory studies, Eisenberg (1963) found that both *Peromyscus crinitus* and *P. californicus* showed a pronounced nest site attachment and defense. Aggressive behavior aside from the nest defense was confined almost entirely to the males. Balph and Stokes (1960) reported that females of *P. m. rufinus* in captivity defended their nest boxes.

Travels Outside the Home Range

Dispersal from the birth site represents a common type of travel for the young. Adults also leave their home ranges, sometimes to establish new ones, but sometimes apparently to make explorations.

When mice are removed from an area, they are replaced by near neighbors, or by wanderers. Most trapping studies reveal a certain proportion of non-resident individuals, which are caught once or twice as they move across the trapped plot. It is not known whether there is indeed a wandering surplus population or whether the apparent wanderers are merely exploring out from their own homes. Dispersal from home ranges, invasion of depopulated areas, and exploratory trips are expressions of such similar phenomena that they cannot well be separated. They are used as categories in the discussion that follows primarily for convenience in arranging examples.

DISPERSAL OF JUVENILES.—For some time after birth, young mice occupy the same home ranges as their parents, but as maturity approaches they leave their nest sites and in time establish home ranges of their own. This behavior is only indirectly indicated by extensive travels of young mice. However, certain studies utilizing nest boxes have made it possible to determine the location of birth and subsequently to measure dispersal from that site.

Dispersal distances found in several of these studies are shown in Table 2. In all these studies, males dispersed farther than females. Of the many offspring of *P. m. bairdi,* whose dispersal behavior was recorded in the study by Howard (1949), only one young mouse was known to have returned. This one, a male, moved 2050 feet

TABLE 2

DISPERSAL OF YOUNG *Peromyscus* FROM NEST SITES

Distance (feet)	*P. m. bairdi* (Howard 1949) Number	*P. m. bairdi* (Dice and Howard 1951) Number*	*P. maniculatus* (Blair 1958) Number*	*P. l. noveboracensis* (Nicholson 1941) Number	*P. leucopus* (Blair and Kennerly 1959) Number
0–500	119	88	110	1	19
550–1000	22	27	1	3	6
1050–1500	6	4	5	0	0
1550–2000	2	8	4	0	0
2050–2500	3	3	1	0	0
2550–3000	1	2	1	0	0
3050–3500	2	0	1	0	0
3550–4000	0	3	0	0	0
Total	155	135	123	4	25

* These numbers are approximate, since they were read from published graphs in which data were grouped and presented differently from the tabulation here.

away from his birth site and then returned to breed in the same nest box where he was born.

P. l. noveboracensis dispersed by swimming from island to island in Lake Opinicon, Ontario (Sheppe, 1965b). Ten mice from natural populations crossed to or from Cow Island, swimming distances of 140–410 feet. All of these mice were young of the year; six were males and four were females.

DISPERSAL OF ADULTS.—Thirty adult *P. m. bairdi* shifted the locations of their home ranges on the areas studied by Blair (1940). Twelve males moved an average of 1168 feet, or a distance equal to about seven times the average width of the home range. Eighteen females moved an average of 758 feet, or about five times the average width of the female home range. The greatest distance that an adult male moved was equal to about 22.9 times the average width of the male home range. Most females moved less than 10 times the width of the average female home range. One female and one male adult *P. l. tornillo* moved from one plot to another about one-half mile away (Blair, 1943). One adult male *P. gossypinus* that was marked in burned thicket was retaken seven days later 2800 feet away (Pournelle, 1950). If he had traveled by the shortest route he would have had to cross two strips of unburned thicket and an extensive area of swamp in order to arrive at the place of second capture. Among *P. m. gambeli* in California, a large number of the movements beyond 400 feet were made by adult animals (Brant, 1962).

Blair (1940) followed the dispersal of *P. m. bairdi* from a 2.9-acre area into which 45 mice (22 adult males, 19 adult females, and 4 immature females) were introduced. Most of them came from a plot about two-thirds of a mile north. Only a few of the released mice remained, and none returned to the original plot. Ten mice (three males and seven females) were recovered at distances from 825 to 1815 feet from the release point. The average distance was 1228 feet, or almost one-fourth mile.

Yeatman (1960) reported that *P. l. leucopus* shifted their home ranges considerably. For example, one male was captured five times on certain trap lines during July, but was absent until January, when it reappeared and was captured five times on a different line, indicating a considerable shift of range. Two other changes, or suspected changes, of home range occurred, one by a male and one by a female.

One adult female *P. maniculatus* caught four times within a 75-foot radius in February and March was caught in May at a station 1000 feet from her original range (Fitch, 1958). This was by far the longest movement recorded. Several other instances of shorter movements were recorded at stations that seemed to be outside the main home range of the animal involved.

The flexibility of movement patterns has been shown by changes that occurred when the environment was altered. Certain minor adjustments of this type were described above in relation to stability of range. Some more pronounced changes will be described here.

P. boylei in California showed seasonal migrations, spring and fall, apparently moving into certain areas after the snow had melted and again before winter (Storer *et al.*, 1944).

P. m. rubidus were trapped before and after a timber burn that followed lumbering (Gashwiler, 1959). While the fire was burning, ten of 16 *Peromyscus* originally marked on the experimental grid were recaptured in the adjacent timber and eight of these were subsequently caught on the original grid after the fire had died down. Fourteen *Peromyscus* traveled from 601 to 1700 feet between the experimental grid and the timbered lines. The travels of one adult were of special interest. He was caught each month from

July to November in the same general vicinity on both the experimental grid and the trap lines outside. His maximum travel distances were 1266, 1254, 1204, 1154, and 1410 feet.

In a Wisconsin study of *P. m. bairdi,* unusual population movements appeared to have been produced by farming operations (LoBue and Darnell, 1959). Twenty-seven mice were taken before alfalfa was cut and 32 afterwards. Twenty-three of 32 post-disturbance individuals were new; thus 75 to 80 per cent were taken for the first time in the post-disturbance period. The preponderance of new individuals indicates unusual population movements. These new individuals may have been forced to move from other portions of the alfalfa field or they may have come from outside areas. The post-disturbance sex ratio in favor of females was the reverse of that obtained in the pre-disturbance period, and indicates abnormal population movements or differential sex mortality as a result of farming activities.

INVASION OF DEPOPULATED AREAS.—Forty-seven *P. m. bairdi* were removed from a 4.46-acre plot in six days in a Michigan bluegrass area (Blair, 1940). The plot was left undisturbed for a week, then trapped again for a week. Invasion was quite rapid and most of it had occurred during the two weeks after the original population had been depleted. In the following two weeks very little change took place. The population after the invasion of the depleted area was only about one-half as great as originally.

In a longer removal study of *P. m. bairdi,* invasion rates followed the seasonal trends of reproduction and population density (Linduska, 1950). Mice were trapped and removed for four days of each week from August 25 to March 31. After the removal of the resident population in August, the numbers of invading animals remained fairly constant for a period of about one month. There was evidence of an increased rate of drift during early October and peak numbers were taken during the first two weeks in November. It is likely that this period coincided with the annual high in population density. The numbers of deer mice caught after November declined gradually and none was taken during four trapping periods in January and February. Two periods of trapping in March showed that invasion of the plots was again taking place. Probably movements during this time were associated with the beginning of sexual activity. The resident population at the time trapping began in August consisted largely of adult animals. The relative numbers of adults taken from the plot declined rapidly and after early fall the catch involved mostly young animals.

Similar seasonal influences on invasion rates were found for *P. maniculatus* and *P. truei* in California (Tevis, 1956). A 15-acre area trapped 261 days from September 1 to May 19 was invaded by at least 730 mice. In the fall, until the first big influx of new animals in mid-November, the rate averaged less than one per night and in the spring it averaged more than four.

In a population of *P. gossypinus* whose reproductive rate was decreased by x-irradiation, invasion by outside animals was sufficient to maintain the overall size of the population (McCarley, 1959a).

In one study, *P. l. noveboracensis* were continuously removed from a one-acre plot for 35 consecutive nights (Stickel, 1946). The mice that first invaded the depopulated area were the nearest neighbors, and these were in turn supplanted by those farther away. Yet not all of the population took part in this movement; many apparently remained in their original areas.

The extent of invasion has been taken to be a measure of the portion of a population that is transient. This was expressed by Townsend (1935) as a wandering index. A variation in the resident/transient ratio has been shown in other studies, notably by Storer *et al.* (1944), who found that the distinction between resident and non-resident groups was more definite in *P. boylei* than in *P. maniculatus*. The non-resident population of *P. boylei* equaled or exceeded the resident population during much of the trapping period. In the first half of September the non-residents made up 70 per cent of the total population. Among *P. maniculatus,* however, the non-resident group never exceeded more than 42 per cent of the total population.

Blair (1943) found that *P. l. tornillo* stayed shorter lengths of time or appeared to have less permanent ranges than *P. m. blandus* in the same area. Additional examples of capture trends in areas subjected to long periods of removal trapping are given by Calhoun (1963).

Resident mice occasionally make short journeys away from their established home ranges and later return. This sort of movement was common among male *P. maniculatus* during the breeding season (Storer *et al.,* 1944). *P. m. blandus* also have been reported to make similar trips (Blair, 1943). For example, one female made two round trips between the two study plots, traveling a distance of about 2000 feet each time she moved from one plot to the other. One-way trips were made by an adult female who moved 2300 feet and another 1200 feet. An adult male made a two-way trip 1200 to 1300 feet each way. Such data indicate that journeys of more than one-third mile are not uncommon.

HOMING.—Homing records for *Peromyscus* provide evidence that home ranges are not strictly the result of mechanical forces, such as those producing the spatially limited distribution of dust motes in a Brownian movement. Displaced mice return home from various distances, but fewer return from greater distances than from nearby, and speed of return increases with successive trials. The consensus from present evidence is that homing is made possible by a combination of random wandering and familiarity with a larger area than the day-to-day home range. Acquisition of this familiarity seems achievable as a result of juvenile wanderings during the dispersal phase and of adult explorations. Recorded distances for these activities very nearly encompass the distances over which any substantial amount of successful homing occurs.

Murie and Murie (1931, 1932) studied the homing behavior of *P. m. artemisiae* in an area where the normal crusing range was about 100 yards. One of 23 mice homed a distance of two miles, six of 49 homed more than a mile, one of eight homed 0.7 mile and 29 of 34 homed 100–1100 yards. In another study of homing in *P. maniculatus* in Wyoming, releases were made over a period of several years (Murie, 1963). Mice were released 200–2500 yards from their capture sites. None of 14 released at 2500 yards returned. The proportion homing decreased with increased distance. Of those released at various distances, 75 per cent (3) homed 200 yards, 28 per cent (7) homed 550 yards, 19 per cent (22) homed 800 yards, 20 per cent (6) homed 900 yards, 9 per cent (8) homed 1400–1500 yards, and 7.8 per cent (3) homed 1600 yards. Some mice returned more rapidly after the second and subsequent releases.

Homing of *P. l. noveboracensis* was studied by Burt (1940) in Michigan. The average cruising range of *Peromyscus* in the area studied was 53 yards. The first trial was made on a 900×240-foot strip of woodland. Each day for four days the captured mice were released at the center of the plot. Thirty-seven mice were transferred and 28 (76 per cent) of them traveled 10 to 155 yards to return home. A second trial was made on a tongue of woodlands surrounded by fields on three sides. Mice were transferred greater distances, 160 to 365 yards. Each day mice from the two halves of the experimental plot were removed to the extreme opposite ends. To return home, some mice went from east to west and others from west to east. Fifty-one mice were transferred and 17 (33 per cent) returned to their homes.

P. l. *noveboracensis* also was studied in a Maryland bottomland forest (Stickel, 1949). Mice were transferred 500 to 700 feet from the periphery to the center of a 22.5-acre area. The observed range length of *Peromyscus* in this area was about 100 feet (females) and 150 feet (males). Of 27 males transferred, 22 (82 per cent) homed, two (7 per cent) possibly homed, one (4 per cent) failed to home, and two (7 per cent) were not recaptured. Of the 21 females transferred, 13 (62 per cent) homed, one (5 per cent) possibly homed, one (5 per cent) failed to home, and six (28 per cent) were not recaptured. Other mice were released at the site of capture. Of 20 males, 15 (75 per cent) would have been considered to have homed, three (15 per cent) would have been considered possibly to have homed, and two (10 per cent) were not retaken. Of the 19 females released where captured, 18 (95 per cent) would have been considered to have homed, one (5 per cent) would have been considered possibly to have homed, and none failed to be retaken.

P. l. *noveboracensis* in New York homed repeatedly from distances of a quarter of a mile (Hamilton, 1937). Sheppe (1965*b*) reported that a P. l. *noveboracensis* swam 765 feet to return to its home island.

Griffo (1961) made extensive studies of homing of P. *gossypinus* in Florida. Mice released in unnatural habitat, a golf course, adjacent to their home area homed less successfully than did those released in natural habitats. In natural areas, males appeared to home more successfully than did females, but this did not appear to be true in the unnatural areas. Mice which had homed a number of times and were then held in captivity for periods of more than 12 weeks before being liberated at former release sites showed no decline in their homing ability. Animals released in their former home ranges after extended periods of laboratory isolation remained and maintained their ranges. Animals not given previous homing experience returned to and maintained their former home ranges when released at various distances from their capture sites after prolonged laboratory isolation. Homing success and speed appeared to improve with repetitive liberations even though these were not made at previous release points. Mice released on the golf course not only homed less successfuly but required greater lengths of time than those released from similar distances in natural areas. Initial orientation in unfamiliar territory could not be correlated with the direct route home and apparently was not influenced by the distance of the home site from the release area.

Broadbooks (1961) released *P. maniculatus* away from their home areas in Oregon sagebrush. Four males and one female were released one-fourth mile from home and two males returned. Four mice were released one-half mile from home and none returned. Home ranges in this area averaged 583 feet, ranging from 342 to 761 feet.

In New Hampshire, an adult male *P. m. gracilis* returned home from a release point 200 meters southwest of its nest and from a second release point 300 meters north of its nest (Rawson and Hartline, 1964). A female transported approximately twice as far did not return. Two *P. m. gracilis* returned home approximately 800 feet within a day (Kendeigh, 1944). The mice had to find their way out of a barn, across a medium-sized lawn, across a stream, and through a dense thicket of shrubs and trees back to the woods. The prompt return of the mice over what would seem to be a difficult route indicates a good knowledge of the terrain over an area probably wider than their usual home range. One *P. m. bairdi* in Illinois returned 20 yards and 150 yards on successive releases (Johnson, 1926). Another returned 200 yards and a third failed to return from 100 yards and was recaptured at the point of release several times.

In South Carolina, *P. polionotus* were released from the center of a nine-acre plowed field 340–640 feet from trap sites in occupied habitats surrounding the field (Gentry, 1964). Of the 26 males released, 23 (81 per cent) returned to their home sites. One became resident in a new area and four were never recaptured. Of the 13 females released, ten (77 per cent) returned to their home sites, two set up new home ranges, and one was never recaptured. The animals were captured the night before the experiment and were kept in traps until the following morning. They were released at 8:00 A. M. and later. Most of them returned home between 9:00 P. M. and midnight of the day they were released.

The Home Range Concept, and Methods Of Measuring Home Range

The concept of home range was expressed by Seton (1909) in the term "home region" with the explanation that, "No wild animal roams at random over the country: each has a home region even if it has not an actual home." Heape (1931) elaborated the concept, but termed it "nomadism." Burt (1940, 1943) clarified the definition, applying it widely to animal species and demonstrating the nature of the home range among *Peromyscus* in a Michigan oak-hickory woods.

The concept is fundamental in understanding the behavior of *Peromyscus* or other wild mammals. The outposts of travel of *Peromyscus* in its home range are ever-changing, in response to both external and internal forces, as pointed out by Burt (1943) and affirmed by many investigators thereafter. The patterns of travel have been shown by tracking in snow and sand, by use of dyes (New, 1958), by tracking studies on smoked paper in artificial shelters (Sheppe, 1965a) and by following released animals to holes and refuges. Telemetry was used with *Peromyscus* in a brief homing study (Rawson and Hartline, 1964) and may prove adaptable to the study of travel patterns within the range.

However, quantitative comparisons of sizes of home ranges of mice of different species, age, or sex, and of mice living in different areas or habitats have been possible only through trapping. The data thus assembled are subject to many biases and possible biases. For example: (1) The traps could interfere with the travels of the mice, incarcerating them near their homes, and so preventing them from canvassing their home ranges. The trap-revealed ranges would be smaller than the true ranges. At the same time, with neighbors safely enclosed, other mice could wander freely and widely over areas outside their normal domains, and their trap-revealed ranges would be larger than their true ranges. (2) Mice could adjust both the sizes and positions of their ranges in response to the shelter and food provided by traps. (3) Some mice may be more prone to enter traps than others; if these also are mice with larger or smaller ranges than the average, trap-revealed ranges will be biased. It has generally been assumed that these biases are not important in relation to the gross nature of the measurements obtained by trapping.

Two general procedures have been employed for expressing home-range data obtained by trapping. In the first of these, data are assembled separately for each individual, and population data are derived by determining averages and variability of the individual measurements. Results may be expressed as linear units or as areas. These methods include the Observed Range Length and Adjusted Range Length (or their translation to area by using these distances as diameters), the Minimum Area Method, and the Boundary Strip Method (Inclusive and Exclusive). These terms are more fully described by Stickel (1954).

In the second general procedure, data are assembled for the population, without separate computations for individuals. Results may be expressed as linear units, with averages and variability measurements (or translated to area, using the average distance as a radius), or as probabilities of records at different distances. These measurements include Distances from Initial Capture, and Distances between Successive Captures. They also include Distances from the Center of Activity (a geometrical center) as suggested by Hayne (1949) and developed by Dice and Clark (1953). A procedure also has been suggested for approximating sizes of home ranges by comparing trap-success trends at different spacings (Stickel, 1965). These approximations can be made in relatively short periods and do not depend on tabulation of individual records.

In data assembled on an individual basis, the trap-revealed size of the home range increases to a point as the number of captures increases, and hence a minimum number of captures ordinarily is required before the records of an individual mouse are included. Rarely, however, are sufficient captures available that no further increase would be expected. Blair (1942) found that the trap-revealed size of the home ranges of adult males of *P. m. gracilis* increased little, if any, beyond 19 captures. Using artificial populations, with home ranges of specified area and shape, Stickel (1954) showed that the number of captures required to indicate average size of home range depended on the trap spacing and that this number was substantially greater than the number achievable in most trapping studies. Requirement of a minimum number of captures reduces the magnitude of the bias associated with number of captures, however, for the trap-revealed size increases more slowly after the first few captures. The long travels of transient individuals also are largely excluded by requirement of a minimum number of captures per individual. Time also will affect the trap-revealed size, for as time passes more individuals will shift their home ranges or areas of principal use and others will be captured during their exploratory trips. The methods of assembling data by individual records are not based on an assumption of fixed boundaries nor of any particular pattern of home-range use in relation to single centers or nest sites. They do assume an area which can be grossly defined in the large units imposed by trap spacing.

Assembling data for all individuals in a group provides records for a much larger proportion of the population than is possible when limitations are set on the number of captures per individual. A bias is introduced by the fact that different numbers of captures are available for different mice and hence all individuals are not equally represented. Time for which data are assembled affects the proportion of long movements produced by transients and by other mice shifting their home ranges or making trips of exploration. If records are removed for travels that obviously are exploratory or that indicate transient behavior, results lose their objectivity in terms of the population, although they probably are closer to expressing the real sizes of home ranges of residents.

Home ranges calculated in terms of probabilities of captures at different distances from the center of activity are based on the assumption that there are no home range limits, but instead a declining frequency of use away from a geometric center. The validity of this tenet is self evident if trapping continues for a long period, because some individuals shift their ranges and areas of use, but some will make exploratory trips, and transients will add their records. However, even if home ranges had finite boundaries and traps were visited strictly at random, the same decreasing probabilities of capture with distance will appear when the records of different individuals are combined (Stickel, 1954). Odum and Kuenzler (1955) assembled data from observations on birds in this manner, using records from individuals separately. They reached the conclusion that the method indicated relatively well-defined territorial limits. Individual data on mice obtained by trapping have not been examined in this way. Tester and Siniff (1965), however, have assembled telemetry data for a raccoon in terms of probabilities and have shown the shifts of home-range use that occurred with time.

Trapping of small mammals, despite its limitations and diverse methodology, has permitted the development of basic concepts of behavior of these animals in the field. These have been supported and extended by other methods of observation and by experimental studies in the laboratory.

Summary

The concept of home range was expressed by Seton (1909) in the term "home region," which Burt (1940, 1943) clarified with a definition of home range and exemplified in a definitive study of *Peromyscus* in the field. Burt pointed out the ever-changing characteristics of home-range area and the consequent absence of boundaries in the usual sense—a finding verified by investigators thereafter.

In the studies summarized in this paper, sizes of home ranges of *Peromyscus* varied within two magnitudes, approximately from 0.1 acre to ten acres, in 34 studies conducted in a variety of habitats from the seaside dunes of Florida to the Alaskan forests.

Variation in sizes of home ranges was correlated with both environmental and physiological factors; with habitat it was conspicuous, both in the same and different regions. Food supply also was related to size of home range, both seasonally and in relation to habitat. Home ranges generally were smallest in winter and largest in spring, at the onset of the breeding season. Activity and size also were affected by changes in weather. Activity was least when temperatures were low and nights were bright. Effects of rainfall were variable. Sizes varied according to sex and age; young mice remained in the parents' range until they approached maturity, when they began to travel more widely. Adult males commonly had larger home ranges than females, although there were a number of exceptions. An inverse relationship between population density and size of home range was shown in several studies and probably is the usual relationship. A basic need for activity and exploration also appeared to influence size of home range.

Behavior within the home range was discussed in terms of travel patterns, travels in relation to home sites and refuges, territory, and stability of size of home range. Travels within the home range consisted of repeated use of well-worn trails to sites of food, shelter, and refuge, plus more random exploratory travels. *Peromyscus* generally used and maintained several or many different home sites and refuges in various parts of their home ranges, and frequently shifted about so that their principal activities centered on different sets of holes at different times.

Once established, many *Peromyscus* remained in the same general area for a long time, perhaps for the duration of their lives. Extent of their travels in different directions and intensity of use of different portions of their home ranges varied within a general area in response to habitat changes, loss of neighbors, or other factors.

Various authors have obtained both direct and indirect evidence of territoriality, in some degree, among certain species of *Peromyscus*.

Young mice dispersed from their birth sites to establish home ranges of their own. Adults also sometimes left their home areas; some re-established elsewhere; others returned after exploratory travels. Most populations contained a certain proportion of transients; these may have been wanderers or individuals exploring out from established home ranges or seeking new ones. When areas were depopulated by removal trapping, other *Peromyscus* invaded. Invasion rates generally followed seasonal trends of reproduction and population density.

Peromyscus removed from their home areas and released elsewhere returned home from various distances, but fewer returned from greater distances than from nearby; speed of return increased with successive trials. The consensus from present evidence is that homing is made possible by a combination of random wandering and familiarity with a larger area than the day-to-day range. Records of juvenal wanderings during the dispersal phase and of adult explorations very nearly encompassed the distances over which any substantial amount of successful homing occurred.

Methods of measuring sizes of home ranges and the limitations of these measurements were discussed in brief synopsis. It was concluded that trapping data, despite their limitations, have been the source of valid biological concepts of the behavior of small mammals in the field.

Literature Cited

ALLRED, D., AND D. E. BECK. 1963. Range of movement and dispersal of some rodents at the Nevada atomic test site. Jour. Mamm., 44:190–200.

BALPH, D. F., AND A. W. STOKES. 1960. Notes on the behavior of deer mice (*Peromyscus maniculatus rufinus*). Proc. Utah Acad. Sci., Arts and Letters, 37:55–62.

BEER, J. R. 1961. Winter home ranges of the red-backed mouse and white-footed mouse. Jour. Mamm., 42:174–180.

BENDELL, J. F. 1959. Food as a control of a population of white-footed mice, *Peromyscus leucopus noveboracensis* (Fischer). Can. Jour. Zool., 37: 173–209.

BLAIR, W. F. 1940. A study of prairie deer-mouse populations in southern Michigan. Amer. Midland Nat., 24:273–305.

——— 1941. The small mammal population of a hardwood forest in northern Michigan. Contr. Lab. Vert. Genet. Univ. Mich., 17:1–10.

——— 1942. Size of home range and notes on the life history of the woodland deer-mouse and eastern chipmunk in northern Michigan. Jour. Mamm., 23:27–36.

——— 1943. Populations of the deer-mouse and associated small mammals in the mesquite association of southern New Mexico. Contrib. Lab. Vert. Biol. Univ. Mich., 21:1–40.

——— 1951. Population structure, social behavior, and environmental relations in a natural population of the beach mouse (*Peromyscus polionotus leucocephalus*). *Ibid.*, 48:1–47.

——— 1958. Effects of x-irradiation on a natural population of the deer mouse (*Peromyscus maniculatus*). Ecology, 39:113–118.

BLAIR, W. F., AND T. E. KENNERLY, JR. 1959. Effects of x-irradiation on a natural population of the wood-mouse (*Peromyscus leucopus*). Texas Jour. Sci., 11:137–149.

BOWDITCH, ADRIENNE J. 1965. Terrestrial mammals of Martha's Vineyard, Massachusetts, with special reference to *Peromyscus*. Smith College, Northampton, Mass., 64 pp.

BRAND, R. H. 1955. Abundance and activity of the wood mouse (*Peromyscus leucopus noveboracensis*) in relation to the character of its habitat. Ph.D. Thesis, Univ. Mich., 168 pp.

BRANT, D. H. 1962. Measures of the movements and population densities of small rodents. Univ. Calif. Publ. Zool., 62:105–184.

BRANT, D. H., AND J. L. KAVANAU. 1965. Exploration and movement patterns of the canyon mouse *Peromyscus crinitus* in an extensive laboratory enclosure. Ecology, 46:452–461.

BROADBOOKS, H. E. 1961. Homing behavior of deer mice and pocket mice. Jour. Mamm., 42:416–417.

BROWN, L. N. 1964a. Dynamics in an ecologically isolated population of the brush mouse. Jour. Mamm., 45:436–442.

———— 1964b. Ecology of three species of *Peromyscus* from southern Missouri. *Ibid.*, 45:189–202.

BURT, W. H. 1940. Territorial behavior and populations of some small mammals in southern Michigan. Misc. Publ. Mus. Zool. Univ. Mich., 45:1–58.

———— 1943. Territoriality and home range concepts as applied to mammals. Jour. Mamm., 24:346–352.

CALHOUN, J. B. 1963. The social use of space. *In*: Physiological Mammalogy. Vol. 1. New York: Academic Press. Pp. 1–187.

DAVENPORT, L. B., JR. 1964. Structure of two *Peromyscus polionotus* populations in old-field ecosystems at the AEC Savannah River plant. Jour. Mamm., 45:95–113.

DAVIS, D. E. 1958. The role of density in aggressive behaviour of house mice. Anim. Behav., 6:207–210.

DICE, L. R., AND P. J. CLARK. 1953. The statistical concept of home range as applied to the recapture radius of the deermouse (*Peromyscus*). Contr. Lab. Vert. Biol. Univ. Mich., 62:1–15.

DICE, L. R., AND W. E. HOWARD. 1951. Distance of dispersal by prairie deer mice from birthplaces to breeding sites. Contr. Lab. Vert. Biol. Univ. Mich., 50:1–15.

DRAKE, J. J. 1958. The brush mouse *Peromyscus boylii* in southern Durango. Mich. State Univ. Publ. Mus., Biol. Ser., 1:99–132.

EISENBERG, J. F. 1963. The intraspecific social behavior of some cricetine rodents of the genus *Peromyscus*. Amer. Midland Nat., 69:240–246.

EVANS, F. C., AND R. HOLDENRIED. 1943. Double captures of small rodents in California. Jour. Mamm., 24:401.

FITCH, H. S. 1958. Home ranges, territories, and seasonal movements of vertebrates of the Natural History Reservation. Univ. Kans. Publ. Mus. Nat. Hist., 11:63–326.

GASHWILER, J. S. 1959. Small mammal study in west-central Oregon. Jour. Mamm., 40:128–139.

GENTRY, J. B. 1964. Homing in the old-field mouse. Jour. Mamm., 45:276–283.

GRIFFO, J. V., JR. 1961. A study of homing in the cotton mouse, *Peromyscus gossypinus*. Amer. Midland Nat., 65:257–289.

HAMILTON, W. J., JR. 1937. Growth and life span of the field mouse. Amer. Nat., 71:500–507.

HAYNE, D. W. 1949. Calculation of size of home range. Jour. Mamm., 30:1–18.

HEAPE, W. 1931. Emigration, migration and nomadism. Cambridge: W. Heffer and Son. Pp. xii + 369.

HIRTH, H. F. 1959. Small mammals in old-field succession. Ecology, 40:417–425.

HOOVEN, E. 1958. Deer mouse and reforestation in the Tillamook Burn. Oreg. Forest Lands Res. Center, Res. Note, 37:1–31.

HOWARD, W. E. 1949. Dispersal, amount of inbreeding and longevity in a local population of prairie deer mice on the George Reserve, southern Michigan. Contr. Lab. Vert. Biol. Univ. Mich., 43:1–50.

HOWELL, J. C. 1954. Populations and home ranges of small mammals on an overgrown field. Jour. Mamm., 35:177–186.

JACKSON, W. B. 1953. Use of nest boxes in wood mouse population studies. Jour. Mamm., 34:505–507.

JOHNSON, BEATRICE W. 1927. Preliminary experimental studies of mice of Mount Desert Island, Maine. Jour. Mamm., 8:276–284.

JOHNSON, M. S. 1926. Activity and distribution of certain wild mice in relation to biotic communities. Jour. Mamm., 7:245–277.

KENDEIGH, S. C. 1944. Homing of *Peromyscus maniculatus gracilis*. Jour. Mamm., 25:405–406.

LINDUSKA, J. P. 1942. Winter rodent populations in field-shocked corn. Jour. Wildl. Mgt., 6:353–363.

———— 1950. Ecology and land-use relationships of small mammals on a Michigan farm. Mich. Dept. Cons., Fed. Aid Project 2-R, ix + 144 pp.

LoBUE, J., AND R. M. DARNELL. 1959. Effect of habitat disturbance on a small mammal population. Jour. Mamm., 40:425–437.

MANVILLE, R. H. 1949. A study of small mammal populations in northern Michigan. Misc. Publ. Mus. Zool. Univ. Mich., 73:1–83.

McCABE, T. T., AND BARBARA D. BLANCHARD. 1950. Three species of *Peromyscus*. Rood Associates: Santa Barbara, Calif., v + 136 pp.

McCARLEY, W. H. 1959a. A study of the dynamics of a population of *Peromyscus gossypinus* and *P. nuttalli* subjected to the effects of x-irradiation. Amer. Midland Nat., 61:447–469.

———— 1959b. The effect of flooding on a marked population of *Peromyscus*. Jour. Mamm., 40:57–63.

McGREGOR, R. C. 1958. Small mammal studies on a southeast Alaska cutover area. Station Paper No. 8. Alaska Forest Res. Center, U. S. Forest Serv., Juneau, Alaska, 9 pp.

MORRIS, R. F. 1955. Population studies on some small forest mammals in eastern Canada. Jour. Mamm., 36:21–35.

MURIE, M. 1963. Homing and orientation of deermice. Jour. Mamm., 44: 338–349.

MURIE, O. J., AND A. MURIE. 1931. Travels of *Peromyscus*. Jour. Mamm., 12: 200–209.

———— 1932. Further notes on travels of *Peromyscus*. *Ibid.*, 13:78–79.

NEW, J. G. 1958. Dyes for studying the movements of small mammals. Jour. Mamm., 39:416–429.

NICHOLSON, A. J. 1941. The homes and social habits of the wood-mouse (*Peromyscus leucopus noveboracensis*) in southern Michigan. Amer. Midland Nat., 25:196–223.

ODUM, E. P., AND E. J. KUENZLER. 1955. Measurement of territory and home range size in birds. Auk, 72:128–137.

ORR, H. D. 1959. Activity of white-footed mice in relation to environment. Jour. Mamm., 40:213–221.

PEARSON, P. G. 1953. A field study of *Peromyscus* populations in Gulf Hammock, Florida. Ecology, 34:199–207.

POURNELLE, G. H. 1950. Mammals of a north Florida swamp. Jour. Mamm., 31:310–319.

RAWSON, K. S., AND P. H. HARTLINE. 1964. Telemetry of homing behavior by the deermouse, *Peromyscus.* Science, 146:1596–1598.

REDMAN, J. P., AND J. A. SEALANDER. 1958. Home ranges of deer mice in southern Arkansas. Jour. Mamm., 39:390–395.

RUFFER, D. G. 1961. Effect of flooding on a population of mice. Jour. Mamm., 42:494–502.

SETON, E. T. 1909. Life histories of northern animals. Charles Scribner's Sons. Vol. 1, Pp. xxx + 673; Vol. 2, Pp. xii + 677–1267.

SHADOWEN, H. E. 1963. A live-trap study of small mammals in Louisiana. Jour. Mamm., 44:103–108.

SHEPPE, W. 1965a. Characteristics and uses of *Peromyscus* tracking data. Ecology, 46:630–634.

————— 1965b. Dispersal by swimming in *Peromyscus leucopus.* Jour. Mamm., 46:336–337.

STEPHENSON, G. K., P. D. GOODRUM, AND R. L. PACKARD. 1963. Small rodents as consumers of pine seed in east Texas uplands. Jour. Forest., 61:523–526.

STICKEL, LUCILLE F. 1946. The source of animals moving into a depopulated area. Jour. Mamm., 27:301–307.

————— 1948a. Observations on the effect of flood on animals. Ecology, 29:505–507.

————— 1948b. The trap line as a measure of small mammal populations. Jour. Wildl. Mgt., 12:153–161.

————— 1949. An experiment on *Peromyscus* homing. Amer. Midland Nat., 41:659–664.

————— 1954. A comparison of certain methods of measuring ranges of small mammals. Jour. Mamm., 35:1–15.

————— 1960. *Peromyscus* ranges at high and low population densities. *Ibid.,* 41:433–441.

————— 1965. A method for approximating range size of small mammals. *Ibid.,* 46:677–679.

STICKEL, LUCILLE F., AND O. WARBACH. 1960. Small-mammal populations of a Maryland woodlot, 1949–1954. Ecology, 41:269–286.

STORER, T. I., F. C. EVANS, AND F. G. PALMER. 1944. Some rodent populations in the Sierra Nevada of California. Ecol. Monogr., 14:165–192.

TERMAN, C. R. 1961. Some dynamics of spatial distribution within semi-natural populations of prairie deermice. Ecology, 42:288–302.

————— 1962. Spatial and homing consequences of the introduction of aliens into semi-natural populations of prairie deer-mice. *Ibid.,* 43:216–223.

TESTER, J. R., AND D. B. SINIFF. 1965. Aspects of animal movement and home range data obtained by telemetry. Trans. N. Amer. Wildl. Conf., 30:379–392.

TEVIS, L., JR. 1956. Behavior of a population of forest mice when subjected to poison. Jour. Mamm., 37:358–370.

THOMSEN, H. P. 1945. The winter habits of the northern white-footed mouse. Jour. Mamm., 26:138–142.

TOWNSEND, M. T. 1935. Studies on some of the small mammals of central New York. Roosevelt Wild Life Ann., 4:1–120.

WHITE, J. E. 1964. An index of the range of activity. Amer. Midland Nat., 71:369–373.

WILL, R. L. 1962. Comparative methods of trapping small mammals in an Illinois woods. Trans. Ill. State Acad. Sci., 55:21–34.

WILLIAMS, O. 1955. Home range of *Peromyscus maniculatus rufinis* in a Colorado ponderosa pine community. Jour. Mamm., 36:42–45.

YEATMAN, H. C. 1960. Population studies of seed-eating mammals. Jour. Tenn. Acad. Sci., 35:32–48.

11

POPULATION DYNAMICS

C. Richard Terman

Introduction

POPULATIONS OF *Peromyscus* have been studied by numerous workers under natural, semi-natural, and laboratory conditions. Although the rationale and techniques of these studies have varied widely, the data obtained show remarkable similarities suggestive of common basic mechanisms influencing the dynamics of populations.

While most of the literature on populations of *Peromyscus* has been reviewed during preparation of this chapter, an exhaustive survey will not be presented here. Rather, the emphasis of this chapter will be to outline existing information exposing gaps in our knowledge. Special attention will be given to factors involved in the regulation of population numbers in nature and in controlled laboratory conditions.

Comparative Population Size and Equilibria

NATURAL POPULATIONS.—Estimating the size of the population is the foremost objective of any census. Unfortunately, in nature this task is extremely difficult and subject to errors because of variability of the physical environment, measurement techniques, and response of the animals being censused.

Table 1 summarizes the studies of natural populations of several species of *Peromyscus*; it includes the species studied, the literature citation, spacing of traps, the location of each study, number of acres involved, the years of the study, the maximum numbers of animals trapped per acre during the Spring (Jan.–June) and Fall (July–Dec.), and the yearly variation in numbers.

The data from different studies, as presented in Table 1, should not be combined because of differences in procedures, traps, length of trapping periods, home range, and density calculation. Comparison among other studies should be made with care for the same reasons.

TABLE 1

SURVEY OF STUDIES OF POPULATIONS OF *Peromyscus* IN NATURE*

Species	Citation	Trap spacing (feet)	Plot	Area (acres)	Years	Spring	Var.	Fall	Var.
P. gossypinus	McCarley (1954)	?	1	20	1951	2.7		1.5	
	Pearson (1953)	25–100	1	9	1949			4.8	2.5
					1950	1.8		2.3	
					Mean				2.5
P. leucopus noveboracensis	Burt (1940)	30 and 60	1 and 2	5.52	1936	5.8	1.5	10.87	
					1937	7.3			
	Howell (1954)	50	1	14.6	1951			0.4	
	Klein (1960)	45	S. W.	6.6	1955			0.9	0.3
				10.6	1956			0.6	
			C. H	6.8	1955			1.0	0.2
				12.0	1956			0.8	
			A. F.	6.0	1955			2.0	1.3
				15.3	1956			3.3	
	Snyder (1956)	60	A	10.0	1949	4.6		5.7	1.8
					1950	1.8	2.8	3.9	
					1951				
			B	12.2	1950			4.8	
					1951	1.7			
	Stickel (1960)	30 and 60	1	17.2	1954	1.9		0.8	
			2	5.3	1957	6.5		7.2	

TABLE 1 (Continued)

Species	Citation	Trap spacing (feet)	Plot	Area (acres)	Years	Mice trapped/acre			
						Spring	Var.	Fall	Var.
P. l. noveboracensis (cont.)	Stickel and Warbach (1960)	60	1	4.4	1949	3.0		3.9	2.9
					1950	5.7	2.7	6.8	0.5
					1951	3.2	2.5	7.3	3.0
					1952	8.9	5.7	4.3	3.0
					1953	2.0	6.9	8.6	4.3
					1954				
					Mean		3.7		1.8
P. leucopus tornillo	Blair (1943)	45	A and B	35.76	1940	0.63			
P. maniculatus	Storer, Evans, and Palmer (1944)	30	B. L.	2.3	1938			10.3	
					1939	17.0			
P. maniculatus bairdi	Blair (1940)	45 and 60	1	7.6	1938	2.0	1.7	1.8	1.6
					1939	0.3		0.2	
			3	9.0	1938	1.7		2.2	
			4	3.8	1938	3.7		9.2	6.3
					1939			2.9	
					Mean		1.7		4.0
P. maniculatus blandus	Blair (1943)	45	A.	35.76	1940	0.63			
P. maniculatus gracilis	Blair (1941)	45	1	18.2	1940			3.8	
	Klein (1960)	45	S. W.	6.6	1955			1.8	
				10.6	1956			2.4	0.6
			C. H.	6.8	1955			0.3	
				12.0	1956			2.5	2.2
			A. F.	6.0	1955			0.0	
				15.3	1956			1.0	1.0

TABLE 1 (Continued)

Species	Citation	Trap spacing (feet)	Plot	Area (acres)	Years	Mice trapped/acre			
						Spring	Var.	Fall	Var.
P. m. gracilis (cont.)	Manville (1949)	30	1	2.1	1940	0.17		0.51	
					1941	0.85			
					1942	0.34			
			2	2.1	1940				
					1941	6.1		6.1	
					1942				
			3	2.1	1940			1.9	
					1941			6.3	4.4
					1942	5.5			
			4	2.1	1940			4.3	
					1941	6.6		7.0	2.7
					1942	11.0	4.4		
			5	2.1	1940			2.2	
					1941			5.0	2.8
					1942	2.2			
			6	2.1	1940			0.9	
					1941	1.9		2.4	1.5
					1942				
			7	2.1	1941			0.5	
					1942	0.2			
			8	2.1	1940	2.4		2.4	
					1941			6.3	3.9
					1942				
					Mean	2.4	4.4	2.4	2.4

TABLE 1 (Continued)

Species	Citation	Trap spacing (feet)	Plot	Area (acres)	Years	Mice trapped/acre			
						Spring	Var.	Fall	Var.
O. nuttalli	Howell (1954)	50	1	14.6	1951	0.5		0.2	
	McCarley (1958)	75	A	13.0	1955	1.7	1.2	0.5	0.4
					1956	2.5	0.8	0.9	0.3
					1957			0.6	
			B	12.9	1955	0.0	0.4	0.0	0.3
					1956	0.4	0.0	0.3	0.3
					1957	0.4		0.0	
	Pearson (1953)	25–100	1	9.0	1949			0.4	
					1950	1.2		1.1	0.7
					Mean		0.6	1.0	0.4
P. polionotus	Blair (1951)	60 and 80	1	65.1	1941	1.1			
					1942				
	Caldwell (1964)	50	1	3.4	1957	10.6	8.2	20.0	8.8
					1958	18.8		11.2	
			2	3.4	1957	6.2	4.1	7.9	5.3
					1958	10.3		2.6	
			3	3.4	1957	8.8	4.1	9.4	6.8
					1958	4.7		2.6	
			4	3.4	1957	9.7		8.5	4.1
					1958			4.4	
			5	3.4	1957	7.4		10.9	
					1958				
			6	3.4	1957	8.2		8.8	5.6
					1958			3.2	

TABLE 1 (Continued)

Species	Citation	Trap spacing (feet)	Plot	Area (acres)	Years	Mice trapped/acre				
						Spring	Var.	Fall	Var.	
P. polionotus (cont.)	Davenport (1964)	40	409	3.6	1953	16.0	6.9	6.3	2.0	
					1954	9.1	3.9	8.3		
					1955	5.2				
			410	3.6	1953	14.3		10.0	10.0	
					1954			0.0		
					Mean		5.4		6.1	

Var. = Annual variation in population numbers.
* All studies listed utilized live trapping techniques.

TABLE 2
DIMENSIONS OF TRAP LINES A AND B OF THE NORTH
AMERICAN CENSUS OF SMALL MAMMALS*

Line	Length of trap line (feet)	Distance between center of trapping stations (feet)	Radius of trapping station (feet)	Distance between the two lines (feet)
A	475	25	2.5	200
B	950	50	5.0	400

* Modified from Calhoun, 1948; 60-trap trap-lines.

The differences in population levels supported by each area, evident from the data, no doubt reflect basic ecological differences of the areas. As might be expected, seasonal variations were evident in most studies with the fall populations being the largest. One of the most outstanding characteristics of these data, however, is the small number of animals recorded per acre. This occurs with great consistency and is suggestive of efficient population controlling mechanisms. Further, while annual variations in population size occurred at the same site during the same season, they were generally not extensive. *P. polionotus* exhibited generally larger populations per acre (Caldwell, 1964; Davenport, 1964; both studies in the same general geographical area) than the other species of *Peromyscus* and the mean yearly variation of 6.1 per acre was the greatest variation of the populations tabulated.

It is clear from the preceding data that in spite of the numerous studies of natural populations of *Peromyscus,* relatively little information is available concerning population growth and equilibrium levels. This is particularly true if one wishes to examine information obtained by identical sampling techniques in the same area for a period longer than a few weeks or months.

The North American Census of Small Mammals (Calhoun, 1948, 1949, 1950, 1951, 1956; Calhoun and Arata, 1957a, 1957b, 1957c, 1957d) is one of the best sources available presently for information of population growth and equilibrium levels for long periods of time in the same locality using the same measuring techniques. During this census, population data were collected by numerous investigators following procedures outlined in "Release No. 2, North American Census of Small Mammals" edited by John B. Calhoun (1948). Generally the suggested trapping periods consisted of three

TABLE 3
PARTIAL TRAPPING RECORDS OF THE NORTH AMERICAN CENSUS OF SMALL MAMMALS

Year / Season	1949 S	1949 F	1950 S	1950 F	1951 S	1951 F	1952 S	1952 F	1953 S	1953 F	1954 S	1954 F	1955 S	1955 F	1956 S	1956 F	Mean S	Mean F
Peromyscus leucopus																		
Iowa 1: BI & II	—	—	—	—	8	27	—	—	—	—	—	—	—	—	—	—		
Maryland 1: BI & II	13	6	7	6	13	7	10	2	10	11	13	96	11	—	—	—	10.5	44.7
" 2: " " "	9	8	8	4	8	2	7	1	2	13	3	0	9	8	5	5	7.8	5.9
Michigan 3: BI & II	5	2	1	23	13	6	0	16	1	5	0	0	8	4	5	3	5.8	3.4
Virginia 3: BI & II	13	3	7	2	23	11	16	1	5	9	2	6	12	—	—	—	4.3	10.3
Peromyscus maniculatus																		
Alberta 1: BI & II	—	—	16	18	48	51	18	33	10	44	11	18	16	35	6	23	14.1	4.3
Brit. Col. 1: BI & II	8	42	10	40	—	—	18	30	3	23	7	42	—	—	—	—	17.9	31.7
" 2: " " "	3	17	13	19	9	38	33	18	1	13	12	29	—	—	—	—	9.2	35.4
Calif. 1: BI & II	—	—	—	12	0	15	—	—	2	10	7	4	0	—	0	0	12.4	19.2
" 3: " " "	—	—	—	3	—	0	—	3	8	22	2	0	0	—	0	0	3.6	12.8
" 4: " " "	—	—	—	—	—	49	0	4	1	2	2	8	4	7	6	—	2	8
New York 3: BI & II	—	—	—	—	—	67	—	3	—	—	8	—	1	—	—	—	2.6	4
" 6: " " "	—	—	—	—	—	20	—	3	—	2	—	—	—	—	—	—	4.5	26.5
" 9: " " "	—	—	—	—	—	17	—	3	3	—	8	—	—	—	—	—		35
" 10: " " "	—	—	—	—	—	16	—	1	—	2	—	—	—	2	1	1	5.5	6.5
" 11: " " "	—	—	—	—	—	10	—	0	15	0	—	—	—	—	—	—	1	5
" 12: " " "	—	—	—	—	15	20	—	3	4	5	15	—	3	—	—	1	11	5.5
Oregon 2: BI & II	8	24	23	25	10	—	13	26	2	2	38	33	0	2	—	4	2	4.8
Wyoming 2: BI & II	12	—	32	52	—	14	—	39	3	12	—	34	—	—	—	—	23.5	23.3
" 13: " " "	—	—	—	—	0	—	7	24	—	37	—	14	—	18	12	73	14.2	38.1
" 14: " " "	—	—	—	—	—	7	—	—	2	11	—	6	0	11	—	3	4.6	11.7
Washington 1: AI & II	—	—	—	—	—	—	—	—	—	—	—	—	2	—	—	—		
" 2: AIII & IV	—	—	—	—	—	—	—	—	—	5	—	—	—	18	—	34		15.8

S = Spring, F = Fall; — indicates no trapping.

TABLE 4

MEAN YEARLY VARIATIONS IN NUMBER OF ANIMALS TRAPPED
DURING THE NORTH AMERICAN CENSUS OF SMALL MAMMALS

| Species | P | Mean number | | Mean variation $\pm$ S. E. | |
		Spring	Fall	Spring	Fall
Peromyscus					
maniculatus	18	8.1	17.7	9.0 $\pm$ 1.84	17.7 $\pm$ 15.9
Peromyscus					
leucopus	5	8.5	13.7	5.3 $\pm$ 1.32	22.4 $\pm$ 1.6
Blarina					
brevicauda	10	1.7	10.2	1.48 $\pm$ 0.57	7.6 $\pm$ 1.42
Clethrionomys					
gapperi	17	13.7	28.2	10.6 $\pm$ 3.7	28.5 $\pm$ 5.73
Microtus					
californicus	2	6.9	22.3	6.0 $\pm$ 4.0	34.9 $\pm$ 10.9
Microtus					
montanus	4	27.2	33.6	36.1 $\pm$ 1.85	65.3 $\pm$ 1.57
Microtus					
pennsylvanicus	12	14.1	19.1	11.2 $\pm$ 2.9	24.3 $\pm$ 5.66
Reithrodontomys					
megalotis	3	34	62.6	4.5 $\pm$ 2.8	46.2 $\pm$ 12.58

P = Number of populations averaging more than four animals.

nights of trapping in the spring and fall of each year. Groups of three snap traps were placed within a suggested radius of each of 20 trap stations in each of two lines. Table 2, modified from Calhoun (1948), gives the suggested dimensions of the traplines and stations for both the "A" and "B" techniques even though most of the data analyzed here were collected by the "B" technique.

Because of the variability of trapping procedures and frequencies, only a small fraction of the data collected in the census could be used in the present analysis. Further, data reflecting population fluctuations caused by extreme environmental changes such as floods, fire, etc., were not included in these considerations.

Table 3 lists the data of the NACSM for populations of *P. leucopus* and *P. maniculatus*. In addition, the average size of spring and fall populations for each locality is indicated. As in Table 1, the data suggest that population levels are typically low for *Peromyscus*. The population levels measured by the NACSM techniques are higher than those shown in Table 1, but direct comparisons are not valid because of the different techniques utilized in obtaining

TABLE 5

TABULATION OF STUDIES OF NATURAL POPULATIONS

Species	Citation	Capture method	Males	Females	Per cent males	X^2
Peromyscus sp.	Brown (1945)	L	220	100	69	$P < 0.001$
P. gossypinus	McCarley (1959*a*)	L	8	3	73	
	McCarley (1959*b*)	L	73	36	67	
	Pearson (1953)	L	78	80	49	
	Pournelle (1952)	S	506	382	57	
	Total		665	501	57	$P < 0.001$
P. leucopus noveboracensis	Bendell (1959)	L	16	23	41	
	Burt (1940)	L	827	724	53	
	Jackson (1952)	S	532	452	54	
	Jackson (1953)	NB	42	36	54	
	Klein (1960)	L	199	102	66	
	Manville (1956)	S	70	23	75	
	Nicholson (1941)	NB	152	136	53	
	Snyder (1956)	L	315	244	56	
	Stickel and Warbach (1960)	L	137	114	55	
	Townsend (1935)	?	194	97	67	
	Total		2484	1951	56	$P < 0.001$
P. oreas	Sheppe (1963)	S	126	148	46	NS
P. maniculatus	Catlett and Brown (1961)	S	46	51	47	
	Hays (1958)	S & L	70	49	59	
	Sheppe (1961)	S	199	168	54	
	Total		315	268	54	$P < 0.1; P > 0.05$

TABLE 5 (Continued)

Species	Citation	Capture method	Males	Females	Per cent males	X²
P. maniculatus bairdi	Blair (1940)	L	170	163	51	
	Howard (1949)	NB	539	514	51	
	Linduska (1942)	S	41	41	50	
		Total	750	718	51	NS
P. maniculatus gracilis	Blair (1942)	L	83	68	55	
	Klein (1960)	L	156	88	64	
	Manville (1949)	L	302	238	56	
		Total	541	394	58	P < 0.001
O. nuttalli	Goodpaster and Hoffmeister (1954)	NB	21	23	48	
	McCarley (1959a)	L	4	9	31	
	McCarley (1959b)	L	32	22	59	
	Pearson (1953)	L	14	8	63	
		Total	71	62	53	NS
P. polionotous	Blair (1951)	L	100	72	58	P < 0.05; P > .025
		Grand Total	4917	4024	54	P < 0.001

L = Live Trap.
S = Snap Trap.
NB = Nest Boxes.

the data. Table 3 contains records of animals trapped at the two lines of trap stations and these records were not expanded or adjusted on a per acre basis as is true for Table 1.

An analysis of the above data (Table 3) with respect to variation in population size is presented in Table 4. Here, for each species, the mean population variation from the previous year is presented by season. In the same table the mean population size and mean variation is presented for other species of small mammals sampled via NACSM techniques.

The data in Table 4 indicate that populations of *Peromyscus* on the average exhibit smaller mean variations than populations of the other small mammals sampled, with the exception of *Blarina brevicauda*. Because of the many variables associated with the collection of these data, statistical tests for the significance of these differences were not made. Further analyses of these and other NACSM records are available elsewhere (Terman, 1966).

Population Structure; Sex Ratio

NATURAL POPULATIONS.—Information concerning the sex ratios of natural populations was obtained by analysis of separate studies from the literature and of data collected during the North American Census of Small Mammals.

Table 5 lists sex ratio information obtained from the literature survey for the species studied. While some variability in the results exists between species, more males than females were recorded for most species and the ratio of males to females was different from 1 : 1 with a chance probability equal to or less than 0.1. Males likewise exceeded females in the grand total of animals captured and the sex ratio was different from 1 : 1 at a "P" value less than 0.001 (Terman and Sassaman, 1966).

Sex ratio information collected via the NACSM techniques is summarized in Table 6 for *Peromyscus leucopus* and *P. maniculatus*. These data exhibit a similar excess of males, but not significantly so for *P. leucopus*.

The explanation most often given for the preponderance of males in records of natural populations is that males "wander" more or travel over larger areas than females and thus have greater trap exposure. This suggestion was first made by Townsend (1935) and has been cited by several other workers since. For example, Burt

TABLE 6

SEXES OF ANIMALS TRAPPED DURING THE NORTH
AMERICAN CENSUS OF SMALL MAMMALS

Species	Localities	Trapping periods	Males	Females	Per cent males	X^2
Peromyscus leucopus	5	61	271	260	51	NS
Peromyscus maniculatus	19	134	1034	826	56	$P < 0.001$

(1940), citing Townsend, explained the surplus of males in his study as resulting from the same factor. There have been numerous studies attempted to ascertain size of home range and dispersal distances related to sex. The conclusions from these studies vary from greater size of male home range and greater dispersal distances to no difference between the sexes (Blair, 1953a). The majority of the studies, however, do suggest larger home ranges for males than for females; thus implying the influence of these differences in movement on the recorded sex ratio.

LABORATORY COLONY.—As a source of additional information, the sexes of young *Peromyscus maniculatus bairdi* born into a laboratory colony were noted and are presented in Table 7. The parents of these young were isolated bisexual pairs composed of mice that were neither siblings nor first cousins. Analysis of the data collected from 435 litters during 30 months revealed a significantly greater production of males than females ($P < 0.02$; Terman and Sassaman, 1966).

The laboratory data support the field data and indicate a secondary sex ratio of more males than females. Any differences produced by males moving more than females would serve to accentuate the differences in sex ratio. The mechanism by which production of males exceeds that of females is not understood in *Peromyscus* although there is evidence for mechanisms producing similar effects in other organisms (Hanks, 1965; Poulson, 1962; Schulten, 1964; Yanders, 1965).

Forces for Population Change

Changes in population numbers depend upon three factors or forces operating upon them. These "forces for change" (Davis,

TABLE 7

SEXES OF YOUNG IN A LABORATORY COLONY OF
Peromyscus maniculatus bairdi

Months	Litters	Males	Females	Per cent males	X²
30	435	1082	969	53	.02 > P

1953) are reproduction, mortality, and movement. Reproduction has an additive effect on population numbers, mortality a damping effect, and movement either effect.

REPRODUCTION.—Table 8 is a summary of the reproductive information from the literature for several species of *Peromyscus*. Included are the number of adult females or litters examined, the percentage of adult females pregnant, the number of young in the uterus (gravidum; Snyder and Christian, 1960) or the immediate post-partum litter size, and estimates of the number of litters per year per female.

The data are few and the means of collection variable, however, the evidence does suggest two distinct groups of species based on litter size. Litters averaging four young or less were recorded for *P. boylei, P. californicus, P. leucopus leucopus,* and *P. truei,* whereas *P. leucopus noveboracensis, P. maniculatus, P. maniculatus bairdi, P. maniculatus gracilis,* and *P. oreas* exhibited a mean litter size greater than four. The significance to population dynamics of such consistent differences in litter size between species or subspecies is not clear, but one would expect the mortality rates to be markedly different, related to these differences.

Clearly evident from Table 8 is the shortage of information available for the percentage of females pregnant (prevalence of pregnancy) and the numbers of litters produced per year. Valid analyses of the reproductive and growth dynamics of populations of *Peromyscus* cannot be made until additional reliable information is available.

MORTALITY AND LONGEVITY.—The second major force determining size of population is mortality. This can logically and perhaps most appropriately, in terms of population dynamics, be studied with reference to the time of its occurrence, namely, whether deaths of young occur prior to birth or at some post-natal time.

TABLE 8

FORCE FOR INCREASE IN POPULATIONS OF *Peromyscus*

Species	Citation	Method	Location	Number of adult females or litters	Per cent pregnant	Young per litter	Litters per year
P. boylei	Jameson (1953)	Embryo	Field	42		3.1	
P. californicus	McCabe and Blanchard (1950)	Embryo	Lab	34		1.79	3.25
			Field	19		2.11	
P. leucopus leucopus	Svihla (1932)	Litter	Lab	15		1.87	
	Svihla (1932)	Litter	Lab	21		3.43	
P. leucopus noveboracensis	Bendell (1959)	Embryo	Field	27		5.19	
			Field	23		5.52	
			Lab	82		4.22	
	Burt (1940)	Embryo	Field	39		4.26	
			Field	596	31		4–5
	Coventry (1937)	Embryo	Field	50		5.04	
	Jackson (1952)	Embryo	Field	174	56	4.14	4.1
			Field	172	47		3.4
	Townsend (1935)	Embryo		18		4.7	
	Svihla (1932)	Litter	Lab	53		4.36	
P. maniculatus	Jameson (1953)	Embryo	Field	96		4.60	
	McCabe and Blanchard (1950)	Embryo	Field	52		5.06	
			Lab	37		4.89	4.0
	Scheffer (1924)	Embryo	Field	48		5.10	
	Sheppe (1963)	Litter	Field	53		5.5	

TABLE 8 (Continued)

Species	Citation	Method	Location	Number of adult females or litters	Per cent pregnant	Young per litter	Litters per year
P. maniculatus bairdi	Biair (1940)	Litter	Field	43		4.0	
			Field	192	55		4
	Howard (1949)	Litter	Field	25		4.28	
	Svihla (1932)	Litter	Lab	21		3.05	
	Terman and Sassaman (1966)	Litter	Lab	521		4.8	
P. maniculatus gracilis	Manville (1949)	Lac. or Preg.	Field	131	74		
	Coventry (1937)	Embryo	Field	50		5.4	
P. oreas	Sheppe (1963)	Litter	Field	21		6.1	
P. polionotus	Caldwell (1964)	Embryo	Pen	63		3.1	
			Pen	205	31		
P. truei	McCabe and Blanchard (1950)	Embryo	Field	27		3.52	3.4
			Lab	22		3.32	
	Svihla (1932)	Litter	Lab	19		2.84	

Prenatal Mortality: In many species of animals, the number of young born is frequently less than the number of ova ovulated. Mortality of ova before fertilization, failure of blastocysts to implant in the uterus, and mortality resulting from various causes prior to parturition all contribute to birth of fewer young than the number of eggs ovulated. Information on the frequency of occurrence of prenatal mortality, the developmental stage involved, the proportion of young lost in each gravidum as well as in the total population, and factors precipitating such mortality in natural populations is almost completely lacking. In the laboratory, however, Helmreich (1960) demonstrated an increase in intra-uterine mortality in *P. m. bairdi* as a result of "crowding" four males and four females in individual cages. The significance of such findings to the dynamics of natural populations is questionable, particularly in view of the present paucity of information.

Post-natal Mortality and Longevity: While Dice (1933) reported individuals of *P. m. gracilis* which lived eight years in his laboratory, and were fertile at four years of age, no such longevity is thought to occur in nature. There is a marked shortage of information on post-natal mortality, but the evidence available indicates that very few *Peromyscus* live one year (Blair, 1953a, 1953b; Burt, 1940; Snyder, 1956). Estimation of percentage mortality for one year ranges from 99 per cent for *P. maniculatus bairdi* (Howard, 1949), 96 per cent for *P. leucopus* (Blair, 1953b; Burt, 1940) to a range of 63–94 per cent for *P. m. gracilis* in different studies in Michigan (Blair, 1941; Manville, 1949). These figures are undoubtedly too high owing to the difficulty of separating loss of animals because of emigration from loss attributable to mortality. The same difficulty existed in Snyder's (1956) study of *P. leucopus* in which dissappearance rates (mortality plus emigration) were found to be constant with age for mice over five weeks old.

More recently, Bendell (1959) studied populations of *P. leucopus* on islands where it was possible to control emigration more adequately. One of the conclusions of his study was that food supply below "some low level" influenced the death rate of mice in excess of one month of age, but when food was available above this low level it had "no" effect on death rate of animals from approximately one month of age. Bendell further concluded that if food is plentiful intraspecific strife for space may become a limiting factor to populations of *P. leucopus* even though the mechanism by which this operates is not clear.

The above conclusions are clearly controversial. Studies of other animals provide evidence that many variables, both biotic and abiotic, other than food and age may influence mortality. Information on these relationships must be obtained. Urgently needed are data on mortality rates related to the stage of population growth and the age and sex of the component animals.

MOVEMENT.—The effect of movement on the size of a local population depends on whether it is emigration or immigration. If the former is excessive, then the population decreases in size. If the latter is greater the population will increase unless compensatory reactions occur.

In many studies of natural populations of *Peromyscus,* it has been useful to subdivide the population into transients and residents, based upon sampling records. Transients are usually individuals recorded only once or a very few times and for which further information is lacking. Transient animals are usually continually moving through an area. The numbers of such animals probably reflects inversely the availability of home sites and home ranges in the local area or directly the expulsion rate caused by density related pressures in adjacent local populations. Depending upon the time of year, as many as 50 per cent of the individuals captured during some studies have been recorded as transients (Blair, 1951; McCarley, 1959*a*; Nicholson, 1941; Stickel and Warbach, 1960). Evidence is available for few species of *Peromyscus* and not abundant for any one species, but that which is available indicates that transient animals are predominantly young or young adults (Blair, 1951; Howard, 1960; McCarley, 1959*b*).

The differentiation of individuals into transients and residents has implications of significance to the dynamics of populations. In several studies it has been suggested that an individual selects a home site or area at approximately the same time it attains sexual maturity, following which it remains relatively sedentary the rest of its life (Howard, 1949 and 1960, for *P. m. bairdi*; McCarley, 1959*b*, for *P. gossypinus*; Nicholson, 1941, and Burt, 1940, for *P. l. noveboracensis*; McCarley, 1959*a*, 1958, for *Ochrotomys nuttalli*; and Caldwell, 1964, and Blair, 1951, for *P. polionotus*).

McCarley (1959*a*) cites an example of the tendency of individual *Ochrotomys nuttalli* to remain within their home area once they have become established. During one of his studies, a flood occurred after the population had been marked and studied for a sufficient time to denote residents and transients. The water was eight feet deep in some areas and "completely disturbed the habitat, washing traps, logs and brushpiles away and depositing piles of drift in other places." Eight days later the water had receded sufficiently to permit trapping. The trapping results showed that 84 per cent of the resident *O. nuttalli* and 60 per cent of the resident *P. gossypinus* were re-trapped within the boundaries of their originally noted home ranges!

The above is not to suggest that movement does not occur among established resident individuals. Several studies have provided evidence suggesting that established individuals continue to move around their home ranges and may frequently change their resident sites (Blair, 1951; Howard, 1949; Nicholson, 1941; Terman, 1961 and 1962). Moves and changes in nest sites are typically of a short distance, however, and are considered local as opposed to the longer moves and complete disappearance of most transient animals.

There is good evidence from varied sources that an individual may maintain several nest sites or refuge sites in a local area. Blair (1951), studying *P. polionotus,* found that the average home range contained a mean of 20 mouse holes of which a mean of five were entered by each mouse upon release from live traps. Studies of other species in nature have revealed that a mouse may flee to one of several different holes upon release from a live trap. Arnold Nicholson (1941), using nest boxes to study *P. leucopus* in nature, found that mice seldom resided long in the same boxes but frequently disappeared from the boxes and returned from a few days to several weeks later. Terman (1961), studying populations of *P. maniculatus bairdi* in half-acre enclosures containing subterranean nest boxes, found that each mouse utilized an average of 4.3 of the 24 nest boxes available. Figure 1 illustrates these data and indicates that between 55 and 65 per cent of the mice moved to different nest boxes following establishment of the population (day 7) in the field. These movements, however, were increasingly to nest boxes in which each mouse had previously been recorded. Figure 2 illustrates this phenomenon even more emphatically. In

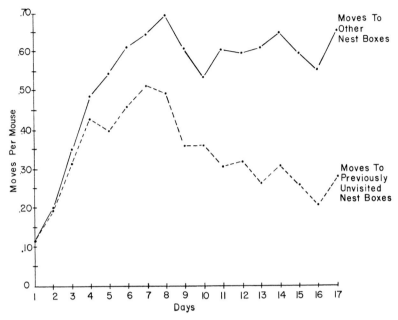

FIG. 1. Proportion of mice moving to other nest boxes and to previously unvisited nest boxes on successive days during the period of population establishment (From Terman, 1961).

this figure the daily records of mice occurring in the same nest box as on the previous day were combined with those indicating movement to nest boxes previously occupied by each animal. This combination was regarded as an index of "localization" behavior and is plotted and compared with the curve reflecting moves per mouse to previously unvisited boxes (Fig. 2). From day eight onward localization behavior was significantly more frequent than the frequency of moves per mouse to previously unvisited nest boxes. Therefore, each mouse localized and used a few nest boxes (mean of 4.3) among which it continued to move. These data indicate that each mouse maintained several refuges and/or nest sites rather than a single one around which its activity centered.

Dispersals, when they occur, are not necessarily extensive. For example, Howard (1949) found that 76 per cent of 115 *P. m. bairdi* moved less than 550 feet from their places of birth before establish-

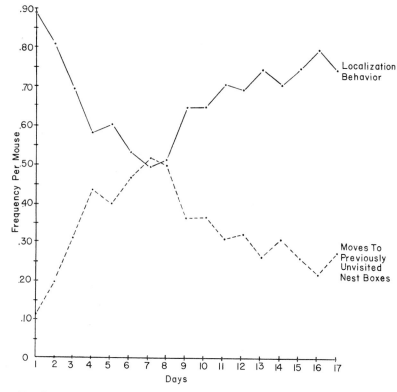

Fɪɢ. 2. Proportion of mice moving to previously unvisited nest boxes or exhibiting localization behavior on successive days during the period of population establishment (From Terman, 1961).

ing themselves in home ranges. There are numerous complexities which are not clearly understood in such studies. Evidence suggests that the density of the population in relation to the requisites of the environment is important to movement behavior. Stickel (1960) found that the density of populations of *Peromyscus* during successive years is inversely related to size of individual home ranges in the same area. Pearson (1953) found the distance moved from the point of release from live traps to that of recapture was inversely related to population density. Under "dense" population conditions, the proportion of moves less than 125 feet was significantly greater than the proportion of movements more than 125 feet by animals in less dense populations. Bendell (1959) also recorded shorter average distances moved by animals in dense populations on islands than by animals in sparce populations.

The cause for such movement behavior has not been clearly demonstrated. The evidence suggests that movements are chiefly by juvenal or young adult animals approaching sexual maturity or recently sexually mature. Movement behavior may, therefore, reflect a complex relationship between endocrine factors associated with reproductive maturity and behavioral factors related to establishment of a home site or range. Both may be influenced by social interaction with other members of the population. The territorial behavior of established residents may act as a stimulus to movement; however, territorial defense has not been demonstrated for *Peromyscus*. Evidence suggestive of a mutual avoidance mechanism has been obtained from semi-natural populations of *P. m. bairdi*. Moving animals, therefore, may avoid prior residents of the same sex and move in relation to the social stimuli received from established individuals. Such behavior occurred in homing experiments with *P. m. bairdi* in half-acre field enclosures (Terman, 1962). In these experiments, populations established themselves in distinct spatial patterns of distribution among the subterranean nest boxes present in the field. Later, the animals of each population were removed from the field and held in isolation in the laboratory for 36 hours. Following the isolation period, the mice were released into the fields at locations distant from their "home" areas after which they demonstrated "homing" to specific previously occupied nest boxes significantly more frequently than by chance. During subsequent trials, 21-day-old male mice were retained in half the nest boxes of the field. Homing success to these "alien-occupied" boxes dropped significantly from previous levels. Further, homing to "alien-occupied" boxes was significantly less frequent than to empty nest boxes during the same trials. Additional indication of avoidance behavior was the fact that during the time the young alien mice were retained in half the nest boxes, the resident mice were found in empty nest boxes significantly more often than expected by chance. Thus, the returning residents appeared to avoid boxes temporarily occupied by alien, 21-day-old mice. These data suggest that the spatial distribution exhibited was influenced by social interaction not based on fighting or defensive ability per se.

An additional dispersal variable recently suggested by Howard (1960) is that young mice may move from the location of their birth (nest site) as a result of an "innate dispersal drive" which has selective value in terms of the population and species. Dispersal by these individuals, therefore, may not be directly related to environmental causes such as mentioned above. Evidence cited by Howard to support the "dispersal drive" hypothesis is that distances of dispersal have been found not to be random in some studies; many recently introduced species spread too rapidly to be caused by population pressures; and the rate of such movements appears to be density independent. Additional experimental evidence testing this hypothesis is needed.

It is apparent from the preceding discussion that movements are important to the regulation of the size of a local population. When prevented because of barriers, etc., the population must utilize other mechanisms to regulate its numbers. Yet, in spite of the spatial and biological importance of movements, factors influencing such behavior in populations of *Peromyscus* are virtually unknown.

Population Control

No discussion of population dynamics is complete without an examination of factors controlling growth and thereby regulating size of population. Forces acting on the control of populations have generally been discussed as density independent and density dependent with a great deal of heat being generated as to the relative importance of each. While it is well recognized that many factors of the physical and biotic environment may be influential directly or indirectly in maintaining population numbers, no attempt will be made in this discussion to enumerate these. Rather, I shall examine mechanisms which involve reactions among the animals themselves with respect to requisites of the physical and biotic environment.

EVIDENCE FOR POPULATION REGULATION AND CONTROL.—*Natural Populations*: Some of the clearest and most direct evidence for growth control of populations of *Peromyscus* in nature is the rather consistent low density of these populations. A discussion of density was presented in the early portion of this chapter and the data in Tables 1 and 4 are pertinent. These consistently low population densities occur and are maintained under conditions in which there is no apparent shortage of food or other requisities of the environment. These data are, therefore, suggestive of controlling mechanisms functioning when there are no obvious environmental shortages or stresses.

A further analysis of the North American Census of Small Mammal data is presented in Figure 3. In this presentation, trapping results were regarded as indicative of population levels and the *maximum* numerical fluctuations (ranges) were noted for the several populations representing each species. Figure 3 is taken from Terman (1966) and compares populations of *Peromyscus* with selected species of small mammals.

In attempting to evaluate the potential for fluctuation of population numbers, consideration was given not only to the maximum fluctuation exhibited by populations of a species but also to the time required for each fluctuation to occur. The relationship between the mean range of population fluctuation for each species and the mean time for these changes to occur was referred to as the "Fluctuation Index" calculated in the following manner:

$$\text{Fluctuation Index} = \frac{\text{Mean range of population fluctuations}}{\text{Mean years for range to occur}}$$

The F. I. value, therefore, reflects the potential for a species to change in numbers radically in a short period of time as indicated by the NACSM data.

Of interest in the present discussion is the fact that while populations of *Peromyscus* did exhibit yearly fluctuations as measured by trapping results, these fluctuations were of a lesser degree than the same measurement techniques revealed for most other small mammals (Fig. 3). Further, the average size of population and average maximum fluctuations are also lower for *Peromyscus* than for most other species of small mammals.

Of the species of *Peromyscus, P. maniculatus* exhibited larger mean maximum fluctuation (30.7) than *P. leucopus* (28.6). The fluctuations of the populations of *P. maniculatus,* however, occurred over a longer period of time (2.7 years) than those of *P. leucopus* (1.0 years) so the fluctuation indexes for each species suggest that

SPECIES	P#	X̄N	X̄ YEARS TRAPPING	MAX. POP. CHANGE	X̄ RANGE IN FLUCTUATIONS OF POPULATIONS	F.I.
Blarina brevicauda	10	15.5	5	1.8		9.6
Microtus montanus	4	45.6	4	1.5		52.3
Microtus californicus	2	22.3	6	1.0		80.0
Microtus pennsylvanicus	12	29.1	5	2.6		18.3
Clethrionomys gapperi	17	28.4	4	1.6		23.8
Peromyscus maniculatus	18	18.0	5	2.9		10.6
Peromyscus leucopus	5	15.9	7	1.0		28.6
Reithrodontomys megalotis	3	62.6	5	3		22.1

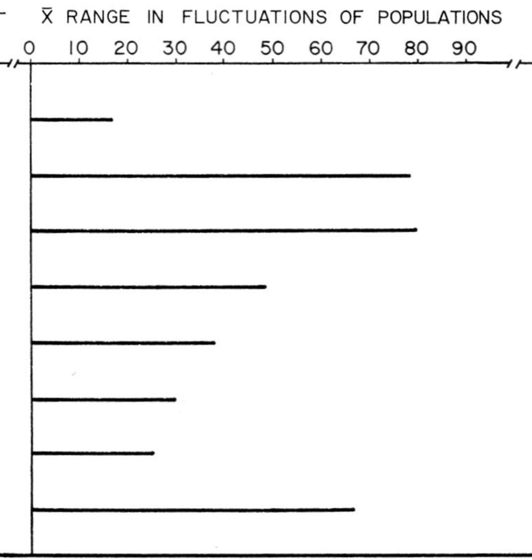

F.I. = Fluctuation Index = $\dfrac{\text{Mean range of population fluctuations}}{\text{Mean years for range to occur}}$

FIG. 3. Variations in the number of animals trapped during the North American Census of Small Mammals. P# = the number of different populations in which a mean of at least four animals were trapped per trapping period; X̄N = the mean number of animals trapped per trapping period; F.I. = the Fluctuation Index (From Terman, 1966).

populations of *P. leucopus* have a greater potential for fluctuation. These data are only indices and relative. The need for more reliable measurements is apparent.

The data analyzed above are in general agreement with other, shorter-term, studies which indicate the low levels of mean yearly fluctuations in populations of *Peromyscus*. High population densities (supposedly eruptions or maximum fluctuations) have been reported for *Peromyscus maniculatus sonoriensis* (Hoffmann, 1955) and for *Peromyscus maniculatus* (Catlett and Brown, 1961, using NACSM techniques).

Hoffmann's study involved 1204 trap-nights in the White Mountains, Mono County, California, in June, 1954. During this period, 528 mice were trapped for a mean of 0.44 mice per trap-night. Catlett and Brown's records are from near Gothic, Colorado, at an elevation of 9500 feet. During three nights of trapping with 60 snap traps, 97 *P. maniculatus* were taken for a mean of 0.54 mice per trap-night. Unfortunately, there is little information available immediately prior to or after these records of a dense population and the typical population levels of the area cannot be ascertained.

With the exception of the above papers, evidence from the literature suggests that outbreaks of species of *Peromyscus* rarely occur (Blair, 1940, 1943; Christian, 1961; Howard, 1949; McCarley, 1954, 1958; Pearson, 1953), and to my knowledge there is no evidence for outbreaks of the type exhibited by *Lemmus, Mus musculus,* and various species of *Microtus*. This information is indicative of population controlling mechanisms which are presently unknown (Terman, 1966).

Laboratory Populations: As a supplement to "field" information, freely growing laboratory populations of *P. m. bairdi* have been studied to reveal growth forms and population regulation mechanisms (Lidicker, 1965; Terman, 1965*a*). In one study (Terman, 1965*a*), food and water were available in excess of utilization in each of several populations housed in identical enclosures. Population growth followed a sigmoid-like curve culminating in a distinct asymptote. Asymptote was arbitrarily determined on the basis of reproductive performance and survival of offspring. A population was regarded as having reached asymptote if, during a minimum period of 15 weeks following the birth of the last surviving litter, no young were born or survived 21 days. This time period permitted

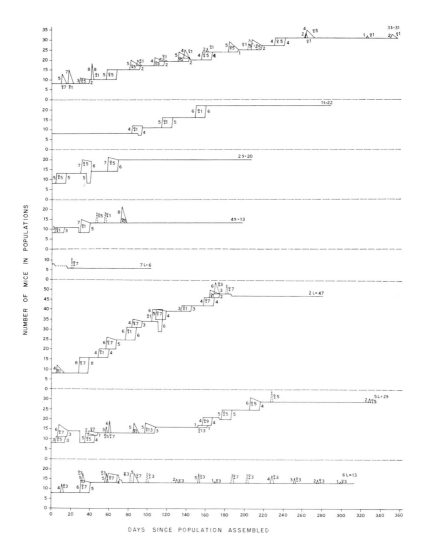

Fig. 4.   Growth curves of laboratory populations. The number of animals was plotted against the time since the population was assembled. Each triangle or quadrilateral records the birth of a litter, the female parent, and the number of young born and surviving ten days. Dotted triangles refer to litters known to be born to females even though the young were never found (Modified from Terman, 1965a).

the surviving young of previous litters to reach a chronological age at which they normally would be reproductively mature.

As illustrated by the growth curves of the populations presented in Figure 4, upper levels of growth were distinct. Further, control of growth, once achieved, was maintained for extensive periods of time. Specifically, in the above study, cessation of growth was maintained for an average of 169 days prior to sacrifice of the populations. How long such stability of numerical levels in the population would have continued is unknown.

MECHANISMS AND CHARACTERISTICS OF POPULATION CONTROL.— *Field Populations*: The demonstration of factors characteristic of or associated with population control in nature is extremely difficult and almost non-existent in the literature of *Peromyscus*.

Movements have been discussed earlier as an important force for population change. To my knowledge, however, there is no experimental information for populations of *Peromyscus* exhibiting change in either frequency of animals moving or in the proportion of animals moving related to increased density.

Practically the same statement may be made for mortality rates with respect to density changes. Evidence presented by Pearson (1953) suggests increased mortality of young *P. gossypinus* associated with a decrease in food and in population numbers. Bendell (1959), working with island populations of *P. leucopus,* presents evidence implicating failure of young to survive as directly influencing population growth.

There is some evidence for a decrease in reproductive rate functioning as a controlling mechanism. Pearson (1953) associated an almost total absence of production of young with a sudden reduction in a population of *P. gossypinus.* Jameson (1953) suggested that changes in numbers of *P. boylei* and *P. maniculatus* could be related solely to observed varying rates of reproduction. Brown (1964), studying *P. boylei* in isolated cedar glades, presented evidence that failure of females to produce litters was more influential than increased mortality of young in producing population decline. McKeever's data (1964) for populations of *P. maniculatus* in California indicate a moderate inverse relationship between population density and sexually active females.

The most striking data reflecting reproductive function inversely related to population density comes from Catlett and Brown's (1961) and Hoffmann's (1955) studies of populations which appear to be the closest to "eruption" of any *Peromyscus* populations recorded. As mentioned previously, Catlett and Brown trapped 97 *P. maniculatus* in 180 trap nights. Sixty-six of these showed sub-adult pelage. No juveniles were taken and *no* females carried embryos even though the time of trapping was within the normal breeding season.

Hoffmann (1955) reported a very dense population of *P. maniculatus sonoriensis* in the White Mountains of California during June, 1954. In this study, 528 animals were captured in 1204 trap-nights. Only 30 per cent of the sample were fully adult animals and there was no external evidence of reproductive activity among those trapped. Ten adults (5 males and 5 females) were autopsied; placental scars but no embryos were found in the females and the testes of the males appeared to be regressing.

While the densities of the above populations are atypical in comparison with most of those reported in the literature, suggestions of regulation of reproduction in these and other more moderately dense populations differ from population theory developing from laboratory studies. However, the need for adequately controlled and systematically collected field data is apparent and necessary before further progress in this area of investigation can be made.

Endocrine Relationships: Evidence from natural populations of *Peromyscus* of characteristic responses of the endocrine glands to changing population densities is variable (Christian, 1963; Christian *et al.,* 1965). McKeever's (1964) work with *P. maniculatus* showed no consistent correlation of population density with adrenal size. Likewise, no correlations of size of thyroid and pituitary were found with population density. Christian (1959) demonstrated a decline in adrenal weight associated with a population decline in *P. leucopus*. Evidence for inhibition of maturation of young born late in the breeding season has been presented for *P. maniculatus* and *P. oreas* by Christian (1961) and Sheppe (1963) and may be related to increasing densities during the breeding season.

Laboratory Studies: Experiments with laboratory populations of *Peromyscus* have yielded information regarding characteristics of population control. Studies by Lidicker (1965) for *P. truei* and

P. maniculatus and by Terman (1965a) for *P. m. bairdi* with freely growing populations supplied with excess food and water revealed that distinct and long lasting levels of equilibrium or asymptote were attained by each population. Cessation of population growth was achieved by one or both of two means, namely, cessation of reproduction or failure of young to survive (Lidicker, 1965; Terman, 1965a.)

Contributing to the determination of asymptote and apparently characteristic of the population situation was the fact that a significantly lower pregnancy rate was recorded for females in a population than for female controls maintained as bisexual pairs (Terman, 1965a). For example, in each population an average of only 1.9 of the mean of ten females present produced young. Of further significance is the fact that approximately 95 per cent of the females born into these freely growing populations never produced young even though their age at death was a minimum of 125 days and averaged 168 days. Inhibition occurred in females of litters born early as well as later along the growth curve of each population. The significance of these findings is apparent when one recalls that Clark (1938) showed that the average age of sexual maturity for this species is 48 days.

Of further interest in the analysis of forces controlling populations is the fact that under the identical conditions of the physical environment of each enclosure, populations stopped growing at radically different numerical levels (Terman, 1965a). Specifically, density of these populations at asymptote varied from 2.35 to 0.3 animals per square foot—a range of 2.05 per square foot. Such variability of asymptotic levels under identical conditions of the physical environment has been demonstrated for other animals as well (Brown, 1953; Christian, 1956; Petrusewicz, 1957; Southwick, 1955).

Endocrine Relationships: Christian *et al.* (1965) have presented a great deal of evidence for the involvement of the endocrine system in regulation of numbers of animals in response to increased density. Evidence for such mechanisms operating in populations of *Peromyscus* is limited.

The analysis of data from the freely growing laboratory populations cited above (Terman, 1965a) is incomplete and largely unpublished. The data presently available, however, indicate that the testes and vesicular glands of animals from populations at asymptote were significantly smaller than those from physically isolated bisexual pairs of comparable age (Terman, 1963*b*). Comparison of the weights of adrenal glands of male experimental and control animals did not reveal significant differences in either absolute or relative (mgs per 100 gs of body weight) weight values (Terman, 1965*b*).

Paired ovary weights of females in a population were significantly less than those from physically isolated bisexual pairs. Absolute adrenal weights of females in a population, whether nulliparous or parous, were not significantly different from those of controls. Relative adrenal weights (mg per 100 gs body weight) of nonparous females from populations were significantly heavier than those of controls (Terman, 1965*b* and unpublished data).

The numbers of animals present at asymptote varied greatly among populations even though conditions of the physical environment were identical. At the same time, weights of glands were similar between animals of asymptotic populations irrespective of the different numerical levels. This suggests that each population was an independent entity, that factors intrinsic to each influence control of growth and are not directly related to numbers of animals in each population per se (Terman, 1963*b*, 1965*a*). Further, these controlling mechanisms were activated in the presence of surplus food and water and at varying population densities. The data presently available from studies of laboratory populations of *Peromyscus* generally agree with studies of other species and suggest that intrinsic regulatory mechanisms are basically behavioral and operate via physiological mechanisms to regulate populations (Christian, 1963; Christian *et al.*, 1965).

Discussion

In spite of numerous published studies, basic information essential to understanding the dynamics of populations of *Peromyscus* is insufficient. There is need for almost all types of measurements. To be useful, these measurements must be made under conditions in which the variables of the physical and biotic environment are controlled or adequately determined. Further, data on individual local populations must be collected with consistent techniques for longer periods of time than has been the case in the past. This is imperative if information of the kind needed is to be obtained.

The specifics of reproduction, mortality, and movement within individual populations are areas of information woefully lacking. Needed are life history data beginning with young at birth, noting their parents, survival, movements, spatial dynamics, and eventual localization. Also needed is information on reproduction at maturity. This appears to be particularly important to an understanding of intrinsic factors controlling population growth.

The above information cannot generally be obtained for *Peromyscus* without long-term studies employing techniques which will permit accurate and sensitive measurement of the population. Studies on islands, in isolated vegetation types, or in large seminatural enclosures, may permit collection of some of the needed data. The development of nest-box techniques or other means of locating nests so that young can be measured shortly after birth is a very important "next" step. Most of the needed techniques must still be developed, however, and these represent some of the greatest challenges in the study of populations in the *field* today.

Information on social interaction within natural populations of *Peromyscus* is essential for understanding their dynamics. Almost nothing is known on this subject. For example, what is the organizational framework of the population? Are the basic units the family, the neighborhood, or something else? Do populations in nature function as entities as laboratory studies suggest. If so, do neighborhoods or series of neighborhoods respond collectively as an entity?

These and many other questions regarding social and spatial organization are unanswerable presently.

Nicholson (1955, 1957) has suggested that basic to factors influencing populations in a density dependent way is intraspecific competition for requisites of the environment—not the requisites themselves. It is evident that population regulation or response by such a mechanism involves communication. Little is known of such variables in *Peromyscus* although Wynne-Edwards (1962, 1965) has recently discussed the concepts for other forms. Neither is it known how animals space themselves in nature and maintain their spacing. Communication is essential to the accumulation of "population pressure" related to space utilization. Any "feed-back" mechanism in nature or in the laboratory must utilize mechanisms of communication not presently demonstrated. The importance to population biology of information about this "frontier" is apparent.

In conclusion, it appears that young mice are the most directly affected by population pressures. While both sexes respond to these pressures, the response of the young females seems of greater significance to population regulation. Experiments with laboratory populations of varying densities have demonstrated rather drastic susceptability of young females to inhibition of reproductive maturation (Terman, 1965a). Of equal interest and probable significance to natural population dynamics is the recent finding that injection of ACTH into intact female *Mus* and *Peromyscus* inhibits reproductive maturation for several weeks (Christian 1964a, 1964b; Christian *et al.*, 1965). Thus, it may be hypothesized that the period of life prior to sexual maturation may be particularly vulnerable to the influence of noxious stimuli of various types. Such stimuli may produce long term inhibition of reproductive maturation and modifications of the behavior of females with a consequent influence on population growth and dynamics.

Summary

Populations of *Peromyscus* exhibit relatively constant low numerical levels with little evidence of violent fluctuations or eruptions. These phenomena are suggestive of sensitive and effective controlling mechanisms.

Data from natural populations and a laboratory colony indicate the production of significantly more males than females in certain species of *Peromyscus*. The greater movement and larger home ranges of males than of females, as suggested in the literature, accentuates what appears to be a basic difference in secondary sex ratio.

Review of the literature revealed a lack of basic information on reproduction, mortality, and movement within natural populations of *Peromyscus*.

The data suggest that young mice move varying distances from the sites of their birth before selecting home areas and establishing themselves. Once an individual has become a resident of an area, it continues to move over the area making use of several nest sites or refuges rather than a single one.

The data presently available on populations are suggestive of sensitive controlling mechanisms which are by no means understood. Prerequisite to understanding these mechanisms is the accumulation of additional information on the social and spatial dynamics of populations and the variables influencing reproduction, mortality and movement.

Literature Cited

BENDELL, J. F. 1959. Food as a control of a population of white-footed mice, *Peromyscus leucopus noveboracensis* (Fischer). Can. Jour. Zool., 37:173–209.

BLAIR, W. F. 1940. A study of prairie deer-mouse populations in southern Michigan. Amer. Midland Nat., 24:273–305.

———— 1941. The small mammal population of a hardwood forest in northern Michigan. Contr. Lab. Vert. Biol. Univ. Mich., 17:1–10.

———— 1942. Size of home range and notes on the life history of the woodland deer-mouse and eastern chipmunk in northern Michigan. Jour. Mamm., 23:27–36.

———— 1943. Populations of the deer-mouse and associated small mammals in the mesquite association of southern New Mexico. Contr. Lab. Vert. Biol. Univ. Mich., 21:1–40.

———— 1951. Population structure, social behavior, and environmental relations in a natural population of the beach mouse (*Peromyscus polionotus leucocephalus*). *Ibid.*, 48:1–47.

———— 1953*a*. Factors affecting gene exchange between populations in the *Peromyscus maniculatus* group. Tex. Jour. Sci., 5:17–33.

———— 1953*b*. Population dynamics of rodents and other small mammals. Adv. Gen., 5:2–37.

BROWN, H. L. 1945. Evidence of winter breeding in *Peromyscus*. Ecology, 26:308–309.

BROWN, L. N. 1964. Dynamics in an ecologically isolated population of the brush mouse. Jour. Mamm., 45:436–442.

BROWN, R. Z. 1953. Social behavior, reproduction, and population changes in the house mouse (*Mus musculus L.*). Ecol. Monogr., 23:217–240.

BURT, W. H. 1940. Territorial behavior and populations of some small mammals in southern Michigan. Misc. Publ. Mus. Zool. Univ. Mich., 45:1–58.

CALDWELL, L. D. 1964. An investigation of competition in natural populations of mice. Jour. Mamm., 45:12–30.

CALHOUN, J. B. 1948. North American census of small mammals. Announcement of Program. Release No. 1, Rodent Ecology Project, Johns Hopkins Univ., 8 pp.

———— 1949. North American Census of Small Mammals. Annual report of census made in 1948. Release No. 2, *Ibid.*, 67 pp.

———— 1950. North American Census of Small Mammals. Annual report of census made in 1949. Release No. 3, Roscoe B. Jackson Memorial Lab., 90 pp.

———— 1951. North American Census of Small Mammals. Annual report of census made in 1950. Release No. 4, *Ibid.*, 136 pp.

———— 1956. Population dynamics of vertebrates. Compilations of research data. 1951. Annual Report—North American Census of Small Mammals, Release No. 5, Admin. Publ. of U. S. Dept of Health, Education and Welfare; Nat. Inst. Mental Health, 164 pp.

CALHOUN, J. B., AND A. A. ARATA. 1957a. Population dynamics of vertebrates. Compilations of research data. 1952. Annual Report—North American Census of Small Mammals. Release No. 6, Admin. Publ. U. S. Dept. of Health, Education and Welfare; Nat. Inst. Mental Health, 155 pp.

———— 1957b. Population dynamics of vertebrates. Compilations of research data. 1953. Annual Report—North American Census of Small Mammals. Release No. 7, *Ibid.*, 167 pp.

———— 1957c. Population dynamics of vertebrates. Compilations of research data. 1954. Annual Report—North American Census of Small Mammals. Release No. 8, *Ibid.*, 96 pp.

———— 1957d. Population dynamics of vertebrates. Compilations of research data. 1955 and 1956. Annual Reports—North American Census of Small Mammals and certain summaries for the years 1948 and 1956. Release No. 9, *Ibid.*, 132 pp.

CATLETT, R. H., AND R. Z. BROWN. 1961. Unusual abundance of *Peromyscus* at Gothic, Colorado. Jour. Mamm., 42:415.

CHRISTIAN, J. J. 1950. The adreno-pituitary system and population cycles in mammals. Jour. Mamm., 31:247–259.

———— 1956. Adrenal and reproductive responses to population size in mice from freely growing populations. Ecology, 37:258–273.

———— 1959. The roles of endocrine and behavioral factors in the growth of mammalian populations. *In*: Gorbman, A. (ed.) Comparative Endocrinology. New York: John Wiley and Sons, Pp. 71–93.

———— 1961. Phenomena associated with population density. Proc. Nat. Acad. Sci., 47:428–449.

———— 1963. Endocrine adaptive mechanisms and the physiologic regulation of population growth. *In*: Mayer, W. V., and R. G. Van Gelder (eds.) Physiological Mammalogy. New York: Academic Press, Pp. 189–353.

———— 1964a. Effect of chronic ACTH treatment on maturation of intact female mice. Endocrinology, 74:669–679.

———— 1964b. Actions of ACTH in intact and corticoid-maintained adrenalectomized female mice with emphasis on the reproductive tract. *Ibid.*, 75:653–669.

CHRISTIAN, J. J., J. A. LLOYD, AND D. E. DAVIS. 1965. The role of endocrines in the self-regulation of mammalian populations. Recent Progr. Hormone Res., 21:501–578.

CLARK, F. 1938. Age of sexual maturity in mice of the genus *Peromyscus.* Jour. Mamm., 19:230–234.

COVENTRY, A. F. 1937. Notes on the breeding of some Cricetidae in Ontario. Jour. Mamm., 18:489–496.

DAVENPORT, L. B. 1964. Structure of two *Peromyscus polionotus* populations in old-field ecosystems at the AEC Savannah River Plant. Jour. Mamm., 45:95–113.

DAVIS, D. E. 1953. The characteristics of rat populations. Quart. Rev. Biol., 28:373–401.

DICE, L. R. 1933. Longevity in *Peromyscus maniculatus gracilis.* Jour. Mamm., 14:147–148.

GOODPASTER, W. W., AND D. F. HOFFMEISTER. 1954. Life history of the golden mouse, *Peromyscus nuttalli,* in Kentucky. Jour. Mamm., 35:16–27.

HANKS, G. D. 1965. Are deviant sex ratios in normal strains of *Drosophila* caused by aberrant segregation? Genetics, 52:259–266.

HAYS, H. A. 1958. The distribution and movement of small mammals in central Oklahoma. Jour. Mamm., 39:235–244.

HELMREICH, R. L. 1960. Regulation of reproductive rate by intra-uterine mortality in the deer mouse. Science, 132:417–418.

HOFFMANN, R. S. 1955. A population high for *Peromyscus maniculatus.* Jour. Mamm., 36:571–572.

HOWARD, W. E. 1949. Dispersal, amount of inbreeding, and longevity in a local population of prairie deermice on the George Reserve, southern Michigan. Contr. Lab. Vert. Biol. Univ. Mich., 43:1–50.

——— 1960. Innate and environmental dispersal of individual vertebrates. Amer. Midland Nat., 63:152–161.

HOWELL, J. C. 1954. Populations and home ranges of small mammals on an overgrown field. Jour. Mamm., 35:177–186.

JACKSON, W. B. 1952. Populations of the woodmouse (*Peromyscus leucopus*) subjected to the applications of DDT and Parathion. Ecol. Monogr., 22:259–281.

——— 1953. Use of nest boxes in wood mouse population studies. Jour. Mamm., 34:505–507.

JAMESON, E. W., JR. 1953. Reproduction of deer mice (*Peromyscus maniculatus* and *P. boylei*) in the Sierra Nevada, California. Jour. Mamm., 34: 44–58.

KLEIN, H. G. 1960. Ecological relationships of *Peromyscus leucopus noveboracensis* and *P. maniculatus gracilis* in central New York. Ecol. Monogr., 30:387–407.

LIDICKER, W. Z. 1965. Comparative study of density regulation in confined populations of four species of rodents. Res. on Pop. Ecology, 7:57–72.

LINDUSKA, J. P. 1942. Winter rodent populations in field-shocked corn. Jour. Wildl. Mgt., 6:353–363.

MANVILLE, R. H. 1949. A study of small mammal populations in northern Michigan. Misc. Publ. Mus. Zool. Univ. Mich., 73:1–83.

———— 1956. Unusual sex ratio in *Peromyscus*. Jour. Mamm., 37:122.

McCabe, T. T., and B. D. Blanchard. 1950. Three species of *Peromyscus*. Rood Assoc., Santa Barbara, 136 pp.

McCarley, W. H. 1954. Fluctuations and structure of *Peromyscus gossypinus* populations in eastern Texas. Jour. Mamm., 35:526–532.

———— 1958. Ecology, behavior and population dynamics of *Peromyscus nuttalli* in eastern Texas. Tex. Jour. Sci., 10:147–171.

———— 1959a. The effect of flooding on a marked population of *Peromyscus*. Jour Mamm., 40:57–63.

———— 1959b. A study of the dynamics of a population of *Peromyscus gossypinus* and *P. nuttalli* subjected to the effects of X-irradiation. Amer. Midland Nat., 61:447–469.

McKeever, S. 1964. Variation in the weight of the adrenal, pituitary and thyroid gland of the white-footed mouse, *Peromyscus maniculatus*. Amer. Jour. Anat., 114:1–15.

Nicholson, Arnold J. 1941. The homes and social habits of the wood-mouse (*Peromyscus leucopus noveboracensis*) in southern Michigan. Amer. Midland Nat., 25:196–223.

Nicholson, A. J. 1955a. Compensatory reactions of populations to stresses, and their evolutionary significance. Aust. Jour. Zool., 2:1–8.

———— 1955b. An outline of the dynamics of animal populations. *Ibid.*, 2:9–65.

———— 1957. Self-adjustment of populations to change. Cold Spring Harbor Sym. Quant. Biol., 22:153–173.

Pearson, P. G. 1953. A field study of *Peromyscus* populations in Gulf Hammock, Florida. Ecology, 34:199–207.

Petrusewicz, K. 1957. Investigation of experimentally induced population growth. Ekologia Polska, Ser. A., Polska Acad. Nauk., 5:281–309.

Poulson, D. F. 1962. Cytoplasmic inheritance and hereditary infection in *Drosophila*. *In*: Methodology in Medical Genetics. San Francisco: Holden Day, Vol. 3.

Pournelle, G. H. 1952. Reproduction and early post-natal development of the cotton mouse, *Peromyscus gossypinus gossypinus*. Jour. Mamm., 33:1–20.

Scheffer, T. H. 1924. Notes on the breeding of *Peromyscus*. Jour. Mamm., 5:258–260.

Schulten, G. G. M. 1964. A heritable aberrant sex-ratio in *Drosophila melanogaster* reducing the number of females. Genetica, 35:182–196.

Sheppe, W. 1961. Systematic and ecological relations of *Peromyscus oreas* and *P. maniculatus*. Proc. Amer. Phil. Soc., 105:421–446.

———— 1963. Population structure of the deer mouse, *Peromyscus*, in the Pacific Northwest. Jour. Mamm., 44:180–185.

Snyder, D. P. 1956. Survival rates, longevity, and population fluctuation in the white-footed mouse, *Peromyscus leucopus*, in southeastern Michigan. Misc. Publ. Mus. Zool. Univ. Mich., 95:1–33.

Snyder, R. L., and J. J. Christian. 1960. Reproductive cycle and litter size of the woodchuck. Ecology, 41:647–656.

Southwick, C. H. 1955. The population dynamics of confined house mice supplied with unlimited food. Ecology, 36:212–225.

Stickel, L. F. 1960. *Peromyscus* ranges at high and low population densities. Jour. Mamm., 41:433–441.

Stickel, L. F., and O. Warbach. 1960. Small mammal populations of a Maryland woodlot. Ecology, 41:269–286.

Storer, T. I., F. C. Evans, and F. G. Palmer. 1944. Some rodent populations in the Sierra Nevada of California. Ecol. Monogr., 14:165–192.

Svihla, A. 1932. A comparative life history study of the mice of the genus *Peromyscus*. Misc. Publ. Mus. Zool. Univ. Mich., 24:1–39.

Terman, C. R. 1961. Some dynamics of spatial distribution within semi-natural populations of prairie deermice. Ecology, 42:288–302.

———— 1962. Spatial and homing consequences of the introduction of aliens in semi-natural populations of prairie deermice. *Ibid.*, 43:216–223.

———— 1963a. The influence of differential early social experience upon spatial distribution within populations of prairie deermice. Anim. Behav., 9:246–262.

———— 1963b. Cessation of population growth and the sex organs of male prairie deermice. (Abstract). Bull. Ecol. Soc. Amer., 44:123.

———— 1965a. A study of population growth and control exhibited in the laboratory by prairie deermice. Ecology, 46:890–895.

———— 1965b. Adrenal gland weights of prairie deermice from populations with a long duration of asymptote. (Abstract) Amer. Zool., 5:656.

———— 1966. Population fluctuations of *Peromyscus maniculatus* and other small mammals as revealed by the North American Census of Small Mammals. Amer. Midland Nat., 76:419–426.

Terman, C. R., and J. F. Sassaman. 1966. Sex ratio in *Peromyscus* populations. (Abstract) Vir. Jour. Sci., 17:284.

Townsend, M. T. 1935. Studies on some of the small mammals of central New York. Roosevelt Wildl. Ann., 4:1–120.

Wynne-Edwards, V. C. 1962. Animal dispersion in relation to social behaviour. New York: Hafner Publ. Co., 653 pp.

———— 1965. Self-regulating systems in populations of animals. Science, 147:1543–1548.

Yanders, A. F. 1965. A relationship between sex ratio and paternal age in *Drosophila*. Genetics, 51:481–486.

ADDENDUM

Beer, J. R., and C. F. MacLeod. 1966. Seasonal population changes in the prairie deer mouse. Amer. Midland Nat., 76:227–290.

Brant, D. H., and J. L. Kavanau. 1965. Exploration and movement patterns of the canyon mouse, *Peromyscus crinitus* in an extensive laboratory enclosure. Ecology, 46:452–461.

Brown, L. N. 1966. Reproduction of *Peromyscus maniculatus* in the Laramie basin of Wyoming. Amer. Midland Nat., 76:183–189.

Gentry, J. B. 1966. Invasion of a one-year abandoned field by *Peromyscus polionotus* and *Mus musculus*. Jour Mamm., 47:431–439.

HEALEY, M. C. 1967. Aggression and self-regulation of population size in deer-mice. Ecology, 48:377–392.

HNATIUK, J., AND S. IVERSON. 1965. Habitat distribution and morphological variations of the deer mouse (*Peromyscus*) complex of northwestern Minnesota and northeastern North Dakota. Proc. North. Dakota Acad. Sci., 19:147–148.

MOORE, R. E. 1965. Ethological isolation between *Peromyscus maniculatus* and *Peromyscus polionotus*. Amer. Midland Nat., 74:341–349.

ORR, H. D. 1966. Behavior of translocated white-footed mice. Jour. Mamm., 47:500–506.

SADLEIR, R. M. S. F. 1965. The relationship between agonistic behaviour and population changes in the deermouse *Peromyscus maniculatus*. (Wagner). Jour. Anim. Ecol., 34:331–352.

SANDERSON, G. C. 1966. The study of mammal movements—a review. Jour. Wildl. Mgt., 30:215–235.

SHEPPE, W. 1965. Island populations and gene flow in the deer mouse, *Peromyscus leucopus*. Evolution, 19:480–495.

———   1966a. Social behavior of the deer mouse, *Peromyscus leucopus*, in the laboratory. Wasmann Jour. Biol., 24:49–65.

———   1966b. Exploration by the deer mouse, *Peromyscus leucopus*. Amer. Midland Nat., 76:257–277.

———   1966c. Determinants of home range in the deer mouse, *Peromyscus leucopus*. Proc. Calif. Acad. Sci., 34:377–418.

TERMAN, C. R. 1966. Population fluctuation of *Peromyscus maniculatus* and other small mammals as revealed by the North American Census of Small Mammals. Amer. Midland Nat., 76:419–426.

12

BEHAVIOR PATTERNS

John F. Eisenberg

Introduction

ETHOLOGY INVOLVES the study of the expression
and the ontogenetic development of the
postures, sounds, movements, and activity patterns of an animal. In general, an animal's
behavioral repertoire consists of relatively stereotyped units of muscular coordination organized
into sequences and elicited by a predictable

stimulus situation. The first task in the study of behavior is the
identification of these behavioral units which may be orienting
movements or fixed action patterns (Lorenz, 1957). Quantitative
studies to determine species differences or temporal patterning can
be undertaken after these preliminary descriptions have been made.
The behavioral units do not occur randomly in time, but rather
are organized into functional sequences with a specific temporal
patterning related to endogenous changes in the physiological state
of the animal, and the presence or absence of various types of
stimulus input. These sequences may be quite stereotyped, but
they may also be labile and less predictable. The determination
of the order of the behavioral units in a sequence may involve very
sophisticated statistical techniques (e.g., Nelson, 1964; Isaac and
Marler, 1963), but before such analyses can even be considered the
units must be identified. In a similar fashion the study of the
stimuli necessary to elicit a behavioral response presupposes a
knowledge of the animal's behavior patterns, and all physiological
studies concerning the neural or endocrine mechanisms underlying
an animal's activity are dependent on a thorough knowledge of
behavioral units and their temporal patterning.

The description of behavior does not end with the study of the
activity of the single animal, and even more complex analyses are
demanded when the actions of two or more interacting individuals
are considered. The behavior patterns must also be studied in
relation to the life phases of the individual. Problems of learning
and maturation must be considered together with the problem of

seasonal change in an adult's behavioral inventory. In addition to the short-term temporal considerations involving the ontogenetic aspect of behavior, one must consider the long-term, phylogenetic development of behavior. Relatively stereotyped behavior patterns, typifying a species, reflect the sustained selective pressures which have molded each species' lineage throughout time, so that in any point in time, for a given environmental circumstance, a population is physiologically, morphologically, and behaviorally adapted.

This chapter does not propose to be definitive, but rather to point out what we do know about the behavior of some selected species of *Peromyscus*. In a previous paper (Eisenberg, 1962) I have published a preliminary analysis of the behavioral patterns of *Peromyscus maniculatus gambeli* and *P. californicus parasiticus*. A detailed description of these patterns will be presented in the appendix and the body of the text will be reserved for a discussion of the functional organization of the major behavioral units.

Adult Behavior Patterns

THE SOLITARY ANIMAL.—*Locomotion*: The locomotor patterns of an animal on a plane surface are generally stereotyped. Two basic movement patterns may be distinguished: (1) the crossed extension pattern of locomotion where the contralateral limbs are in synchrony and (2) the quadrupedal ricochet where the forelimbs and hindlimbs strike the ground alternately. These two patterns may alternate in the normal forward progression of the animal, but when species are compared one mode of progression may definitely predominate in one and not in the other (Eisenberg, 1962). Tail postures may be quite variable during locomotion and the rigidity of the tail often indicates the relative amount of muscular tension in the animal's body (see below, *Exploration*).

In climbing, two similar patterns of limb coordination are discerned. The tail functions in maintaining balance (Horner, 1954). Climbing ability and the spontaneous tendency to climb are closely related and correlate well with the morphology of the animal. Long-tailed forms with a long hindfoot tend to climb readily and dexterously (Horner, 1954).

Swimming is readily accomplished by *Peromyscus*. *P. maniculatus austerus* swims by a strong, alternate kicking movement with the hindfeet. The forefeet are occasionally paddled, evidently to maintain the head above water. The body is submerged except for the head. The ears are depressed, and the tail is wriggled slightly from side to side. Upon leaving the water the animal shakes, and begins to extend and flex its body while lying on its side. Interspersed with these rubbing movements are bouts of washing movements with the forepaws and mouth, and bouts of scratching with the hindfeet. The fur may be dried in this fashion within ten minutes.

The locomotor activity of *Peromyscus* is organized into definite periods that are synchronized by light, but maintain themselves in the absence of any obvious environmental cues. Patterned activity with a periodicity of approximately 24 hours is referred to as a circadian rhythm and has been widely studied in recent years (e.g., Aschoff, 1960, and Kavanau, 1962). *Peromyscus* is active during the dark cycle and all vital activities such as foraging, hoarding, reproduction, and dispersal seem to occur during this period (Behney, 1936; Orr, 1959).

Exploration and Utilization of the Living Space: Laboratory observations in open field testing arenas indicate that *Peromyscus* will respond to a novel area by actively exploring. This exploratory behavior occurs after a brief initial period of hesitancy when the animal is prone to freeze and moves with a tense, elongate body posture, including expanded forward-directed pinnae and a rigid tail (Fig. 1). The body posture and gait alter as an animal continues to explore in a novel environment and gradually becomes familiar with it. During this period the animal may pause from time to time to rise on its hind legs and test the air. The body contours become more rounded and less elongate until the animal pauses to examine a new object, then the tail and body may exhibit a tenseness with the usual elongate configuration. Similarly, the degree of folding and the position of the pinna also vary during exploration. The pinna may be fully expanded and directed forward if the animal has been startled by some motion or sound. If the animal crouches and freezes, the ears are sometimes folded slightly and depressed. Similar variations in the position of the pinna may be discerned during social interactions.

Objects placed in the arena, such as small branches, stones, or lengths of pipe, are actively investigated, climbed on, or crawled into. The frequency of responses to different test objects may well be species specific and an inviting area of research could be pursued along these lines (see Harris, 1952, and Foster, 1959).

Field studies have repeatedly shown that the individuals of a population of *Peromyscus* tend to utilize a rather consistent area for foraging and seeking mates (Blair, 1940; McCabe and Blanchard, 1950; Blair, 1951; Howard, 1948; Nicholson, 1941). This living space has been termed the home range (Burt, 1940, 1943), within which a nest site and often several subsidiary nests or "rest stations" are located. The choice of a nest site within the home range may vary from species to species in a predictable fashion. A species such as *P. maniculatus* may show a great lability in its selection of a nest site, since abandoned burrows are often appropriated, and the nests of *maniculatus* have even been found in abandoned bird nests. Other forms may be much more specific. Species such as *P. polionotus* (Sumner and Karol, 1929; Hayne, 1936) and *P. floridanus* (Blair and Kilby, 1936) may excavate or utilize burrows as a dwelling place. *P. californicus* (McCabe and Blanchard, 1950) builds a complex ground nest of grasses and twigs, whereas *Ochrotomys nuttalli* (Goodpaster and Hoffmeister, 1954; Barbour, 1942) builds grass and leaf nests in trees.

In the laboratory special areas of the cage may be set aside for defecation and urination (Eisenberg, 1962), and there is a marked tendency to keep the nest free from feces and urine; however, field evidence indicates that nests of some species may become fouled during the rearing of litters (Howard, 1948).

Habitat Selection and Homing: There would appear to be a species-specific preference for a habitat of a certain distinctive character. Harris (1952) was able to demonstrate significant preferences in a free choice experiment by laboratory-raised stocks of two races of *P. maniculatus* for artificial habitats characteristic for each race. Further studies using *P. maniculatus bairdi* have demon-

→

Fig. 1. Behavior patterns of *Peromyscus*. Top: *Peromyscus maniculatus* exploring in their home cage. Note the extreme elongate posture of the right animal. The tail is rigid, the ears directed forward, and the eyes are open. This posture occurs in social contexts also. Middle: *Peromyscus maniculatus* washing the fur of its ventrum. Bottom: *Peromyscus californicus* in the sitting posture feeding on a sunflower seed. Note the use of the forepaws in manipulating and holding the seed.

strated the genetic basis for habitat preference which is little modified by early experience in an alien habitat (Wecker, 1963).

Numerous experiments have been performed to test the ability of *Peromyscus* to home when displaced from its home range (Murie and Murie, 1931; Burt, 1940; Stickel, 1949; Terman, 1962; Murie, 1963). Although there is some evidence from work on European murids for spontaneous direction preferences (Lindenlaub, 1955), the evidence for *Peromyscus* points to homing by means of wandering until a familiar part of its range is encountered (Murie, 1963).

Rest and Sleep: *Peromyscus* sleeps in its nest intermittently during the daylight hours. It adopts a curled body posture by sitting on its hindlegs and tucking the head under its body while the tail is generally curled around the feet. This same curled posture may be adopted while the animal lies on its side; both positions are typical for the Rodentia (Eibl-Eibesfeldt, 1958). An individual may also rest in a variety of postures, such as sitting on all fours or lying on its ventrum in a prone posture. During sleep the eyes are closed and the ears may be folded. The body temperature falls several degrees but remains steady at about $3°$ C below the normal, active temperature. These mice are not known to hibernate, but they may become torpid and hypothermic if subjected to severe cold. This hypothermic condition is a critical state, and the animals will die if they are not warmed sufficiently to produce movement, shivering, and a metabolic reestablishment of their normal body temperature. *P. eremicus* is known to become torpid and enter an aestivating state if it experiences water deprivation. This ability would appear to aid water conservation by this desert-adapted species (Macmillan, 1964).

Marking: Exudates from the anal, clitoridal, preputial, and specialized epidermal glands are frequently deposited through special movements by many mammals (Bourlière, 1956; Hediger, 1950). This is a form of chemical communication, since the exudates could theoretically yield information concerning the sex, age, physiological state, species, and individual identity of the "marker" to any passing animal. Studies by Godfrey (1958) with voles have shown that subspecies identification and preference based on odor alone are demonstrable. Similarly the importance of odors from the flank glands and clitoridal glands of hamsters in their sexual behavior has been demonstrated by Lipkow (1954). The work by

Magnen (1951) and Parkes and Bruce (1961) with the white rat and laboratory mouse has yielded further confirmation of the importance of olfaction in the social life of rodents.

Several species of *Peromyscus* show a special marking movement consisting of dragging the perineal area over the substrate (Eisenberg, 1962) ; however, the exact significance of this movement has yet to be elucidated. In addition to marking movements involving glandular areas, the urine and feces can be potential vehicles for chemical communication. Specific local depositions of these substances could serve in chemical communication.

Care of the Body Surface: Several different patterns of movement are employed in washing, scratching, and sandbathing. These patterns are discussed together since they are functionally related to the common problem of dressing the pelage, cleaning the digits and genitalia, and removing parasites.

Washing involves a relatively stereotyped sequence of events utilizing the forepaws, forelimbs, tongue, and teeth. In a typical sequence the animal licks the forepaws employing a rolling movement of the paws under the mouth. The face, head, and ears are then cleaned by bringing up the paws and wiping down simultaneously on the sides of the head. The strokes begin with the nose and gradually increase in amplitude while the head is bobbed up and down to facilitate the final reach behind the ears. The animal may then turn and comb downward with its forelimbs and paws, lick with its tongue, and nibble with its teeth at the flanks or ventrum (Fig. 1). The genitalia and perineal region may be licked during the ventrum wash. In a complete sequence the tail is grasped with the forepaws, brought up to the mouth, and licked from the base to the tip. This general sequence of activity is widespread in the Rodentia (Bürger, 1959).

Scratching consists of the stereotyped pendular movements of one hindleg directed at various points on the animal's body. The toenails are generally cleaned during a scratching bout by bringing the hindfoot up to the mouth, holding it with the forepaws, and nibbling at the claws. Washing and scratching increase in frequency during encounters with other mice. The washing pattern may be abbreviated under these circumstances and consists of the nose-wipe portion alone.

Sandbathing is of rare occurrence in the genus *Peromyscus*, but the pattern has been observed in studies of *P. crinitus*. This species has a long, dense pelage which tends to become oily and matted in the absence of soil or sand. The sandbathing movement consists of a series of extensions and flexions on alternate sides of the body, which seem to be functional in pelage dressing. The same pattern of movement is displayed by *P. maniculatus* after it has become wet, and it is probable that this behavior is common to all members of the genus. However, *P. crinitus* is unique in the circumstances in which the behavior is shown.

One may consider that three major "cleaning organs" are involved in the care of the body surface: the mouth (tongue and incisors), the forelimbs, and the hindfeet (Bürger, 1959). With the washing and scratching patterns, the animal is able to reach all of the body surface in a given time. The incisor teeth and toenails are probably quite functional in removal of ectoparasites and the adaptive value of washing and scratching cannot be overestimated (Murray, 1961).

Nest Building and Burrowing: As indicated previously in the discussion of utilization of the living space, species of *Peromyscus* have a variety of nest-site preferences; however, digging, gnawing, and nest-building movements tend to be typical for all species studied regardless of their actual nest site requirements.

Digging involves the rapid forward and backward forepaw movements until a pile of earth accumulates under the animal. Generally this pile is kicked back by one or two simultaneous thrusts of the hindfeet. Gnawing comes into play when the animals are enlarging cavities in old tree stumps or similar situations. The animal braces itself with its hindfeet and bites rapidly at the wood surface with the incisors. Backward jerks of the head are effective in dislodging splinters.

Once a suitable site has been selected, the animal begins transporting nesting material. This may consist of dry grass, hair, feathers, seed hulls, or the soft parts of plants. Gathering involves searching movements and the eventual transport by carrying quantities of the nest material in the mouth. The assembled material is generally arranged by a variable combination of five basic movements: (a) boring into the pile, (b) combing the material to the side and above with the forepaws, (c) molding the nest cup by turning movements of the body, (d) pushing movements with the forepaws and the nose, and (e) stripping stalks of plant material by holding it in the forepaws, nibbling along one edge with the incisors, and jerking the head back.

Assembly of the Foodstuffs: Gathering foodstuffs involves foraging and transport. The foodstuffs assembled are variable depending in part on the species and the seasonal availability. Seeds, berries, nuts, green vegetation, and insects predominate. *P. truei* has been seen springing at house flies, employing both the forepaws and the mouth to aid in prey capture.

The foraging animal typically sniffs at the substrate, digs at cracks and crevices, and climbs in the underbrush. The relative frequency of each of these patterns is in part species specific and thus varies from race to race or species to species. Seeds hanging above the ground are grasped in the incisors while standing on the hindlegs, whereas seeds scattered on the substrate are picked up in the forepaws and transferred to the mouth or picked up directly with the incisors. The arboreal species will readily climb small bushes to retrieve seeds with the incisors.

Transport of seeds is apparently facilitated by the internal cheek-pouches present in some species (Hamilton, 1942), although these pouches are rather weakly developed compared with the internal pouches of the sciurids. The food is either carried in the mouth or dragged by holding on with the incisors while pushing with the limbs against the substrate.

Caching is common in all species that have been studied. The cache may be of a subsidiary nature including only a mouthful of seeds buried within the home range, or quite often the cache is extensive, including a quart or more of material (McCabe and Blanchard, 1950). Concentrated caches may be made next to the nest chamber or at several preferred loci in the vicinity. The caching movements include digging, placing, and covering. The covering movements include pushing and patting back the earth with the forepaws.

Feeding involves a variety of movements, but all are quite stereotyped. Basically the animal sits in a crouch and manipulates or holds the food in its forepaws (Fig. 1). The incisors are employed to bite off small pieces that are ground between the molars and then swallowed. Seeds are shelled by biting through one edge of the hull then jerking back the head to strip away the edge of the shell. The shell may then be peeled off by pushing with the forepaws while gripping the torn edge of the hull with the incisors. Nuts are gnawed open by starting from a rough or projecting point on the shell's surface.

Individuals of *P. maniculatus austerus* reared in the laboratory do not immediately open nuts presented to them. In one experiment, hazel nuts were left in a cage with two mice for three weeks before they were opened. The nut was opened by gnawing into its side as it lay on the cage floor, and three to four holes were made before the kernel was extracted from the shell. A second presentation of nuts resulted in an immediate response by the animals, and within 24 hours two nuts had been opened; however, in this case the nuts were both opened more efficiently by gnawing one hole at the tip. These observations were duplicated on another pair of *P. m. austerus* with similar results. One could conclude that some practice is necessary to effect an efficient nut-opening technique.

Defensive and Offensive Behavior: These mice are preyed upon by a variety of predatory vertebrates, thus defensive patterns have a genuine survival value. In addition, offensive and defensive behaviors may be displayed towards competitive species (King, 1957; Brown, 1964) or conspecifics (Eisenberg, 1962, 1963*a*). Although it can deliver a serious bite if it is disturbed in the nest, the animal is more prone to run to cover if surprised by humans. Lactating females tend to remain in their nests (King, 1958; Huestis, 1933), but they can and do flee, dragging their litter members which attach firmly to the teats. The nest is defended by means of warding with the forepaws, swift bites, and nest-defense squeaks. These sounds are quite intimidating to strange mice seeking to enter an occupied nest (Eisenberg, 1962; Balph and Stokes, 1960).

Freezing is commonly displayed if the animal is disturbed by a sudden sound or movement. Inconspicuousness is enhanced during freezing not only because of the immobility but also because of the camouflaging character of the coat color which tends to match the shade of the prevailing soil (Benson, 1933; Dice, 1947). After a variable interval the animal may resume its activity or if further startled it will flee or freeze again. The mode of fleeing is variable from species to species. Arboreal forms tend to climb in bushes or shrubs, whereas the more terrestrial species and races will run along established trails in the grass or underbrush. They can and do utilize the pathways constructed by such community members as *Microtus* (Pearson, 1959); however, *Peromyscus* do not construct such runways. They are adept at seeking out routes among the relatively clear areas under the stems and twigs of their micro-habitat. The familiarity with such definite routes that an individual has within its home range must be of great survival value. Indeed it would appear that their greatest vulnerability occurs when they are juveniles dispersing through strange areas (Howard, 1948).

SOCIAL BEHAVIOR.—Social behavior is defined as all behavior involving two or more interacting, conspecific individuals. "Inter-acting" implies that the animals in question are mutually influenc-ing one another through some form of communication system. Communication is thus objectively defined by correlating a visual, tactile, auditory, or chemical signal with a response by another animal. A discrete behavior pattern of one animal causing a response on the part of a second animal permits the first animal to be designated a transmitter with the second animal designated as a receiver. Subjective questions of meaning and intent are eliminated and the task for the ethologist is the establishment of correlations among the behavior patterns of two or more animals (Marler, 1961).

Communication: A variety of behavior patterns in *Peromyscus* are of potential communicatory value given the above definition. Auditory communication involves sounds produced by the animal as it moves, sounds resulting from the physical manipulation of its environment, or actual vocalizations. Ultrasonic vocalizations have been determined for several species of rodents (Zippelius and Schleidt, 1956), but little along this line has been done with *Peromyscus* as an experimental subject. Sounds produced within the range of the human ear have been described for several species (Eisenberg, 1962) and Tables 1 and 2 list the vocal and non-vocal sounds for *P. maniculatus, P. crinitus, P. eremicus,* and *P. califor-nicus.* Figures 6 through 8 portray a sonagraph analysis of several adult and neonatal vocalizations. (See also the section on Vocaliza-tions under Ontogeny of Behavior.)

Tactile communication occurs during all phases of interaction ranging from the grooming behavior during an initial encounter to biting during an aggressive interaction. Initial contact generally

TABLE 1

AUDITORY COMMUNICATION IN ADULT *Peromyscus*

(Non Vocal Communication)

Name of sound:	Pattering	Tooth-chattering
Circumstances of occurrence:	Situations when startled	Aggressive arousal
P. maniculatus gambeli	Rarely displayed	Present
P. maniculatus austerus	Frequently displayed	Present
P. californicus parasiticus	Rarely displayed	Present
P. truei	Frequently displayed	Present
P. crinitus stephensi	Rarely displayed	Present
P. eremicus eremicus	Frequently displayed	Present

TABLE 2

AUDITORY COMMUNICATION IN ADULT *Peromyscus*

(Vocal Communication)

Name of sound:	Squeal	Defense squeal or "chit"	Mew
Circumstances of occurrence:	Response to bites or injury	Warding or during nest defense	Withdrawal from an opponent
P. maniculatus gambeli	Present	Single chit or bursts of 3–4	Absent
P. californicus parasiticus	Present	Single or in twos	Present
P. crinitus stephensi	Present	Given in bursts of 1–10	Absent
P. eremicus eremicus	Present	Single chits	Absent

involves naso-nasal and naso-anal control postures (Figs. 2 and 3). The vibrissae, mouth, and tongue are employed during this interaction. If grooming follows, the tongue, teeth, and forepaws are functional in the role of promoting tactile input.

Chemical communication has been discussed under "Marking" where the need for future research was indicated. That redback voles (*Clethrionomys*) discriminate local populations from the odor of females in estrus (Godfrey, 1958) suggests that individual recognition could be based on chemical cues also. It is possible that individual recognition necessary for the formation of a pair bond and the facilitation of mating is also based on olfactory cues. Certainly inter-individual familiarity was instrumental in inducing successful copulation in the laboratory study by Clemens (1966), and it seems reasonable to assume that chemical cues had some relationship to this recognition. Pair recognition may persist for as long as fifteen days in experiments with the murid *Apodemus* (Zimmerman, 1952).

Visual signals in nocturnal species are probably of minor importance, but we are in no position to say with any certainty just what the limits of visual discrimination are for *Peromyscus* at low light intensities. They are quite able to perceive movement at 0.5 ft-c and the use of visual communication cannot be ruled out. The upright postures (Eisenberg, 1962) are often employed during an encounter, and the assumption of this posture includes both movement and the sudden exposure of the white ventrum. This posture could induce a similar movement in an encounter, since a mutual upright with warding movements of the forepaws is a common form of interaction (Fig. 5).

The Encounter: During an encounter between two strange animals, the perceptual worlds of the interacting members are temporarily altered. A host of new stimuli are being received and a number of alternative behavior patterns may be undertaken. The animal may remove itself from the situation by moving away or fleeing, or it may stay in one position apparently doing nothing. On the other hand it may actively initiate contact, sexual, or agonistic behaviors. The outcome of an encounter is somewhat predictable to an observer if some additional information is known.

The locus of the encounter quite often determines the outcome (Eisenberg, 1962). All things being equal, an animal in familiar surroundings will initially dominate, or attack and displace an intruder. It would appear that an individual is most prone to attack or defend in the vicinity of or actually in its nest. Field evidence based on live-trapping and marking studies clearly indicates that some species of *Peromyscus* are able to maintain an area in the home range from which conspecifics of the same or opposite sex are excluded (Burt, 1940; McCabe and Blanchard, 1950). Whether this phenomenon results from an active defensive or offensive effort by the occupant or by avoidance behavior on the part of intruders is not known. Laboratory studies have shown, however, that the propensity to dominate or attack is profoundly influenced by the proximity to the nest (Eisenberg, 1962).

The physiological states of the interacting animals also influence

the expressed behavior. Aggressive interaction among males increases with the onset of breeding condition and presumably with increased levels of circulating androgen (Eisenberg, 1962; Sadleir, 1965). The nest defense by females increases markedly during their lactation phase (King, 1958). The estrous behavior of a female is instrumental in arousing the male to begin sexual behavior (Clemens, 1966). Such observations on *Peromyscus* are amply confirmed by similar studies with *Mus* and *Rattus*. Beeman (1947) found that androgen affected the aggressive behavior of *Mus*. The replacement-therapy studies of Beach and Holz (1946) and other workers have demonstrated the dependence on circulating androgen for the complete expression of male sexual behavior in the rat. Kuehn and Beach (1963) also demonstrated the shifts in female rat behavior accompanying the natural course of the estrous cycle. These same behavioral changes can be duplicated in part with estrogen and progesterone treatment.

Numerous experiments on *Mus* and *Rattus* have demonstrated the effect that learning can have on the outcome of an encounter. Allee (1942) summarized the diverse effects of experience on the behavior of *Mus musculus* in staged encounter situations, and Crowcroft and Rowe (1963) and Calhoun (1962) have demonstrated the stability of learned relationships and their effects on spatial distribution in population studies of *Mus musculus* and *Rattus norvegicus*. Thus, stable social relationships among groups of rats and mice can be created in the laboratory based in part on learned relationships resulting from encounters. Similarly, experience based on individual recognition can result in a stable rank order being established among males of *Peromyscus* run in successive encounters in the laboratory (Balph and Stokes, 1960; Sadleir, 1965; Healey, 1967). It should be added that early experience can profoundly

---

←

FIG. 2. Behavior patterns of *Peromyscus*. Top: Threat (proper) by the left animal. Note the expanded pinna and open mouth as the animal emits a threat squeal. The right animal is elongate with the ears slightly folded. Middle: Nasonasal contact by two *Peromyscus californicus*. The left animal is in the relaxed sitting posture holding a seed. The right animal is in the tense, elongate posture and has partially folded back its ears as a first sign of potential submission. Bottom: The same two animals as before. The left mouse is directing its ears forward and will advance. The right mouse shows signs of withdrawal or submission by its closed eyes and partially folded ears.

affect adult social behavior (Terman, 1963). In Terman's experiments, young *Peromyscus* isolated at weaning showed a pronounced disinclination for contact and aggregation in subsequent field tests.

It is well established that the propensity for an animal to exhibit aggressive behavior is in part a result of its heredity (Lagerspetz, 1961). Most of the research on the genetic basis for aggressive behavior has been done with *Mus musculus,* but species and race differences in the expression of agonistic behavior by *Peromyscus* are demonstrable (Eisenberg, 1963*a*), and these differences are probably related to general hereditary differences in the endocrine and metabolic physiology among species. Although the frequency and intensity of aggressive behavior varies, the basic motor patterns are similar.

A final variable influencing the outcome of an encounter is the sex of the animals encountering one another. The observed patterns of behavior are not confined solely to one sex or one circumstance. The sexes differ only in the relative frequency of the occurrence of the various behavior patterns, and seem to be potentially capable of exhibiting aspects of all patterns common to the species (Eisenberg, 1962). In general the non-parturient female is less aggressive, more prone to tolerate contact, or shows defensive patterns of behavior. On the other hand, the male is more prone to initiate contact behavior and is more apt to exhibit agonistic behavior, especially toward other males.

Forms of Social Interaction: It is possible, somewhat arbitrarily, to classify the behavior patterns into functional groups based, in part, on their temporal contiguity. *Initial contact* generally involves naso-nasal or naso-anal control, and this may be followed by *contact-promoting* behaviors such as grooming (Figs. 2 through 4). Behavior patterns of this type generally occur in adult encounters staged in the laboratory in a neutral testing arena (Eisenberg, 1962), and such contact-promoting behaviors as grooming may lead to the

→

FIG. 3. Male-female interactions of *P. maniculatus.* Top: Naso-nasal encounter stance with the mouths opposed. Note the rigid tails and folded ears. Middle: Nose to nose encounter stance between the middle and right animals. Note the submissive aspect of the center animal with its closed eye and folded pinnae. Bottom: Nose to nose encounter between a male (right) and female *P. truei.* Compare with the preceding photograph. The vibrissae only are in contact.

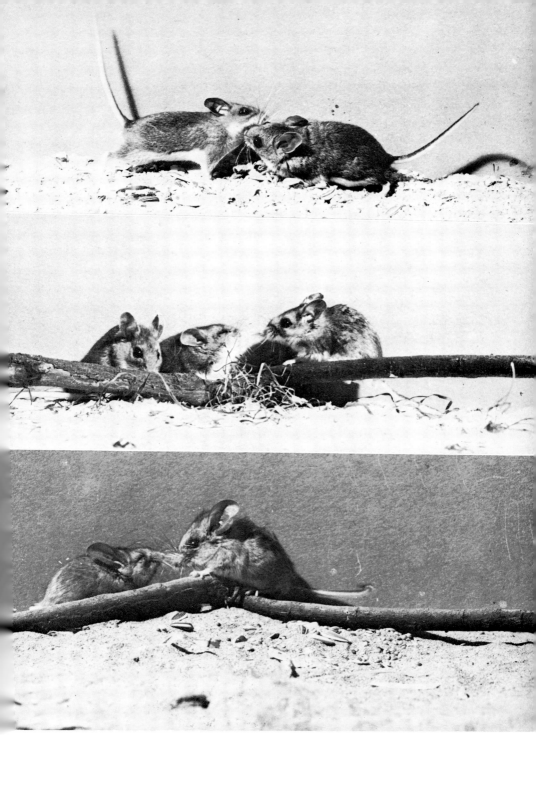

peaceful establishment of a dominance relationship. Grooming involves pushing and patting with the forepaws, licking with the tongue, and nibbling the fur with the incisors. Grooming as a precursor to sexual behavior is often qualitatively different from grooming in the dominance context. The latter grooming type is often rapid, jerky, and apt to grade into nips on the submissive animal's body.

In a *dominance-subordination* relationship one animal is able, by grooming or actual attack, to establish a superior social status with respect to another. The superior status is reflected by freer movement in a testing arena while the subordinate withdraws or moves less. Established groups of mice having a dominant member show little difference in individual behavior, but the dominant member can displace subordinates if there are competitive interactions over the possession of food (Eisenberg, 1962).

Agonistic patterns of behavior involve two sub-categories. The first includes the behaviors which precede or follow a fight, and these may be divided into upright and quadrupedal or horizontal postures. Uprights generally involve warding or striking with the forepaws; this pattern involves no real damage to either animal, but rather is defensive in character. In an upright crouch, warding is often shown during nest defense. Horizontal postures run a spectrum from slow, elongate approaches to actual rushes with bites. If the second animal flees, a chase will generally ensue. A variant of the preceding horizontal postures involves a side approach with the flank exposed to the opponent and this seems to be employed in low intensity agonistic situations (Balph and Stokes, 1960). An actual fight includes locking with the ventrums pressed together and rolling about while biting and kicking at one another (Eisenberg, 1962). (For pictures of agonistic behavior patterns, see Figure 2, top; and Figure 5.)

Sexual behavior includes those patterns associated with copulation. It is generally displayed in a male-female encounter, although mounting among laboratory stocks of male groups is common. The

←

Fig. 4. Behavior patterns of *Peromyscus*. Top: Grooming and attempted mount by *P. maniculatus*. Middle: Naso-anal encounter type in *P. maniculatus*. Bottom: Naso-anal encounter type in *P. californicus*.

estrous cycle of the female *P. maniculatus* (Clark, 1936) and *P. gossypinus* (Pournelle, 1952) shows an approximate five-day periodicity and at the time of estrus she is prone to adopt lordosis and tolerate mounting by the male. Sexual behavior involves grooming, driving, mounting, intromission, and ejaculation by the male (Eisenberg, 1962). At the conclusion of a mount the male and female separate temporarily and wash their genital areas. The mounts are less than one second in duration for *P. m. gambeli,* and several mounts with intromission precede a mount with ejaculation (Clemens, 1967). This temporal patterning is typical for many rodents (Eisenberg 1963*b*; Beach and Jordan, 1956).

Parental care patterns are most frequently displayed by the female, but in those species where the male is tolerated in the nest throughout the rearing of the young, the potentiality for the expression of male parental care exists. Males have been known to retrieve young, groom them, and huddle over them (Horner, 1947). At the onset of parturition the female shows an increased tendency to build a nest, and her sleeping nest may be greatly enlarged at this time. Nest defense increases strongly in the first week following the birth but wanes during the remainder of the rearing phase (King, 1958).

The *parturition* behavior has been studied by several workers (Clark, 1937; Pournelle, 1952; Svihla, 1932). Prior to the birth the female is lethargic and may rest on all fours, pausing to stretch occasionally. The posture during the birth generally consists of squatting on the hindlegs with the head between the legs while the female licks at the vaginal orifice. She may use her incisors or forepaws or both to pull the young from the vagina although there is individual variability in the performance of these activities. Clark (1937) reports that some females of *P. maniculatus nebrascensis* did not pull out the young with their incisors during birth, but rather squatted with their heads raised. Generally, as the young

←

FIG. 5. Agonistic patterns of *Peromyscus*. Top: Side display by *P. maniculatus*. Note the partial exposure of the white ventrum as the left animal turns away. Middle: Mutual upright displays by two *P. maniculatus*. Note the forepaws which may be used to ward off an approach. Bottom: Fighting by two male *P. maniculatus*. The body axes are at right angles as the animals roll about on the substrate.

emerges from the vagina, the female begins to lick at its head. The cord may be broken by the movements of the female or neonate or as the female devours the placenta.

After the birth is completed, the *maternal behavior* consists of grooming, nursing, brooding, and retrieving. Grooming includes manipulation of the young with the forepaws; during this behavior the mother may push or pat at the young or hold them down. The actual grooming involves licking while bobbing her head up and down. The body of the young is licked primarily in the anal-genital area. Licking the genitalia is vital for the litter during the first few days of life, since this stimulates normal urination and defecation. This reflex may be artificially elicited from the young by gently stroking the genitalia with a soft brush moistened in isotonic saline.

Retrieving and transfer to a new nest site has been observed in the wild (Hall, 1928; Smith, 1939) and studied intensively in the laboratory (King, 1963). The female is aroused by the abandoned cry of the young, and will search until she finds it. She will grasp the young by the nape or ventrum and carry it back to the nest. This response is well developed after the young are about four days old, but wanes when the young reach two to three weeks (King, 1958, 1963).

Brooding involves crouching over the young with the back slightly humped (Huestis, 1933), and the young generally suckle at this time. The young are able to cling to the teats quite strongly while suckling and the female is able to transport them by dragging them behind her. Teat transport is not strongly developed in *P. maniculatus* (King, 1963), but very characteristic of *P. gossypinus* (Pournelle, 1952), and of *P. crinitus, P. eremicus,* and *P. californicus.* It is interesting to note that this trait is well developed in the murid counterpart of *Peromyscus,* the Old World *Apodemus* (Eibl-Eibesfeldt, 1958). The length of time spent in the nest by the female begins to wane as the young become independent and take solid food (King, 1958, 1963).

Social Organization.—Some species of rodents such as the common hamster, *Cricetus cricetus,* show a dispersed distribution with the adults dwelling separately except at the time of pairing. This may be termed the closed-dispersed system. Other species, including the beaver, the prairie dog, and the Norway rat, have semi-permanent groupings of either a male and female, a male and several females, or a group of mixed males and females. These social groupings generally are relatively impermeable and may also be considered closed to outsiders; these will be referred to here as closed, clumped systems. However, one must recognize the fact that many natural social systems of rodents have a loose structure (Eisenberg, 1963*a, b*). That is to say that in contradistinction to the relatively fixed composition of the closed social systems, a loose social system is capable of organizational modification with different population densities or different seasonal changes in the physiological state of the constituent members. The loose social structures are really variations on a common theme. This theme involves a temporary pairing between a male and a female; the birth of the young; the constitution of the female-young social unit; and the subsequent separation of the littermates and the mother through the development of agonistic behaviors or through passive dispersion by movement from the home nest, or a combination of these two activities.

Social Groupings: For many species of *Peromyscus,* live trapping, marking, and releasing studies have shown that the male and female home ranges overlap to a great extent, including in part *P. m. gracilis* (Blair, 1942), *P. m. bairdi* (Howard, 1948), *P. truei* (McCabe and Blanchard, 1950), *P. m. blandus* (Blair, 1943), and *P. californicus* (McCabe and Blanchard, 1950). Some parts of the home range may be exclusively occupied by one animal especially when the ranges of the same sex are compared and in particular the ranges of the females (*P. leucopus*—Burt, 1940; Nicholson, 1941, *P. californicus*—McCabe and Blanchard, 1950, *P. truei*—McCabe and Blanchard, 1950).

From the available field evidence it appears that most species of *Peromyscus* exhibit a loose social structure during the breeding season. The duration of pairing is variable and some females will tolerate the presence of the male through parturition, as in *P. maniculatus bairdi* (Howard, 1948) and presumably in *P. m. gracilis* (Blair, 1940). Under some circumstances two females may share the same nest and apparently collectively nurse their litters. This latter behavior has been reported for *P. maniculatus* (Hansen, 1957; Howard, 1948). In contradistinction to the preceding some species exhibit a more closed social system from virtual monogamy and family group formation in *P. polionotus* (Blair, 1951) to dispersion and intolerance shown to the male by the parturient female in *P. leucopus* (Nicholson, 1941; Burt, 1940). It should be noted that outside the breeding season during the temperate zone winters larger aggregations of adults and juveniles can occur in *P. leucopus* populations.

Studies on confined populations have shown similar species differences in social tolerance (Eisenberg, 1963*a*). In territorial encounters, males of *P. crinitus, P. californicus,* and *P. maniculatus* are very aggressive. On the other hand, males of *P. eremicus* are much more tolerant. With the species showing a high female tolerance the pairing of a male and female can be prolonged in the laboratory and the male may remain with the female through parturition. This is especially true for *P. californicus* and *P. eremicus,* and to a lesser extent for females of *P. maniculatus gambeli.* In contrast, females of *P. crinitus* are relatively intolerant of the male while they are rearing young (Egoscue, 1964; Eisenberg, 1963*a*).

In summary, differences in aggressiveness and tolerance for conspecifics do exist when one compares the various species of *Peromyscus.* This is reflected in the patterns of adult spacing in the field which ranges from relatively closed, dispersed tendencies to the formation of closed family groups. On the other hand, some species seem to exhibit a lability in social structure that varies with the season and population density.

Modification of Grouping Tendencies: The loose social structure has an obvious adaptive value for species which are exploiting a relatively unstable niche. Such species usually have a high reproductive potential and in an oscillating temperate zone environment the ability to reproduce rapidly and tolerate crowding during favorable conditions could allow a maximum utilization of their environment. Current research is aimed at clarifying the behavioral mechanisms responsible for the observed differences in sociality which may be shown during different population densities of a given species.

It has been proposed that the grouping tendencies and resultant social structure may be modified in part by early experience (Terman, 1963). Thus, the increased rate of contact that juveniles may be subjected to at high densities or the effects of wintering together as littermates may profoundly affect the adult social behavior and subsequent clumping tendencies. At the same time it is important to realize that the ability to tolerate proximity at high densities may be a species-specific characteristic with a high survival value and thus be favored in selection. Howard (1950) has suggested the possible benefits accruing from clumping during the winter when heat conservation could be effected with several animals in the same nest. The defining characteristic of the loose social structure is the ability to form subgroups of mixed sexes of compatible individuals, and Frank (1957) has proposed that this social characteristic, the *Verdichtungspotential,* may be responsible in part for the population cycles of the vole, *Microtus arvalis.*

Adult aggressiveness in those species with more dispersed social systems may result in decreased survivorship by the dispersing juveniles, especially if nest sites are in short supply (McCabe and Blanchard, 1950; Sadleir, 1965). In addition, the isolation resulting from this type of spacing may serve as a form of early experience which affects the subsequent social behavior of surviving juveniles when they are adults (Terman, 1963). In those species where such early experience is important, the social structure of a population could fluctuate from season to season as a result of the different types of social experience endured by the juveniles.

*Ontogeny of Behavior**

Several excellent studies of growth and morphological development have been undertaken with various species of *Peromyscus.* The morphological aspects of development are discussed fully in Chapter 6, and I include only a brief physical description which is necessary for an understanding of the behavioral maturation.

The young of *Peromyscus* are born blind, naked, toothless, and with partially fused toes on the fore- and hindpaws. In general the incisors appear during the first week and the toes separate by the eleventh day. The naked young are pink at birth except for the haplomylomids, *P. eremicus* and *P. californicus,* which have a pigmented dorsum. The first appearance of a dorsal pelage results

---

* Editor's note: Although this subject is discussed in Chapter 6, it is retained here because of the author's original observations on vocalizations, which make up the majority of this section.

in a dark brown color, and the dorsum is generally covered by the end of the first to second week of age. The white ventral pelage is slightly slower in initial appearance. Shortly after the brown phase, the juvenal gray pelage manifests itself with the completed growth of the hair. This gray pelage may be completed in as short a time as fourteen days in *P. eremicus* or as long as 18–25 days in *P. californicus.* The gray dorsal pelage is lost with the molt to the subadult brown. At birth the young have vibrissae, but the eyes and auditory meatus are closed. No data exist concerning the neonate olfactory sense. Shortly after birth the young will respond to temperature changes and touch. The auditory meatus opens a day or two before the eyes do. Times of eye opening vary from species to species, but two weeks is an average for many forms (Eisenberg and Kuehn, unpubl. MS.).

VOCALIZATION.—The following discussion is based in part on unpublished data from a current study by the author, and includes the following species: *P. crinitus, P. eremicus, P. maniculatus,* and *P. californicus.*

There are four basic types of sounds made by the young. The *tic* sound, which occurs when the young is removed from the teat, is a concomitant of sucking activity and thus is not a true vocalization. The other sounds result from forced expirations and may be considered as true vocalizations. These include: *comfort sounds,* low, rapid "peeps" made by the neonate while the female grooms it; the *pain squeak,* a sharp sound accompanying rough handling; and the *abandoned cry,* which consists of a burst of squeals repeated at intervals and given when the young mouse has been displaced from the nest. Sounds of the neonates seem to grade into each other and appear to result from the change in the intensity of the stimulus input. Thus, when the female grooms them they produce a short, soft note. When they are displaced from the nest and subjected to cooling and alien odors, the prolonged, rhythmic, and intense abandoned cry is given. A sudden stimulus contrast produces the pain squeak or startle squeak, which is intense and short.

The abandoned cry has been studied in some detail using a tape recorder and sonagrams (sound spectographs). This cry is often scarcely audible in young *P. crinitus* and *P. eremicus,* but the rhythmic pulsating of their sides allows one to determine the num-

TABLE 3
SUMMARY OF THE PHYSICAL CHARACTERISTICS OF THE ABANDONED CRY

Species	Duration (sec) of cry		Frequency (Hz) of fundamental	
	Type 1	Type 2	Type 1	Type 2
P. maniculatus gambeli	.15–.21	.20–.40	2500–12000	12000–20000
P. crinitus stephensi	.08–.14	.11–.14	1700–2500	12000–24000
P. eremicus eremicus	——	.10–.17	——	15000–24000
P. californicus parasiticus	.40–.50	.40+	4000–12000	12000–22000

ber of cries in a burst. In general, young *P. crinitus* and *P. eremicus* have a range of one to eight and two to eight with an average of 4.5 and 4.2 squeaks to a burst, respectively. *P. m. gambeli* and *P. californicus* have shorter bursts of one to five with averages of 3.4 and 3.0, respectively. The cries of the young *P. californicus* are clearly audible.

The abandoned cry has been analyzed on the sonagraph for *P. californicus, P. maniculatus, P. crinitus,* and *P. eremicus* (Table 3). It is known from other studies (Zippelius and Schleidt, 1956) that the cries of young mice may have energy distributions including harmonics at 70,000 Hz. The greatest energy may not be carried at the fundamental frequency and the true communication channel may indeed be ultrasonic. The data discussed here do not reflect a complete analysis of the signal; however, they do give a clear picture of the sound types and the energy distributions at the fundamental frequencies. The abandoned cries are complex harmonic systems with a fundamental of around 3000 to 25,000 Hz. Each squeak may be roughly classified as one of two types: (1) a low sound (fundamental approximately 3000 Hz), rising and falling over a range of about 500 cycles and often beginning or ending in a click; (2) a high sound (fundamental approximately 25,000 Hz) ; this cry has a similar duration but is generally repeated several times with a very abbreviated low syllable (Type 1) inserted between two successive high-pitched cries.

The abandoned cry changes in its structure as a function of two variables, the age of the animal and the length of time it has been displaced from the nest. Young of one to three days produce more of the complex sounds of Type 1, with a lower fundamental, and as they mature the typical, high-pitched, repeated cry (Type 2) is

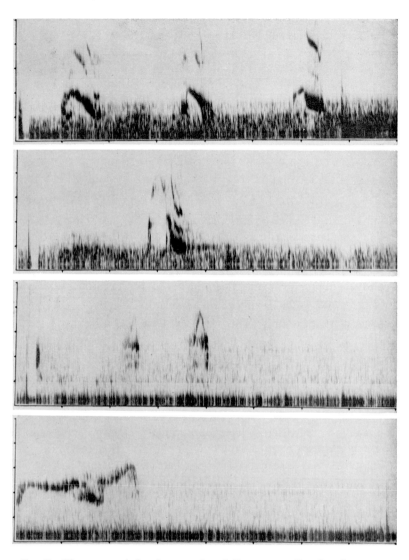

Fig. 6. Threat squeals by three species of *Peromyscus*. Reading from top to bottom: The first sonagram is from *P. maniculatus*; the second from *P. crinitus*; and the third and fourth from *P. californicus*. The ordinate is in 6000 Hz increments and the abscissa in 0.05–sec increments for the first two sonagrams. The ordinate is in 3000 Hz increments and the abscissa is in 0.1-sec increments in the bottom two sonagrams. Note that the threat sounds are brief with a rather low fundamental. The fourth sonagram illustrates two fused squeals.

TABLE 4

SUMMARY OF THE PHYSICAL CHARACTERISTICS OF
SOME *Peromyscus* VOCALIZATIONS*

Species	Warning squeak or "chit"		Comfort sound	
	Duration (sec)	Frequency of fundamental (Hz)	Duration (sec)	Frequency of fundamental (Hz)
P. maniculatus gambeli	.03–.05	4000–18000	.03	8000–9000
P. crinitus stephensi	.04–.05	3600–21000	—	——
P. californicus parasiticus	.20	4000–10000	—	——

* Exclusive of the abandoned cry.

produced with greater frequency. All pups tend to produce the complex, lower-frequency sounds on initial removal from the nest and then after a period of exposure begin to adopt the high-frequency pulses (Fig. 7). The rhythmic patterning of the cry, rather than the harmonic structure of a given syllable, seems to be its most distinctive feature. This cry wanes and disappears when the eyes are open and the pups can move about in a coordinated fashion. The times of waning for the various species are: *P. californicus,* 17–20 days; *P. eremicus,* 15–17 days; *P. crinitus,* 16–17 days; *P. maniculatus,* 14–15 days.

The adult defensive squeak or "chit," which accompanies warding and upright postures, begins to appear at about 12 days in *P. maniculatus,* and is present in all forms at three weeks of age (Table 4). The adult sounds appear to be a graded series of squeaks, culminating in the chit or defensive squeal, that are a series of expirations directly related to the intensity of the stimulus input. Species specificity may be present, but I believe this is the case only when selection has favored a discrete sound for communication in a specific circumstance such as the mewing cry of *P. californicus* which inhibits attack.

BEHAVIORAL DEVELOPMENT.—Unless specified, the following description of behavioral development is taken from *P. maniculatus gambeli,* and may be considered typical. The motor patterns develop according to an anterior to posterior gradient. The hindlimbs lag behind the forelimbs in all patterns of coordination. At birth, *P. maniculatus* can push with the forepaws, extend and flex

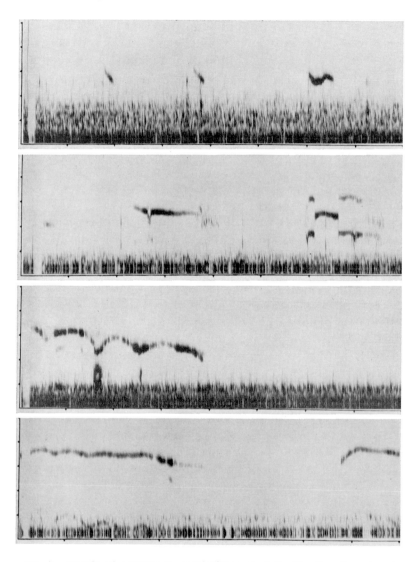

Fig. 7. Sounds of neonate *P. maniculatus gambeli*. Reading from top to bottom: The first sonagram shows the sounds emitted by the neonate while the female grooms it. These vocalizations are faint and very brief. The second portrays the abandoned cry of sound Type 1. These sounds were recorded from two-day-old young and indicate the low fundamental and complex harmonics. The ordinate is in 3000 Hz increments and the abscissa in 0.1-sec increments for these first two recordings. The third sonagram indicates syllable Type 2 of the

the body, suck, vocalize weakly, bob the head up and down, swing the head from side to side, and move the forelimbs forward and backward.

The newborn will generally rest on its side or back, but after its second day it will attempt to right itself, except for *P. californicus* and *P. eremicus* which move very little and usually lie quietly until about six days of age. *P. maniculatus* can right itself at four to six days. The animals sprawl during the first seven to ten days but gradually attempt to bring the feet under the body and crawl, using their forelimbs, as maturation proceeds. Only gradually, at eight to ten days, are the hindlimbs coordinated in locomotion. During the first six days they back up or turn rather than crawl forward on a plane surface. From 11 to 14 days all four limbs are coordinated in a crossed extension pattern and the ventrum can be held off the substrate.

Wiping the face with the forepaws has been noted as early as eight and ten days. The complete wash, including body and tail, is gradually added and appears completed at 21 days. Scratching appears by day 12.

DEVELOPMENT OF SOCIAL BEHAVIOR.—The young attach to the teats and push at the female's underside with their nose and fore-paws. This attachment is pronounced in the genus *Peromyscus,* and the female can transport the litter when she is frightened from the nest by allowing the young to cling to her teats. When leaving the nest under relaxed circumstances, the female has no difficulty in causing the young to release the teat; however, her movements during this action have not been analyzed. The young are weaned at three to four weeks. Recorded dates include: *P. eremicus,* 21 days; *P. crinitus,* 24–26 days.

During the first week of life the littermates huddle when they contact one another. They push and crawl over one another in the nest, but defensive warding does not appear until eight to ten days.

---

abandoned cry which is prolonged with a high fundamental. Note that two cries are bridged by a very short syllable resembling the Type 1 component. This tracing and the next were recorded from six-day-old young. The fourth sonagram represents two portions of syllable Type 2 of the abandoned cry. The ordinate is in 6000 Hz and the abscissa in 0.05–sec increments for sonagrams 3 and 4.

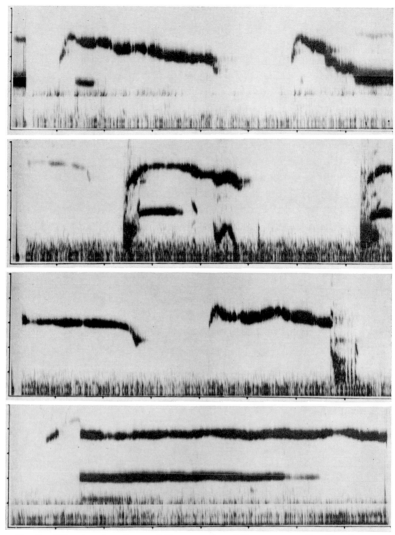

FIG. 8. The abandoned cry of three species of *Peromyscus*. Reading from top to bottom: The first sonagram represents syllable Type 2 for *P. maniculatus gambeli* at 12 days of age. The second represents syllable Type 2 for *P. crinitus* at 12 days of age. The third represents syllable Type 2 for *P. maniculatus gambeli* at 12 days of age. The fourth represents syllable Type 2 for *P. californicus* at 4 days of age. The ordinates are in 6000 Hz and the abscissae are in 0.05-sec increments for all sonagrams. Note the long duration of the *P. californicus* cry.

At about 17 days mutual grooming begins to appear, and huddling on contact becomes less frequent.

The onset of sexual maturity is variable. It may range from 30 to 51 days for females of *P. polionotus* and *P. boylei,* respectively (Clark, 1939). In general the males lag some ten days behind the females in the attainment of reproductive maturity (Clark, 1939). Littermates follow, chase, and groom one another during their first nights out in the rearing cages. In the wild, dispersal would generally be effected at the time of reproductive maturity (Dice and Howard, 1951; Howard, 1948). Prior to dispersal, the young remain in the home range of the parent and may travel as a group of littermates or with the parent (Howard, 1948). Dispersal in the wild is generally at the age of greatly increased activity on the part of laboratory-reared litters.

Adaptation and Evolution of Behavior

In a recent review, King (1961) stressed the role of behavior in directing evolution; also, the importance of genetically controlled maturation rates in the regulation of the types of experience instrumental in directing the development of behavior. Dispersal and other spacing mechanisms also affect the genetic balance in a population. Sexual isolation by mechanisms of behavioral and hybrid sterility also controls gene flow (Blair and Howard, 1944; Dice, 1933; Harris, 1954). However, these aspects of the behavioral influence on evolution are outside the realm of our consideration here. The role of early experience may be important in directing the form of the particular behavior pattern in question. What I am seeking to establish in this chapter is the recognition that the behavior patterns themselves, treated as describable and quantifiable entities, may be either taxon specific or exhibit taxon specific differences in intensity or frequency.

Within any population of animals, a morphological or behavioral character will exhibit some variability when it is measured for several individuals. Statistically, this can be expressed as the variance for a given measurement. It is conventional to assign all trait variance not resulting from environmental variance as that variance caused by genetic variability from one individual to another. The existence of genetically controlled variance can be exploited in selection experiments so that the trait variance is reduced within a selected line and the two descendant stocks differ by greater amounts with respect to the measurable trait one is selecting for. Geneticists thus define the heritability of a trait (behavioral, morphological, etc.) as the amount of trait variance not accounted for by environmental variance (Fuller and Thompson, 1960).

Selection and crossing experiments with *Peromyscus* have shown the genetic basis for differences in habitat selection (Harris, 1952), temperament, and activity (Foster, 1959). The evidence indicates that behavioral traits may have a heritability within a given population. The constant selection pressure of the environment must, however, tend to maintain a degree of population specificity to account for the observed behavioral differences when one compares various races and species. Given that some, if not all, behavior has a genetic basis, it is obvious that the behavioral differences between or among populations are the result of natural selection tending to produce an adapted population, i.e., a population selected to exhibit appropriate behavioral responses to specific stimulus situations.

Lorenz operationally defines innate behavior as that behavior appearing in the presence of biologically appropriate stimuli without a learning process by imitation or practice. The experimental condition defining the innate behavior patterns is the isolation experiment where the animal is reared in an environment which permits normal growth and muscular and sensory development, but denies the animal the sensory experience necessary to release many of the stereotyped behavior patterns to the stimulus objects which are normally encountered. If an isolate-reared animal responds in a species-specific manner to relevant stimulus situations, the behavior is termed innate. For Lorenz there are two sources of information open to the organism. One is the information stored in the central nervous system as a result of sensory experience. The second is the information contained in the genetic code which presupposes certain environmental conditions because this genetic information has been coded through the millennia of selection processes in the phylogeny of the organism (Lorenz, 1961). In this way the behavior, morphology, and physiology are genetically adapted to certain environmental conditions. According to this definition some of the species-

or subspecies-specific behavior patterns of laboratory-reared stocks could be termed innate. Or again they could be termed "species specific" in the sense of Frank Beach (1960). Whatever terminology is employed, species differences in activity (Foster, 1959), maternal behavior (King, 1958), climbing ability (Horner, 1954), habitat selection (Harris, 1952), and social behavior (Eisenberg, 1962 and 1963a) reflect behavioral adaptations resulting from selection for survivorship in differing environments for vastly differing niches.

Most behavioral units such as the motor patterns and orientating responses employed in foraging, mating, maternal care, care of the body surface, or fighting are common not only to all species of *Peromyscus* but to other rodents as well. For example, the behavioral units listed in the appendix may be found in many other genera and families of rodents (Eisenberg, 1963b; Eibl-Eibesfeldt, 1958). This is not to imply that these movements and postures are homologous from one rodent family to another; they may well be, but only ontogenetic studies can clarify this question.

The uniformity of the behavioral units found among closely related species or genera has led some workers to propose higher behavioral categories such as genus-specific or family-specific behavior. When the relevant ontogenetic data are available to permit such a decision this type of behavioral classification is permissible, but often such a system is based solely on the form of the movement pattern rather than a consideration of its temporal patterning such as the frequency with which the movement is displayed. Analyses concerned with the frequency of occurrence permit a "behavioral median" for a species to be established (Leyhausen, 1965).

In this review the author wishes to emphasize that the study of behavioral units can proceed without reference to higher categories of classification. One could list several behavior patterns that are species specific within the genus *Peromyscus,* but not necessarily unique within the Rodentia. These would include the mewing cry of the subordinate typified by *P. californicus* and the sandbathing behavior of *P. crinitus.* In addition many discrete or complex behavior patterns from several species or subspecies can be compared and definite species- or race-specific variations in intensity or frequency can be discerned, including: the relative frequency of the modified fighting technique (Eisenberg, 1963a), the number of squeals in a single burst of the abandoned cry by young (Eisenberg and Kuehn, unpubl. MS.), the intensity of nest defense by lactating females (King, 1963), and the relative excitability in a standard testing situation (Foster, 1959). It would appear that the behavior patterns of *Peromyscus* are quite uniform, and one does not often find the definite loss or acquisition of discrete motor units when species are compared. One finds instead a basic set of postures, movements, and vocalizations that in a species-specific manner vary in their frequency of expression and vary in the strength and character of the stimuli adequate to elicit them.

Summary

The behavior patterns for several species of *Peromyscus* were compared in order to define the basic movement patterns which comprise the behavioral repertoire of this genus. The ontogenetic development as well as the adult patterns of behavior were considered. The behavior patterns were discussed in terms of their functional organization for both the isolated and the socially interacting mouse. Certain behavior patterns are considered to be species specific in their form or temporal patterning and thus reflect the process of natural selection in producing an individual behaviorally adapted to a specific ecologic niche. The evolution of behavior is discussed in the light of several current, theoretical positions.

Appendix

List with Descriptions of the Postures and Movement Patterns for Adult *Peromyscus*

I.   The isolated animal
  A. Sleep and resting
     1. Curled—Animal has tucked its head under its body; this posture may be assumed either with the weight resting on the hindlegs and head or on the side
     2. Stretched—Animal has not tucked its head under the body but rests with the body stretched out
        a. On ventrum (prone)
        b. On back (supine)
  B. Locomotion (Units discussed in the text)
     1. On plane surface
        a. Diagonal (crossed extension)
        b. Quadrupedal saltation

2. Climbing
   a. Diagonal
   b. Alternate gripping with fore- and hindfeet
*3. Swimming
C. Care of the body surface and comfort movements
   1. Washing—A sequence of separate subunits (a–d) ; the pattern of these units is discussed in the text
      a. Mouthing the fur
         (1) Licking (with tongue)
         (2) Nibbling (with incisors)
      b. Wiping with forepaws and forelimbs
      c. Bobbing the head
      d. Nibbling the toenails
   *2. Scratching
   3. Sneezing or coughing
   *4. Sandbathing
   5. Stretch
   6. Yawn
   7. Shake
   8. Defecation—Fecal pellets may be dropped as the animal moves; however, on certain occasions pellets are removed with the incisors as they begin to emerge from the anus. Fecal pellets may be eaten.
   9. Urination—The animal stands still for a moment on all fours. The perineum may be depressed slightly at the conclusion as the animal starts to walk away again
   *10. Marking (perineal drag)
D. Ingestion
   1. Manipulation (of the food with the forepaws)
   2. Drinking (lapping)
   3. Gnawing—The incisors are used during this activity and the jaw articulation is different from the case for chewing
   4. Chewing—The animal uses the molars for grinding
   5. Swallowing
E. Gathering foodstuffs and caching (see also section H)
   1. Picking up
      a. With forepaws
      b. With incisors
   2. Carrying in mouth
   3. Dragging with incisors
   4. Digging (see section F)
   5. Placing—The food stuff is dropped
   6. Pushing with forepaws—The foodstuff is packed at the site
   7. Covering—Soil or material is placed over the assembled foodstuffs
      a. Pushing with forepaws
      b. Patting with forepaws

---

* See text for more complete description.

F. Digging activity
   1. Forepaw movements (forward and backward movements of forepaws with power on the back stroke)
   2. Kick back—The substrate material is collected under the body and displaced to the rear with the hindfeet
   3. Turn and push—The accumulated substrate is pushed with the fore paws to the rear

G. Nest-building
   1. Gathering (see under E, above)
   *2. Stripping
      a. Biting with incisors
      b. Jerking back with the head
   3. Pushing and patting with forepaws
   4. Combing (lateral movements of forelimbs)
   5. Molding—The nest cup is shaped by the turning movements of the animal

H. Exploring and foraging (various degrees of ear-folding and ear position may be noted)
   1. Elongate posture—The animal shows rigidity and muscular tension as it moves. The body is stretched out and elongated as it contacts or approaches a new object or conspecific. The tail may be rigid and the ears and vibrissae extended forward
   2. Upright—The animal rears on its hindlegs with varying degrees of inclination to the substrate
   3. Testing the air—In the upright or elongate posture the animal bobs its head horizontally and/or vertically
   4. Sniffing the substrate—Bobbing movements of the head with the nose close to the substrate are often accompanied by movements of the vibrissae (whiskering)
   5. Alarm postures (typified by tensing and remaining immobile)
      a. Rigid upright—The animal stiffens in a motionless upright posture. This is to be distinguished from an investigatory upright where the animal actively moves its head and is not temporarily immobilized
      b. Freezing (on all fours)
   6. Relaxed exploratory posture (lack of rigidity and tension as the animal moves about)

II. Social behavior
   A. Initial contact
      1. Naso-nasal—Animals touch noses, vibrissae, or mouths
      2. Naso-anal—This may involve sniffing and licking the anal or genital region
      3. Grooming—This behavior pattern involves patting with the forepaws while licking and nibbling the body and fur of the partner. The head is bobbed up and down
   B. Contact-promoting and sexual behavior
      1. Circling (mutual naso-anal)

2. Follow and driving—While the female moves slowly ahead the male follows behind
3. Male patterns
   a. Mounting—The male grips with the forelimbs anterior to the female's pelvis
   b. Thrusting (pelvic movements while mounted)
   c. Intromission (insertion of the penis into the vagina)
   d. Ejaculation
4. Female patterns
   a. Raising the tail
   b. Lordosis (elevation of hindquarters while assuming a frozen posture)
5. Post-copulatory washing—Animals separate to wash, chiefly the ventrum and genital area

C. Approach, flight, attack, and other agonistic patterns
1. Turn toward
2. Approach
   a. Slow approach (body contours relaxed)
   b. Elongate approach (see I, H, 1)
3. Threat (proper) —An animal remains on all fours with incisors bared and ears erect. It may dart its head at an opponent
4. Rush (generally stems from the preceding category)
5. Chase
6. Flight
7. Move away (not a fleeing movement; rather, an orientated avoidance)
8. Bite
9. Locked fight—The animals lock together with ventrums pressed together while rolling about. Occasionally the locked fight develops into a situation where one animal is defending by warding with its forepaws
10. Modified fight—A rush, and attack leap with clash and immediate separation
11. Side display—The animal presents its side as it approaches
12. Uprights (an orientation of the body with the forepaws off the substrate). The body may be inclined forward or almost vertical. The forepaws are often held out to ward, strike, or spar with the opponent
    a. Upright class 1—The subject is inclined forward with the body axis at about thirty degrees to the substrate. In this posture the animal is prone to attack
    b. Upright class 2—The body axis approximates verticality and the subject is prone to adopt warding movements rather than to rush and attack
13. Submission (withdrawal) —The animal moves little, and assumes rounded body contours with the ears folded and pressed back and the eyes half-closed

14. Defeat—The animal lies immobile on its back under an opponent
15. Kicking
16. Attack leap—An aggressor leaps into the air directly at an opponent striking the opponent with his limbs or body
17. Escape leap—An attacked animal avoids by wild and erratic leaps
18. Tooth-chattering
*19. Pattering (with the forefoot)
D. Miscellaneous patterns seen in a social context
   *1. Sandbathing
   *2. Marking
    3. Pilo-erection
    4. Trembling of the body
E. Maternal patterns
   *1. Parturition crouch (squatting with the head between the hind legs)
    2. Grasping the neonate with the incisors
    3. Manipulating the neonate with the forepaws
    4. Patting and pushing the neonate with the forepaws
    5. Grooming (see II, A, 3)
   *6. Retrieving
    7. Pulling under (with forepaws)
   *8. Nursing (brooding) posture

Literature Cited

Allee, W. C. 1942. Social dominance and subordination among vertebrates. Biol. Symp., 8:139–162.

Aschoff, J. 1960. Exogenous and endogenous components in circadian rhythms. Cold Spring Harbor Symposia on Quantitative Biology, 25:11–28.

Balph, D. F., and A. W. Stokes. 1960. Notes on the behavior of deer mice. Proc. Utah Acad., 37:55–62.

Barbour, R. W. 1942. Nests and habitat of the golden mouse in eastern Kentucky. Jour. Mamm., 23:90–91.

Beach, F. A. 1960. Experimental investigations of species-specific behavior. Amer. Psychol., 15:1–18.

Beach, F. A., and A. M. Holz. 1946. Mating behavior in male rats castrated at various ages and injected with androgen. Jour. Exper. Zool., 101: 91–142.

Beach, F. A., and L. Jordan. 1956. Sexual exhaustion and recovery in the male rat. Quart. Jour. Exper. Psychol., 8:121–133.

Beeman, E. A. 1947. The effect of male hormone on aggressive behavior in mice. Physiol. Zool., 20:373–404.

Behney, W. H. 1936. Nocturnal exploration of the forest deer mouse. Jour. Mamm., 17:226–229.

Benson, S. B. 1933. Concealing coloration among some desert rodents of the southwestern United States. Univ. Calif. Publ. Zool., 40:1–70.

Blair, W. F. 1940. A study of prairie deer mouse populations in southern Michigan. Amer. Midland Nat., 24:273–305.

————— 1942. Size of home range and notes on the life history of the woodland deer-mouse and eastern chipmunk in northern Michigan. Jour. Mamm., 23:27–36.

————— 1943. Populations of the deer-mouse and associated small mammals in the mesquite association of southern New Mexico. Contrib. Lab. Vert. Biol. Univ. Mich., 21:1–40.

————— 1951. Population structure, social behavior, and environmental relations in a natural population of the beach mouse. *Ibid.*, 48:1–47.

BLAIR, W. F., AND W. E. HOWARD. 1944. Experimental evidence of sexual isolation between three forms of mice of the cenospecies *Peromyscus maniculatus*. Contrib. Lab. Vert. Biol., Univ. Mich., 26:1–19.

BLAIR, W. F., AND J. D. KILBY. 1936. The gopher mouse—Peromyscus floridanus. Jour. Mamm., 17:421–422.

BOURLIÈRE, F. 1956. The natural history of mammals. New York: Knopf, xxi + 364 pp.

BROWN, L. N. 1964. Ecology of three species of *Peromyscus* from southern Missouri. Jour. Mamm., 45:189–202.

BÜRGER, M. 1959. Eine vergleichende Untersuchung über Putzbewegungen bei Lagomorpha und Rodentia. Zool. Gart. Lpz., 23:434–506.

BURT, W. H. 1940. Territorial behavior and populations of some small mammals in southern Michigan. Misc. Publ. Mus. Zool. Univ. Mich., 45:1–58.

————— 1943. Territoriality and home range concepts as applied to mammals. Jour. Mamm., 24:346–352.

CALHOUN, J. B. 1962. The ecology and sociology of the Norway rat. U. S. D. H. E. W., Publ. Health Service, viii + 288 pp.

CLARK, F. H. 1936. Estrus cycle of the deer mouse. Contrib. Lab. Vert. Gen. Univ. Mich., 1:1–7.

————— 1937. Parturition in the deer-mouse. Jour. Mamm., 18:85–87.

————— 1939. Age of sexual maturity in mice of the genus Peromyscus. *Ibid.*, 19:230.

CLEMENS, L. 1966. Mating behavior of *Peromyscus maniculatus gambelii*. Doctoral thesis, Univ. Calif., Berkeley.

————— 1967. Effect of stimulus female variation on sexual performance of the male deermouse, *Peromyscus maniculatus*. Proc. 75th Ann. Conv., Amer. Psychol. Assoc. Pp. 119–120.

CROWCROFT, P., AND F. P. ROWE. 1963. Social organization and territorial behavior in the wild house mouse (*Mus musculus* L.). Proc. Zool. Soc. Lond., 140:517–532.

DICE, L. R. 1933. Fertility relationships between some of the species and subspecies of mice in the genus Peromyscus. Jour. Mamm., 14:298–305.

————— 1947. Effectiveness of selection by owls of deer mice (*Peromyscus maniculatus*) which contrast in color with their background. Contr. Lab. Vert. Biol. Univ. Mich., 34:1–20.

DICE, L. R., AND W. E. HOWARD. 1951. Distance of dispersal by prairie deermice from birth places to breeding sites. Contr. Lab. Vert. Biol. Univ. Mich., 50:1–15.

EGOSCUE, H. J. 1964. Ecological notes and laboratory life history of the canyon mouse. Jour. Mamm., 45:387–396.

EIBL-EIBESFELDT, I. 1958. Das Verhalten der Nagetiere. Handb. Zool. Berl., Band 8, 12:1–88.

EISENBERG, J. F. 1962. Studies on the behavior of *Peromyscus maniculatus gambelii* and *Peromyscus californicus parasiticus*. Behaviour, 19:177–207.

———— 1963*a*. The intraspecific social behavior of some cricetine rodents of the genus *Peromyscus*. Amer. Midland Nat., 69:240–246.

———— 1963*b*. The behavior of heteromyid rodents. Univ. Calif., Publ. Zool., 69, iv + 100 pp.

EISENBERG, J. F., AND R. E. KUEHN. 1963. Reproduction, parental care and the ontogeny of behavior in four species of *Peromyscus*. Unpubl. MS.

FOSTER, D. D. 1959. Differences in behavior and temperament between two races of the deer mouse. Jour. Mamm., 40:496–512.

FRANK, F. 1957. The causality of microtine cycles in Germany. Jour. Wildlife Mgt., 21:113–121.

FULLER, J. L., AND W. R. THOMPSON. 1960. Behavior Genetics. New York: Wiley, ix + 396 pp.

GODFREY, J. 1958. The origin of sexual isolation between bank voles. Proc. Royal Physical Soc. Edinburgh, 27:47–55.

GOODPASTER, W. W., AND D. F. HOFFMEISTER. 1954. Life history of the golden mouse, *Peromyscus nuttalli*, in Kentucky. Jour. Mamm., 35:16–27.

HALL, E. R. 1928. Note on the life history of the woodland deer mouse. Jour. Mamm., 9:255–256.

HAMILTON, W. J., JR. 1942. The buccal pouch of Peromyscus. Jour. Mamm., 23:449–450.

HANSEN, R. M. 1957. Communal litters of *Peromyscus maniculatus*. Jour. Mamm., 38:523.

HARRIS, VAN T. 1952. An experimental study of habitat selection by prairie and forest races of the deermouse, *Peromyscus maniculatus*. Contrib. Lab. Vert. Biol. Univ. Mich., 56:1–53.

———— 1954. Experimental evidence of reproductive isolation between two subspecies of *Peromyscus maniculatus*. *Ibid.*, 70:1–13.

HAYNE, D. W. 1936. Burrowing habits of Peromyscus polionotus. Jour. Mamm., 17:420–421.

HEDIGER, H. 1950. Wild animals in captivity. London: Butterworths, ix + 207 pp.

HORNER, E. 1947. Parental care of young mice of the genus Peromyscus. Jour. Mamm., 28:31.

———— 1954. Arboreal adaptations of *Peromyscus* with special reference to use of the tail. Contrib. Lab. Vert. Biol. Univ. Mich., 61:1–84.

HOWARD, W. E. 1948. Dispersal, amount of inbreeding, and longevity in a local population of prairie deer-mice on the George Reserve, southern Michigan. Contr. Lab. Vert. Biol. Univ. Mich., 43:1–50.

———— 1950. Relation between low temperature and available food to survival of small rodents. Jour. Mamm., 32:300–312.

HUESTIS, R. R. 1933. Maternal behavior in the deer mouse. Jour. Mamm., 14:47–49.

ISAAC, D., AND P. MARLER. 1963. Ordering of sequences of singing behavior of mistle thrushes in relationship to timing. Animal Behaviour, 11: 179–188.

KAVANAU, J. L. 1962. Automatic multi-channel sensing and recording of animal behavior. Ecology, 43:161–166.

KING, J. A. 1957. Intra- and interspecific conflict of *Mus* and *Peromyscus*. Ecology, 38:355–357.

———— 1958. Maternal behavior and behavioral development in two subspecies of *Peromyscus maniculatus*. Jour. Mamm., 39:177–190.

———— 1961. Development and behavioral evolution in *Peromyscus. In*: W. F. Blair (ed.) Vertebrate speciation, Austin: Univ. Texas Press.

———— 1963. Maternal behavior in *Peromyscus. In*: H. Rheingold (ed.) Maternal behavior in mammals. New York: Wiley.

KUEHN, R. E., AND F. A. BEACH. 1963. Quantitative measurement of sexual receptivity in female rats. Behaviour, 21:282–299.

LAGERSPETZ, K. 1961. Genetic and social causes of aggressive behaviour in mice. Scand. Jour. Psychol., 2:167–173.

LEYHAUSEN, P. 1965. Über die Funktion der Relativen Stimmungs-hierarchie. Zeit. für Tierpsychol., 22:412–494.

LINDENLAUB, E. 1955. Über des Heimfindenvermögen von Säugetieren II Versuch an Mäusen. Zeit. für Tierpsychol., 12:452–458.

LIPKOW, J. 1954. Über das Seitenorgan des Goldhamsters (*Mesocricetus auratus auratus*). Zeit Morph. Ökol. Tiere, 42:333–372.

LORENZ, K. Z. 1957. Methoden der Verhaltensforschung. Handb. der Zool. Berl., Band 8, 8:1–22.

———— 1961. Phylogenetische Anpassung und Adaptive Modification des Verhaltens. Zeit. für Tierpsychol., 18:139–187.

McCABE, T. T., AND B. D. BLANCHARD. 1950. Three species of *Peromyscus*. Rood Associates, Santa Barbara, Calif., v + 136 pp.

MACMILLAN, R. E. 1964. Aestivation in the cactus mouse, and its significance as a water-conserving mechanism. (Abstract). Amer. Zool., 4:304–305.

MAGNEN, J. LE. 1951. Étude des phénomènes olfacto-sexuels chez le rat blanc. C. R. Soc. Biol. Paris, 145:851–860.

MARLER, P. 1961. The logical analysis of animal communication. Jour. Theoret. Biol., 1:295–317.

MURIE, M. 1963. Homing and orientation of deermice. Jour. Mamm., 44: 338–348.

MURIE, O. J., AND A. MURIE. 1931. Travels of Peromyscus. Jour. Mamm., 12: 200–209.

MURRAY, M. D. 1961. The ecology of the louse *Polyplex serrata* Burmeister on the mouse *Mus musculus* (L). Aust. Jour. Zool., 9:1–13.

NELSON, K. 1964. The temporal patterning of courtship behaviour in the glandulocaudine fishes. Behaviour, 24:90–146.

NICHOLSON, ARNOLD J. 1941. The homes and social habits of the woodmouse *Peromyscus leucopus noveboracensis* in southern Michigan. Amer. Midland Nat., 25:196–223.

ORR, H. D. 1959. The activity of white-footed mice in relation to environment. Jour. Mamm., 40:213–220.

PARKES, A. S., AND H. M. BRUCE. 1961. Olfactory stimuli in mammalian reproduction. Science, 134:1049–1054.

PEARSON, O. P. 1959. A traffic survey of *Microtus-Reithrodontomys* runways. Jour. Mamm., 40:169–179.

POURNELLE, G. 1952. Reproduction and early post-natal development of the cotton mouse, *Peromyscus gossypinus gossypinus.* Jour. Mamm., 33:1–20.

SADLEIR, R. M. F. S. 1965. The relationship between agonistic behaviour and population changes in the deermouse, *Peromyscus maniculatus* (Wagner). Jour. Anim. Ecol., 34:331–352.

SMITH, W. P. 1939. The transfer of a Peromyscus family. Jour. Mamm., 20:108.

STICKEL, L. F. 1949. An experiment in *Peromyscus* homing. Amer. Midland Nat., 41:659–664.

SUMNER, F. B., AND J. J. KAROL. 1929. Notes on the burrowing habits of Peromyscus polionotus. Jour. Mamm., 10:213–215.

SVIHLA, A. 1932. A comparative life history study of mice of the genus *Peromyscus.* Misc. Publ. Mus. Zool. Univ. Mich., 24:1–39.

TERMAN, C. R. 1962. Spatial and homing consequences of the introduction of aliens into semi-natural populations of prairie deermice. Ecology, 43:216–223.

————— 1963. The influence of differential early experience upon spatial distribution within populations of prairie deermice. Anim. Behav., 11:246–262.

WECKER, S. C. 1963. The role of early experience in habitat selection by the prairie deer mouse, *Peromyscus maniculatus bairdii.* Ecol. Monogr., 33:307–325.

ZIMMERMAN, K. 1952. Das Verhalten verpaarter Feldmäuse bei Begegnung nach Trennung. Zeit für Tierpsychol., 9:1–11.

ZIPPELIUS, H. M., AND W. SCHLEIDT. 1956. Ultraschall-Laute bei jungen Mäusen. Naturwissenschaften, 43:502.

ADDENDUM

EISENBERG, J. F. 1967. A comparative study in rodent ethology with emphasis on evolution of social behavior, I. Proc. U. S. Natl. Mus., 122:1–51.

HEALEY, M. C. 1967. Aggression and self-regulation of population size in deermice. Ecology, 48 (3) :377–392.

KAVANAU, J. L. 1967. Behavior of captive white-footed mice. Science, 155:1623–1639.

SCUDDER, C. L., A. G. KARCZMAR, AND L. LOCKETT. 1967. Behavioural developmental studies on four genera and several strains of mice. Anim. Behav., 15:353–363.

SMITH, M. H. 1967. Effects of social behavior, sex, and ambient temperature on the endogenous diel body temperature cycle of the old field mouse, *Peromyscus polionotus*. Physiol. Zool., 40:31–39.

13

PSYCHOLOGY

John A. King

Introduction

BEHAVIOR PERVADES the entire biology of an animal as an expression of its morphology, physiology, ecology, and distribution. Because of this pervasiveness, behavior has been mentioned in many chapters of this book, particularly Chapters 12 and 14. The preceding chapter by Eisenberg presented an ethogram, which is a comprehensive and initial analysis of behavior. Quantitative analyses of particular behavior patterns should follow the critical leads provided by the ethogram. Since scientific progress is rarely so orderly, the behavior of *Peromyscus* has been quantitatively examined for various purposes long before an adequate ethogram was prepared for any single species. The current chapter reviews quantitative laboratory studies of behavior from the psychological point of view. The following chapter by Falls deals exclusively with activity, which has been investigated quantitatively more than any other behavioral pattern.

Quantitative studies of the behavior of *Peromyscus* can be classified into three categories: motor patterns, sensory capacities, and learning. This classification essentially asks what the animal does, how it perceives environmental stimuli, and how does it modify its motor patterns in response to what it perceives? Questions of emotionality, temperament, and motivation are essentially answered by analysis of motor and sensory patterns. Indeed, learning and even sensory capacities are studied by measurements of motor responses. The classification employed here is somewhat arbitrarily based on convenience and areas of most study.

An adequate analysis of motor patterns, sensory capacities, and learning has not been made for any species, but the combined results of several experiments enable a few general conclusions. Many studies cited in this chapter have been discussed previously, but here the aim is to provide quantitative results and evaluate them in respect to their contribution to comparative behavior within the genus *Peromyscus*.

Motor Patterns

The raw materials for any behavioral study of animals are the motor patterns produced by the contractions of muscles or groups of muscles in particular temporal and sequential patterns. Four levels of analysis of these motor patterns can be recognized. First is the *muscular* level, in which the muscles, their origins, insertions, size, and patterns of contractions are studied. This level is exemplified by the functional anatomy of locomotion in the horse and cheetah (Hildebrand, 1959) and the facial expressions of primates (Andrew, 1963). Such detailed studies on the functional anatomy of *Peromyscus* have not been undertaken. The second level is that of *action* patterns or the overt behavior we identify as locomotion, swimming, grooming, climbing, digging, and eating. At this level, the particular muscles involved are ignored and the amount, frequency, or duration of the action pattern is measured. Quantitative studies of these action patterns in *Peromyscus* will be presented in this section. At the third level of analysis, the *resultant* level, measures are made of the results of action patterns, rather than the action patterns themselves. For example, nest building, food hoarding, and water consumption can be measured without observing the actions that shred the fibers into a nest, that bury the food in a cache, or that lap the water in drinking. This level of behavioral analysis has also been applied to *Peromyscus* and will be treated in this section. The fourth level is the *abstract* level of analysis. Abstract concepts of emotion, motivation, and learning can be derived from motor patterns. For example, an animal running rapidly about an open field or freezing motionlessly may indicate an emotional state; or an animal that repeatedly makes the same turns in a maze may reveal learning. In these instances, the action patterns provide the means for studying abstract characteristics that cannot be directly observed.

All four levels can be used to study causal mechanisms of behavior, sensory capacities, and preferences. Hormonal, genetic, and neural control of locomotion, food consumption, or learning can be examined, although the lower levels are preferable to those less direct and more abstract. Abstractions of learning and emotion require study in their own right. However, if tests of visual acuity are desired, the training of animals to discriminate distances between lines involves more confounding variables than measuring oculo-motor muscle contractions when a striped drum is rotated about the animal. For comparisons of genetic differences among species, the lower levels of analysis also have the advantage of fewer inter-vening and confounding variables.

At each level of analysis, the nomenclature and taxonomy of motor patterns is complex. Usually operational definitions involve the fewest assumptions. Thus, "climbing" can be defined by the frequency that a mouse ascends a tree in the wild or a rope sus-pended in its cage. The same motor patterns could be called "loco-motion," "vertical ambulation," or "alternate grasping and releas-ing." Climbing is not an entity apart from the measures employed and nothing is gained by making it an abstraction, despite the advantage of measuring it in many different contexts. The nomen-clature and classification used in this section will follow the defini-tions provided by the investigator reviewed.

LOCOMOTION.—Locomotion is a form of spatial propulsion: crawl-ing, walking, running, climbing, swimming. Climbing and swim-ming will be discussed separately. Locomotion is here defined as limb propulsion on a horizontal, solid surface, which includes crawling, stepping, walking, running, and bounding. All of these action patterns require orientation to gravity, which is achieved primarily through the vestibular apparatus.

Quantitative differences in locomotion measured by activity wheels occurred between species, between individuals, and within individuals from one day to the next (Dice and Hoslett, 1950). Among the eleven stocks of *Peromyscus* compared, *P. polionotus* had the fewest wheel rotations and *P. m. bairdi* and *P. m. austerus* the most. Other stocks were intermediate. The amount of running during a 24-hour period averaged about seven miles, which is amplified by coasting with the mouse both in and out of the wheel. The actual amount of locomotion under natural conditions is considerably less, although quantitative measures are rare. Trails made by *P. polionotus* in the sand, for example, averaged 262 feet for the longest trails from 65 sets of tracks, with a maximum of 789 feet (Blair, 1951). Even if these distances are doubled for a return trip to the burrow, they are considerably less than those recorded in activity wheels.

The action pattern of locomotion in activity wheels is primarily running or bounding, but with practice the mice exhibit many skills in the wheels. They hop in and out of a rapidly rotating wheel; they coast on the rim both inside and outside; they leap over the axle to the opposite side; they change their running speed, which allows them to run down the incline or up the incline until they are upside down. Many mice actually prefer square wheels or wheels with low barriers inside the rim, which requires split-second leaping (Kavanau and Brant, 1965). The speed of running has been calibrated by tachometer generators attached to the wheel axle (Kavanau, 1963*a*). After a "warm up" period, one *Peromyscus maniculatus* maintained a remarkably constant speed of 70 revolutions per minute for several hours in an activity wheel 11 inches in diameter. Although the mouse frequently stopped the wheel, most of seven continuous hours were run at approximately 2.3 miles per hour. The wheel probably restricts the mouse from attaining greater speeds, which in short bursts may go to four or even five miles per hour.

Locomotion is frequently used to measure activity and periodicity, which are treated in Chapter 15.

SWIMMING.—Like most other terrestrial mammals, *Peromyscus* can swim. They do so by using the locomotory pattern of running, sometimes employing their tails in a sculling motion. The mice do not readily enter the water, although the propensity to do so may be species specific. They have been observed to enter the water when pursued by predators or other mice (Teeters, 1946; Orr, 1933). Sheppe (1965) released *P. leucopus* into the water from a boat and found they could swim one-half hour without difficulty, although the swimming was interrupted by floating rest periods. Swimming distances up to 765 feet were recorded between islands and mainland (Sheppe, 1965).

Young *Peromyscus maniculatus bairdi* and *P. m. gracilis* exhibited coordinated swimming motions at 10, 12, and 14 days of age when placed in water 27° to 30° C (King, 1961). *P. m. bairdi* at ten days kept their heads out of the water, but *P. m. gracilis* arched their backs until their heads became submerged. The young mice swam without direction until approximately 16 days of age, at which age *P. m. bairdi* was capable of swimming across a tub of water to an elevated platform in significantly less time than *P. m. gracilis.* This difference reflects the more rapid maturation of *P. m. bairdi.*

The readiness with which mice of several species entered water from an exposed wire mesh platform located in the center of a tub of water differed (King, Price, Weber, 1968). *Peromyscus polionotus* and *P. leucopus* entered the water significantly more often and *P. m. gracilis* significantly less often than *P. floridanus, P. californicus, P. eremicus,* or *P. crinitus,* which were intermediate. In general, those species entering the water most frequently were also less hesitant to do so, since their latency to enter was shorter. The test situation probably measured the relative aversiveness of the wire platform and the water: those species most aversely affected by sitting exposed on the platform and least aversely affected by entering the water tended to swim most often. The relative aversion to water by the species tested did not conform to any phylogenetic or ecologic pattern.

CLIMBING.—Many species and subspecies of the genus can be separated into two forms: long-tailed, forest, brush, or rock-inhabiting forms and short-tailed, grassland-inhabiting forms. Intermediate forms are common, but the extremes are readily distinguished. In general, the long-tailed, semi-arboreal types are more common than the short-tailed, terrestrial types. Apparently the genus readily responds to the selection pressures of forest and grassland habitats in terms of morphology and behavior. Since the morphological differences are easily discernible, some attempts have been made to measure climbing behavior. Field observations, derived primarily from the behavior of mice released from live traps, have provided some insight into the climbing adaptations of various species (McCabe and Blanchard, 1950), but laboratory investigations have been limited.

The development of clinging as measured by the duration young mice hung to an inverted wire screen was examined in two semi-arboreal forms, *P. m. gracilis* and *P. eremicus,* and two terrestrial forms, *P. m. bairdi* and *P. polionotus* (Price and King, 1966). One semi-arboreal form, *P. eremicus,* hung inverted longer than the other three from six to twenty days of age. The two terrestrial forms, *P. m. bairdi* and *P. polionotus,* hung for significantly shorter periods during this age period, but the rate of development

of this response was similar to that of *P. eremicus*. In contrast to
the other three forms, *P. m. gracilis,* a semi-arboreal form, was slow
in developing the clinging response, but at the last day tested,
25 days of age, it clung longer than the others. Cross-sectional
groups (see Early Learning) of the stocks reveal different developmental patterns of clinging in some, indicating that prior experience can contribute to this response. These results indicate that
clinging patterns emerge differently in each stock and that their
subsequent development may be affected by prior clinging experiences.

Adult climbing performance has been studied in a variety of test
situations with four terrestrial forms, *P. m. bairdi, P. m. nebrascensis, P. m. blandus,* and *P. polionotus,* five semi-arboreal forms,
P. m. gracilis, P. m. oreas, P. leucopus, P. nasutus, and *P. truei,* and
one intermediate form, *P. m. rubidus* (Horner, 1954). The mice
were tested for their inclination to climb, reaction to high places,
vertical climbing, crossing of gaps, climbing on small branches, and
climbing after their tails had been amputated. All were capable
climbers, and made similar body and tail motions in climbing, but
the semi-arboreal forms performed better and used their tails more
effectively than did the terrestrial kinds. That the semi-arboreal
forms relied upon their tails for climbing was demonstrated by
amputating their tails, which impeded their climbing more than
it did that of the terrestrial forms. The behavior of the semi-
arboreal forms suggested caution or deliberation in climbing as
though they were aware of a more precarious position. Terrestrial
forms appeared more reckless and tended to fall more frequently.
Semi-arboreal forms apparently are characterized by a constellation
of behavior patterns, morphological adaptations, and temperaments,
which distinguish them from terrestrial forms. A descriptive evaluation of such possible constellations are well presented in McCabe
and Blanchard (1950) for three species of *Peromyscus: truei, maniculatus,* and *californicus.*

The semi-arboreal *P. m. gracilis* climbed small boxes in an open-
field situation more frequently than the terrestrial *P. m. bairdi*
(Foster, 1959); a result which further documents the polarity of
climbing responses in these two subspecies. In a situation resembling one of Horner's (1954) tests, eight stocks were examined
for their frequency to leave an elevated platform and climb out on

a horizontal limb constructed of 1½-inch wooden dowelling, which tapered to ⅛ inch at the distal end (King, Price, Weber, 1968). *P. floridanus* was least likely to venture out on the limb, and *P. crinitus* initially went out more often than *P. eremicus, P. leucopus, P. m. bairdi,* or *P. polionotus,* but not more than *P. californicus* or *P. m. gracilis.* When the test was repeated on the following day, many more individuals of *P. polionotus* went out on the limb, which is a response not associated with their terrestrial habits. The results of the test only approximated predictions of climbing behavior based on the habitat of each species. Many other parameters of the behavior of these species were measured in addition to the tendency to climb.

The choice of elevated versus surface nest boxes has been offered to several species. In a large enclosure with nest boxes located on the surface and on top of logs, *P. m. bairdi* almost invariably selected the elevated nest boxes in contrast to what might be expected from the terrestrial habits of this subspecies (Evans, 1957). The two semi-arboreal forms, *P. leucopus* and *P. gossypinus,* similarly selected elevated nest boxes more frequently than those at the surface when either species occupied the cage alone (Taylor and McCarley, 1963). When the two species were combined in one situation, *P. gossypinus* used the surface nest boxes more frequently than did *P. leucopus.*

DIGGING.—Fossorial habits in *Peromyscus* are not well developed; however, members of a few species like *P. polionotus* regularly excavate burrows and nest chambers in sandy soils (Hayne, 1936). Others, like *P. floridanus,* often live in the burrows of other animals and probably do little digging themselves (Blair and Kilby, 1936). Most species of *Peromyscus* probably make minor excavations for their refuges under rocks, logs, or tree stumps. The propensity to dig should relate to their utilization of the habitat: terrestrial, grassland forms being more inclined to dig than semi-arboreal, forest forms.

The only quantified study of digging reported does not unconditionally support this hypothesis (King and Weisman, 1964). When *P. leucopus, P. maniculatus bairdi, P. m. gracilis,* and *P. floridanus* were given an opportunity to dig an unlimited quantity of sand out of a tunnel, only *P. m. gracilis* dug significantly less sand than the other species. Since *P. leucopus* is also a semi-arboreal species, it should not have dug more sand than *P. m. gracilis*. It is possible that the digging situation in this experiment (removing sand from a plastic tunnel) measured some type of behavior other than the propensity to dig. Motivation to escape is a possible alternative behavior. That digging itself can provide reinforcement, however, was demonstrated in a subsequent experiment reported in the same paper. *P. leucopus* learned a simple spatial discrimination with the opportunity to dig sand as the only reinforcement. This learning suggests that sand digging was sufficient motivation for the large quantities removed from the tunnel by this species.

Sand removal from a tunnel was also tested in eight stocks of *Peromyscus* for a 15-minute period on each of two days (King, Price, and Weber, 1968). The mice were placed in a small plastic cage (6 × 6 × 12 inches) joined to a similar cage by a tunnel filled with sand. Approximately nine pounds of sand had to be removed before the mouse could traverse the tunnel. The number of mice digging at least 1½ ounces of sand was used for species comparisons. Approximately 25 mice of both sexes were tested in each stock. In contrast to their 24-hour digging performance, fewer *P. leucopus* dug than any other stock suggesting that the 15-minute period was not sufficiently long for them to habituate to the situation. The other species fell roughly in two groups—diggers and non-diggers. These groups were significantly different from each other. *P. polionotus*, *P. floridanus*, and *P. crinitus* were the digging species, with at least 60 per cent of the animals digging. *P. m. bairdi*, *P. m. gracilis*, *P. leucopus*, *P. californicus*, and *P. eremicus* were the non-digging species, with fewer than 40 per cent of the animals digging. This dichotomy is consistent with the habitat of these species, with the possible exception of *P. m. bairdi* whose terrestrial habits should have included them among the diggers. More *P. m. bairdi* dug than any of the other non-digging species and they did not differ significantly from the digging *P. crinitus*. Of the mice that dug, *P. polionotus*, *P. floridanus*, and *P. m. bairdi* dug the most sand, which is consistent with the numbers of mice digging. However, 15 of the 24 digging *P. crinitus* removed the least amount of sand and ten of the 32 digging *P. m. gracilis* individuals removed almost as much sand as *P. m. bairdi*. These results, while not definitive, do suggest that the propensity to dig sand is related to the habitat occupied by the species.

NEST BUILDING.—The action patterns in nest building may include locomotion to the source of nest material, climbing, gathering, transporting, shredding the material, and shaping the nest. On occasion one or more of these action patterns have been analyzed separately. Thorne (1958), for example, quantified the amount of paper shredded. Usually the resultant behavior is measured, such as, the size, shape, and location of the nest.

Nest sites and nest building of *Peromyscus* have been discussed in a recent review (King, 1963), which pointed out the opportunism of most species in their utilization of any available site and material for their nests.

The effects of light, temperature, and humidity on the nest-building behavior of *P. m. osgoodi* has been analyzed by Thorne (1958). Paper towels were provided each day and the amount (percentage) shredded was measured. The quality of the nest was rated on a scale of 0 to 9. The quality and size of the nests were inversely related to temperature, which suggests the function of the nest as a thermoregulatory device for these small rodents. Light was a less critical determinant, but tended to reduce nest building at the three temperatures examined. The combination of high temperature and light produced the situation in which mice built their smallest nests. Humidity apparently did not affect nest building and was not systematically investigated.

A comparative study of nest building among four stocks of *Peromyscus* originally from wide-spread geographic localities suggested a north-south geographic cline in the amount of nest material used (King, Maas, and Weisman, 1964). Stocks of *P. polionotus* and *P. floridanus* obtained from Florida pulled significantly less cotton from a dispenser for their nests than did *P. m. gracilis* from northern Michigan. A stock of *P. m. bairdi* from southern Michigan used an intermediate amount of cotton. These results suggested that the propensity to build nests of suitable size under constant light and temperature conditions of the laboratory is related to the climatic necessity for each genotype to do so in the field. That acclimatization to summer and winter temperatures can also contribute to nest building was indicated by the quality of nests built by *P. l. noveboracensis* caught during the summer and winter and subjected to temperatures of approximately –30° C (Sealander, 1952*a*). Summer-caught mice built poorer nests than those caught in winter, at these low temperatures.

The factors known to control nest building in *Peromyscus* are temperature, light, genotype (species differences), and acclimatization. Hormonal factors related to reproductive activity and metabolism (thyroid) have not been investigated as they have been in laboratory rats (Richter, 1937) and mice (Koller, 1962).

FOOD AND WATER CONSUMPTION.—The action patterns in eating, such as the gnawing patterns used by squirrels in opening nut shells (Eibl-Eibesfeldt, 1957), have not been systematically studied. Some information on the frequency of eating and drinking as well as their durations is available (French, 1956; Kavanau, 1962*a*, 1963*a*). However, most quantitative measures are on the resultant behavior —the amount of food and water consumed.

Food consumption was measured in *P. leucopus, P. m. bairdi,* and *P. m. austerus* as a function of air temperature and previous thermal acclimation (Sealander, 1952*b*). In general, food consumption decreased with increasing temperature, but at the same temperature, mice with a warm thermal history consumed more food that those with a cold thermal history. Since the previous thermal environment of mice may contribute to the quantity of food consumed, species comparisons are valid only when food consumption is measured at the same temperature and after mice have acclimated to that temperature for at least ten days. On a diet of Purina Laboratory Chow, twelve *P. leucopus* consumed an average of 4.10 grams (0.16 g/g body wt.), over a ten-day period, eight *P. m. bairdi* consumed 3.55 grams (0.20 g/g body wt.), and seven *P. m. austerus* consumed 2.32 grams (0.14 g/g body wt.) after 19 days at 20.5° C (Sealander, 1952*b*). The differences in food consumption per gram body weight may be a function of activity or digestion and assimilation which were not measured.

By recording the frequency and duration of trips to separate chambers containing either food or water, French (1956) found that ten *P. m. sonoriensis* on an *ad libitum* schedule visited the feeder a mean of 23.7 times for a total duration of 49 minutes, and consumed 2.54 grams of Purina Chow. The mean number of drinker visits was 8.9 times, for a total duration of 12.6 minutes, with 2.65 cc of water consumed. Patterns of frequency and duration to the food and water chambers changed when the mice were deprived to 30 per cent of the normal water intake. The amount of food consumed was also proportionally reduced with the mice on a water deprivation schedule.

Frequency and duration measures of mice entering water and food chambers do not provide actual drinking or eating data because the mouse may go to the chamber without eating or drinking. A more reliable measure of drinking behavior can be obtained by requiring the mouse to hold aside a shutter covering the water tube (Kavanau, 1962*a*) or by using a drinkometer (Schaeffer and Premack, 1961). An eatometer has also been devised to measure the frequency and duration of eating (Fallon, 1965). Any operant response by which the animal presses a bar or lever and receives food or water can also adequately measure eating and drinking responses.

A single female *P. maniculatus* drank at the rate of 1.42 cc per minute in drinking bouts of three to six seconds, with few bouts longer than 15 seconds, as measured by the mouse pushing aside a shutter over a water bottle (Kavanau, 1962*a*). Another mouse of the same species drank at the rate of 0.90 cc per minute with a total consumption of 3.82 cc per day (Kavanau, 1963*c*). The number of 97 mg Noyes food pellets consumed per day were also measured in two *P. maniculatus,* first living in isolation and later together. The number of pellets consumed by these mice varied from about 33 to 59, depending on the individual, the types of behavior necessary to obtain the food, and the social situation (alone or together). Another record of eating and drinking in a single male *P. maniculatus* provided by Kavanau (1963*c*) shows 4.17 grams of food per day and 4.53 cc of water; the latter being consumed in 54.3 drinking bouts for a total of 2.92 minutes. These studies suggest that *Peromyscus* is primarily a nibbler and a sipper, that is, it eats and drinks in small quantities, and does so frequently in the laboratory. To what extent these habits prevail in the field is not known, but they probably exist whenever both food and water are readily available.

The quantity of water consumed as related to the physiology of water metabolism has been thoroughly reviewed in *Peromyscus* and

TABLE 1

MEAN DAILY WATER CONSUMPTION IN *Peromyscus,* CC PER GRAM BODY WEIGHT

Species	No.	Mean body weight, grams	cc/g/day	Source
P. l. tornillo	20	31.1	.06	Lindeborg (1952)
P. l. noveboracensis	40	21.9	.12	Lindeborg (1952)
P. l. noveboracensis	5	22.9	.11	Odum (1944)
P. l. noveboracensis	12	—	.13	Dice (1922)
P. m.bairdi	12	—	.15	Dice (1922)
P. m. bairdi	80	18.6	.22	Lindeborg (1952)
P. m. blandus	60	24.1	.11	Lindeborg (1952)
P. m. nebrascensis	38	19.6	.10	Lindeborg (1952)
P. m. gracilis	30	21.5	.12	Lindeborg (1952)
P. m. epigus	20	23.8	.13	Lindeborg (1952)
P. m. sonoriensis	10	19.1	.16	French (1956)
P. m. gambeli	10	20.4	.11	MacMillen (1964)
P. m. rufinus/osgoodi*	26	21.0	.16	Williams (1959a)
P. m. rufinus/osgoodi†	28	17.6	.25	Williams (1959a)
P. eremicus	40	21.4	.11	Lindeborg (1952)
P. eremicus	10	18.9	.12	MacMillen (1964)
P. californicus	10	34.3	.12	MacMillen (1964)
P. truei	20	32.4	.08	Lindeborg (1952)
P. comanche	18	22.6	.16	Lindeborg (1952)
P. floridanus**	4	33.0	.20	Fertig & Layne (1963)
P. floridanus††	4	35.4	.11	Fertig & Layne (1963)

* Low relative humidity
† High relative humidity
** Field caught
†† Laboratory

other mammals (Chew, 1965). Many studies have attempted to relate water consumption to the habitats occupied by each species. The relationships between arid habitats and water consumption are not well established largely because of considerable variability in water consumption and the limitations of the *ad libitum* measure. Physiological tolerances indicated by maintenance of body weight during water deprivation or with hypertonic saline solutions may provide more suitable comparisons than *ad libitum* water consumption (Chew and Hinegardner, 1957).

Table 1 presents water consumption for 24 hours calculated on the basis of cc consumed per gram body weight. The table must be read with reservation because not all mice were treated the same in respect to diet, ambient temperature, relative humidity, and techniques of measurement. For example, water consumption differed in *P. floridanus* as a result of being maintained in the laboratory (Fertig and Layne, 1963). These authors also used the technique of providing water with various concentrations of salt, which enabled them to conclude that the physiological tolerances of *P. floridanus* resemble those of mice from xeric regions. The effect of diet and relative humidity has also been thoroughly investigated by Williams (1959*a*).

A comparative study of behavior patterns for eating and drinking, particularly during periods of deprivation or when placed on lean fixed ratio schedules of reinforcement, might reveal behavioral adaptations to various types of habitats as well as physiological adaptations. Unfortunately, such studies have not yet been undertaken.

GROOMING.—All species of *Peromyscus* devote considerable time to grooming, scratching, and maintenance of the body surface. The bodily movements in grooming are presented in Chapter 13. Quantitative measures of the frequency of grooming responses are given by Foster (1959) in two situations. During a ten-minute observation period in an open-field situation, *P. m. bairdi* groomed significantly less than *P. m. gracilis*; however, the subspecies did not differ during a 2.5-minute period in a two-compartment apparatus. Species differences in this response probably are less a function of the relative care given to the body surface than a measure of the propensity to engage in this response over another (such as freezing) in specified situations. For example, an animal in a strange open field can run about, sit motionless, or sit and groom. In the open field used by Foster, apparently *P. m. gracilis* had a higher probability of grooming than did *P. m. bairdi*. Reference to grooming as "displacement" probably contributes little in the absence of knowing exactly what it is displacing. Laban (1966) also found the amount of grooming activities to vary in accordance with the situation in which *P. maniculatus rubidus* were confined.

VOCALIZATION.—Vocal patterns offer many quantitative measures for species comparisons, but studies of vocalizations in *Peromyscus* have just started. Descriptions of vocalizations (Eisenberg, 1962) are being replaced by quantitative measures (see Eisenberg, Chapter 12). More audible squeaks were uttered by infant (3 to 20 days old) *P. m. gracilis* than by *P. m. bairdi* during a five-minute test period (King, 1963). Sonagraphic analyses of three-day-old young of these two subspecies and *P. floridanus* also revealed considerable differences in the pattern of their vocalizations (King, 1963). In a recent study (Hart and King, 1966) vocalizations in *P. m. bairdi* and *P. m. gracilis* from three to 15 days old were examined, using sets of squeaks per 30 seconds, number of pulses per set, maximum frequency, pulse duration, and interpulse interval for analysis. The subspecies differed in several of these parameters and changes occurred with increasing age of the mice. In general, *P. m. gracilis* squeaked more than *P. m. bairdi* and the maximum frequency of *P. m. bairdi* was higher, at 25.3 kHz; *P. m. gracilis* at 20.5 kHz. Future analyses of adult vocalizations may provide evidence for phylogenetic relationships and species isolating mechanisms.

CONCLUSIONS.—Since motor patterns are the basis for the analysis of all other behavior, it is important that they be thoroughly examined and quantified. Comparative studies of learning and motivation, for example, are worthless if motor patterns vary among the species. If one species jumps or runs more than the other, learning may depend upon the frequency of these responses more than on the capacity to learn. Quantifying motor patterns, however, is no simple task. Each test situation may differentially affect the responses of each species. For example, in one 24-hour test, *P. leucopus* dug more sand than the other species, but in a somewhat different test which lasted 15 minutes, *P. leucopus* dug less sand than other species. Here the critical variable was apparently duration of the test, which completely reversed the digging performance of *P. leucopus*. Many other examples could be selected to illustrate how the type of response depends upon the test situation.

The test may actually measure some characteristic of the mice other than the motor pattern under examination. Fear, timidity, activity, motivation to escape, as well as other temperamental characteristics may determine the motor pattern, but these characteristics are abstractions and are farther removed from the observations than is the motor response itself. This recurring dilemma of all behavior studies, in which the motor pattern depends upon the motivational state of the animal and measurement of the motivational state depends upon some pattern of motor response, makes a choice difficult for the investigator. The safest choice, however, is the most parsimonious, which in this dilemma is to conclude that the motor patterns differ in their frequency, duration, intensity, or propensity. The advantage of this choice is that the motor patterns can be examined in many different situations, the most valid being those prevailing under natural conditions. After the motor patterns have been measured in a variety of situations, a reasonable comparative statement regarding them can be made, such as, "In most situations *P. m. gracilis* climbs more than *P. m. bairdi,* which is compatible with behavioral adaptations to their habitats and not with the phylogenetic proximity of the two subspecies."

The review of motor patterns in this section obviously does not permit many such statements. Many studies reviewed were not comparative, since only one species or subspecies was examined. Other studies did not provide a sufficient number of situations to permit a valid assessment of the behavior. This is not criticism of the studies, because the investigators were often not attempting comparisons or behavioral validations, regardless of the aims of this book. Therefore, any attempt to cast these heterogeneous studies under one framework is hazardous and tentative.

In the absence of studies at the muscular level of motor behavior, we must conclude that species do not differ in the topography of their motor patterns. That is, they will use the same muscles in roughly the same sequence when running, climbing, digging, eating, etc. Probably all members of the genus are physically capable of performing all the motor patterns described. The primary difference between species is the frequency, intensity, duration, or propensity to perform each pattern. When a mouse is placed in a particular test situation, which of all the measurable response patterns does it exhibit? Is it most likely to exhibit running, jumping, climbing, grooming, or sitting motionless? In the terms of the dependent variables used to measure these patterns, we can compare species and subspecies.

For predicting the behavior of an individual, the habitat occupied is generally better than its phylogenetic position. Closely related subspecies from different habitats are more likely to diverge behaviorally than are species of different subgenera occupying the same or similar habitats. This is another way of saying that convergence in the behavior of the genus is more common than homology, which in turn indicates that behavioral characteristics readily respond to selection pressures of the habitat. When the behavior tested does not conform to the expected utilization of the habitat, phylogenetic relationships usually offer no better alternative. Either we have not adequately tested the behavior or our concepts about habitat utilization are wrong for the species concerned.

Climbing or the constellation of patterns associated with a semi-arboreal life illustrates behavioral convergence to habitat utilization best, perhaps because it has been most thoroughly investigated. Water consumption, which has also been carefully examined, tends to show a similar relationship, but suitable tests are lacking in many of the studies. Most of the other motor patterns have not been investigated in a sufficient number of species or situations to support this conclusion. Indeed, several studies suggest the opposite, but these studies often used only one test situation, which may not have been a valid test of the motor pattern. Obviously, more research is required before the above conclusions are justified.

Perception

Perception, as distinguished from sensation, pertains to the neural organization of sensory stimuli and is often measured by behavioral responses. In contrast, sensation refers to the stimulation of a sense organ and can be measured by physiological responses. For example, two species may have the same sensory capacity to be stimulated by light of a given intensity, but one species may perceive it as a noxious stimulus and avoid it, whereas the other species perceives it as a positive stimulus and approaches it. In this section, we are concerned with perception and how an animal responds behaviorally to stimuli.

In many respects, perception is the basis of behavior because it determines what stimuli an animal will respond to and often how it will respond. The same stimulus, a tree, for example, may be perceived as an obstacle in the path, a source of cover or food, a refuge, or a predator's lair. Habitat selection and utilization, food resources, interspecific competitors, predators and social associations are determined by an animal's perception of its environment.

The usual physical and chemical stimuli to which organisms respond will provide the organization for this section: light, sound,

TABLE 2
MEAN WEIGHT OF BODY AND LENS IN *Peromyscus*

Species	Body weight (grams)	Lens weight (milligrams)	Lens/body (mg/g)
P. polionotus	13.3	21.8	1.64
P. leucopus	21.3	27.1	1.28
P. floridanus	37.8	46.9	1.24
P. m. gracilis	24.0	26.5	1.10
P. m. bairdi	17.2	18.2	1.06
P. eremicus	20.0	17.4	.87
P. californicus	53.1	29.4	.29
Mus musculus	28.4	8.2	.29

chemical odors and tastes, temperature, humidity, and gravity. Each stimulus has its receptor organs and response patterns, which will be dealt with here as vision, audition, olfaction, gustation, thermoregulation, and equilibrium.

LIGHT.—One of the most outstanding morphological characteristics of members of the genus are their large eyes. Compared with the house mouse of approximately equal body weight, the weight of the lens per unit body weight of *Peromyscus* may be more than five times that of *Mus*. Within the genus, eye size, measured by lens weight, varies considerably from one species to the other as shown in Table 2 (King, 1965). This divergence cannot be accounted for phylogenetically, nor does any single environmental factor explain the differences. One would expect that large eyes would increase the amount of light and give more resolution of the image on the retina (Walls, 1942). Even if the large eyes do enable better vision with minimal quantities of light, the adaptive significance of this ability is not understood. One might hypothesize that arboreal locomotion, spatial orientation, or the presence of aerial predators are significant features for the large eyes in some species.

Very low intensities of light can be perceived by at least one species of *Peromyscus*. *P. maniculatus gracilis* could discriminate 1.13×10^{-6} candle power per square centimeter of surface from darkness when the incorrect response was punished by electric shock (Moody, 1929). When discrimination between two intensities of light were tested, the brighter the light, the greater the differences in light intensity necessary for discrimination, which conforms to the Weber law. Thus, at 36.11×10^{-6} Candle/cm^2 the just noticeable difference (jnd) was 60×10^{-6} C/cm^2 or 83 per cent of the difference; at 40.84×10^{-6} C/cm^2 the jnd was 87.8×10^{-6} C/cm^2 or 87 per cent; and at 493×10^{-6} C/cm^2 the jnd was 1031.3×10^{-6} C/cm^2 or 109 per cent of the difference. Since these absolute values are somewhat esoteric, the comments of Moody provide an anthropomorphic measure: "The mice at no time discriminated differences so small as to rival the ability of the writer's own eyes" (p. 399).

That the ability to discriminate light intensity may be important for survival is indicated in an experiment which measured activity at various intensities (Blair, 1943). Individuals of *P. m. blandus* were housed in a large room with sand on the floor, which enabled recording of activity by tracks left in the sand. An owl was kept in the room, but confined at night, in order to simulate natural conditions with a predator. As the light intensity was increased from no light, through about 12 intervals up to one ft-c, mouse activity decreased from running through all the quadrats to no activity; the mice remained in the nest box. These observations confirm those of many mouse trappers who find little evidence of mouse activity on clear, moonlit nights (Blair, 1951; Burt, 1940). Although no comparative studies have been undertaken, one might speculate that the threshold of light intensity for activity varies with increasing eye size.

Tests of daily activity cycles (Rawson, 1959; Johnson, 1939; Kavanau, 1962*b*, 1962*c*) have shown that light entrains the cycles—activity starts when light goes off and stops when light is on. *Peromyscus* is a distinctly nocturnal animal, although during long summer days, mice are frequently caught in traps during the last half hour or so of twilight. Records of *Peromyscus* being caught or observed in the wild during full daylight are rare, unless the mice have been disturbed (McNab and Morrison, 1963).

When *Peromyscus* is given some control of light by turning it off or on, or by increasing or decreasing the intensity, they select dim lights (8×10^{-4} ft-c) during active periods and very dim light (dark to 1.1×10^{-4} ft-c) during inactive periods (Kavanau, 1966*a*). *P. crinitus* will repeatedly turn off a light that automatically comes on every half hour by pressing a bar nine times through a series of decreasing intensities (Kavanau, 1963*a*). They will also turn on a light to a low intensity, if the light goes off automatically every half hour. Mice will alternate between manipulanda, one of which turns on the light and the other turns off the light (Kavanau, 1963*a*, 1966). These experiments, in general, indicate that *Peromyscus* has a preferred light intensity for its activities, but the absolute values of these intensities have not been rigidly established for any species. They also suggest that the preferred intensity is less critical than any change in the stimulus value of light (McCall, 1965). A very dim light similar to celestial light of early dawn, late dusk and moonlight, may also function as a navigational cue, since *P. crinitus* will run in an activity wheel towards the light, changing direction whenever the position of the light is changed (Kavanau, 1967).

A strong 500-watt photoflood lamp mounted 15 inches above the mice has been used as an aversive stimulus in a shuttle-box, but light was confounded with heat (Dice and Clark, 1962). This comparative study of *P. polionotus, P. m. bairdi,* and *P. m. gracilis* revealed that all stocks quickly shuttled and escaped the light, that all stocks reduced the latency of their escape responses, and that on the second day with ten trials each stock had a different latency. *P. m. bairdi* responded the slowest ($\overline{X} = 14.7$ sec), *P. m. gracilis* intermediate ($\overline{X} = 11.5$) and *P. polionotus* fastest ($\overline{X} = 8.7$ sec), which correspond well with their relative lens weights.

In another comparative study light was used as the discriminative stimulus for learning (King and Weisman, 1966). Mice had to press a bar in order to obtain a small drop of water, which was presented only when a small circle of light was in the test chamber. All of the five species examined learned the discrimination equally well, indicating that they all attended to the appropriate stimulus. In an unpublished study, both *P. m. bairdi* and *P. m. gracilis* learned to discriminate a small circle of light from a small cross of light in order to obtain water reinforcement.

In summary, within the genus there is wide variability in eye size, but the function of the eyes other than simple light receptors is not known. The mice are most active at minimal light intensities but will turn on relatively bright lights in darkness and they will escape bright lights. They attend to visual cues in learning situations and are capable of simple pattern discriminations. Comparative studies which might reveal the adaptive significance of the diversity in eye size and vision are needed (Rahmann, Rahmann, and King, in press).

SOUND.—Size of the external ear, which is often positively correlated with size of auditory bullae, is one of the principal measurements of taxonomists and unlike lens size, there is a plethora of data which has not been systematically examined. In wide-ranging species, ear size may be quite variable, but in species with restricted ranges, ear size may have narrower limits. However, size of ear is not a suitable phylogenetic trait. Clines in size of ear are not conspicuous and where they do exist, they do not follow Allen's rule of decreasing size from south to north (Hoffmeister, 1951). Desert-dwelling species tend to have larger ears than their grassland and forest inhabiting relatives (Dice, 1940). Semi-arboreal forms also tend to have larger ears than terrestrial forms. These ecological correlates to ear size apparently have many exceptions, although a systematic study has not been undertaken. Once size of external pinnae and auditory bullae have been related to auditory receptivity, then perhaps the functional significance of ear size can be understood.

The only evidence that size of external ear is related to hearing ability is supplied by Dice and Barto (1952), who found that large-eared hybrids of *P. m. artemisiae* and *P. m. blandus* responded to lower intensities of sound between 5 and 60 kHz, than the smaller eared *P. m. bairdi*. They also noted ear movement responses in the large-eared *P. nasutus* at 100 kHz, which is approximately 5 to 10 kHz above the response of *P. maniculatus*. These investigators used the ear twitch, a conditioned response, and audiogenic seizures to confirm the auditory reception of tones up to 65 and 95 kHz. Since mice can hear in the ultrasonic ranges used by bats in echolocation, it is possible that mice generate high frequency tones and use them for this purpose, as Gould, Negus, and Novick (1964) found for shrews. Mice with the EP convulsive mutation become deaf at about four months of age and no longer exhibit audiogenic seizures (Dice, Barto, and Clark, 1963). Deafness is apparently caused by degeneration of parts of the auditory nerve (Ross, 1962).

The highest frequency recorded for the vocalization of *Peromyscus* is about 24 kHz (King and Hart, 1966), which is well within the range of their hearing. However, ultrasonic microphones and recording equipment probably placed the limits on these frequencies more than did the capacity of the mice to make them.

Peromyscus are attentive to sounds. One need only watch them move their ears about and respond with a twitch to many low intensity sounds. Sounds at high intensity, such as key jingling, have produced audiogenic seizures (Barto, 1956; Watson, 1939). Buzzer sounds, so commonly used for auditory cues in conditioning rats, usually tend to make *Peromyscus* freeze.

In addition to the possible use of audition for echolocation, the auditory responsiveness of *Peromyscus* has been suggested for communication (Eisenberg, 1962, Chapter 12) and for detecting predators (Dice, Barto, and Clark, 1963). Certainly the same receptors can be used for a variety of functions, but the uniformity of ear size within similar environments suggests a primary selective advantage operating in each environment.

ODOR.—In contrast to the eyes and ears which can be measured externally, olfaction has no obvious external morphology that may be correlated with function. All of our knowledge for olfaction therefore comes from behavioral studies, most of which are sexual isolation studies with olfactory cues. Additional indirect evidence comes from studies on the Bruce effect, in which olfactory cues affect the reproductive physiology of female mice.

Sexual behavior of mice is characterized by considerable sniffing of the sexual partner in the genital area (Tamsitt, 1961a, 1961b), which suggests that olfactory cues may differentiate sexes, reproductive condition, and species. Tests for species discrimination, in which olfactory cues are implicated, usually require a male to select an area adjacent to or previously occupied by a female of either the same or a different species (or subspecies). Males of *P. maniculatus* selected the area previously occupied by females of their own species more often than the area occupied by females of *P. polionotus,* however, males of *P. polionotus* failed to make the discrimination (Moore, 1965). Since *P. polionotus* were originally from an area without other species of *Peromyscus,* Moore concluded that species discrimination was not necessary, unlike *P. maniculatus* which was sympatric with other species. A similar experiment with *P. eremicus* and *P. californicus* from sympatric and allopatric locations also revealed that males selected the proximity (but not contact) of females of their own species (Smith, 1965). Males of *P. eremicus* from an allopatric population were the only exceptions. Discrimination between species for an appropriate mate would be necessary only in regions inhabited by more than one species. Olfactory discrimination could function prior to a non-reproductive encounter between sexes of different species. After the encounter other isolating mechanisms may operate.

Olfactory stimuli have been implicated in the inhibition and synchronization of estrous cycles; and in blocking pregnancy among laboratory mice (Parkes and Bruce, 1961). A series of experiments (Bronson, Eleftheriou, and Garick, 1964; Bronson and Marsden, 1964) demonstrate the same phenomena in *P. m. bairdi,* although the critical operation of rendering their subjects anosmic has not been performed. However, the presence of male urine in the female's cage prior to mating can reduce the peak time of insemination from two to three days. Besides the confirmation of laboratory mouse studies (Bruce and Parrott, 1960; Whitten, 1956) with deermice, these investigators have some evidence to suggest that pregnancy blocks are most likely when females are exposed to odors of the same species and subspecies (Bronson and Eleftheriou, 1963).

Mice certainly depend upon olfaction for many other necessary discriminations in their visual day-to-day activity. The detection and selection of food, the recognition of the home site and land marks in their home range, the discrimination of different conspecifics (parent, offspring, mate, sibling, stranger), and the identification of predators are probably aided by olfaction. Adequate tests of these functions have not been undertaken.

Mice with the CV convulsive mutation differed from wild type *P. m. bairdi* in maze running and other tests possibly because of a modification of the olfactory sense in the mutants (Dice, Barto, and Clark, 1963). This mutant strain is susceptible to osmogenic convulsions, which may account for a different response to odors in the test apparatus than that exhibited by the wild types.

TASTE.—The primary distinction between taste and smell is that taste results from stimulation of contact chemoreceptors by soluble materials and smell results from stimulation of distant chemoreceptors by volatile substances. The distinction is often difficult to make from the behavior of *Peromyscus,* since they nibble or gnaw objects at the same time they smell them. Olfactory cues probably replace many gustatory cues during associations which occur early in the life of the mice.

Preferred concentrations of glucose solutions were tested in *P. eremicus, P. leucopus,* and *P. maniculatus* by offering individuals two different concentrations simultaneously for a two-hour period each day for 15 days (Wagner and Rowntree, 1966, pers. comm.). The mice were given food and water *ad libitum* at all times except during the two-hour tests. When pairs of 10, 25, and 37½ per cent glucose solutions were presented in all combinations, all three species consumed larger amounts of the sweeter solution of the pair despite the fact that an index of relative sweetness varies from 0.5 (in solutions of 25 vs. 37½ per cent) to 2.7 (in solutions of 10 vs. 37½ per cent). The only noticeable species difference was that *P. eremicus* tended to consume less of the 37½ per cent solution when paired with 10 and 25 per cent solutions than did *P. leucopus* or *P. maniculatus.* Since this species difference is perceptable (not tested statistically) with 10 per cent as well as 25 per cent concentrations, *P. eremicus* probably has a weaker preference for 37½ per cent concentrations than the other two species, but it can discriminate the glucose concentrations as well as *leucopus* and *maniculatus.*

Systematic studies of taste discrimination other than that of Wagner and Rowntree (1967) have not been undertaken, although types of foods eaten have been extensively explored. When *P. leucopus noveboracensis, P. maniculatus bairdi, P. m. gracilis,* and *P. m. sonoriensis* were presented 52 kinds of seeds, nuts, and fruits, 15 kinds of buds and bark, and 19 kinds of invertebrates, all mice ate all types of food in addition to an adequate laboratory diet (Cogshall, 1928). The only noticeable difference among the stocks tested was the tendency for *P. m. bairdi* to prefer sprouting grains, leaves, and bulbs, and the tendency for *P. m. gracilis* and *P. leucopus* to prefer bark and buds. Stomach analyses of foods eaten by wild caught *P. leucopus, P. maniculatus,* and *P. boylei* indicate an assortment of seeds, fruits, nuts, insects, leaves, fungi, annelids, crustacea, molluscs, insects, and vertebrates consumed (Hamilton, 1941; Jameson, 1952; Williams, 1959*b*; Brown, 1964). The types of seeds stored by *P. m. bairdi* depends largely on availability, with the seeds

of ragweed, oak, bush clover, and panic grass making up the largest
volume of the caches (Howard and Evans, 1961; Criddle, 1950).

In the absence of comparative and systematic studies on taste
discrimination or food preferences the conclusion can be tentatively
accepted that species of *Peromyscus* are opportunistic feeders and
will eat almost anything that provides nourishment. However, the
observations of Cogshall (1928) suggest that some species have a pre-
dilection for fruits and others for insects, and that food preferences
may be related to the availability of foods normally present in the
habitats occupied by each species. This hypothesis deserves testing
in the absence of such confounding variables as previous experience,
current diet, and nutritional value.

TEMPERATURE AND HUMIDITY.—Mice frequently abandon intolerable conditions of temperature and humidity by moving about until
suitable conditions are encountered. While this mobility is advantageous, it can expose them to even less favorable conditions, and
it provides no assurance that a suitable refuge will be encountered.
A mouse driven from its nest because of solar heat may enter direct
sunlight and may fail to locate a cool site before a lethal exposure.
Selection or construction of suitable refuges prior to encountering
intolerable conditions would be more advantageous than escaping
from them. Probably many stimuli provide the cues necessary for
the selection of suitable refuges, but temperature and humidity
themselves may be discriminated by the mouse before they become
physiologically intolerable. Studies of temperature and humidity
discrimination and selection are scarce, although changes in activity,
periodicity, food and water consumption, nest building, and huddling have been shown to be influenced by these physical conditions.

Peromyscus maniculatus bairdi were allowed to select a tem-
prature along a gradient from 0° C at one end to 60° C at the
other end (Stinson and Fisher, 1953). Humidity was not controlled.
The gradient was established in an aluminum tube ten and one-half
feet long, and provided with windows for observation. Positions
of the mice along the horizontal gradient were recorded each
minute for an hour. The mice spent most time between 20° C
and 30° C, with a mean of 24.1° C and a standard deviation of
4.3° C. A vertical temperature gradient used with four *P. leucopus*
indicated slightly lower (21° C to 26° C) temperature preferences,

but selection was not as sharp as in the horizontal tube. Preliminary tests for the effects of acclimation on temperature selection were also made after mice were maintained at 14° C, 22° C, and 33° C for a period of two to five weeks. In general, mice kept at high temperatures tended to select low temperatures and those acclimated at low temperatures selected high temperatures. The suggestion that *P. maniculatus* and *P. leucopus* differ in temperatures they selected was obscured by the use of two different techniques. A recent experiment (Ogilvie and Stinson, 1966) demonstrated that *P. m. bairdi* and *P. m. gracilis* significantly selected lower floor temperatures than *P. leucopus* and *Mus*.

The effect of temperature and humidity on activity in the field and laboratory is reviewed in Chapter 14 by Falls, who concluded that *Peromyscus* is most active at intermediate temperatures and at low humidities for the season. Torpor, characterized by hypothermia, has been described in *P. m. bairdi* (Howard, 1951) and *P. eremicus* (MacMillen, 1965) as a response to yet unspecified conditions of temperature and humidity, as well as to food and water deprivation.

The usual nocturnal behavior of *P. maniculatus* may be changed to diurnal activity by high temperatures (Hatfield, 1940). This departure from nocturnalism with increased ambient temperature may explain the diurnal activity observed in *P. eremicus* and *P. crinitus* (McNab and Morrison, 1963), although personal observations of captive animals indicate considerable diurnal activity in these two species.

Food consumption tends to be inversely related to temperature and a function of previous thermal history (Sealander, 1952*b*; see Food Consumption). When the temperature was raised from 21° C to 28° C, *P. m. bairdi* consumed an average of 22 per cent less wheat and *P. leucopus* consumed 24 per cent less (Dice, 1922). A female *P. m. bairdi* given 45 sunflower seeds per day at laboratory temperature gained weight, but lost weight on the same ration at freezing temperatures (Howard, 1951). Although systematic studies relating caloric input to temperature are few, there is no reason to suspect that *Peromyscus* differs from other mammals in this respect. Humidity, except as it contributes to thermoregulation, probably has little effect on food consumption.

Water consumption tends to increase at high temperatures and decrease with increased relative humidity, which corresponds to the physiological requirements of thermoregulation and water balance. At three temperatures of 21° C, 28° C, and 32°–34° C in both dry and humid air, the water consumption of *P. m. bairdi* was apparently higher than that of *P. leucopus* (Dice, 1922). Intergrades of *P. m. osgoodi* × *rufinus* increased their water consumption an average of 36 per cent (calculated from cc/g body wt.) when the relative humidity was decreased from 70–80 per cent to 10–20 per cent at 21° C (Williams, 1959*a*). On a diet of bird seed and without drinking water, *P. eremicus* could not maintain its body weight and survive at low relative humidities (40–60 per cent), but could do so at relative humidities of 60 to 80 per cent (Mac-Millen, 1964, 1965).

Thermoregulatory behavior in *Peromyscus* primarily consists of nocturnal periodicity, nest building, and social huddling. Mice avoid high diurnal temperatures by remaining in cool, moist, subterranean chambers during the day and by becoming active at night (Chapter 14). The selection, location, and construction of burrows and nest sites have not been thoroughly examined in the field or in the laboratory (King, 1963). One laboratory study on nest building and the shredding of paper for nests tested *P. m. osgoodi* at temperatures of 8° C, 20.5° C, and 33° C (Thorne, 1958). The best nests and most paper shredded occurred at the lowest temperature and they were almost eliminated at 33° C. The amount of nesting material (cotton) removed from a dispenser at a constant temperature of 21° C varied according to the geographical origins of four stocks of mice tested (King, Maas, and Weisman, 1965; see Motor Patterns).

In addition to building nests for heat conservation, *P. m. bairdi* and *P. leucopus* will often huddle together in communal conspecific and interspecific nests during the winter (Howard, 1949). The effect of huddling was demonstrated by the longer survival of *P. leucopus* when caged in a group of four than in smaller groups or alone (Howard, 1951).

Temperature and humidity have often been regarded as the significant factors determining the geographic distribution of the species and subspecies of *Peromyscus*. However, neither field nor laboratory studies have demonstrated that these factors may be responsible for the distribution of these mice. Most species have wide limits of tolerance in environments with controlled temperature and humidity, and the species studied do not differ to any great extent within these limits (McNab and Morrison, 1963; MacMillen, 1965). Although the habitat of each species differs in temperature and humidity, these factors by themselves are probably relatively unimportant in affecting the distribution of species, except possibly among those inhabiting extreme environments. If mice select their habitat, temperature or humidity are probably not the critical cues for their discrimination. Rather, the biota associated with temperature and humidity most likely provide the critical discriminatory cues (Harris, 1952; Wecker, 1963). The ability of a single species to occupy many different habitats when not competing with other species (Sheppe, 1961; Pruitt, 1959) further suggests that dynamic features of the environment are more important for distribution than the physical features of temperature and humidity.

GRAVITY.—The geotactic orientation of young rats (Crozier and Pincus, 1933) stimulated a comparative study of this response in six stocks of *Peromyscus* (Clark, 1936). Young mice between 11 and 12 days of age were placed on a wire screen incline at three angles: 20°, 40°, and 60°. The orientation of the mice on the incline at the three angles showed that the semi-arboreal forms (*P. m. artemisiae, P. boylei,* and *P. californicus*) generally oriented toward a greater vertical angle than did the terrestrial forms (*P. m. bairdi, P. m. osgoodi,* and *P. m. nebrascensis*). Since the geotactic "laws" proposed by Crozier and Pincus (1926) are open to doubt (Hovey, 1929; Hunter, 1931), and since Clark's study included a number of uncontrolled variables (temperamental differences, stimulation), the conclusions that terrestrial and semi-arboreal forms show a difference in geotactic orientation are, at most, suggestive.

CONCLUSIONS.—Although comparative studies of receptor organs and behavioral responses to physical and chemical stimuli have been undertaken for many years, they have usually been at the genetic and phyletic levels rather than the specific level of comparison (King and Nichols, 1960). Comparisons of species within a genus have not been fully exploited for the details of adaptive radiation, which species comparisons can illustrate. The genus

Peromyscus is particularly valuable for such studies because of its wide ecologic and geographic distribution and the readily apparent divergence in the external morphology of its sense organs, primarily its eyes and ears. Whether these external structures are related to visual and auditory sensitivity and whether these are related to habitat utilization is not known (Rahmann *et al.,* 1967). None of the studies reviewed in this section provides evidence for or against these relationships.

Knowledge of the perceptual world of the animals is fundamental to an understanding of their behavior. Selection and utilization of habitats may depend upon their ability to see, smell, or sense temperature. Their daily periodicity, homing ability, and arboreal or terrestrial habits may again depend upon their sensory capacities. Social relationships in selecting and finding mates, discriminating conspecific friend from foe, and parental care are limited by the animals' sensory capacities. The detection of predators and the responses given to them also depends upon the ability to perceive them. Perhaps, even more important than the sensory ability itself is whether the animal attends to or perceives the stimuli. Ethologists have stressed that animals respond to only a small fraction of the stimuli they are capable of sensing. A mouse that can readily see the stars, may not respond to them in any manner. Therefore, it is essential to examine the mouse's responses to the stimuli and not just its ability to sense them. Electrophysiological examinations are at most preliminary to psychological tests of sensory discriminations.

Comparisons at the species level can also reveal significant features of the sense organs themselves. The eye and vision, for example, have been the subjects of many studies (i.e., Walls, 1942), but minor differences in size of eye, number of retinal cells versus optic nerve fibers, differential density of cells in retina, and visual areas of the cortex have not been systematically examined within a genus. The same type of comparisons could be applied to any of the sensory modalities of *Peromyscus* with the assurance of significant contributions. Until the perceptual world of *Peromyscus* is more thoroughly studied and understood, our knowledge of its behavior will be largely superficial.

Learning

The modification of behavior as the result of prior experience is one of the most appealing areas for study; it has occupied the efforts of psychologists for decades. Comparative studies of learning, however, are surmounted with many difficulties, which too frequently have led to their abandonment. One primary difficulty, that of equating perceptual and motor capacities, is less within the genus *Peromyscus* than between families or orders because members of the genus have relatively similar perceptual and motor capacities. Despite this advantage, comparative studies, indeed, any learning studies with *Peromyscus,* are few.

Learning ability or the types of tasks learned could be a significant variable in the utilization of the environment by each species. An arboreal species, for example, could find learning the kinesthetic cues associated with running along a tree branch critical to its survival in forests. Whereas, a terrestrial species might better learn light intensity cues from the horizon which could direct it to its nearest refuge. Again, some species may find learning the location of water sources more important than learning where to locate food; or learning stimuli associated with aerial predators more important than those associated with terrestrial predators.

The classification of learning studies reviewed here is an expedient one, based upon the types of learning studies undertaken. Habituation, maze learning, avoidance, discrimination, and early learning will be examined in that order.

HABITUATION.—*Peromyscus,* like many other animals, become familiar with (learn) the features of their environment. When changes in the environment are perceived, the mouse may be aroused into activity or withdraw and reduce its activity depending on the properties of the new stimuli (Berlyne, 1966). A mouse placed in a strange, but not noxious situation, for example, may first sit quietly in a corner, later move about and explore, and finally reduce its activity once more. Changes in the level of activity or other behavior subsequent to the introduction of a stimulus, thus provides a measure of the animal's learning the stimulus. A familiar neutral stimulus presumably fails to elicit a response.

Changes in the level of activity and latency to enter an open field test apparatus were recorded over a two-day period for three stocks: *P. polionotus, P. m. bairdi, P. m. gracilis* (Dice and Clark, 1962). That the mice were stimulated by the new situation is indicated by their short latencies to enter and by their high activity scores. A reduction in both scores on the second day suggests that they had become familiar with the situation and were less stimulated by it than they were on the first day. In an activity box, these investigators found that *P. m. bairdi* reduced its activity from the first to the second day, *P. m. gracilis* did not change appreciably, and *P. polionotus* increased in activity. The same tests were used for the examination of behavioral differences among three mutant strains of *Peromyscus* (Dice, Barto, and Clark, 1963). This situation may not have been a suitable measure of habituation. In general, changes of activity over shorter time periods are more useful for the study of habituation than those over several days.

In another open-field test, habituation of the first laboratory-born generation of wild-caught *P. m. bairdi* were compared with those that had been bred in the laboratory for approximately 15 years (Price, 1967). The semidomestic mice reduced their latency to enter scores on the second day of exposure to the situation more than the first laboratory-reared wild-caught generation, indicating that the semidomestic mice habituated to the open field.

MAZES.—Maze tests of learning were developed for the laboratory rat and only in a few other species have they been found useful. *Peromyscus* is not one of those species, because the necessary handling between trials interferes with their maze performance. Many investigators have tried and abandoned mazes for the study of learning in *Peromyscus*. Successful use of mazes has necessitated allowing the mice a free run of them.

In an automated, six-choice maze, three stocks (*P. polionotus, P. m. bairdi,* and *P. m. gracilis*) were given one trial per day for ten days (Dice and Clark, 1962). The mice were free to run the maze at any time by leaving the home cage and traversing the maze to another cage at the end. Motivation for running the maze was apparently exploration. The error curves for *polionotus* and *bairdi* were similar, with a reduction to one error by the third and fourth days. *P. m. gracilis* started with fewer errors and showed less of a reduction than the other two stocks. Since motivational levels were low, the maze-running time is not a suitable measure of learning and only *P. polionotus* showed an orderly reduction of time over the ten-day period. The time curves for *P. m. bairdi* and *P. m. gracilis* were almost straight and both stocks ran slower than *polionotus*, with *bairdi* running slower than *gracilis*. Conclusions regarding the comparative learning ability of these stocks cannot be made from these results, although it is evident that they could all readily learn the maze. Convulsive (EP and CV strains) and waltzing (WZ strain) mutants on a *P. m. bairdi* background also exhibited modifications in their maze performance in comparison with wild-type *P. m. bairdi* (Dice, Barto, and Clark, 1963).

One of the most complex mazes ever used in the study of animal behavior was designed by Brant and Kavanau (1964, 1965). Several mazes were arranged in horizontal, inclined, and vertical positions, and constructed of different materials—glass tubes, Plexiglas, acetate film, and aluminum strips. The mazes were connected to six different cages provided with dirt, activity wheels, food, and water. The purpose of this elaborate system was an attempt to duplicate in the laboratory a situation resembling the natural habitat of the mice. Movements of the mice were automatically recorded or observed with closed circuit television with infrared sensitivity. Mice with different laboratory and field histories included *P. maniculatus sonoriensis, P. m. gambeli, P. m. clementis, P. crinitus,* and wild and domestic *Mus musculus* for a total of 23 individuals of both sexes. One maze was completely explored by 12 mice, two mazes by seven mice, and all three mazes by two mice. Once maze exploration began, it was usually completed within a day. Seven mice apparently became lost, since they did not return to the food and water for 24 hours. The outgoing trips generally took longer than the return trips and repeated trips were usually much shorter than the initial trip through the maze. Although no reliable comparisons can be made among the different species tested, the *Peromyscus* stocks performed somewhat alike and were more active than *Mus* in the mazes. The complexity and extent of these mazes, involving 96 feet for the shortest path, 20 feet of climbing, 148 blind alleys, and 313 turns of 90°, reveal that *Peromyscus* has little difficulty in learning mazes, even without tissue needs. One *P. crinitus* made 13 round trips during its 10 days in the apparatus.

When an activity wheel was unlocked, maze running was much reduced.

The willingness of *Peromyscus* to run a maze was also illustrated when a maze was placed in a large field enclosing *P. m. bairdi*. The mice learned this maze to obtain sunflower seeds exposed to but not obtainable from outside. One female eventually used the goal chamber for a nest, running the maze each time she returned to the nest.

If mazes could be used for *Peromyscus* as they have been for laboratory rats, there is every reason to suspect that their ability would be comparable and perhaps superior to rats.

AVOIDANCE.—Two types of avoidance learning are recognized—active and passive. In active avoidance, an animal must perform a specific response upon perceiving a more or less neutral stimulus in order to prevent receiving a noxious stimulus, usually electric shock. A typical testing procedure is to make the animal run from an electric grid to a safe area when it hears a buzzer. If it does not run within a brief period (about 5 sec) it receives a shock. In passive avoidance, the animal avoids the noxious stimulus by remaining motionless. For example, an animal placed on a small shelf must step down to another shelf to get off, but by stepping from one shelf to another, it completes an electric circuit and receives a shock. It will soon learn to remain on one shelf and not attempt to step to the other. Escape learning usually precedes avoidance learning, but it can be used alone. In escape learning, the animal must perform a specific response when it receives the noxious stimulus.

In a Sidman avoidance procedure, a brief electric shock (0.2 sec) was given to the mice every ten seconds unless they pressed a lever, which delayed the shock for 20 seconds (King and Eleftheriou, 1959). Repeated lever presses within the 20 seconds enabled the mice to avoid all shocks. *P. m. bairdi* and *P. m. gracilis* were placed in this situation one hour each day for five days. Both subspecies increased the frequency of lever presses and frequency of shock avoidance during the five-day period. *P. m. gracilis* significantly avoided more shocks during the five days than did *P. m. bairdi*. This difference in performance could be explained by the fact that when shocked *P. m. gracilis* leaps about more than *P. m. bairdi* (Foster, 1959) and thus, accidently hits the lever more often.

An escape and passive avoidance learning test was given to the same two subspecies with opposite results, since *P. m. bairdi* performed better than *P. m. gracilis* (King, 1961). Mice were tested on alternate days from 10 to 30 days of age and another group was tested at 20 days only. The mice were placed on a grid which was electrified as soon as a door was raised, permitting them to run to an insulated platform. They learned to escape the shock by running to the platform and learned to avoid it passively by remaining there throughout the one minute duration of each trial. Since the mice ran on and off the pan many times during the test, the longest continuous time they remained on the platforms was used as the measure of learning. *P. m. bairdi* remained on the safe platform longer than *P. m. gracilis,* both as a function of age and as a function of trials within a given age group.

A similar procedure of avoidance conditioning to that above was used on *P. m. gracilis* to determine the effects of CNS depressant drugs and the nature of the response (Wolf, Swinyard, and Clark, 1962). One group of mice avoided shock by running to an insulated pan on an electrified grid and the other group avoided shock by climbing a pole. The animals in the pole-trained group learned faster, reached a higher level of performance, extinguished slower and were less affected by chlorpromazine and pentobarbitol doses than the pan-trained group. These results were interpreted in terms of the natural escape patterns of this semi-arboreal subspecies. In comparison with the previous study (King, 1961), this interpretation seems valid because *P. m. bairdi,* a terrestrial mouse, performed better than *P. m. gracilis* when given only the pan for an escape. The critical test with *P. m. bairdi* in a pole situation has not been made.

That learning performance depends upon the test situation is illustrated by the conflicting comparative results of these two experiments. *P. m. bairdi* appeared to be the better learner in one test and *gracilis* better in the other. Factors other than or including learning ability are involved in these tests, indicating that learning may depend on emotionality, pain thresholds, response patterns, and motivational levels characteristic of each species.

DISCRIMINATION.—When an animal associates one stimulus with a reinforcer among other non-reinforced stimuli, a discrimination between the stimuli has occurred. The stimuli may be presented either simultaneously or sequentially. Under natural conditions, mice presumably make many discriminations between food and non-food objects, between sexes, between species, and between predators and non-predators. The stimuli may be olfactory, auditory, visual, tactile, chemical (taste) or other capable of being perceived by the mice.

Position discrimination involves learning the location of a particular site or reinforcer, or the simple discrimination of left from right. Position discriminations are usually one of the easiest for animals to make and once a position habit is established, it is difficult to change. Many mazes involve position discriminations, however, cues other than position may be used.

The only published study on position discrimination in *Peromyscus* used a weak reinforcer, which was not sufficiently strong to break the habit when the reinforcing positions were changed (King and Weisman, 1964). Providing *P. leucopus* with the opportunity to dig sand was the reinforcer. The mice could press either of two levers, only one of which delivered a small quantity of sand. Within 12 days, the mice learned which of the two levers delivered the sand, but when the position of the reinforcing levers was reversed, they failed to change their position within the next 12 days. This study demonstrated that sand could act as a reinforcer, but was not strong enough to bring about a reversal.

Water is a strong reinforcer for mice deprived of it and the strength of the position preference is indicated by several unpublished studies with *P. m. bairdi*. When a visual cue for water reinforcement was alternated from right to left, the mice persisted in going to one side. Thus, they received only half of the reinforcements they could have received by attending to the visual cue instead of the position cue. In another situation, the mice were provided two water bottles. They persistently drank from one position, even when the bottles were switched, or when a light flashed on and off with each lap of the tongue.

Discrimination learning has been used to study light intensity thresholds in *P. leucopus* (Moody, 1929). The mice were taught the discrimination with electric shocks as a negative reinforcer, but few comments about learning appear in the study. The rate of learning was discontinuous and often took as many as 600 trials before a strong discrimination was formed. Variability in learning among individuals was great.

A sequential light-on, light-off discrimination for water reinforcement was tested on five stocks of *Peromyscus*—*P. m. bairdi, P. m. gracilis, P. leucopus, P. floridanus,* and *P. eremicus* (King and Weisman, 1965). The mice could obtain a small drop of water at a fountain by pressing a single lever only when a disc of light was projected on a frosted glass window which was lit about one minute for a total of 20 minutes of each hour. Pressing the lever during the dark 40 minutes did not produce water at the fountain. Thus, the mice had to learn to press only when the light was on. All stocks of mice learned this discrimination at the same rate and at the same level of performance over the 20 test days. The stocks did differ significantly, however, in the number of lever presses during the hour, which indicated that learning was independent of response rate. Unlike the avoidance studies, in which the type or the rate of responses affected learning, this study indicates that this type of learning ability does not differ among the species and subspecies tested.

EARLY LEARNING.—Usually learning refers to a specific task for which the animal has been trained. However, less specific experiences obtained throughout the life cycle of the organism may modify its later behavior. In some instances the relationship between the early experiences and the later behavior is quite clear, such as imprinting. In other instances, this relationship is obscured and the early experiences apparently contribute to an entire organismic change, which is expressed in a variety of different behavior patterns. Both imprinting and the effects of early experience are included in this section without a commitment to learning theory.

Studies of the ontogeny of behavior frequently include two categories for comparison—longitudinal and cross-sectional groups of subjects. Longitudinal groups repeat the test on the same group of animals at each age examined. For example, a litter of mice may be weighed each day to obtain their growth rates. Cross-sectional studies use a separate group of animals at each age. If the growth of the brain is studied, for example, a group of animals at each age must be killed in order to weigh the brain. Although longitudinal studies have the advantage of following an individual from day to day or week to week, they have the disadvantage of possibly modifying later measures by the previous examination. Even the weight of growing mice may be affected by removing them from their mothers and handling them. Behavioral tests are particularly prone to the effects of prior treatment, which necessitates the use of both longitudinal and cross-sectional groups.

When development of locomotion was examined in *P. m. bairdi* by measuring the duration that young mice remained on an elevated platform, the longitudinal groups remained significantly longer than cross-sectional groups at 18 and 25 days of age (King and Shea, 1959). This difference in duration may have resulted from the longitudinal mice habituating to the exposed platform without attempting to locomote off. Cross-sectional groups encountering the mildly noxious situation for the first time attempted to escape by moving about and consequently fell off. If this interpretation is correct, the longitudinal mice apparently learned to remain motionless prior to 18 days of age.

A swimming test for *P. m. bairdi* and *P. m. gracilis* revealed no significant differences between longitudinal and cross-sectional groups of both subspecies (King, 1961). In the same study, differences were found between the two groups of *P. m. bairdi* in the longest continuous time the mice passively avoided an electrified grid. Since the longitudinal groups avoided the grid for longer periods, they apparently learned the appropriate response during the preceding days. That the young mice could learn this response is illustrated by the longer duration of avoidance by cross-sectional groups over repeated trials at 15 and 18 days of age. Since longitudinal and cross-sectional groups of *P. m. gracilis* did not differ, the early learning of this response is apparently species specific.

The differences between longitudinal and cross-sectional groups in a learning situation and the lack of difference in swimming may mistakingly lead to the conclusion that prior experience affects only learned performances. A study on clinging behavior in *P. m. bairdi, P. m. gracilis, P. eremicus,* and *P. polionotus* refutes this conclusion (Price and King, 1966). Since the semi-arboreal and terrestrial stocks differed in clinging during development, clinging performance is associated with the genotype. However, the modification of the genotypic expression of the clinging response is apparent from the differences between the longitudinal and cross-sectional groups of some stocks at certain ages.

Another approach to early learning is to give two groups of animals different early experiences at a specified age period and then later test their performance. For example, King and Eleftheriou (1959) mechanically manipulated groups of *P. m. bairdi* and *P. m. gracilis* between the ages of 3 and 25 days and later tested their activity and active avoidance performance at 70 days of age. Control groups not receiving the manipulation were used for both subspecies. The treated groups of both subspecies differed significantly from the control groups in activity and avoidance. Furthermore, the subspecies responded differently to the treatment, particularly in the avoidance test. *P. m. gracilis* had higher avoidance scores than those of the control group and *P. m. bairdi* had lower scores, indicating that the early experience had a differential effect on each genotype.

The possibility that early social experience may affect later spatial distribution of *P. m. bairdi* in seminatural conditions was examined by Terman (1963). Although significance levels for each of his measures varied considerably, the general conclusion was that mice raised in groups were more socially oriented and less spatially oriented in the field than mice raised in isolation. Isolated mice had fewer social combinations in the field, they stayed farther from other mice, and they took longer to unite with other mice, when they did unite, than group-raised mice. When the isolated mice were temporarily removed from the field and later returned, they homed to their prior area of residence more often than socially raised mice. Isolated mice were also more disturbed upon encountering an alien mouse in their home nest box than were the socially raised mice. These results are significant for the study of population dynamics of this genus.

Habitat selection may also be influenced by early experience (Wecker, 1963). A semidomestic stock of *P. m. bairdi* selected a grassland habitat over a forest habitat, only after they had been raised in a grassland environment; not if they were raised in the laboratory or forest habitat. Offspring of recently captured *P. m. bairdi,* however, selected the grassland habitat regardless of where they had been raised. These results suggest that semidomestic mice could be imprinted on the habitat of their parental stock, whereas the offspring of wild caught mice could make the selection without the prior "imprinting" experience. Apparently during the 12-year period that the semidomestic stock were bred and raised in the laboratory, selection pressure for habitat preference had relaxed, but only to the point where the genotype was still vulnerable to the effects of early experience.

CONCLUSIONS.—The primary conclusion derived from the preceding review is that many more comparative studies with many different learning tests are needed before any other conclusions can be firmly drawn. A start on learning in some species has begun, but the genus has not been exploited to the extent warranted by the wealth of comparative material. The problem of control by equation of response tendencies, perceptual or sensory capabilities, and motivational levels among the species may interfere with comparative conclusions. Some control by systematic variation as proposed by Bitterman (1965) may be more applicable, although it is unlikely that species differences in *Peromyscus* will appear in probability learning or repeated reversals. With a systematic variation of several learning situations, demanding different perceptual cues, different responses, and different motivational levels applied to a number of species, some systematic pattern of learning should emerge. Not that one species is more capable of learning than another, but that each species performs best in situations most nearly resembling those for which it has been selected (King, 1967). While such a conclusion is not new, it does lack supporting evidence, which could be supplied by comparative studies of *Peromyscus*.

Discussion

The material for this chapter on psychological parameters in the behavior of *Peromyscus* has come from a wide variety of investigations, each with its own specific aims and methods. With some difficulty, the studies were pulled apart and reorganized in a manner not entirely compatible with their aims. The reasons for this reorganization are that it corresponds with comparable investigations in psychology and that it provides a means for comparing species. The techniques and results of psychological studies on many different animals are obviously useful to the investigator studying *Peromyscus* behavior. This organization further extends the comparisons possible with others in the psychological literature by stressing species-level comparisons, which are uncommon both in psychology and zoology.

It is difficult to compare all aspects of behavior in several species, particularly if all are to be quantified. Thus, some selection is necessary and the organization provided here indicates types of behavior which have been systematically investigated. Since the choice of behavior to be studied does not entirely depend upon types of behavior which have been previously investigated, how does one choose a behavior to study?

The answer demands an inquiry into the aims of studies in comparative behavior. One of the principle aims is to discover the adaptive function of behavior, which involves identification of the selection pressures affecting the behavior of each species (Tinbergen, 1965). Comparative studies provide insight into the nature of selection pressures because each species has behavioral adaptations appropriate to its environment. An analysis of behavior among several species from different habitats can provide correlational evidence for the type of selection affecting the behavior. The correlation of behavior with habitat or with a particular phylogeny does not reveal the type of selection, but it does provide preliminary material for experimental studies, such as those suggested by Tinbergen (1965). Experimental isolation of the selection pressure for any particular behavior is difficult in *Peromyscus* and few patterns of behavior can be so thoroughly analyzed. Therefore, many behavioral studies will be limited to the correlation phase of comparative analysis.

The comparative technique can be used with precision within a diverse genus like *Peromyscus*. For example, one can determine if all species develop the same responses or if each species, depending on its phylogeny, develops different behavior patterns to the same types of selection. Behavioral convergence is expected, but the extent of convergence depends upon the genetic substrate provided by each species. The gene pool of each species or subgenus is different and selection may bring about similar patterns of behavior with a different constellation of genes. Before comparative analyses with this degree of precision are possible, the phylogenetic relationships of the genus must be well established.

Fortunately the phylogeny of *Peromyscus* is relatively well understood for a genus with so many species and such a wide, complex geographic distribution. The solution of remaining systematic problems may be furthered by behavioral data. Although behavior has not been used in mammalian taxonomy, the fact that primitive behavior patterns derived from a common gene pool are often masked by phenotypically modifiable behavior need not prevent their use in taxonomy. Systematic comparative studies may yet reveal patterns of behavior common to a particular species group or subgenus of *Peromyscus*. Fertility relationships within a species group, which have contributed to the solution of taxonomic problems, may involve species-specific behavior.

Like other scientific disciplines, behavioral comparisons require quantification. Quantified measures are not sufficient in themselves and a number representing a behavioral response does not make it inviolable. Indeed, quantification may be used too early in the analysis of behavior and the resulting numerical values are meaningless. If behaviorists have erred in this respect, however, the errors tend to be in the direction of too little quantification, rather than too much. Descriptive terms often convey the message more clearly than numerical data, but accuracy and reliability are sacrificed. For example, terms like "nervousness," "timidity," and "docility" have been used in describing *Peromyscus* behavior and they do suggest behavioral characteristics, but each reader is left to his own interpretation of them. The advantage of even premature quantification is that it provides a definite start, which can be revised and perfected by future studies.

The review of behavioral studies presented in this chapter reveals that the aims of the comparative technique have rarely been achieved. Most studies involve only one species or a haphazard sample of several species. Controls for age, previous experience, sex, and sample size are seldom incorporated. Suitable quantification and statistical analyses are attempted without meeting basic assumptions. The most telling criticism, however, is that comparative questions have not been asked. Again, this criticism is not aimed at the investigators, who are at liberty to ask their own questions, but it is aimed at the area of comparative behavior, which is too often an amorphous conglomeration of data. One exception to this criticism is the study of arboreal behavior in ten stocks of *Peromyscus* by Horner (1954). Although the study is open to criticism for other reasons, it is oriented towards the basic comparative question—what is the adaptive function of a specific behavioral pattern?

Summary

Quantitative laboratory studies of adult behavior in the genus *Peromyscus* were reviewed in an attempt to reveal evolutionary or ecologic patterns of behavior. Motor patterns, perception, and learning were presented and examined for systematic trends.

The motor patterns of locomotion, swimming, climbing, digging, food and water consumption, grooming, nest building, and vocalization have been studied more thoroughly than the perception of the physical environment and learning. However, ecologic or phylogenetic relationships were not discernible except possibly for climbing and water consumption, which indicate ecologic rather than phylogenetic relationships. These motor patterns are also those which have been most systematically studied.

Some isolated values for the perception of light and sound have been obtained, but adequate comparative material is not available. *Peromyscus* can discriminate odors, particularly those of other mice, but olfaction has not been quantitatively studied. The first quantitative study of taste for sugar solutions was examined. Perception of temperature and humidity have been studied in respect to thermoregulation, which does suggest ecologic relationships.

Although species differences have been found in habituation, maze, avoidance, and discrimination learning, the number of species tested and the number of techniques used are not sufficient for drawing any conclusions regarding the comparative learning ability of species within the genus. Early learning studies have indicated the importance of this variable in habitat selection and in population dynamics, but again comparative studies are lacking.

The most apparent conclusion derived from this review is that the excellent comparative material offered by this genus has not been experimentally exploited. Systematic studies of even a few carefully selected species will certainly reveal some consistent patterns in the way different species of this genus perceive their environment, respond to it, and modify their behavior accordingly.

Literature Cited

ANDREW, R. J. 1963. Evolution of facial expression. Science, 142:1034–1041.

BARTO, E. 1956. Tests for independence of waltzer and EP sonogenic convulsive from certain other genes in the deer mouse (*Peromyscus maniculatus*). Contr. Lab. Vert. Biol. Univ. Mich., 74:1–16.

BERLYNE, D. E. 1966. Curiosity and exploration. Science, 153:25–33.

BITTERMAN, M. E. 1965. Phyletic differences in learning. Amer. Psychol., 20: 396–410.

BLAIR, W. F. 1943. Activities of the Chihuahua deer-mouse in relation to light intensity. Jour. Wildl. Mgt., 7:92–97.

———— 1951. Population structure, social behavior and environmental relations in a natural population of the beach mouse (*Peromyscus polionotus leucocephalus*). Contr. Lab. Vert. Biol. Univ. Mich., 48:1–47.

BLAIR, W. F., AND J. D. KILBY. 1936. The gopher mouse—*Peromyscus floridanus*. Jour. Mamm., 17:421–422.

BRANT, D. H., AND J. L. KAVANAU. 1964. "Unrewarded" exploration and learning of complex mazes by wild and domestic mice. Nature, 204:267–269.

———— 1965. Exploration and movement patterns of the canyon mouse *Peromyscus crinitus* in an extensive laboratory enclosure. Ecology, 46:452–461.

BRONSON, F. H., AND B. E. ELEFTHERIOU. 1963. Influence of strange males on implantation in the deermouse. Gen. and Comp. Endocrin., 3:515–518.

BRONSON, F. H., B. E. ELEFTHERIOU, AND E. I. GARICK. 1964. Effects of intra- and inter-specific social stimulation on implantation in deermice. Jour. Reprod. Fertil., 8:23–27.

BRONSON, F. H., AND H. M. MARSDEN. 1964. Male-induced synchrony of estrus in deermice. Gen. and Comp. Endocrin., 4:634–637.

BROWN, L. N. 1964. Ecology of three species of *Peromyscus* from southern Missouri. Jour. Mamm., 45:189–202.

BRUCE, H. M., AND D. M. V. PARROTT. 1960. Role of olfactory sense in pregnancy block by strange males. Science, 131:1526.

BURT, W. H. 1940. Territorial behavior and populations of some small mammals in southern Michigan. Misc. Publ. Mus. Zool. Univ. Mich., 45:1–58.

CHEW, R. M. 1965. Water metabolism of mammals. *In*: W. V. Mayer and R. G. Van Gelder (eds.) Physiological Mammalogy, New York: Academic Press, 43–178.

CHEW, R. M., AND R. T. HINEGARDNER. 1957. Effects of chronic insufficiency of drinking water in white mice. Jour. Mamm., 38:361–374.

CLARK, F. H. 1936. Geotropic behavior on a sloping plane of arboreal and non-arboreal races of mice of the genus *Peromyscus*. Jour. Mamm., 17:44–47.

COGSHALL, A. S. 1928. Food habits of deer mice of the genus *Peromyscus* in captivity. Jour. Mamm., 9:217–221.

CRIDDLE, S. 1950. The *P. m. bairdii* complex in Manitoba. Canad. Field-Nat., 64:169–177.

CROZIER, W. J., AND G. PINCUS. 1926. The geotropic conduct of young rats. Jour. Gen. Physiol., 10:257–269.

———— 1933. Analysis of the geotropic orientation of young rats. VIII. *Ibid.*, 16:883–893.

DICE, L. R. 1922. Some factors affecting the distribution of the prairie vole, forest deer mouse, and prairie deer mouse. Ecology, 3:29–47.

———— 1940. Ecologic and genetic variability within species of *Peromyscus.* Amer. Nat., 74:212–221.

DICE, L. R., AND E. BARTO. 1952. Ability of mice of the genus *Peromyscus* to hear ultrasonic sounds. Science, 116:110–111.

DICE, L. R., E. BARTO, AND P. J. CLARK. 1963. Modifications of behavior associated with inherited convulsions or whirling in three strains of *Peromyscus.* Anim. Behav., 11:40–50.

DICE, L. R., AND P. J. CLARK. 1962. Variation in measures of behavior among three races of *Peromyscus.* Contr. Lab. Vert. Biol. Univ. Mich., 76: 1–28.

DICE, L. R., AND S. A. HOSLETT. 1950. Variation in the spontaneous activity of *Peromyscus,* as shown by recording wheels. Contr. Lab. Vert. Biol. Univ. Mich., 47:1–18.

EIBL-EIBESFELDT, I. 1957. Technik des Nüsseöffnens. Zeit. Säugetierkunde, 21: 132–134.

EISENBERG, J. F. 1962. Studies on the behavior of *Peromyscus maniculatus gambeli* and *Peromyscus californicus parasiticus.* Behav., 19:177–207.

EVANS, F. C. 1957. Utilization of resources by experimental populations of *Peromyscus.* Bull. Ecol. Soc. Amer., 38:66 (abstract).

FALLON, D. 1965. Eatometer: a device for continuous recording of free-feeding behavior. Science, 148:977–978.

FERTIG, D. S., AND J. N. LAYNE. 1963. Water relationships in the Florida mouse. Jour. Mamm., 44:322–334.

FOSTER, D. D. 1959. Differences in behavior and temperament between two races of the deer mouse. Jour. Mamm., 40:496–513.

FRENCH, R. L. 1956. Eating, drinking, and activity patterns in *Peromyscus maniculatus sonoriensis.* Jour. Mamm., 37:74–79.

GOULD, E., N. C. NEGUS, AND A. NOVICK. 1964. Evidence for echolocation in shrews. Jour. Exp. Zool., 156:19–38.

HAMILTON, W. J., JR. 1941. The food of small forest mammals in eastern United States. Jour. Mamm., 22:250–263.

HARRIS, V. T. 1952. An experimental study of habitat selection by prairie and forest races of the deermouse, *Peromyscus maniculatus.* Contr. Lab. Vert. Biol. Univ. Mich., 56:1–53.

HART, F. M., AND J. A. KING. 1966. Distress vocalizations of young in two subspecies of *Peromyscus maniculatus.* Jour. Mamm., 47:287–293.

HATFIELD, D. 1940. Activities and food consumption in *Microtus* and *Peromyscus.* Jour. Mamm., 21:29–36.

HAYNE, D. W. 1936. Burrowing habits of *Peromyscus polionotus.* Jour. Mamm., 17:420–421.

HILDEBRAND, M. 1959. Motions of the running cheetah and horse. Jour. Mamm., 40:481–495.

HOFFMEISTER, D. F. 1951. A taxonomic and evolutionary study of the piñon mouse, *Peromyscus truei*. Illinois Biol. Monogr., 24 (4) :1–104.

HORNER, B. E. 1954. Arboreal adaptations of *Peromyscus,* with special reference to use of the tail. Contr. Lab. Vert. Biol. Univ. Mich., 61:1–84.

HOVEY, H. B. 1928. The nature of the apparent geotropism of young rats. Physiol. Zoöl., 1:550–560.

HOWARD, W. E. 1949. Dispersal, amount of inbreeding, and longevity in a local population of prairie deermice on the George Reserve, southern Michigan. Contr. Lab. Vert. Biol. Univ. Mich., 43:1–50.

——— 1951. Relation between low temperatures and available food to survival of small rodents. Jour. Mamm., 32:300–312.

HOWARD, W. E., AND F. C. EVANS. 1961. Seeds stored by prairie deermice. Jour. Mamm., 42:260–263.

HUNTER, W. S. 1931. The mechanisms involved in the behavior of white rats on the incline plane. Jour. Gen. Psychol., 5:295–310.

JAMESON, E. W., JR. 1952. Food of deer mice, *Peromyscus maniculatus* and *P. boylei,* in the northern Sierra Nevada, California. Jour. Mamm., 33:50–60.

JOHNSON, M. S. 1939. Effect of continuous light on periodic spontaneous activity of white-footed mice (*Peromyscus*). Jour. Exp. Zool., 82: 315–328.

KAVANAU, J. L. 1962a. Precise monitoring of drinking behavior in small mammals. Jour. Mamm., 43:345–351.

——— 1962b. Twilight transitions and biological rhythmicity. Nature, 194: 1293–1295.

——— 1962c. Activity patterns on regimes employing artificial twilight transitions. Experientia, 18:382–384.

——— 1963a. Compulsory regime and control of environment in animal behavior. I. Wheel-running. Behav., 20:251–281.

——— 1963b. Continuous automatic monitoring of the activities of small captive animals. Ecology, 44:95–110.

——— 1963c. The study of social interaction between small animals. Anim. Behav., 11:263–273.

——— 1966. Automatic monitoring of the activities of small mammals. *In*: K. E. F. Watt (ed.) Systems Analysis in Ecology, New York: Academic Press, 99–146.

——— 1967. Behavior of captive white-footed mice. Science, 155:1623–1639.

KAVANAU, J. L., AND D. H. BRANT. 1965. Wheel-running preferences of *Peromyscus.* Nature, 208:597–598.

KING, J. A. 1961. Swimming and reaction to electric shock in two subspecies of deermice (*Peromyscus maniculatus*) during development. Anim. Behav., 9:142–150.

——— 1963. Maternal behavior in *Peromyscus*. *In*: H. E. Rheingold (ed.) Maternal Behavior in Mammals, New York: Wiley, 58–93.

———— 1965. Body, brain, and lens weights of *Peromyscus*. Zool. Jharb. Anat., 82:177–188.

———— 1967. Behavioral modification of the gene pool. *In*: J. Hirsch (ed.) Behavior-Genetic Analysis, New York: McGraw-Hill.

KING, J. A., AND B. E. ELEFTHERIOU. 1959. Effects of early handling upon adult behavior in two subspecies of deermice, *Peromyscus maniculatus*. Jour. Comp. Physiol. Psychol., 52:82–88.

KING, J. A., D. MAAS, AND R. G. WEISMAN. 1964. Geographic variation in nest size among species of *Peromyscus*. Evolution, 18:230–234.

KING, J. A., AND J. W. NICHOLS. 1960. Problems of classification. *In*: R. H. Waters, D. A. Rethlingshafer, and W. E. Caldwell (eds.) Principles of Comparative Psychology, New York: McGraw-Hill, 18–42.

KING, J. A., E. O. PRICE, AND P. G. WEBER. 1968. Behavioral comparisons within the genus *Peromyscus*. Papers Mich. Acad. Sci. Arts, and Letters, 53:113–136.

KING, J. A., AND N. J. SHEA. 1959. Subspecific differences in the responses of young deermice on an elevated maze. Jour. Hered., 50:14–18.

KING, J. A., AND R. G. WEISMAN. 1964. Sand digging contingent upon bar pressing in deermice (*Peromyscus*). Anim. Behav., 12:446–450.

———— 1966. Visual discrimination in deermice. Psychon. Sci., 4:43–44.

KOLLER, G. 1952. Der Nestbau der weissen Maus und seine hormonale Auslösung. Verh. Dtsch. Zool. Ges., Freiburg, 1952:160–168.

LABAN, C. 1966. A study of behavior in the deer mouse *Peromyscus maniculatus rubidus* Osgood during its 24-hour cycle of activity in a simulated natural habitat. Unpublished Ph.D. dissertation, Oregon State Univ.

LINDEBORG, R. G. 1952. Water requirements of certain rodents from xeric and mesic habitats. Contr. Lab. Vert. Biol. Univ. Mich., 58:1–32.

McCABE, T. T., AND B. D. BLANCHARD. 1950. Three species of *Peromyscus*. Rood Associates, Santa Barbara.

McCALL, R. B. 1965. Stimulus change in light-contingent bar pressing. Jour. Comp. Physiol. Psychol., 59:258–262.

MacMILLEN, R. E. 1964. Population ecology, water relations, and social behavior of a southern California semidesert rodent fauna. Univ. Calif. Publ. Zool., 71:1–66.

———— 1965. Aestivation in the cactus mouse, *Peromyscus eremicus*. Comp. Biochem. Physiol., 16:227–248.

McNAB, B. K., AND P. MORRISON. 1963. Body temperatures and metabolism in subspecies of *Peromyscus* from arid and mesic environments. Ecol. Monogr., 33:63–82.

MOODY, P. A. 1929. Brightness vision in the deer-mouse, *Peromyscus maniculatus gracilis*. Exper. Zool., 52:367–405.

MOORE, R. E. 1965. Olfactory discrimination as an isolating mechanism between *Peromyscus maniculatus* and *Peromyscus polionotus*. Amer. Midland Nat., 73:85–100.

ODUM, E. P. 1944. Water consumption of certain mice in relation to habitat selection. Jour. Mamm., 25:404–405.

OGILVIE, D. M., AND R. H. STINSON. 1966. Temperature selection in *Peromyscus* and laboratory mice, *Mus musculus.* Jour. Mamm., 47:655–660.

ORR, R. T. 1933. Aquatic habits of *Peromyscus maniculatus.* Jour. Mamm., 14:160–161.

PARKES, A. S., AND H. M. BRUCE. 1961. Olfactory stimuli in mammalian reproduction. Science, 134:1049–1054.

PRICE, E. O. 1967. Behavioral changes in the prairie deermouse, *Peromyscus maniculatus bairdii,* following twenty-five generations of laboratory breeding. Unpublished Ph.D. dissertation, Mich. State Univ.

PRICE, E. O., AND J. A. KING. 1966. Ontogeny of clinging in four stocks of *Peromyscus.* Amer. Zool., 6:309–310 (abstract).

PRUITT, W. O., JR. 1959. Microclimates and local distribution of small mammals on the George Reserve, Michigan. Misc. Publ. Mus. Zool. Univ. Mich., 109:1–27.

RAHMANN, H., M. RAHMANN, AND J. A. KING. In Press. Comparative visual acuity (minimum separable) in five species and subspecies of deermice (*Peromyscus*). Physiol. Zool.

RAWSON, K. S. 1959. Experimental modification of mammalian endogenous activity rhythms. In Photoperiodism and Related Phenomena in Plants and Animals. Washington: A. A. A. S., 791–800.

RICHTER, C. P. 1937. Hypophyseal control of behavior. Cold Spr. Harb. Symp. Quant. Biol., 5:258–268.

ROSS, M. D. 1962. The auditory pathway of the epileptic waltzing mouse: I. A comparison of the acoustic pathways of the normal mouse with those of the totally deaf epileptic waltzer. Jour. Comp. Neurol., 119: 317–339.

SCHAEFFER, R. W., AND D. PREMACK. 1961. Licking rates in infant albino rats. Science, 134:1980–1981.

SEALANDER, J. A., JR. 1952a. The relationship of nest protection and huddling to survival of *Peromyscus* at low temperature. Ecology, 33 (1) :63–71.

——— 1952b. Food consumption in *Peromyscus* in relation to air temperature and previous thermal experience. Jour. Mamm., 33:206–218.

SHEPPE, W., JR. 1961. Systematic and ecological relations of *Peromyscus oreas* and *P. maniculatus.* Proc. Amer. Phil. Soc., 105:421–446.

——— 1965. Dispersal by swimming in *Peromyscus leucopus.* Jour. Mamm., 46:336–337.

SMITH, M. H. 1965. Behavioral discrimination shown by allopatric and sympatric males of *Peromyscus eremicus* and *Peromyscus californicus* between females of the same two species. Evolution, 19:430–435.

STINSON, R. H., AND K. C. FISHER. 1953. Temperature selection in deermice. Canad. Jour. Zool., 31:404–416.

TAMSITT, J. R. 1961a. Tests for social discrimination between three species of the *Peromyscus truei* group of white-footed mice. Evolution, 15: 555–563.

——— 1961b. Mating behavior of the *Peromyscus truei* species group of white-footed mice. Amer. Midland Nat., 65:501–507.

TAYLOR, J. R., AND H. McCARLEY. 1963. Vertical distribution of *Peromyscus leucopus* and *P. gossypinus* under experimental conditions. Southwestern Nat., 8:107–108.

TEETERS, R. 1945. Swimming ability of a wood mouse. Jour. Mamm., 26:197.

TERMAN, C. R. 1963. The influence of differential early social experience upon spatial distribution within populations of prairie deermice. Anim. Behav., 11:246–262.

THORNE, O. II. 1958. Shredding behavior of the white-footed mouse, *Peromyscus maniculatus osgoodi*, with special reference to nest building, temperature, and light. Thorne Ecol. Res. Sta., Bull., No. 6.

TINBERGEN, N. 1965. Behavior and natural selection. *In*: J. A. Moore (ed.) Ideas in Modern Biology. New York: Natural History Press, 519–542.

WAGNER, M. W., AND J. T. ROWNTREE. 1966. Methodology of relative sugar preferences in laboratory rats and deer mice. Jour. Psychol., 64: 151–158.

WALLS, G. L. 1942. The vertebrate eye and its adaptive radiation. Bloomfield Hills, Mich.: Cranbrook Inst. Sci.

WATSON, M. L. 1939. The inheritance of epilepsy and waltzing in *Peromyscus*. Contr. Lab. Vert. Genet. Univ. Mich., 11:1–24.

WECKER, S. C. 1963. The role of early experience in habitat selection by the prairie deer mouse, *Peromyscus maniculatus bairdii*. Ecol. Monogr., 33:307–325.

WHITTEN, W. K. 1956. Modification of the oestrous cycle of the mouse by external stimuli associated with the male. Jour. Endocrin., 13:399–404.

WILLIAMS, O. 1959a. Water intake in the deer mouse. Jour. Mamm., 40: 602–606.

———— 1959b. Food habits of the deer mouse. *Ibid.*, 40:415–419.

WOLFE, H., E. A. SWINYARD, AND L. D. CLARK. 1962. The differential effects of chlorpromazine and pentobarbital on two different forms of conditioned avoidance behavior in *Peromyscus maniculatus gracilis*. Psychopharmacologia, 3:438–448.

ADDENDUM

HOWARD, W. E., AND R. E. COLE. 1967. Olfaction in seed detection by deer mice. Jour. Mamm., 48:147–150.

RALLS, K. 1967. Auditory sensitivity in mice: *Peromyscus* and *Mus musculus*. Anim. Behav., 15:123–128.

VESTAL, B. M., AND J. A. KING. 1967. Relation of age of eye opening to first optokinetic response in four taxa of *Peromyscus*. Amer. Zool., 7: 216–217.

WHITAKER, J. O., JR. 1966. Food of *Mus musculus*, *Peromyscus maniculatus bairdi* and *Peromyscus leucopus* in Vigo County, Indiana. Jour. Mamm., 47:473–486.

14

ACTIVITY

J. Bruce Falls

Introduction

M ICE OF THE GENUS *Pero-
myscus* are easily main-
tained and bred in captivity
and present an array of related
forms suited to comparative
study. They readily accept activity wheels and other devices, and
exhibit, both in the field and
laboratory, a marked 24-hour
periodicity in activity. These
qualities make them favorite
subjects for studies of activity.

Activity is some form of behav-
ior distinguished from sleep or quiescence. The term denotes
movement, usually locomotion, although feeding, drinking, elimination and other functions are often referred to as activities. The
present chapter will deal primarily with locomotion and only
incidentally with other activities as they help to explain overall
activity patterns. In the natural life of a mouse, locomotor activity
serves in social behavior, obtaining of food and water, exploratory
movements within the home range, dispersal, escape from enemies,
and exercise. It is not within the scope of this chapter to discuss
the many motivating factors involved in all these forms of behavior
or the end results in terms of home range, territoriality, and dispersal movements. Many of these subjects are dealt with in other
chapters. This chapter will be limited to a review of the available
information on the amount and timing of general activity and the
effects of certain environmental and internal factors upon activity.

In reviewing this subject one is limited to those aspects of activity
measured by methods commonly in use. The results of different
methods employed are not necessarily comparable. Thus, suspended
bouncing cages (stabilimeters) respond to a variety of movements

and measure the time spent in all activity, whereas activity wheels record only the time, duration, direction and distance of running. It is unlikely that either of these methods measures activity in proportion to energy expended. This was confirmed in a personal communication from Peter Mann who measured oxygen consumption of *Peromyscus* in these and other activity devices. It is often observed that *Peromyscus* using activity wheels indulge in movements such as jumping which are not recorded as turns and, conversely, that by jumping in and out a mouse can keep a wheel turning without running continuously. Kavanau (1967) found that experienced mice preferred a square wheel or one with hurdles over the standard round type. The preferred types call for split-second timing, coordination, and quick reflex actions at which these mice are adept. These observations indicate the desirability of providing an animal with opportunities for activity commensurate with its inherent abilities.

Measurements of oxygen consumption of active animals can, if combined with data for basal metabolism, provide a measure of energy expended on total activity (Hart, 1952). However, this method combines all forms of activity and is difficult to carry out without seriously limiting the animal's opportunities for activity. Counting squares tracked by mice on a sanded floor measures locomotion directly (Blair, 1943). Less direct methods include the use of treadles and gates to record trips to and from the nest or number of passages from one part of an enclosure to another. Although time out of the nest is often used as a measure of the active period, Kavanau (1963a) observed a period of activity in the nest prior to an animal's departure. Treadles may in some cases be avoided by mice and thus influence the activity they are supposed to measure. The most complex laboratory enclosures described so far are those of Kavanau (1967). Employing electromechanical transducers and electric eyes for second-by-second monitoring, one version of his apparatus has 22 independent channels recording information concerning position, movements, eating, drinking, and eliminating (Kavanau, 1963a). He warns that an adaptation period of several days is required before a stable pattern of activity is established. While Kavanau's approach provides detailed information on the response of mice to environmental manipulations, his papers so far report on the behavior of very few individuals and this may be a limitation of the complexity of his method.

The kinds and amounts of activity which an animal will perform depend upon the possibilities available to it. This was well illustrated in a study by Kavanau (1962*a*), who found that a *P. maniculatus,* which spent only one and one-half hours per day out of the nest, stayed out for ten hours when an activity wheel was provided. In another instance, using *P. crinitus,* Brant and Kavanau (1965) found that use of an activity wheel tended to replace exploration of a complex maze. They refer to exploration, learning, and running as self-rewarding activities and argue that the movement patterns of an animal will represent a balance among the different types of activity available. Kavanau (1967), on the basis of several studies in which *Peromyscus* could manipulate aspects of their environment, states that "Confined animals seize upon and repeatedly exercise opportunities to modify their surroundings. In addition they tend to counteract non-volitional (imposed) and 'unexpected' deviations from the status quo." These tendencies will influence the results of any study of enclosed mice. They may also explain why *Peromyscus* so readily accept and use free-running activity wheels.

It is difficult to find laboratory equivalents of behavior patterns important in field activities. Brant and Kavanau have approached this problem by using a complex series of burrow-simulating mazes in which environmental variables can be manipulated. Their preliminary results indicate several ways in which movement patterns in the apparatus parallel behavior in the field.

The writer believes that, while laboratory studies are necessary to determine effects of individual factors and define relationships which may be significant to animals, it is also necessary to study activity in the field, if it is to be properly understood. Activity wheels and various enclosures have been used to study captive animals under field conditions. For the study of wild mice, treadles and electric eyes have been employed to a limited extent. Live trapping has been the method most frequently used. Disadvantages of trapping are that traps restrain mice, animals may avoid or habitually visit traps for bait, and only a very limited number of events can be recorded. Track counts in sand, where possible, provide a more sensitive measure of changes in activity than captures in baited traps (Falls, 1953) and the use of smoked cards for this purpose may have limited value (Sheppe, 1965*a*). Measures of activity in the field have tended to be rather fragmentary. Recently developed instruments such as automatic cameras or telemetering systems should help to correct this deficiency in future studies.

A further problem in interpreting findings of activity studies, especially where a small sample of animals is used, is that the results refer only to the individuals studied. Laboratory stocks are subject to conditions that may produce genotypic differences from their wild ancestors. Thus, to determine characteristics of wild forms and the range of variation which may be important in adaptation to a variable environment, it is desirable to study samples taken directly from the field.

Factors Affecting the Amount of Activity

GENETIC DIFFERENCES.—Dice and Hoslett (1950) used activity wheels in a comparison of stocks representing seven species and four additional races of *Peromyscus*. Most of the animals were laboratory-bred. Even after a period of about six days in which mice learned to use the wheels, there was still considerable day-to-day variation in the records of individual mice. However, this variation "within individuals" was significantly less than that "between individuals." Heterogeneity among individuals of the same stock made comparisons among stocks unreliable. However, the mean number of turns in 24 hours for *P. polionotus* ($\overline{X} = 5815$) was far below the mean for all stocks ($\overline{X} = 17,521$), while *P. m. austerus* ($\overline{X} = 30,676$) and *P. m. bairdi* ($\overline{X} = 25,881$) had values well above average. The low scores of *P. polionotus* may be the result of their light weight or inability to learn to use the apparatus. The highest single 24-hour record, 58,760 turns for an individual *P. leucopus*, is equivalent to 23 miles of travel. However, operating a wheel cannot be equated to travel in nature. The greatest movement in a short period known in nature is probably still that reported by Murie and Murie (1931) of a mouse that homed two miles in two days.

Some of the difficulties of comparing the activity among different stocks are exemplified by several studies involving *P. m. bairdi* and *P. m. gracilis*. Using the number of crossings of three treadles in an eight-foot alleyway as a measure of activity, King and Eleftheriou

(1959) found *bairdi* ($\overline{X}$ for 24 hours = 1762) to be three times as active as *gracilis* ($\overline{X}$ = 639). Among many measures of behavior, Dice and Clark (1962) measured frequency of treadle crossing in a 17-inch alleyway. They found *gracilis* crossed an average of 803 times in 23 hours, which was significantly more than did *bairdi* ($\overline{X}$ = 205) or *polionotus* ($\overline{X}$ = 176). In other tests involving movements between compartments, these three stocks did not differ. Other studies might be mentioned, but it is obvious that no simple conclusion can be drawn concerning the relative activity of different stocks. Different types of activity or effects of different factors may be involved in the various tests.

Inherited differences certainly occur in the types of locomotor activity used by the various forms of *Peromyscus*. Thus, McCabe and Blanchard (1950) found that *P. truei gilberti* was a better climber than *P. c. californicus,* while *P. maniculatus gambeli* didn't climb at all. Recently, Eisenberg (1962) has compared the behavior of two of these same species. He gives a detailed description of running which is the chief mode of progression of *maniculatus.* *P. californicus,* however, more often leaps or climbs and also employs a quadrupedal hop. Horner (1954) demonstrated notable differences in climbing ability and behavior of ten forms of *Peromyscus* and related these to body proportions and utilization of natural habitat. *P. m. gracilis, P. m. oreas, P. leucopus noveboracensis, P. n. nasutus,* and *P. t. truei* are semi-arboreal and proved to be better climbers in a series of tests than *P. m. bairdi, P. m. blandus, P. m. nebrascensis* or *P. polionotus leucocephalus,* which live in open habitats. *P. m. rubidus* proved to be intermediate. Foster (1959), in a series of experiments with a laboratory enclosure, found that *bairdi* tended to freeze or make quick darting movements in a strange environment, whereas *gracilis* moved deliberately. Sheppe (1965*b*) described swimming in *P. l. noveboracensis.* These studies represent only a sampling of the types of locomotion to be found in the genus.

SEX AND AGE.—Writing of *P. l. noveboracensis,* Burt (1940) states, "males wander farther than do females, and in this respect are more active." His own data on home range (males, $\overline{X}$ = 0.27 acres; females $\overline{X}$ = 0.208 acres) tend to bear this out, as do those of many other authors for different forms of *Peromyscus* (e.g., Stickel, 1946;

Hirth, 1959; Brown, 1964; White, 1964). Dice and Howard (1951) showed that males of *P. m. bairdi* tended to disperse farther than females from their birthplaces (recorded distances: males, $\overline{X}$ = 339 feet; females, $\overline{X}$ = 188 feet). Howard (1949) also demonstrated that dispersal usually takes place just prior to the time when mice become sexually mature. Thus, subadults presumably make more long distance movements than other age groups. However, it is not known whether any of these observed differences is the result of variation in total activity and there appears to have been no attempt to compare the activity of groups of different sex or age in the laboratory.

OESTROUS CYCLE.—Stinson (1952) associated day-to-day changes in running activity of female *P. maniculatus bairdi* with oestrous cycle. Vaginal smears showed a 4.2-day cycle in eight females, with peaks of activity associated with peaks of cornified cells (ovulation). Significant differences in activity between peaks and troughs of the cornified cell cycle were demonstrated for 24-hour periods but not for periods from 9:00 A.M. to 5:00 P.M., indicating that nocturnal activity was mainly affected. Increased activity during heat may increase the chances of contact with a male. If this phenomenon is essentially similar to that described for white rats (Richter, 1927), activity of young or pregnant females should be at a low level without a cycle of peaks. The same might apply where anestrum occurs during the winter months. However, the activity of anestrous females has apparently not been studied in *Peromyscus*.

OTHER ANIMALS.—Although mice have been studied in groups or adjacent enclosures, there is little information concerning effects of social factors on activity. However, Kavanau (1963*b*) reported on various activities of two mature females of *P. maniculatus,* when isolated and when together. When together they tended to remain in each other's company and their behavior patterns converged. Thus, one that was very active when alone became less active, while the other, relatively inactive alone, became more active. These changes resulted from social interactions since, when returned to a solitary existence, each animal reverted to essentially its previous pattern of activity.

FOOD AND WATER.—Movements of a mouse within its home range are principally associated with feeding and food gathering according to Blair (1951), who tracked individuals of *P. polionotus leucocephalus* at Santa Rosa Island, Florida. In several studies feeding and drinking have been related to activity patterns in captive mice; these will be reviewed in the section on time of activity. A number of investigators have shown that deprivation of food results in increased activity. Stinson (1952) starved 14 *P. maniculatus bairdi* after a two-week period during which food was supplied and running activity in wheels was recorded both for the day (9:00 A.M. to 5:00 P.M.) and night. In most instances, activity rose to a high level and then fell to a low level before the death of the animal. The highest day and night runs for each animal were significantly greater in the starvation period than in the preceding period of feeding, but the difference was considerably greater in the daytime activity. Thus, the nocturnal pattern of activity broke down to some extent. Similarly, Rawson (1960) noted that a male *P. leucopus,* accidentally deprived of food, ran during the normal inactive period. Kavanau (1962*b*) also noted that when an individual *P. maniculatus* was deprived of food its running activity in a wheel increased fourfold. However, with no food and the wheel locked the animal spent most of its time in the nest. One may postulate that the increased activity shown in most cases would increase the probability of a starving animal encountering food.

The only report on the effect of water deprivation on *Peromyscus* activity appears to be that of French (1956). Eating and drinking and activity patterns of *P. m. sonoriensis* were examined by recording the presence of each mouse in a series of compartments. Ten animals given only 30 per cent of daily normal water intake for 30 days showed a decrease of about 50 per cent in total activity although their activity pattern was unchanged.

LIGHT.—Light is evidently an important timing factor for the activity rhythm, since these mice are strictly nocturnal. Light intensity also affects the amount, as distinct from the time, of activity. This has been established both in laboratory and field studies.

The activity of a desert mouse, *P. m. blandus,* at different light intensities was studied by Blair (1943), who used as an index of activity the mean number of squares with tracks in a grid on the sand-covered floor of an enclosure. The mice were from a laboratory stock and in order to "create that fear of enemies which affects the activities of wild mice" an owl was caged in the room when the mice were active. No light intensities between zero and 0.0082 ft-c were used. He found a decline in activity with increasing intensity above about 0.001 to 0.01 ft-c, depending on which part of his data is examined. According to Blair, some reduction in activity took place at about half the intensity of full moonlight (0.012 ft-c) and further decrease occurred at greater light intensity. At 1.28 ft-c the mice did not track a single quadrat.

Records of activity of *P. l. noveboracensis* in continuous light at intensities of 0.25 and 25 ft-c indicate that total activity was less at the higher intensity (Johnson, 1939).

Stinson (1952) measured the running activity at different light intensities of six male *P. m. bairdi* confined in small cages connected with activity wheels. No nesting material was used and the individual units were arranged in a large chamber with a ground glass side so that shadows were minimized and the animals were subjected to diffuse light. Eleven different intensities from zero to 8.4 ft-c were used. Average results showed a gradual increase of about threefold in activity from zero to about 0.015 ft-c, followed by a decline to about the original level at 8.4 ft-c. Thus, activity peaked at an intensity just below that given by Blair (1943) for the full moon (0.023 ft-c). Neither the length of the active period nor the amount of activity in the normal daylight hours were affected. Using the same apparatus, Falls (1953) was able to repeat Stinson's results. However, if nesting material was used, the effect of light was less, although a peak still occurred at an intermediate intensity. When ground glass was not used and the apparatus was oriented so that shadows developed, activity was about the same at all intensities. The fact that neither Blair nor Johnson found reduced activity in very dim light can be explained since the intensities they used were in the upper part of Stinson's range.

Brant and Kavanau (1965), however, state that in a complex enclosure, only two of six *P. crinitus* spent appreciable time in a dark maze. When four *P. maniculatus* were accustomed to the apparatus and were allowed to control illumination in ten steps from dark to 4.0 ft-c, Kavanau (1967) obtained results remarkably parallel to Stinson's. The animals frequently changed the illumination but spent only brief periods in the dark or at the brightest light. Three mice spent 77 per cent of their time at levels between

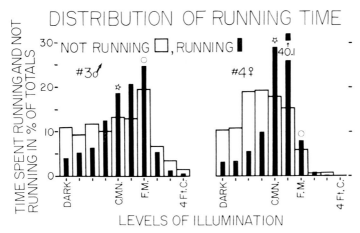

FIG. 1. Time spent running and not running at various light intensities by two *P. maniculatus* given control over illumination (Kavanau, 1967). F. M. = full moon; C. M. N. = clear, moonless night.

half starlight and full moonlight, 8 per cent in the dark, and only 0.5 per cent at 4 ft-c. The fourth animal usually turned the lights off at the beginning of the inactive period and spent 53 per cent of its time in the dark. The data for the other three mice show a peak somewhere between starlight and full moonlight with fewer occasions and less time spent at higher and lower intensities. The parallel with Stinson's data is even more marked when time spent running in a wheel at different intensities is examined (Fig. 1). The peak occurs at a higher intensity than for total time and lies in the same range as Stinson's peak. Animals selected a lower average intensity during rest than when active.

Blair (1951) studied the activity of *P. m. blandus* in an open dune area by live-trapping and following tracks. He states that light was the most potent factor affecting activity, catches being reduced under a half moon and few, if any, tracks occurring under the full moon. On moonless nights the mice were very active whether it was clear or cloudy.

In a study of factors affecting activity of *P. m. bairdi* on a Lake Erie beach in summer, Falls (1953) made hourly observations of light and several different measures of activity. Mice in wheels, which had access to small wire cages with nesting cotton, showed

LIGHT   CONDITIONS

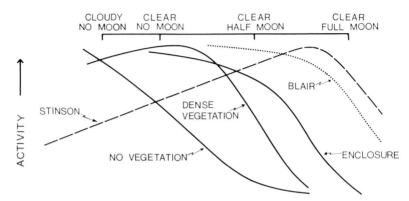

FIG. 2. Effects of light on activity found by Blair (1943) for *P. m. blandus* (tracks on floor), and Stinson (1952) (activity wheels) and Falls (1953) (tracks on beach) for *P. m. bairdi*. Light scale logarithmic.

no variation in activity with light intensity even when wheels were sheltered only under a translucent plastic sheet. However, captures in live-traps and numbers of tracks counted on a mile-long strip of sand and in an 80-square-foot enclosure around a nest box all indicated a marked decrease in activity with increasing light intensity. Track counts showed this effect to a greater degree than captures in traps. On a few very dark nights activity seemed to be slightly reduced. Differences in activity could be demonstrated for the same time on different nights or during the same night if, for example, the moon rose or set.

It is unlikely that these results were caused by any factor other than light. Since tracks could be counted only on dry nights, precipitation could not have influenced the results. Also, during periods of equal light intensity other weather factors failed to have an effect on activity.

In open areas, activity decreased gradually with increasing light, even at low intensities encountered on relatively dark nights, and reached a minimum at half moonlight. In dune grass or near logs, activity remained high at intermediate intensities and did not reach a minimum until full moonlight. In the enclosure, activity was high at half moonlight, but low with a full moon. These results are compared with those of Blair and Stinson in Figure 2. Thus, although even dim light inhibits movement in the open, activity persists near the nest or where shadows occur until much higher intensities are reached. This may account for the rather high intensities required to bring about a decrease in activity in animals in enclosures (Blair, 1943; Stinson, 1952), and the failure to show effects of light where nesting material and shadows are present. It seems likely, as Blair suggests, that this reaction to light enables mice to escape predation in open environments. The author observed *P. m. bairdi* in dune grass on moonlight nights and noted that they remained motionless or zigzagged quickly, a behavior pattern which rendered them inconspicuous. As already noted, this same type of behavior is characteristic of *bairdi* in laboratory enclosures (Foster, 1959).

Using live traps and activity wheels in a mature hardwood forest, Falls (1953) found no effect of light on the activity of *P. m. gracilis*. Similarly, Orr (1959) found no effect of moonlight on the activity of *P. l. noveboracensis* in a woodland enclosure. These results would be predicted from the work in open habitats. However, Hirth (1959) obtained a correlation of +0.74 between degree of cloudiness and combined captures in live traps for three habitats ranging from an old field through forest edge to oak-hickory forest. It is possible that other factors, such as rainfall and temperature, may have been correlated with cloudiness, as in the study of Gentry and Odum (1957).

RAINFALL.—Rain has been found to increase captures of *Peromyscus* in live traps (Burt, 1940; Hays, 1958). Hays found that catches in snap traps fell off in rainy periods and many traps were set off by the rain. However, increased catches in snap traps during light rain were found by Gentry and Odum (1957). Hirth (1959) reported a tendency for catches in live traps to be low when rain fell before midnight but normal if rain fell later. In a hardwood forest, Falls (1953) found that catches of *P. m. gracilis* increased on 16 out of 18 occasions when rain began or increased during the trapping period. Dry routes were present in the forest and captured mice were seldom wet. On an open beach, catches of *P. m. bairdi* increased with moderate rain but were low in heavy rain. On one occasion nightly departure from a nest site under surveillance was delayed until heavy rain stopped. These studies suggest the following: with light rain *Peromyscus* are very active as in other dark periods; with moderate rain, they tend to seek shelter, including live traps; with heavy rain activity is greatly restricted in open areas. The advantage of keeping dry is indicated by the common observation that mice in traps frequently die if their fur becomes wet or matted, presumably owing to loss of insulation and failure to maintain body temperature.

TEMPERATURE AND HUMIDITY.—The most detailed study of effects of temperature and humidity on *Peromyscus* activity is that of Stinson (1952) who maintained laboratory-bred *P. m. bairdi* in wheels under controlled conditions. Some interaction between the effects of the two factors was noted. In general, activity decreased above 80° F and below about 59° F. At temperatures within this range, activity decreased as vapor pressure increased. Stinson explained these results in terms of temperature regulation. Inactivity at high temperatures reduces heat output, and at low temperatures may conserve heat as shown by Hart (1951) for white mice. Increased activity at low vapor pressures might compensate for evaporative heat loss.

Other studies have generally confirmed these results so far as temperature is concerned. Howard (1951) found that caged *Peromyscus* became less active when subjected to below-freezing temperatures. Hart (1953) showed that running capacity of *P. l. noveboracensis* confined to a mechanically driven activity wheel fell off above 68° F and below 23° F if the mice were acclimated to 58° F. At acclimation temperatures of 68° F and 86° F, the lower limit for maximal activity was 32° F and 50° F, respectively. Thus, acclimation may make it possible for mice to be active at lower temperatures in winter than would be possible in summer. Hart's results were also interpreted in terms of temperature regulation. Although based on limits of forced rather than spontaneous activity, they showed a plateau at intermediate temperatures as did Stinson's experiments.

From observations of several desert taxa in metabolism chambers at temperatures up to 38° C, McNab and Morrison (1963) reported that the major behavioral responses to high temperatures were a prostrate form and greatly reduced activity.

Falls (1953) found a tendency for running activity of *P. m. bairdi* to decrease at temperatures of 69° F and higher when wheels were exposed to outdoor conditions in summer. When wheels were exposed in winter, running activity was about a third, and time out of the nest about a half, the values obtained in summer. Running was evidently more concentrated in winter because it occupied only 5 per cent of the time out of the nest compared with 37 per cent in summer. These differences may be partly the result of seasonal effects unrelated to temperature, since five *P. m. gracilis,* kept at temperatures above 58° F, showed twice as much running activity in April as in November and December. However, the mice kept outdoors in winter spent 20 per cent less time out of the nest when minimum temperature was below 20° F than on warmer nights. Similarly, Hatfield (1940) reported that the time spent obtaining food by two captive *P. m. bairdi* was shortened at low temperatures. In other experiments with captive *P. m. bairdi,* the writer found that a larger fraction of time out of the nest was spent feeding at freezing temperatures than at room temperature.

Orr (1959) kept *P. l. noveboracensis* in an outdoor enclosure at temperatures varying from 25° to 75° F, and found that they were more active in crossing treadles between 40° and 50° F than at higher or lower temperatures.

Orr also reported that maximum activity occurred at high relative humidities, although little activity was noted when a point near saturation was reached. Falls (1953) found a tendency for running activity of *P. m. bairdi* exposed outdoors in summer to decline above vapor pressures of 15 mm and relative humidities of 75 per cent.

There is little indication from field studies of effects of temperature and humidity on wild mice. This may result from the small amount of temporal variation in these factors in *Peromyscus* habitats compared with marked microclimatic variation. Gentry and Odum (1959) reported larger catches in snap traps of oldfield mice (*P. polionotus*) on warm cloudy nights than on cold clear nights in winter in Georgia. While it is difficult to sort out the possible factors involved, these authors also state that the trend occurred at the dark of the moon as well as with moonlight, and was less pronounced in warmer parts of the year. Hence temperature may be the most significant factor. The writer has found in several years of live-trapping *P. m. gracilis* that fewer captures are made when the midnight temperature is below 60° F than on warmer nights.

Howard (1951) found that groups of wild *P. m. bairdi* utilizing nest boxes or captives caged outdoors might become torpid when temperatures fell below freezing. At least some mice found torpid during the day were active at night. Torpidity was interpreted as an adaptation reducing food requirements which would otherwise increase at low temperatures. *Peromyscus* may be active at very low temperatures. Beer (1961) captured *P. leucopus* when the temperature was –19° F. He also found that winter movements were limited to restricted travel lanes compared with wide ranging activity in summer.

To summarize, *Peromyscus* activity seems to be greatest at intermediate temperatures. Effects of humidity are not so clear but there is evidence of a decline at high humidities. The actual values at which these effects occur appear to be variable and may depend on acclimation and interactions between factors.

OTHER PHYSICAL FACTORS.—There is little information concerning the influence of other weather factors on activity. Stinson (1952) found no effect of changes in atmospheric pressure when temperature and humidity remained within narrow limits.

Hock (1961) found a significant difference in the endurance times of *P. m. sonoriensis* from two elevations when mice were forced to run on an inclined treadmill. Each group was tested at its own elevation and the same temperatures. Satisfactory performances averaged 15.8 minutes for nine mice at 4000 feet and 7.5 minutes for nine others at 12,470 feet. The difference was attributed mainly to reduced pO_2 and consequent hypoxia at the higher altitude. It would be interesting to compare spontaneous activity of these mice.

Time of Activity

NOCTURNALISM.—*Peromyscus* has been extensively used for the study of 24-hour periodicity in activity, and its nocturnal behavior has been demonstrated in many laboratory and field studies. The extent to which activity is restricted to the dark hours varies among individuals and with the conditions of the environment. Often the entire daylight period is spent in the nest and the writer knows this to be the case with several wild *P. m. bairdi* in artificial and natural nests in summer. Stinson (1952) states that, on the average, *bairdi* in wheels performed less than 5 per cent of the total 24-hour activity outside the 12-hour nocturnal period. Falls (1953) found that for *bairdi* housed outdoors in winter and exposed to daylight, 12 per cent of the total time spent out of the nest occurred during the "inactive" period. The corresponding figure for mice in continuous darkness was higher. Behney (1936) reported no daylight trips outside the nest box for *P. l. noveboracensis* housed outdoors during the winter as long as the cages were bare of snow and plenty of food was supplied. If snow accumulated in the cages, mice came out and tunnelled during the day. Hamilton (1937) states that he has seen indications of daylight activity by deer mice under the snow. However, Getz (1959) made only one of 60 captures of *P. leucopus* in tunnels under the snow during daylight hours although live-traps were set at all times. Evidently, periods of subnivean activity are worthy of further investigation.

McNab and Morrison (1963) found that while *Peromyscus* in southwestern deserts were generally crepuscular, they were often trapped in late afternoon when ambient temperatures were still high (35° C). This was especially true of *P. e. eremicus* near Las Vegas, Nevada. Possibly related to this observation is that of Hatfield (1940) who found that two captive *P. m. bairdi* left their nest boxes at intervals throughout the day and night when subjected to a temperature of 35° C.

Morrison (1948) measured oxygen consumption of *P. m. gracilis* and *P. l. noveboracensis* in a commodious chamber and found that 31 per cent more oxygen was consumed per hour during the 9-hour active period than during the inactive period.

CIRCADIAN RHYTHM.—One of the first systematic studies of periodicity in activity of *Peromyscus* was performed by Johnson (1926), with *P. leucopus* and *P. m. bairdi* in bouncing cages. Mice kept in cages in the field and in the laboratory showed the same periodicity. In continuous darkness the rhythm was maintained, even among mice born and raised in the dark. The activity cycle could be reversed by lighting the cage during the formerly dark period, and this rhythm was maintained after 12 days even in continuous darkness. Later, Johnson (1939) examined the spontaneous activity of *P. leucopus* under continuous light of various intensities (2.5 to

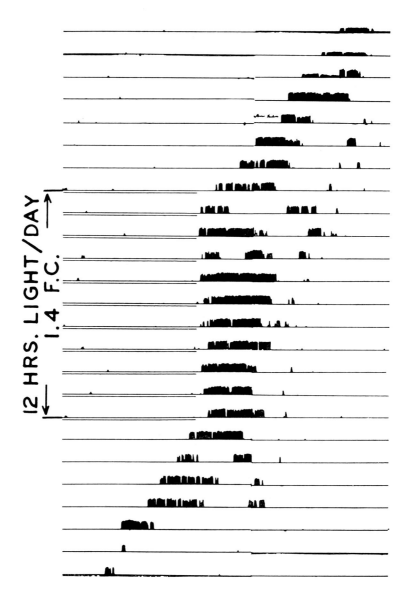

FIG. 3. Twenty-five-day record of running activity of a female *Peromyscus* (Rawson, 1959). Initial rhythm in constant darkness 22 hours, 50 minutes. Twelve-hour light period indicated by double line. Each horizontal base line represents 24 hours.

25 ft-c) and found that activity began at a later time each night. The logarithms of rate of delay of the active period and light intensity were linearly related. The effect of any particular intensity was increased by using diffused instead of concentrated light, and decreased when nesting material was present. Thus, Johnson demonstrated a circadian rhythm in *Peromyscus*, i.e., a rhythm of activity under constant conditions with a free-running period approaching, but not exactly, 24 hours.

It remained for Rawson (1956, 1959) to show for *P. leucopus* that, in continuous darkness, activity in wheels typically begins earlier each day (Fig. 3). Free-running periods of individuals varied from nearly 23 hours to just over 24 hours. Stinson (1960) confirmed these results, using *P. m. gracilis* and *P. m. bairdi.* His animals also showed individual differences and mice in neighboring cages were sometimes completely out of phase.

Although we cannot dismiss the possibility that some unmeasured environmental factor may control activity rhythms under so-called constant conditions (Brown, 1960), in view of the evidence, it is reasonable to regard the circadian activity rhythm of *Peromyscus* as endogenous.

Johnson was unable, by manipulating light cycles, to substitute a 16-hour day (L : D, 8 : 8) for a 24-hour day. However, Kavanau (1962c) succeeded in doing so with *P. maniculatus.* Instead of turning the light on and off abruptly, as is usually done, he gradually altered intensity to conform to the twilight periods of dawn and dusk (Fig. 4). Mice subjected to a 16-hour day (L : D, 8 : 8 and L : D, 10 : 6) adjusted activity accordingly. When placed in continuous darkness, these animals returned to a circadian rhythm but initially showed two activity periods per day.

A number of the foregoing experiments show that light is an effective timing factor (*Zeitgeber*) for activity rhythms of *Peromyscus.* Rawson (1956, 1959) found that 12 hours of light presented during the active period produced a delay in the onset of activity equal to that brought about by continuous light. However, when the light was presented only during the inactive period, there was no delay, even though the animal was aware of the light. Similarly, Kavanau (1967) found that *Peromyscus* responded much less to changed light intensity during the inactive period than in the active period, even though aware of the change. Thus, there appears to

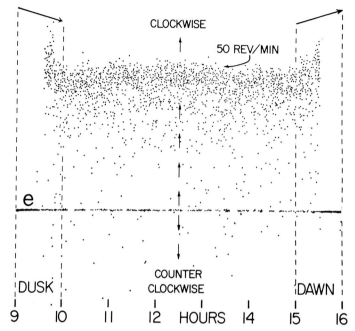

FIG. 4.   Record of running speed of a *P. maniculatus* on a 16-hour periodicity (9 hours daylight) with twilight transitions (Kavanau, 1967). The dim-light source was in the clockwise direction.

be a daily rhythm in responsiveness to light. Rawson also showed how an animal's rhythm can become adjusted to the external light cycle. If the circadian rhythm is less than 24 hours, the animal becomes active earlier each day until its active period encounters the end of daylight, after which activity begins at the same time each night (Fig. 3). In the rare case of a circadian rhythm longer than 24 hours, the animal's activity begins and ends later each day until the end of activity encounters the daylight. It then stops shifting. Evidently, the active period has a characteristic length because the onset of activity as well as the end becomes stabilized. An example of this is also illustrated by Bruce (1960). With four two-hour bursts of light a day spaced at four-hour intervals, a mouse was active for only one four-hour period in 24 hours.

Johnson (1939) showed that under continuous bright light (25 ft-c) the activity of *Peromyscus* was aperiodic. Pittendrigh (1960) cites a case where a *P. maniculatus* after such treatment was placed in continuous darkness. It immediately resumed a circadian rhythm with the onset occurring later each day, as usually occurs in dim light. After several weeks, it reverted to the usual dark cycle of less than 24 hours. It appears that in this case there was an after effect of illumination.

Kavanau (1967) found that when given control over illumination *Peromyscus* maintained a circadian rhythm, selecting a higher intensity for activity than for rest. In line with this result he was able to entrain five mice using a dark : dim, 14 : 10 light cycle. The animals chose the dim period for activity and followed the rhythm when it was shortened. On a light : dim regime they were active in the dim period. It is of interest that when running in dim light they almost invariably ran towards the light source (Fig. 4). He suggested that the moon may be used as a navigational reference. However, in the field, mice seem to be most active in the absence of moonlight, as discussed in the section on effects of light on amount of activity.

No one appears to have found another *Zeitgeber* as effective as light for entraining the activity of *Peromyscus*.

While it is not intended to discuss in detail the physiological basis of endogenous rhythms, a few experiments by Rawson (1959, 1960) with *Peromyscus* will be mentioned. Body temperatures of several mice were lowered by 6° to 17° C for periods of five to eight hours under sodium pentobarbital anesthesia. Temperature coefficients (Q_{10}) for the delay in onset of the following active period ranged from 1.1 to 1.4, which are lower than those usually calculated for biological systems. Thus, it appears that the rhythm is not directly dependent on the rate of metabolism. This was borne out by further experiments in which Rawson failed to alter the rhythm by increasing metabolic rate in a cold environment or by the administration of drugs having metabolic effects. In contrast, activity rhythms affect metabolic rates as illustrated by the more rapid recovery of *P. m. rufinus* from nembutal during the active period than during the inactive period (Emlen and Kem, 1963). Activity rhythms must be delimited and accounted for in any meaningful physiological study.

SEASONAL CHANGES AND PATTERN DETAILS.—The onset of activity in individual mice may be very precise, as Rawson (1956) demonstrated in the laboratory. The same precision occurs in the field, as Falls (1953) found by monitoring the first departure of four *P. m. bairdi* from a natural nest on a beach, using an electric eye. From June 5 to July 17, the average time was 8:50 P.M. EST (standard error, 1.2 min.) or 52 minutes after sunset. On three mornings when the last arrival was noted it was from 3:38 to 3:40 A.M., averaging 65 minutes before sunrise. Both in the morning and evening the light intensity averaged about 0.1 ft-c, a little brighter than the full moon. Kavanau (1962c) found that when twilight transitions were used in the laboratory, *Peromyscus* activity began and ended at intensities of one to ten times full moonlight. However, Falls found no tendency for departure times to be correlated with day-to-day changes in light intensity as a result of cloudiness. Mice in activity wheels, exposed to daylight, were just as precise but started 30 minutes after sunset. In a hardwood forest, *P. m. gracilis* began to enter traps in the evening when the light intensity was about the same as that noted for *bairdi*. Darkness, of course, came earlier in the forest than on the beach and trapping records indicate that the period of activity for *gracilis* was correspondingly longer than for *bairdi*.

Johnson (1926) found that the nocturnal activity in *P. l. noveboracensis* and *P. m. bairdi* lasted longer in winter than in summer. His figures show a 13-hour period of activity for four mice kept in a greenhouse in November and a nine-hour period for two mice kept in the woods in April. Orr (1959) reported the same general trend for *noveboracensis* in an outdoor enclosure but presented no data. Falls (1953) found a gradual increase in tracks counted between 9:00 P.M. and 10:00 P.M. from June through August. Track counts, live trapping, and wheels in which *P. m. bairdi* and *P. m. gracilis* were kept outdoors all showed active periods of seven to eight hours in June, 10 to 12 hours in September, and 14 hours (only nest boxes with activity wheels used) in February (Figs. 5 and 6).

Behney (1936) appears to have been the first to present data on the "shape" of the nocturnal activity pattern, of *P. l. noveboracensis*, based on the time out of the nest. Activity was plotted for 50 equal periods between sunset and sunrise. Exploration began shortly after sunset and rose to a peak about an hour later. Thereafter, there was less activity until about halfway between midnight and

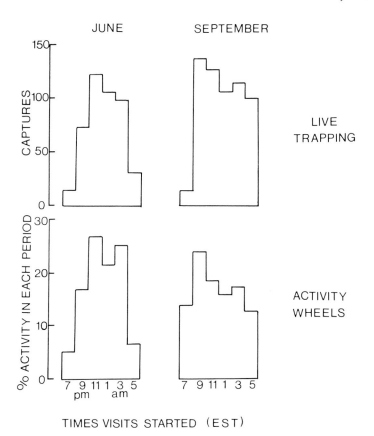

FIG. 5. Summer activity patterns of *P. m. gracilis* in Algonquin Park, Ontario. Five mice in wheels in June and four in September.

sunrise when there was a slight increase followed by a decline to zero. These results were obtained in the winter and it appears from a figure that activity lasted about 12 hours.

The oxygen consumption of four *P. leucopus* and *P. maniculatus* over a period of activity of about nine hours rose to a peak about 1:00 A.M. and then gradually declined, according to Morrison (1948).

Orr (1959) found no essential difference in the activity patterns of individuals or groups of six *P. l. noveboracensis,* measured by inter-compartment treadles in his outdoor enclosure. In what ap-

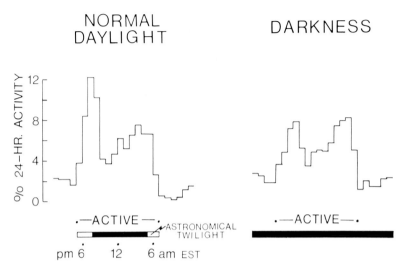

Fig. 6. Activity of *P. m. bairdi* in February measured as time out of nest box in enclosure with food, water, and activity wheel. To bring animals in darkness in phase, 14-hour period of greatest activity for each mouse (determined weekly) lined up in calculating average.

pears to be a 10- or 11-hour period of activity, a major peak occurs at 10:00 P.M. or 11:00 P.M., followed by a rapid drop with a slight plateau at 2:00 A.M. or 5:00 A.M.

On the basis of trapping records, Burt (1940) noted that wild mice (*P. leucopus*) were most active, at least in their search for food, during the early part of the night. On the basis of live-trapping, track counts, time out of nest boxes and running in activity wheels, Falls (1953) presented histograms to show the pattern of activity of *P. m. gracilis* and *P. m. bairdi*. (In open habitats, patterns vary from those described, depending on the occurrence of strong moonlight which suppresses activity. For example, if moonlight occurs early in the night, the major peak may be delayed.) Examples are given in Figures 5 and 6. In the short nights of June, activity rose to a peak between 10:00 P.M. and midnight EST, remained fairly high through the night, and then dropped sharply after 4:00 A.M. In some records a minor peak occurred between midnight and 3:00 A.M. In September, the main peak occurred between 8:00 P.M. and 10:00 P.M., a minor peak was present between 2:00 A.M. and 4:00 A.M., and activity fell sharply after 6:00 A.M. In winter, time out of the nest per hour was used in preference to running activity, since the latter was at a low level. For mice exposed to normal daylight, activity rose to a sharp peak at 7:00 P.M., dropped to a low point at 10:00 P.M., followed by a gradual increase with minor peaks at midnight and 3:00 A.M. A sudden drop occurred about 6:00 A.M. In mice kept outdoors in darkness the same general pattern occurred, but the first peak was lower and times were out of synchrony with daylight. These patterns are averages for five mice in each case over five weeks. Individual days for single mice show a series of short periods out of the nest which tend to occur more often at the peak hours of the average histograms than at other times. To summarize, it appears that on short summer nights activity is sustained at a high level and there is usually a single peak in the evening; on long winter nights, activity is less continuous and the average pattern shows two major periods of activity early and late in the night.

Although an examination of practically all the published original records shows nocturnal activity to be discontinuous, no one has clearly demonstrated a regular short cycle of activity in *Peromyscus,* apart from the tendencies resulting in the major peaks already referred to. Thus, Hatfield (1940), who described a short cycle for *Microtus,* could find no evidence of one in the movements of two *P. m. bairdi* between a nest box and a food chamber. Although Morrison (1948) gives values of 2.2 and 2.7 hours, respectively, for the periods of short cycles in oxygen consumption of *P. l. noveboracensis* and *P. m. gracilis,* he also states that it was difficult to distinguish any regular short cycle in these mice.

The nightly pattern of various activities has been the subject of several studies. French (1956) examined eating, drinking, and activity patterns in *P. m. sonoriensis,* using a bouncing cage with side compartments for food, water, and nesting. Movements of the whole cage as well as presence of a mouse in any compartment were recorded. On *ad libitum* food and water, 90 per cent of visits to the feeder and drinker took place in the dark hours from 6:00 P.M. to 6:00 A.M, 75 per cent occurred before midnight. With water deprivation, activity decreased but the pattern was unchanged.

Kavanau (1963*a*) is critical of these results on the grounds that animals were not given time to adapt to the apparatus and that visits to the appropriate chambers are not necessarily evidence of eating and drinking. In his own studies, mice were required to perform one or two instrumental acts to obtain food or water as well as to leave the nest or operate an activity wheel. Very detailed information is presented concerning a variety of acts but only a few results will be summarized here, based on experiments with three *P. maniculatus* (Kavanau, 1963*a*). Running in the wheel was punctuated by brief intervals in which the animal ate, drank, eliminated, or visited the nest. Food consumption tended to peak both in the first hour or so of the active period and again in the last few hours, with the latter peak being of greater magnitude. Water consumption tended to increase gradually, reaching a pronounced peak in the last hour or two of activity. The longest drinks tended to occur at the end of the activity period and on a few excursions in predawn hours and daylight. Eliminations were frequent throughout the active period but fell off towards the end. Prompt elimination usually took place in the course of brief daylight excursions. These results are based on a 12-hour daily period of activity.

It is now possible to consider explanations for the two main peaks seen in the winter pattern of activity. In view of the tendency for activity to be greatest at an intermediate light intensity (Stinson, 1952), it might seem that peaks are associated with twilight periods. However, the first peak is closer to sunset than the second is to sunrise. Kavanau (1962*c*) found that running in wheels increased both in the evening and morning periods of twilight (Fig. 4); he attributed this to a stimulating effect of twilight transition. He also found that the middle to late phase of dusk stimulated mice to greater running activity during the night than occurred on abrupt bright : dim regimes. While this may account for the precision and sharpness of the first peak in daylight, it does not explain the persistence of two peaks in continuous darkness. It seems possible to the writer that these peaks represent a feeding pattern. An interesting parallel is the observation of Elton *et al.* (1931) that *Apodemus* showed a bimodal nocturnal activity pattern and ate the equivalent of two stomachfuls in a night. Both Kavanau and the writer (unpublished data) have found that feeding does show a bimodal pattern (with an active period of about 12 hours). Perhaps *Peromyscus* activity, as measured by time out of the nest, is bimodal in winter because it is sporadic and mainly associated with feeding whereas, in summer, animals are out more continuously and a smaller fraction of their time is spent feeding. It may be pertinent that on three nights of full moon in summer when activity was low, track counts showed two obvious peaks early and late in the night (Falls, 1953). It is also possible that there is not normally time for two distinct feeding periods in the short nights of early summer. Finally the bimodal pattern of feeding and other activity seen clearly in winter may be a manifestation of a more fundamental periodicity in the organism.

Summary

A number of biotic and physical factors have been shown to affect activity. Of the physical factors it is clear that light, temperature, and moisture influence the activity of *Peromyscus*. Laboratory and field studies have complemented one another in suggesting similar effects, but in some cases at different levels of the factor concerned. These differences may result from confinement of animals in the laboratory and from interactions of factors in the field. At present, there is a dearth of accurate information from field studies. New instrumentation may make possible better measures of both activity and environmental factors. Certainly, variations in activity must be taken into account in sampling techniques, especially those based on the assumptions of homogeneity within a population and constant probability of capture with time.

Activity patterns of *Peromyscus* are typical of nocturnal animals (Aschoff, 1960). Thus, they show a precise circadian rhythm with a free-running period of less than 24 hours in darkness and longer in continuous light. Light appears to be the principal *Zeitgeber* and adjusts the onset of activity to seasonal change. The length of the active period also varies seasonally, which enables mice in the northern part of the range to obtain food over a longer period in winter when it is not so readily available and more is needed. Details of the pattern of activity appear to be influenced by light, time of year, and the need to obtain food and water.

Literature Cited

Aschoff, J. 1960. Exogenous and endogenous components in circadian rhythms. Cold Spring Harbor: Symp. Quant. Biol., 25:11–28.

BEER, J. R. 1961. Winter home ranges of the red-backed mouse and white-footed mouse. Jour. Mamm., 42:174–180.

BEHNEY, W. H. 1936. Nocturnal explorations of the forest deer-mouse. Jour. Mamm., 17:225–230.

BLAIR, W. F. 1943. Activities of the Chihuahua deer-mouse in relation to light intensity. Jour. Wildl. Mgt., 7:92–97.

————— 1951. Population structure, social behavior and environmental relations in a natural population of the beach mouse, *Peromyscus polionotus leucocephalus*. Contr. Lab. Vert. Biol. Univ. Mich., 48:1–47.

BRANT, D. H., AND J. L. KAVANAU. 1965. Exploration and movement patterns of the canyon mouse (*Peromyscus crinitus*) in an extensive laboratory enclosure. Ecology, 46:452–461.

BROWN, F. A., JR. 1960. Response to pervasive geophysical factors, and the biological clock problem. Cold Spring Harbor: Symp. Quant. Biol., 25:57–71.

BROWN, L. N. 1964. Dynamics in an ecologically isolated population of the brush mouse. Jour. Mamm., 45:436–442.

BRUCE, V. G. 1960. Environmental entrainment of circadian rhythms. Cold Spring Harbor: Symp. Quant. Biol., 25:29–48.

BURT, W. H. 1940. Territorial behavior and populations of some small mammals in southern Michigan. Misc. Publ. Mus. Zool. Univ. Mich., 45:1–58.

DICE, L. R., AND P. J. CLARK. 1962. Variation in measures of behavior among three races of *Peromyscus*. Contr. Lab. Vert. Biol. Univ. Mich., 76: 1–28.

DICE, L. R., AND S. A. HOSLETT. 1950. Variation in the spontaneous activity of *Peromyscus* as shown by recording wheels. Contr. Lab. Vert. Biol. Univ. Mich., 47:1–18.

DICE, L. R., AND W. E. HOWARD. 1951. Distance of dispersal by prairie deer-mice from birthplaces to breeding sites. Contr. Lab. Vert. Biol. Univ. Mich., 50:1–15.

EISENBERG, J. F. 1962. Studies on the behavior of *Peromyscus maniculatus gambelii* and *Peromyscus californicus parasiticus*. Behaviour, 19:177–207.

ELTON, C., E. B. FORD, J. R. BAKER, AND A. D. GARDNER. 1931. The health and parasites of a wild mouse population. Proc. Zool. Soc. London:657–721.

EMLEN, S. T., AND W. KEM. 1963. Activity rhythm in *Peromyscus*: Its influence on rates of recovery from nembutal. Science, 142:1682–1683.

FALLS, J. B. 1953. Activity and local distribution of deer mice in relation to certain environmental factors. Unpublished Ph.D. dissertation, Univ. Toronto.

FOSTER, D. D. 1959. Differences in behavior and temperament between two races of the deer mouse. Jour. Mamm., 40:496–513.

FRENCH, R. L. 1956. Eating, drinking, and activity patterns in *Peromyscus maniculatus sonoriensis*. Jour. Mamm., 37:74–79.

GENTRY, J. B., AND E. P. ODUM. 1957. The effect of weather on the winter activity of old-field rodents. Jour. Mamm., 38:72–77.

GETZ, L. L.   1959.   Activity of *Peromyscus leucopus*. Jour. Mamm., 40:449–450.

HAMILTON, W. J., JR.   1937.   Activity and home range of the field mouse, *Microtus pennsylvanicus pennsylvanicus* (Ord). Ecology, 18:255–263.

HART, J. S.   1951.   Calorimetric determination of average body temperature of small mammals and its variation with environmental conditions. Can. Jour. Zool., 29:224–233.

———   1952.   Effects of temperature and work on metabolism, body temperature, and insulation: results with mice. *Ibid.*, 30:90–98.

———   1953.   Energy metabolism of the white-footed mouse, *Peromyscus leucopus noveboracensis*, after acclimation at various environmental temperatures. *Ibid.*, 31:99–105.

HATFIELD, D. M.   1940.   Activity and food consumption in *Microtus* and *Peromyscus*. Jour. Mamm., 21:29–36.

HAYS, H. A.   1958.   The effect of microclimate on the distribution of small mammals in a tall-grass prairie plot. Trans. Kans. Acad. Sci., 61:40–63.

HIRTH, H. F.   1959.   Small mammals in old field succession. Ecology, 40:417–425.

HOCK, R. J.   1961.   Effect of altitude on endurance running of *Peromyscus maniculatus*. Jour. Appl. Physiol., 16:435–438.

HORNER, B. E.   1954.   Arboreal adaptations of *Peromyscus* with special reference to use of the tail. Contr. Lab. Vert. Biol. Univ. Mich., 61:1–85.

HOWARD, W. E.   1949.   Dispersal, amount of inbreeding, and longevity in a local population of prairie deermice on the George Reserve, southern Michigan. Contr. Lab. Vert. Biol. Univ. Mich., 43:1–52.

———   1951.   Relation between low temperature and available food to survival of small rodents. Jour. Mamm., 32:300–312.

JOHNSON, M. S.   1926.   Activity and distribution of certain wild mice in relation to biotic communities. Jour. Mamm., 7:245–277.

———   1939.   Effect of continuous light on periodic spontaneous activity of white-footed mice *(Peromyscus)*. Jour. Exp. Zool., 82:315–328.

KAVANAU, J. L.   1962*a*.   Precise monitoring of drinking behavior in small mammals. Jour. Mamm., 43:345–351.

———   1962*b*.   Automatic multi-channel sensing and recording of animal behavior. Ecology, 43:161–166.

———   1962*c*.   Twilight transitions and biological rhythmicity. Nature, 194:1293–1295.

———   1963*a*.   Continuous automatic monitoring of the activities of small captive animals. Ecology, 44:95–110.

———   1963*b*.   The study of social interaction between small animals. Anim. Behav., 11:263–273.

———   1967.   Behavior of captive white-footed mice. Science, 155:1623–1639.

KING, J. A., AND B. E. ELEFTHERIOU.   1959.   Effects of early handling upon adult behavior in two subspecies of deermice, *Peromyscus maniculatus*. Jour. Comp. Physiol. Psychol., 52:82–88.

McCABE, T. T., AND B. D. BLANCHARD.   1950.   Three species of *Peromyscus*. Rood Associates, Santa Barbara, 136 pp.

McNab, B. K., and P. Morrison. 1963. Body temperature and metabolism in subspecies of *Peromyscus* from arid and mesic environments. Ecol. Monogr., 33:63–82.

Morrison, P. R. 1948. Oxygen consumption in several small wild mammals. Jour. Cell. Comp. Physiol., 31:69–96.

Murie, O. J., and A. Murie. 1931. Travels of *Peromyscus*. Jour. Mamm., 12: 200–209.

Orr, H. D. 1959. Activity of white-footed mice in relation to environment. Jour. Mamm., 40:213–221.

Pittendrigh, C. S. 1960. Circadian rhythms and the circadian organization of living systems. Cold Spring Harbor: Symp. Quant. Biol., 25:159–184.

Rawson, K. S. 1956. The accuracy of the endogenous activity rhythms of small mammals and their response to low body temperatures. Jour. Cell. Comp. Physiol., 48:343.

——— 1959. Experimental modification of mammalian endogenous activity rhythms. *In*: Photoperiodism and related phenomena in plants and animals. Washington, D. C.: A. A. A. S.

——— 1960. Effects of tissue temperature on mammalian activity rhythms. Cold Spring Harbor: Symp. Quant. Biol., 25:105–113.

Richter, C. P. 1927. Animal behavior and internal drives. Quart. Rev. Biol., 2:307–343.

Sheppe, W. 1965a. Characteristics and uses of *Peromyscus* tracking data. Ecology, 46:630–634.

——— 1965b. Dispersal by swimming in *Peromyscus leucopus*. Jour. Mamm., 46:336–337.

Stickel, L. F. 1946. Experimental analysis of methods for measuring small mammal populations. Jour. Wildl. Mgt., 10:150–159.

Stinson, R. H. 1952. Effects of some environmental factors on the behaviour of *Peromyscus*. Unpublished Ph.D. dissertation, Univ. Toronto.

——— 1960. The timing of the activity pattern of *Peromyscus* in constant darkness. Can. Jour. Zool., 38:51–55.

White, J. E. 1964. An index of the range of activity. Amer. Midland Nat., 71:369–373.

INDEX TO AUTHORS

INDEX TO SUBJECTS

INDEX TO TECHNICAL NAMES

(Only pages for the smallest taxon listed. Names of parasites omitted.)

587